Minerals as Advanced Materials II

Sergey V. Krivovichev
Editor

Minerals as Advanced Materials II

Springer

Editor
Sergey V. Krivovichev
Nanomaterials Research Center
Kola Science Center
The Russian Academy of Sciences
14 Fersman Street, 184209 Moscow
Russia
and
Department of Crystallography
Faculty of Geology
St. Petersburg State University
University Emb. 7/9, 199034 St. Petersburg
Russia
skrivovi@mail.ru

ISBN 978-3-642-43912-4 ISBN 978-3-642-20018-2 (eBook)
DOI 10.1007/978-3-642-20018-2
Springer Heidelberg New York Dordrecht London

Softcover re-print of the Hardcover 1st edition 2012

Cover illustration: crimson / fotolia.com

Printed on acid-free paper

Springer is part of Springer Science+Business Media (www.springer.com)

Foreword

This book represents a collection of papers presented at the 2nd international workshop 'Minerals as Advanced Materials II' that was held on 19–25 July 2010 in Kirovsk, Kola peninsula, Russian Federation. Kola peninsula is famous for its natural heritage, both in terms of mineral deposits and its unique mineralogical diversity. Many of the mineral species discovered here are now known as materials used in various areas of modern industry. The most remarkable examples are zorite (natural analogue of the ETS-4 molecular sieve titanosilicate) and sitinakite (natural counterpart of ion-exchanger UOP-910 used for the removal of Cs-137 from radioactive waste solutions). For this reason, Kola peninsula was an excellent locality for the workshop, especially taking into account that the lecture days were followed by field excursions to famous mineral deposits.

Mineralogy is probably the oldest branch of material science, on one hand, and the oldest branch of geology, on the other. For several centuries, mineralogy was dealing with materials that appear in Nature as minerals, and it still continues to provide inspiration to material chemists in synthesis of new materials. The remarkable fact is that there exists a large number of minerals that have not yet been synthesized under laboratory conditions. The good example is charoite, which is famous for its beauty and attractiveness. Recent studies (see contribution by Rozhdestvenskaya et al. in this book) demonstrated that its structure contains nanotubular silicate anions comparable in their external and internal diameters to carbon nanotubes. Charoite occurs in Nature in tons, but it has never been prepared synthetically.

Papers in this book cover a wide range of topics starting from gas release from minerals, microporous minerals, layered materials, minerals and their synthetic analogues with unique physical and chemical properties to biological minerals and microbe-mediated mineral formation. The authors are experts in different fields of science, mainly from mineralogy and material chemistry that provide a special interest from the viewpoint of interaction of scientists with different areas of expertise.

This workshop would not be possible without considerable infrastructure support from the 'Apatit' mining company and personally from Dr. A.V. Grigoriev and his colleagues. It is a pleasure to acknowledge their essential support and collaboration in organization of the workshop.

Sergey V. Krivovichev

Contents

From Minerals to Materials

Wulf Depmeier

1 Introduction

It goes without saying that rocks and minerals have been used as materials ever since the earliest days of mankind. Early usages were certainly restricted to as-found, or at best primitively processed, species, but it did not take long and pre-industrial processes, like ore smelting or sintering of ceramics, were invented, thereby extending the application fields of representatives of the mineral kingdom. A more or less smooth evolution over centuries driven by the great inventions of chemistry and physics has allowed a gradual development of technology, and continues to do so. Furthermore, roughly in the middle of the past century a genuine technical revolution appeared which not only started to change our daily life, but also bore important consequences for culture, economics, life-style and welfare of mankind. The basis of the new technology was the development of tailor-made materials having specific properties and defined functionalities. New scientific disciplines emerged, which became known as materials sciences and nano-science. This development called for materials with hitherto unknown or even unthinkable compositions, often for the making of devices with sizes, shapes, architectures or combination of materials which were never seen before, and which, for sure, do not occur in Nature.

From this one might be tempted to conclude that for our current needs, at least with respect to materials sciences, Nature does not have to offer much more than the raw matter needed for the production of the new advanced materials and devices made thereof. An example would be quartz sand which after several intermediate production steps is eventually transformed into silicon-based microchips. From our point of view this is not entirely true. While it cannot be denied that Nature has its specific limitations – e.g. it is highly improbable that one will ever find a naturally occurring mineral species containing just one single rare earth element, or a

W. Depmeier (✉)
Inst. f. Geowissenschaften, Universität Kiel, Olshausenstr. 40, D–24098 Kiel, Germany
e-mail: wd@min.uni-kiel.de

S.V. Krivovichev (ed.), *Minerals as Advanced Materials II*,
DOI 10.1007/978-3-642-20018-2_1,

multilayer of thin films properly deposited on a substrate and correctly doped for a specific purpose – we propose that there are still many cases where researchers or engineers can get inspiration, if not advice, from Nature. This was the basic motivation for the workshop "Minerals as Advanced Materials II, MAAM II" which was held at Kirovsk, Kola Peninsula, Russia, from July 19–24, 2010. This was a follow-up event after a first one held in 2007 at close-by Apatity, and the results of which were summarized in a book (Krivovichev 2008). Inspired by the 2007 workshop, the present author made some general considerations about the topic (Depmeier 2009). In particular, he suggested that a close relationship exists between the cultural development of early men and his materials, and, furthermore, discussed the question what accounts for a substance to become a good material, at least in those early days of mankind. He proposed that such a substance, in addition to having at least one property which makes it appropriate for a planned application, should meet three requirements, namely (1) availability, (2) processibility and (3) performance. This statement was depicted by a number of examples. Furthermore, the advantages and disadvantages of Nature were discussed in comparison with technique and with respect to certain material characteristics. It turned out that both realms have their particularities which make them partly complementary. In conclusion, it was suggested that a scientist or engineer looking for a new material would be well advised if he or she not only consulted the usual sources for data on chemically pure compounds, but also turned to appropriate databases listing information on the around 4,500 minerals which are actually known. The paper ended by the presentation of a few case studies. It is the purpose of this short contribution to complement this enumeration.

2 Minerals as Materials

A comprehensive treatment of structure – property relationships can be found in Newnham (2005). It is clear that the properties of a crystal depend on its composition, its symmetry, the arrangement of the atoms and on the nature of the bonds between them. In principle, all this information is accessible by a structural analysis. Often a desired macroscopic property depends on the symmetry of the crystal and an appropriate description will rely on the tensor notation. However, usually all of this is not sufficient to characterize a modern functional material. In most cases a given property will also depend markedly on the real structure of the crystal, its size (especially in the nanometre range), the presence and distribution of various defects, substitution and doping, and on external parameters like temperature or pressure. Often it is necessary to fine-tune these variables in order to optimize a desired property, or to impair an adverse one. It is often a cumbersome and, last but not least, expensive undertaking to vary all relevant parameters experimentally, even by some sort of high-throughput combinatorial methods. Computational methods have their limits, too, especially when multi-element substitutions have to be investigated. Therefore, the extreme wealth of Nature with respect to various combinations of these parameters should be exploited whenever possible. For instance, this could be a reasonable strategy for an investigation of multinary complex sulfosalts in view of optimizing their performance, e.g. as absorber material for solar cells. When the long term behaviour of certain materials should be studied, the investigation of natural material can become the method of choice, too. Obvious examples are the investigation of slow processes of diffusion, ordering/disordering, weathering or metamictization.

As-found minerals are only rarely directly applicable as materials. One exception is bentonite which finds wide-spread use for various geo-engineering tasks, mostly because of its impermeability to water and its absorbing properties. Bentonites can also be transformed into materials with higher added value, e.g. by mixing them with natural polymers like polysaccharides or proteins to produce organic-inorganic nano-composites. Such materials are non-toxic and biocompatible, and thus environmentally-friendly, and could serve for biomedical applications, e.g. bone repair (Carrado and Komadel 2009). This work can be considered to be bioinspired by observation of natural pearls. Pearls are the products of biomineralisation. These natural organic-inorganic hybrid nano-composites consist of an oriented assembly of calcite/aragonite nano-crystals agglutinated by conchiolin, a protein. Pearls are much valued as pieces of jewellery and represent one of the (rare) cases where natural stony objects are used without any further finishing (apart from beading or other kinds of attachment for making necklaces, rings or earrings). Other natural gemstones usually have to be finished, i.e. they are cut and polished to produce the final product, for instance brilliants from natural diamonds. These can then also be used for jewellery, or, because of the extraordinary properties of diamond (hardness, thermal conductivity, transparency), be employed as a real high-performance material, finding applications in fields as different as cutting tools, heat dissipators, in diamond anvil cells for high pressure research, or as optical devices at synchrotron radiation sources. The outstanding properties of diamond, and its high prize, have already long time ago led to attempts to synthesize diamond. This technique has nowadays reached a quite advanced level and for many industrial purposes synthetic diamonds are available.

The special venue of both workshops (2007: Apatity; 2010: Kirovsk) in the direct neighbourhood of the Khibiny and Lovozero mountains on Kola peninsula with their particular geochemical situation and resulting unique inventory of minerals, including microporous titano- and zirconosilicates, was probably one of the main reasons, why heteropolyhedral microporous minerals and their possible materials properties represented a major part of the contributions to both programmes. Also, in Krivovichev (2008) several reports were devoted to these materials. The fascinating case of the mineral zorite from Lovozero and its synthetic offsprings ETS-4 and ETS-10 was already presented in some detail (Depmeier 2009). Therefore, this interesting type of minerals/materials will not be further considered here.

The study of multiferroics is currently a very busy field. Multiferroics promise very interesting properties and applications. For instance, multiferroics that couple electrical and magnetic properties would enable to write some information electrically, which could then be read out by a magnetic sensor. This separation of writing and reading properties has certain technical advantages. Other possible fields of application are spintronics. Various aspects are discussed in Fiebig (2005); Eerenstein et al. (2006); Schmid (2008).

Natural boracite with its ideal composition $Mg_3B_7O_{13}Cl$ is in a certain sense the grandfather of multiferroics, as it is simultaneously ferroelectric and ferroelastic. Its synthetic homologue Ni-I-boracite, $Ni_3B_7O_{13}I$, is in addition ferromagnetic and represents the archetype of single phase multiferroics (Ascher et al. 1966). The effect in single-phase materials is rather small and for practical purposes one prefers multiphase composite materials (Eerenstein et al. 2006). The interesting story of the scientific history of boracites is planned to be published by the discoverer of multiferroicity, Prof. Hans Schmid from Geneva, Switzerland, who also named the effect (Schmid 2010). A short description of the discovery of boracite and of the identification of its true nature has already been given in the literature (Schmid and Tippmann 1978). As an aside it is interesting to note that the first (scientific) discoverer, Georg Siegmund Otto Lasius (1752–1833), described boracite as "cubic quartz". He was probably mislead by the fact that the new mineral occurred together with euhedral trigonal quartz crystals in the gypsum cap rock of the salt dome at Lüneburg, not far from Hamburg in Northern Germany, and its outward appearance (hardness, transparency, but not morphology) is not very different from quartz. Soon after, however, it was realized that boracite in fact contains boron and is definitely different from quartz. Lasius was an engineer responsible for the roadwork in the then Kingdom of Hannover. In the course of his activities he was able to build up a quite representative collection of minerals and rocks of the region he worked in. It is highly probable that his collection also comprised boracites and the story has it that in 1821 the collection was sold to the Mining Institute at Saint Petersburg, Russia. A recent search did not prove the evidence of Lasius-boracites in the collection of the Mining Institute despite the fact that it holds several different specimens of boracite. The search is quite difficult because apparently it was not before 1842 that a systematic cataloguing of mineral samples started at the mining institute and, hence, the looked-for samples might well be present, but could not be identified.

In this context it is worth mentioning S. C. Abrahams' work on a systematic search for potential ferroelectric materials in minerals and synthetic compounds (Abrahams 1988). Using this method he and his co-workers were able to identify, for example, fresnoite as a ferroelectric mineral (Foster et al. 1999). A basic property of a ferroelectric is that its symmetry belongs to one of the ten pyroelectric point groups which allow the occurrence of a spontaneous electrical polarisation (1, m, 2, mm2, 4, 4mm, 3, 3m, 6, 6mm). The polarisation can be reversed under the action of an electric field, at least in principle. However, from an application point of view this property is less important than the concurrently occurring optoelectronic and non-linear optical properties.

Such properties are allowed also in other non-centrosymmetric, but non-polar symmetries. Such is the case for the minerals of the melilite family with their basic space group P-42_1m. The general formula can be written $A_2T'T_2O_7$, with A being an 8-fold coordinated cation, and T', T tetrahedrally coordinated cations. The melilite structure type is a very "successful" one in the sense that it shows a great versatility with respect to the chemical composition, i.e. many different chemical elements can occupy the A, T' and T positions. Melilites are also constituents of the calcium and aluminium rich inclusions in chondritic meteorites and, thus, belong to the oldest minerals. With respect to possible applications, it has to be noted that this structure is in a certain sense a "dense" structure, supporting "good" optical properties. Appropriately doped with trivalent rare earth elements on the A position laser properties can be obtained. Recently, the linear and non-linear optical properties of synthetic germanate melilites, e.g. $Ba_2MgGe_2O_7$, doped with rare earth atoms have been studied (Becker et al. 2010). The Czochralski-grown crystals show a broad transmission range and allow the adjustment of linear optical properties by substitution. Efficient phase matching, iso-index points and multi-wavelength generation reveal these melilites as promising optical materials.

Despite the "density" of the melilite structure, it also shows a pronounced layered character as tetrahedral layers $T'T_2O_7$ alternate with layers consisting entirely of cations A. In some cases there is mismatch between the two types of layers and modulated phases occur. It is perhaps worthwhile mentioning that the melilite structure type allows not only for great chemical flexibility, but also for elastic flexibility as discussed by Peters et al. in Krivovichev (2008). Here it was argued that it is most probably the high flexibility of the melilite layers which allows for the observed violation of Loewenstein's rule.

α-Quartz is still one of the most important piezoelectric materials, being able to transform an elastic deformation into an electric signal and vice versa, which explains the wide range of possible applications, for instance in modern communication techniques. Nowadays the great majority of quartz crystals used as impulse generator are of synthetic origin. Tiny quartz crystals were already synthesized in the nineteenth century. During World War II Brazil, then and today the most important supplier of natural quartz crystals, declared a ban on the export of these goods. R. Nacken (1884–1971) in Frankfurt/Main had already successfully grown quartz crystals by the hydrothermal method, and soon he was able to optimise the method and to produce centimetre-sized single crystals. After all, this did not change the history. After the war his experience was exploited and the methods refined on both sides of the then iron curtain. The scientific history of synthetic quartz has been described several times in the literature, e.g. Byrappa (2005); Iwasaki and Iwasaki (2002).

α-Quartz has the disadvantage that its use as efficient piezoelectric material is restricted to relatively low temperatures, because of adverse effects at higher temperatures, like decreasing resistivity. In any case, the absolute upper limit of its applicability would be the α-β phase transition at about 846 K, because the hexagonal symmetry of β–quartz does not allow for piezoelectricity. However, there is strong demand for piezoelectric devices, such as sensors or actuators, for usage in various high temperature technical processes. Therefore, there is much activity going on in the field of the development of high-temperature piezoelectrics. Langasite, $La_3Ga_5SiO_{14}$, is one of the most intensively studied of such materials in this field; another family of compounds with possible application up to 1,500 K is discussed by R. Möckel in this book. A different approach has been proposed by J. Schreuer in his abstract for the application of the MAAM II workshop (eventually, he was unable to attend the workshop). He noticed that one of the oldest known piezoelectric material is of natural origin, namely the frequently occurring mineral tourmaline. Tourmaline is a cyclosilicate of general composition $XY_3Z_6[Si_6O_{18}(BO_3)_3(OH)_3W]$, with, for example but not exclusively, $X = Na^+$, K^+, Ca^{2+}, $Y = Li^+$, Mg^{2+}, Fe^{2+}, Mn^{2+}, Al^{3+}, Fe^{3+}, Cr^{3+}, $Z = Al^{3+}$, Fe^{3+}, Mg^{2+} and $W = OH^-$, F^-. Tourmaline exhibits piezoelectricity, in principle, up to its decomposition at temperatures above, say, 1,100 K. However, adverse effects would probably restrict the use again to considerably lower temperatures. There are, however, several reasons why natural tourmaline is not really in use as material. First of all, the complex structure with different substitution schemes results in chemical compositions which change from crystal to crystal, or even within one and the same crystal as demonstrated by the multicoloured tourmalines which are high valued as gemstones. A possible way out would be the production of synthetic tourmalines of high quality and reproducible composition. However, up to now the usually employed hydrothermal methods have not been able to yield tourmaline crystals of the required gemstone quality and sufficient size (see, e. g. Setkova et al. (2009)).

Mayenite, $Ca_{12}Al_{14}O_{33}$, is a rare mineral from Bellerberg, Mayen, Eifel, Germany, The mineral was found only in 1964 (Hentschel 1964), but the compound has been known as $12CaO{\cdot}7Al_2O_3$, or $C_{12}A_7$, for long time already as a technical product and constituent of calcium aluminate cement. Recently, this compound has met considerable interest in materials science because of its possible applications as ionic conductor, transparent conductive oxide or catalyst for combustion of organic volatiles. A careful analysis has recently solved some relevant open questions with respect to its structure (Boysen et al. 2007). Whereas formerly there was general agreement that the structure should be considered as an open calcium-aluminate framework structure of composition $[Ca_{12}Al_{14}O_{32}]^{2+}$, consisting of AlO_4-tetrahedra and rather irregular Ca-O polyhedra, with the 33rd oxygen being disordered over six cages, Boysen et al. proposed that the structure should be better considered as a framework consisting of corner-connected AlO_4-tetrahedra with the Ca atoms showing considerable degree of disorder in response to that of the "free" oxygen. Note that the more recent perception of the mayenite framework of Boysen et al. is more in agreement with the usual view of zeolitic frameworks than the traditional one, because it considers a negatively charged tetrahedral framework rather than a positively charged heteropolyhedral framework. As a matter of fact, positively charged frameworks are rare, examples are layered double hydroxides (see e.g. the contribution of S. Krivovichev in this work), and a recently prepared thorium borate (Wang et al. 2010). Such cationic layer or framework structures are of considerable interest as they should allow for exchange and/or immobilization of anionic species. With respect to the latter characteristic, i.e. anion exchange, there seems to be a certain entitlement to consider mayenite indeed as a positively charged framework as the "free" oxygen can be replaced partly or fully by other anionic species. Much interest was attracted recently by the possibility of substituting

N^{3-} for the "free" oxygen (Boysen et al. 2008). The "free" oxygen can also be replaced by free electrons e^- (Matsuishi et al. 2003), thus giving rise to the possibility of electronic conductivity in a transparent oxide. The situation is somewhat similar to that in so-called "black sodalite", where formally e^- replaces anions like Cl^-, thus forming periodical arrays of F-centres (see e.g. Trill 2002).

With respect to the general topology of their structure, the examples just given belong to dense, microporous and layered structures. What about one- or zero-dimensional structures and their possible applications? The beneficial, but also the harmful properties of fibrous asbestos are well-known, they are related with the extreme aspect ratio of the fibres. Some silicate minerals, such as canasite or frankamenite, contain tubular structural units which in some cases also leave their imprint on the morphology. For instance, the tubular units in the structure of canasite are formed by joining together four wollastonite-type chains. The tubules can also be considered as consisting of two xonotlite double-chains. Xonotlite is known to crystallize in extremely needle- or hair-like form. The structural particularities of canasite and frankamenite have been described in Rozhdestvenskaya et al. (1996); Rastsvetaeva et al. (2003) and a compilation and comparison with other alkali calcium silicate minerals containing tubular chains can be found in Frank-Kamenetskaya and Rozhdestvenskaya (2004).

One particular member of the family of alkali-calcium silicates is charoite. This high-valued semi-precious gemstone has resisted its definitive structure solution for almost 50 years, before recently newly available instrumentation and advanced methodology made its structure determination possible (Rozhdestvenskaya et al. 2010). The structure of charoite contains a hitherto unknown type of tubular silicate chain. Canasite glass-ceramics have been considered as potential biocompatible substitutes for hard tissues (Miller et al. 2004).

The mere presence of parallel tubular building units in the structures of charoite and canasite is tempting to speculate whether this structural particularity could be useful for some purposes other than strengthening glass ceramics. The most obvious field where one would expect some useful property would be some kind of ion exchange. Note, however, that in the sample studied charoite fibres of about 100 nm diameter were imbedded in an amorphous material which was severely depleted in K and Ca, thus lending support to the idea that charoite does not survive leaching in aqueous environment, and other media have to be looked for.

A quite different way of speculation may come from the observation that domain walls in multiferroics show conduction properties (Seidel et al. 2009). Perhaps an appropriately changed composition of the silicate skeleton of charoite or canasite would allow for similar effects.

Another interesting case of one-dimensional character of a structure-type is the family of cancrinite-type structures. In Nature up to now a dozen, or so, of these structures have been found as minerals. These are the result of periodically changing sodalite (...ABC...) and cancrinite (...AB...) stacking schemes. Recently, a new member of the series, kircherite, has been described which has the highest periodicity found so far, namely not less than 36 (Bellatreccia et al. 2010). In the laboratory intermediate phases between sodalite and cancrinite could also be prepared (Hermeler et al. 1991), however, the products were usually disordered stacking variants, and it seems that it has not been possible to prepare the ordered long-periodic stacking variants found in Nature. In this special case the long time which Nature has available does not seem to play a decisive role, since the natural long-periodic variants are usually found in volcanic ejecta which can safely be supposed to have been cooled quite rapidly. Recently, possible useful zeolite-like behaviour of the nano-crystalline intermediate phases prepared by low-temperature hydrothermal synthesis has been reported (Grader et al. 2010).

In classical mineralogy zero-dimensional cluster-like structures are rare. On the other hand, there is increasing evidence that such structures play an enormous role in environmental chemistry. In particular, the aqueous chemistry of aluminium is governed by large aqueous aluminium hydroxide molecules, the importance of which can be appreciated when it is recalled that aluminium is the third most abundant element in the near-surface areas of the earth. Thus weathering and soil-formation can be expected to be heavily influenced by such clusters. A recent comprehensive review article highlights the importance of aluminium polyoxocation chemistry (Casey 2006). Heteropolymetallates, e.g. the Keggin ion, have been known for almost two centuries. These important cluster structures are interesting for various applications, notably as catalysts, but also for certain physical properties, e.g. as electrooptical materials. A very interesting property relates to the ability of certain heteropolymetallates to bind not only metals, but also to proteins and viruses. In the latter case this could be beneficial for an organism at risk to become infected, because being fixed to bulky clusters the viruses would no longer be able to penetrate cell walls.

The number of known natural heteropolymetallates is quite limited. Only recently the first natural heteropolyniobate, menezesite, of idealized composition $Ba_2MgZr_4(BaNb_{12}O_{42})\cdot 12H_2O$, has been described (Atencio et al. 2008). In another interesting recent finding, the mineral bouazzerite has been described which is built from Bi-As-Fe nanometre-sized clusters of composition $[Bi_3Fe_7O_6(OH)_2(AsO_4)_9]^{11-}$ which, as a big surprise, contain Fe^{3+} not only in the common octahedral coordination, but also in the rare trigonal prismatic coordination. Thus, the knowledge of the structure of this rare mineral might help not only to indicate synthetic pathways to this rare coordination, but also might help to understand the transport of toxic elements, such as arsenic, via the formation of nanoclusters (Brugger et al. 2007).

Superconductivity, since its discovery nearly 100 years ago, has been in the focus of solid state research, and continues to do so. The interest relies not only on the fascinating science behind this effect, but also on the many actual and potential technological applications of this effect. Various classes of materials were found to become superconducting at sufficiently low temperatures, from $Hg_{0.8}Tl_{0.2}Ba_2Ca_2Cu_3O_8$ with a record-high critical temperature of 138 K down to close to 0 K. In this respect, it was amazing that no report on superconductivity on a natural material has appeared in the literature up to 2006, when Di Benedetto et al. published the results of their study on the mineral covellite, CuS (Di Benedetto et al. 2006). Covellite becomes superconducting at 1.63(5) K. The occurrence of superconductivity in covellite has been related with the particularities of its structure with CuS_3 planes alternating with S_2 planes.

In this context it is worthwhile to mention the recent efforts of Liebau and colleagues to relate the occurrence of superconductivity with structural particularities, using crystal chemical arguments and reasoning (Liebau 2011; Liebau et al. 2011). This new approach may have the potential of spotting new superconductors among natural as well as synthetic materials.

Fast ionic conductors are important materials for present day's life. Their use spans wide ranges from various kinds of batteries to fuel cells, information storage, etc. In search for natural ionic conductors, complex silver-copper-sulfosalts minerals belonging to the pearceite-polybasite group have been investigated. In addition to the basic structures, the diffusion path ways of the mobile silver cations could be determined. The complex and variable chemical composition of the minerals of this group allows to study the effects of substitution. It could be realized that copper plays a decisive role, as it stabilizes disorder in the structures and, hence, improves the conductivity (Bindi et al. 2006; Bindi et al. 2007).

$Pb_{2+x}OCl_{2+2x}$ has been identified as a fast ionic conductor, the major charge carriers of which are Cl^- anions (Matsumoto et al. 2001). A recent structure determination of synthetic $Pb_{2+x}OCl_{2+2x}$ enabled us to look into details and to come to an understanding of the ionic conductivity (Siidra et al. 2007). In particular, it could be shown that the structure can be divided into alternating conducting and non-conducting two-dimensional blocks of about 1.5 nm width. The conducting blocks are characterized by atomic positions of low occupancy, whereas the positions in the non-conducting blocks are fully occupied. It has been proposed that the structural details allow considering $Pb_{2+x}OCl_{2+2x}$ tentatively as a nano-capacitor. Indeed, lead oxyhalogenides seem to be promising candidates for potential nano-technological applications. So-called nanobelts with the composition of the mineral mendipite, $Pb_3O_2Cl_2$, could be grown under special conditions which showed an enhancement of the birefringence by an order of magnitude due to the small size and special shape (Sigman and Korgel 2005).

This ends our short contemplation of the relationships between the mineral world and materials sciences. In conclusion, we insist on the fact that Nature, in general, and minerals, in particular, are indispensable sources of inspiration for many fields of solid state research and materials sciences, and should be consulted whenever possible.

Acknowledgements Financial support of the workshop "Minerals as Advanced Materials II" by the Deutsche Forschungsgemeinschaft under contract number DE 412/46-1 is gratefully acknowledged.

References

Abrahams SC (1988) Structurally based prediction of ferroelectricity in inorganic materials with point group *6mm*. Acta Crystallogr B44:585–595

Ascher E, Tieder H, Schmid H, Stössel H (1966) Some properties of ferromagnetoelectric nickel-iodine boracite. J Appl Phys 37:1404–1405

Atencio D, Coutinho JMV, Diriguetto AC, Mascarenhas YP, Ellena J, Ferrari VC (2008) Menezesite, the first natural heteropolyniobate, from Cajati, São Paulo, Brazil: description and crystal structure. Am Mineralog 93:81–87

Becker P, Kaminskii AA, Rhee H, Eichler HJ, Liebertz J, Bohatý L (2010) Linear and nonlinear optical properties of germanate melilites. Acta Cryst A 66:s37

Bellatreccia F, Cámara F, Della Ventura G, Gunter ME, Cavallo A, Sebastiani M (2010) Kircherite, a new mineral of the cancrinite-sodalite group with a 36-layer stacking sequence: occurrence and crystal structure. In: 20th general meeting of the IMA (IMA2010), Budapest, Hungary, 21–27 August 2010, CD of Abstracts: 493

Bindi L, Evain E, Spry PG, Menchetti S (2007) The pearceite-polybasite group of minerals: crystal chemistry and new nomenclature rules. Am Mineralog 92:918–925

Bindi L, Evain M, Pradel A, Albert S, Ribes M, Menchetti S (2006) Fast ionic conduction character and ionic phase-transitions in disordered crystals: the complex case of the minerals of the pearceite-polybasite group. Phys Chem Miner 33:677–690

Boysen H, Kaiser-Bischoff I, Lerch M (2008) Anion diffusion processes in O- and N-mayenite investigated by neutron powder diffraction. Diffus Fundam 8:2.1–2.7

Boysen H, Lerch M, Stys A, Senyshyn A (2007) Structure and oxygen mobility in mayenite ($Ca_{12}Al_{14}O_{33}$): a high-temperature neutron powder diffraction study. Acta Crystallogr B63:675–682

Brugger J, Meisser N, Krivovichev S, Armbruster T, Favreau G (2007) Mineralogy and crystal structure of bouazzerite from Bou Azzer, Anti-Atlas, Morocco: Bi-As-Fe nanoclusters containing Fe^{3+} in trigonal prismatic coordination. Am Mineralog 92:1630–1639

Byrappa K (2005) Growth of quartz crystals. In: Capper P (ed) Bulk crystal growth of electronic, optical & optoelectronic materials. John Wiley & Sons, Ltd., Chichester, England

Carrado KA, Komadel P (2009) Acid activation of bentonites and polymer-clay nanocomposites. Elements 5:111–116

Casey WH (2006) Large aqueous aluminium hydroxide molecules. Chem Rev 106:1–16

Depmeier W (2009, 2010) Minerals as advanced materials. Crystal Res Technol 44:1122–1130; Erratum. Crystal Res Technol 45: 6 / DOI 10.1002/crat.2009006

Di Benedetto F, Borgheresi M, Caneschi A, Chastanet G, Cipriani C, Gatteschi D, Pratesi G, Romanelli M, Sessoli R (2006) First evidence of natural superconductivity: covellite. Eur J Mineral 18:283–287

Eerenstein W, Mathur ND, Scott JF (2006) Multiferroic and magnetoelectric materials. Nature 442:759–765

Fiebig M (2005) Revival of the magnetoelectric effect. J Phys D Appl Phys 38:R123–R152

Foster MC, Arbogast DJ, Nielson RM, Photinos P, Abrahams SC (1999) Fresnoite: a new ferroelectric mineral. J Appl Phys 85:2299–2303

Frank-Kamenetskaya OV, Rozhdestvenskaya IV (2004) Atomic defects and crystal structure of minerals. In: Advances in science and technics, vol 33, Crystal chemistry. Yanus, St. Petersburg, p 187

Grader C, Robben L, Buhl JCh (2010) Synthesis of nanocrystalline intermediate phase between cancrinite and sodalite. In: 26th European crystallographic meeting, ECM26, Darmstadt, Acta Crystallogr A66:s230

Hentschel G (1964) Die Kalksteineinschluesse der Lava des Ettringer Bellerberges bei Mayen (Eifel), Paragenesen seltener und zweier neuer Minerale: Mayenit, 12CaO.7Al2O3, und Brownmillerit, 2CaO. (Al, Fe)2O3. N Jahrb Miner Monatsh 1964:22–29

Hermeler G, Buhl JCh, Hoffmann W (1991) The influence of carbonate on the synthesis of an intermediate phase between sodalite and cancrinite. Catal Today 8:415–426

Iwasaki F, Iwasaki H (2002) Historical review of quartz crystal growth. J Cryst Growth 237–239:820–827

Krivovichev S (ed) (2008) Minerals as advanced materials I. Springer, Berlin/Heidelberg

Liebau F (2011) Nonstoichiometry and bond character in unconventional superconductors. Zeitschr. Kristallogr., 226:319–322

Liebau F, Klein HJ, Wang X (2011) A crystal-chemical approach to superconductivity. I. A bond-valence sum analysis of inorganic compounds. Zeitschr. Kristallogr., 226:309–318

Matsuishi S, Toda Y, Miyakawa M, Hayashi K, Kamiya T, Hirano M, Tanaka I, Hosono H (2003) High-density electron anions in a nanoporous single crystal: $[Ca_{24}Al_{28}O_{64}]^{4+}(4e^-)$. Science 301:626–630

Matsumoto H, Miyake T, Iwahara H (2001) Chloride ion conduction in $PbCl_2$-PbO system. Mater Res Bull 36:1177–1184

Miller CA, Reaney IM, Hatton PV, James PF (2004) Crystallization of canasite/frankamenite-based glass-ceramics. Chem Mater 16:5736–5743

Newnham RE (2005) Properties of materials: anisotropy, symmetry, structure. Oxford University Press, Oxford

Rastsvetaeva RK, Rozenberg KA, Khomyakov AP, Rozhdestvenskaya IV (2003) Crystal structure of F-canasite. Dokl Chem 391:177–180

Rozhdestvenskaya IV, Mugnaioli E, Czank M, Depmeier W, Kolb U, Reinholdt A, Weirich T (2010) The structure of charoite, (K, Sr, Ba, Mn)15–16(Ca, Na)32[(Si70(O, OH)180)](OH, F)4.0. nH2O, solved by conventional and automated electron diffraction. Mineral Mag 74:159–177

Rozhdestvenskaya IV, Nikishova LV, Lazebnik KA (1996) The crystal structure of frankamenite. Mineral Mag 60:897–905

Schmid H (2008) Some symmetry aspects of ferroics and single phase multiferroics. J Phys Condens Matter 20:434201–434225

Schmid H (2010) Personal communication

Schmid H, Tippmann H (1978) Spontaneous birefringence in boracites – measurements and applications. Ferroelectrics 20:21–36

Seidel J, Martin LW, He Q, Zhan Q, Chu YH, Rother A, Hawkridge ME, Maksymovych P, Yu P, Gajek M, Balke N, Kalinin SV, Gemming S, Wang F, Catalan G, Scott JF, Spaldin NA, Orenstein J, Ramesh R (2009) Conduction at domain walls in oxide multiferroics. Nat Mater 8:229–234

Setkova TV, Shapovalov YuB, Marakushev AA, Balitskii VS (2009) Experimental study of stability and crystallization peculiarities of tourmaline in hydrothermal conditions. Dokl Earth Sci 425:490–493

Sigman MB Jr, Korgel BA (2005) Strongly birefringent $Pb_3O_2Cl_2$ nanobelts. J Am Chem Soc 127:10089–10095

Siidra OI, Krivovichev SV, Depmeier W (2007) Structure and mechanism of the ionic conductivity of the nonstoichiometric compound $Pb_{2+x}OCl_{2+2x}$. Dokl Phys Chem 414:128–131

Trill H (2002) Sodalite solid solution systems: synthesis, topotactic transformations, and investigation of framework-guest and guest-guest interaction. Doctorate Thesis, University of Münster

Wang S, Alekseev EV, Divu J, Casey WH, Phillips BL, Depmeier W, Albrecht-Schmitt TE (2010) NDTB-1: a supertetrahedral cationic framework that removes TcO_4^- from solution. Angew Chem 122:1075–1078, Angew Chem Int Ed DOI: 10.1002/ange.200906397

Where Are New Minerals Hiding? The Main Features of Rare Mineral Localization Within Alkaline Massifs

Gregory Yu. Ivanyuk, Victor N. Yakovenchuk, and Yakov A. Pakhomovsky

1 Introduction

Alkaline and alkaline-ultrabasic massifs of the Kola Peninsula are unrestrained world's leaders in mineral diversity. More than 700 mineral species have been found here, and more than 200 of them – for the first time in the world. Discoveries of new minerals within alkaline massifs of the Kola Peninsula started in nineteenth century from W. Ramsay's expeditions in the Khibiny and Lovozero mountains (Ramsay 1890; Ramsay and Hackman 1893) when lamprophyllite and murmanite were described. In twentieth century, quantity of minerals firstly discovered here was increasing exponentially with time, and well-known monograph of A. Khomyakov "Mineralogy of hyperagpaitic alkaline rock" (1995) gave list of 109 new minerals from these massifs. Now list of minerals discovered in the Khibiny and Lovozero massifs includes 198 species and constantly grows on 5–10 minerals per year.

A lot of minerals discovered in these massifs attract a special attention as prototypes of new functional materials. Synthetic analogues of zorite, chuvruaiite, sitinakite, ivanyukite, strontiofluorite and some other minerals are promising materials for a wide range of industrial applications, including gas separation, catalysis, radioactive waste management, pharmacology, optics, laser production, etc. It permits us to found a technology of new mineral prospecting in alkaline massifs for purposes of new functional materials development.

G.Y. Ivanyuk (✉) • V.N. Yakovenchuk • Y.A. Pakhomovsky
Nanomaterials Research Center, Kola Science Center, the Russian Academy of Sciences, 14 Fersman Street, Apatity 184209, Russia
e-mail: ivanyuk@ksc.ru

S.V. Krivovichev (ed.), *Minerals as Advanced Materials II*,
DOI 10.1007/978-3-642-20018-2_2, © Springer-Verlag Berlin Heidelberg 2012

2 Kola Alkaline Province

The Kola Peninsula is a part of the ancient Fennoscandian shield (1.5–3.2 milliard years old) consisting of different metamorphic rocks: granite-gneiss, amphibolite, Banded Iron-Formation, kyanite and mica schists. The Archaean and Proterozoic metamorphic complexes are intruded by granitic, ultrabasic, basic and alkaline massifs (Fig. 1). The Kola alkaline province includes 22 named alkaline and alkaline-ultrabasic massifs and about 60 unnamed massifs, separated pipes and dikes of alkaline rocks. Most of them have Devonian age of about 380 million years (Bayanova 2004).

Distribution of the massifs size is of power kind (Fig. 2), which corresponds to our knowledge about the Kola alkaline province as a unified system with self-organized criticality (SOC) (Ivanyuk et al. 2009). According to the SOC theory (Bak 1997) distribution of all other characteristics, for example, mineral diversity also must be power, and the largest Khibiny massif must be the leader again. Really, the massif size determines both quantity of minerals in whole and quantity of firstly discovered minerals (Fig. 3). It is important that the larger massif contains the greater proportion of new and endemic minerals. If the Kola alkaline province would be larger, it would contain a superlarge massif (about 20000 km^2) consisting of only endemic minerals!

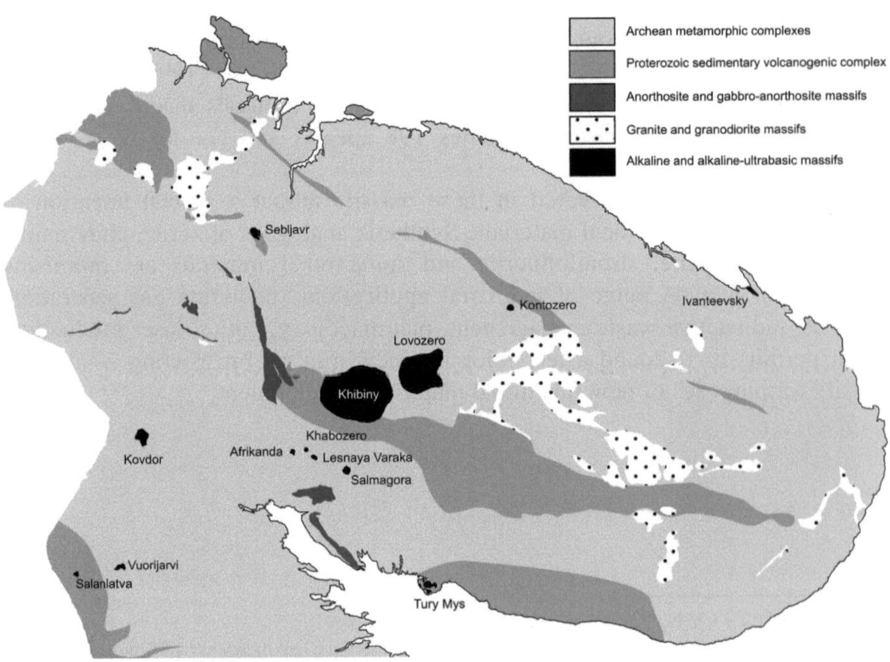

Fig. 1 Simplified geological map of the Murmansk Region

Fig. 2 Number of alkaline massifs with area exceeding S km^2 as a function of S

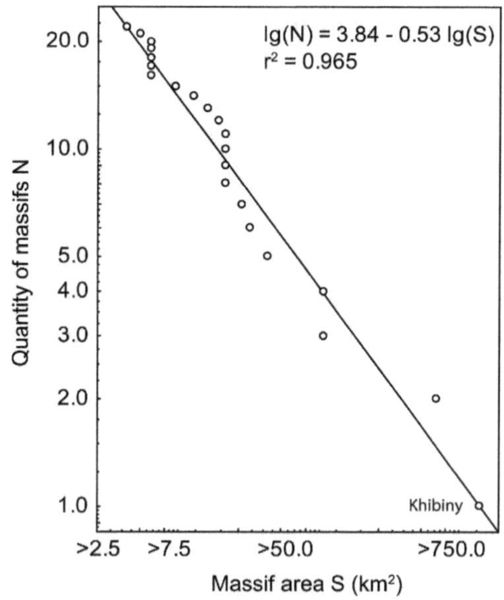

Fig. 3 Relationship between the massif size and number of minerals known in this massif

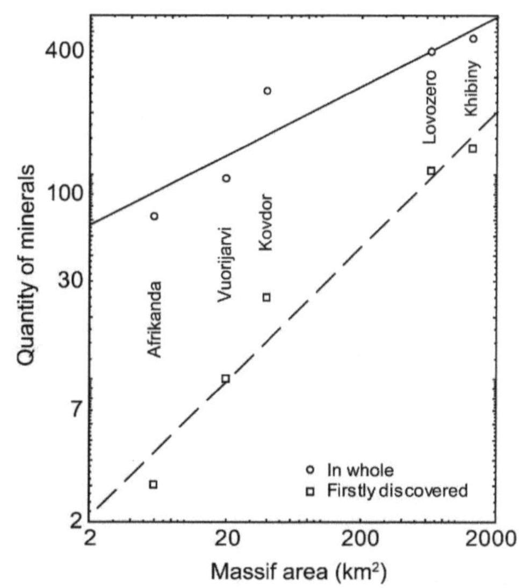

Quantity of known minerals – prototypes of advanced materials – also depends on the massif size, and the Khibiny massif is the most promising again (14 such minerals in comparison with 8 ones in the Lovozero massif and 1 mineral in the Kovdor massif). For this reason it is reasonable to discuss features of rare minerals localization within alkaline complexes on an example of the Khibiny massif.

3 The Khibiny Massif

The world's largest Khibiny alkaline massif occupies the area of about 1327 km^2 in the extreme West of the Kola Peninsula, at the contact of rocks of the Imandra-Varzuga Proterozoic greenstone belt and the Archaean metamorphic complexes of the Kola-Norwegian megablock (see Fig. 1). About 70% of the massif area is occupied by nepheline syenites (foyaite) monotonous in composition which are, in most works, subdivided into two equal parts: foyaite proper (in the center) and "khibinite" (surrounding them), separated from each other by a zone rock complex of the Main Ring (Fig. 4). Within the Main Ring, foidolites (melteigite–ijolite-urtite), high-potassic (leucite normative) poikilitic nepheline syenite (rischorrite) and less widespread malignite, as well as titanite-nepheline, titanite-apatite and

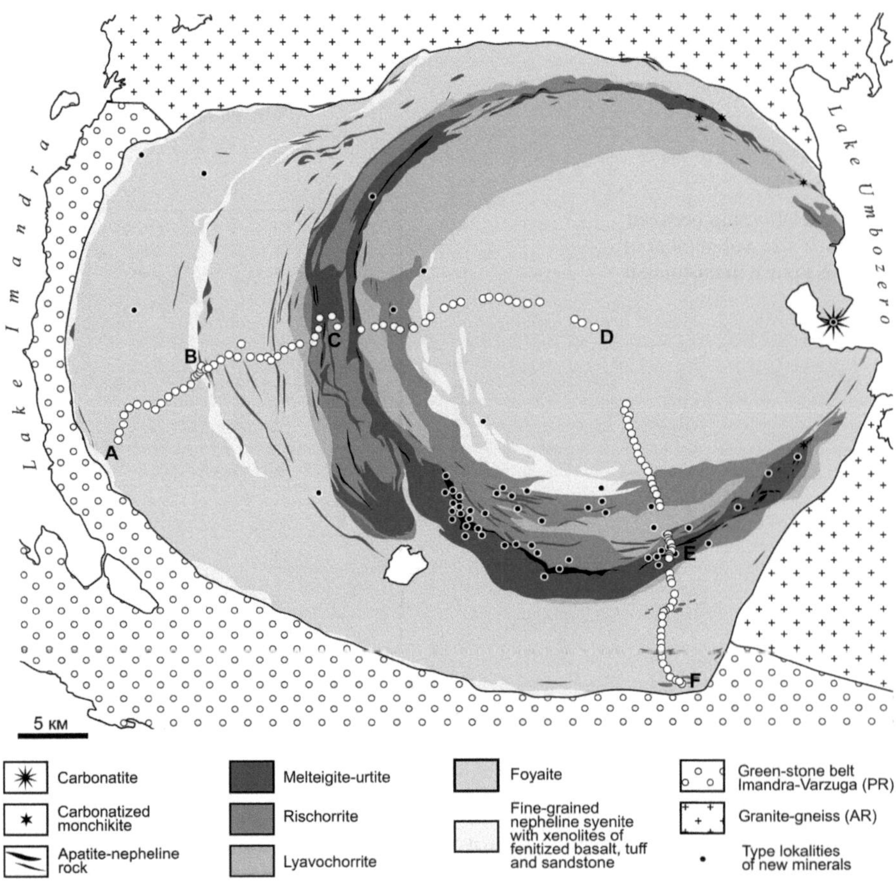

Fig. 4 Simplified geological map of the Khibiny massif (Ivanyuk et al. 2009). *A-B-C-D-E-F* is a profile for the massif zoning study with sampling points

apatite-nepheline rocks are of crucial importance. The same complex of rocks can be related to the so-called irregular-grained nepheline syenite ("lyavochorrite"), transitive to rischorrite in accordance with modal composition, texture-structural features and geological position (Yakovenchuk et al. 2005; Ivanyuk et al. 2009).

The rock complex of the Main Ring fills a conic fault in which the angle between the axis and generatrix varies between 50–70° close to the surface and 10–40° at the depth of more than 1 km. On the day surface, rocks of this complex occupy 30% of the total area of the massif, the share of foidolites, rischorrite and lyavochorrite making up 10 vol.% each. Apatite-nepheline and titanite-apatite-nepheline rocks form ore stockworks in the apical parts of the foidolite ring, being related to it by gradational transitions. The thickness of these deposits, proven only on the basis of isolines of apatite content, ranges from 200 m in the south-western part of the Main Ring up to the first meters in its north-eastern part.

Within the Main Ring and, especially, in the adjoining parts of nepheline syenites (on both sides of the Ring), there are a lot of xenoliths (from half a meter up to several kilometers across) of volcanogenic-sedimentary rocks metamorphosed to hornfels and fenitized. Normative composition of xenoliths varies from practically pure quartzite and olivine basalt to nepheline syenite (i.e. fenite). Xenoliths, though occupying less than 1% of the total day surface of the massif, are in constant association with the much wider spread fine-grained alkaline and nepheline syenites obviously representing the result of a more or less deep fenitization of volcanogenic-sedimentary rocks metamorphosed to hornfels.

Dyke rocks of the Khibiny massif are represented, for the most part, by hypabyssal analogues of its plutonic rocks: alkali-feldspar trachyte, phonolite and melanephelinite, mainly concentrated near the Main Ring, as well as by monchiquite and carbonatite composing veins and explosion pipes in its eastern part (Arzamastsev et al. 1988; Yakovenchuk et al. 2005). Pegmatite and hydrothermal veins, including an unusually great number of mineral species (about 300), are common throughout the massif, with their main concentration within rischorrite and foidolites of the Main Ring. In foyaite, there are ordinary clinopyroxene-nepheline-microcline veins, but, as the Main Ring is approached, their mineral composition becomes more and more varied – up to 80 minerals in a vein (Khomyakov 1995; Yakovenchuk et al. 2005). Almost all new minerals have been found within or nearby the Main Ring, while the rest part of the massif is free from rare minerals (see Fig. 4).

To understand reason of this inhomogeneity we have carried out study of mineral, petrographic and geochemical zonation of the Khibiny massif along the profile from its NW (point *A* at Fig. 4) to SE boundary (*F*) across the Marchenko apatite deposit (*C*), central point of the massif (*D*) and Koashva apatite deposit (*E*). The plot of quantity of rock-forming and accessory minerals in a rock has an intensive minimum in the area of the Koashva deposit and a weak minimum in the area of the Marchenko deposit (Fig. 5). These minimums correspond to the maximal quantity of mineral species known at these intervals. It means that the great mineral diversity of apatite deposits is related to pegmatites and zones of a later mineralization in both of which the impurities were moved during the ore zone

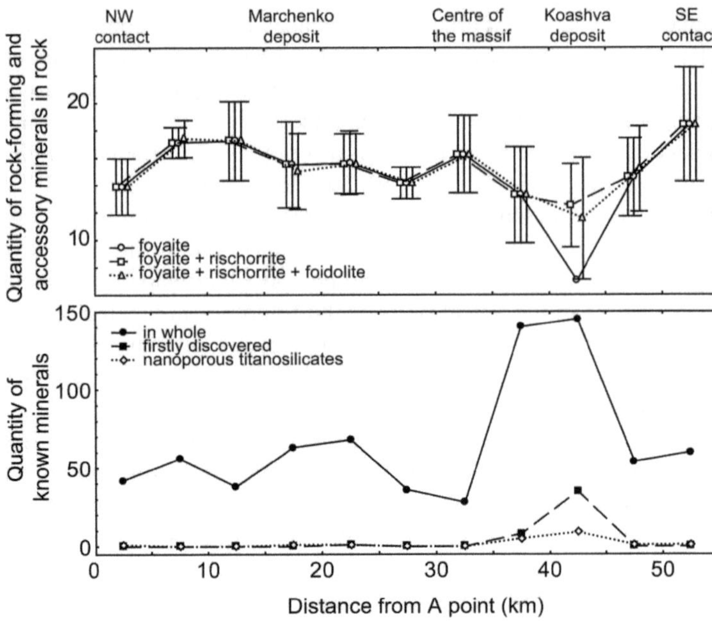

Fig. 5 Variation of quantity of mineral in alkaline rock along the *A-B-C-D-E-F* profile

formation. These impurities can be produced by accessory minerals destruction as well as by rock-forming minerals self-cleaning. The larger thickness of foidolite intrusion in the area of the Koashva deposit causes more long and intensive metasomatic and hydrothermal processes, longer chains of mineral transformations and, finally, larger mineral diversity.

For example, numerous rare minerals of sodalite-aegirine-microcline bulbs in apatitized urtite of the Koashva deposit were produced by foidolite self- cleaning from impurities. Ivanyukite-Na-*T* is a result of lamprophyllite decomposition within one of these bulbs (Yakovenchuk et al. 2009). Ivanyukite-Na-*C* and ivanyukite-K are consecutive products of partial decationization of ivanyukite-Na-*T*. Lastly, ivanyukite-Na-Cu is a result of copper–potassium exchange in ivanyukite-K:

$$Na_2K\left[Ti_4(OH)O_3(SiO_4)_3\right]\bullet 7H_2O \text{ – ivanyukite-Na-T}$$
$$\downarrow \text{-Na}$$
$$NaK\left[Ti_4(OH)_2O_2(SiO_4)_3\right]\bullet 6H_2O \text{ – ivanyukite-Na-C}$$
$$\downarrow \text{-Na}$$
$$K\left[Ti_4(OH)_3O(SiO_4)_3\right]\bullet 9H_2O \text{ – ivanyukite-K}$$
$$\downarrow \text{-K,+Cu}$$
$$Cu\left[Ti_4(OH)_2O_2(SiO_4)_3\right]\bullet 7H_2O \text{ – ivanyukite-Cu.}$$

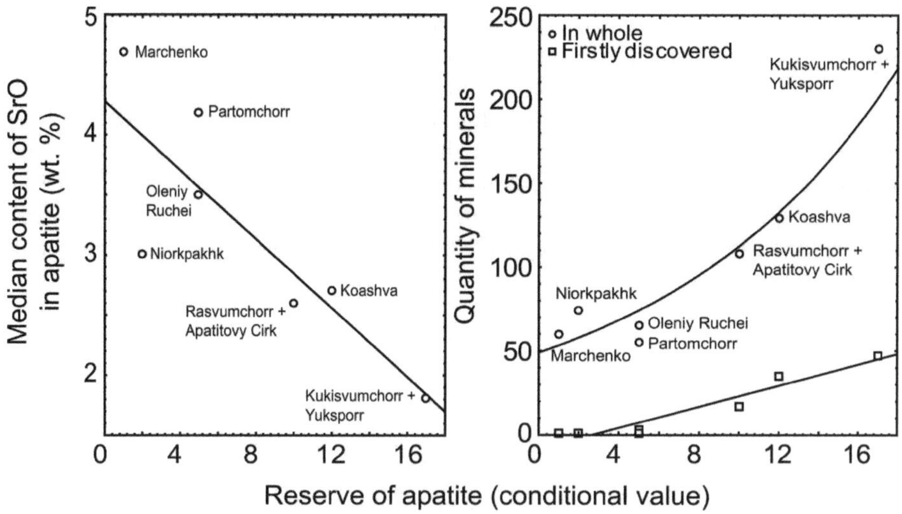

Fig. 6 Relation between size of apatite deposit, composition of apatite and quantity of minerals known in this deposit

Origin of the most of rare minerals by means of self-cleaning of rock-forming minerals causes good correlation between composition of rock-forming minerals and mineral diversity (Fig. 6).

The largest deposit has the simplest mineral composition of ores, closest to ideal composition of rock-forming minerals, highest mineral diversity and longest list of firstly discovered minerals. This rule is true for all subsystems of the SOC system including set of massifs within a province, set of deposits within a massif, set of ore bodies within a deposit and set of zones within an ore body. According to this rule, in the Khibiny massif, Kukisvumchorr-Yuksporr apatite deposit is mostly perspective for new minerals discovering and it is so in reality.

This approach is also effective within a separate deposit, which can be evidently shown on an example of the Kovdor deposit of magnetite, apatite and baddeleyite. Kovdor deposit of magnetite, apatite and baddeleyite is a well-known source of new phosphates (bakhchisaraitsevite, girvasite, gladiusite, juonniite, cattiite, kovdorskite, krasnovite, pakhomovskyite, rimkorolgite and strontiowhitlockite) and quintinite group minerals (quintinite-6*R*, -1*M*, -2*H*, manasseite and karchevskyite) – promising for many industrial purposes layered double hydroxides (Britvin 2008; Krivovichev et al., this book).

4 The Kovdor Deposit of Magnetite, Apatite and Baddeleyite

The Kovdor massif of ultrabasic, alkaline rocks and carbonatites is a central-type, multiphase igneous intrusion emplaced into Archaean granite gneisses and granite-gneiss (Ivanyuk et al. 2002). In plan, the massif has a distinct concentric, zoned structure and contains three pronounced, ring-shaped complexes (from the centre towards the outer part of the massif): olivinite (1), diopside-, phlogopite-, and melilite-rich metasomatic rocks (2), turjaite and melteigite-urtite (3). At the contact of olivinite and foidolite intrusions in the west, the massif is intruded by a vertical concentric zoned pipe of apatite-magnetite-forsterite rock in the outer zone and magnetite-carbonate rock in the central zone (Iron-Ore Complex, Fig. 7). Transformation of apatite-forsterite and magnetite-apatite-forsterite rocks of the outer zone into comparatively late apatite-magnetite-calcite rock of the central zone is gradual.

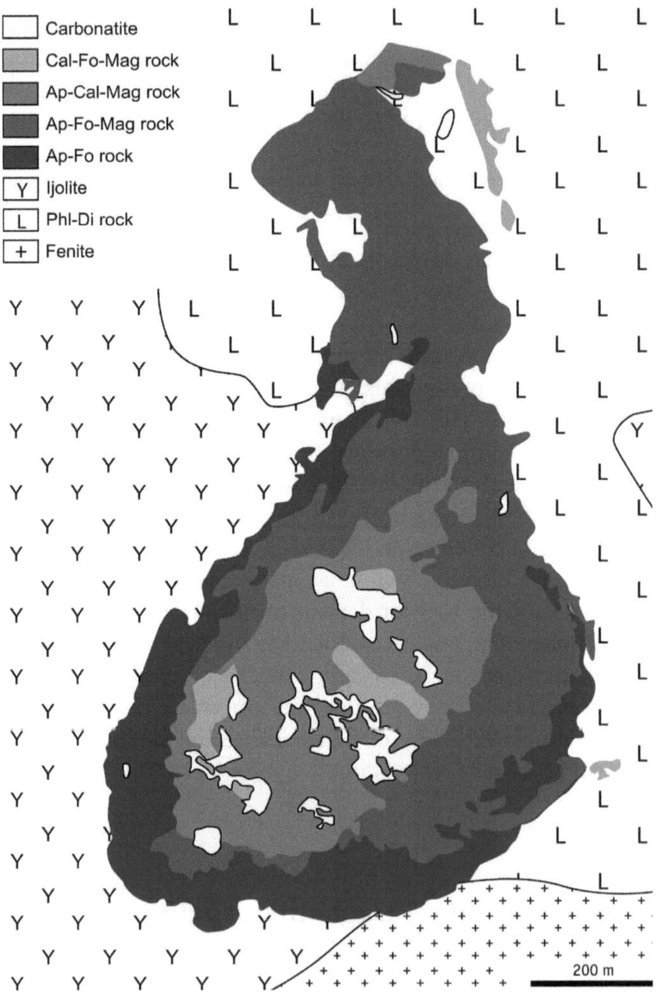

Fig. 7 Simplified geological map of the Kovdor deposit of magnetite, apatite and baddeleyite (Ivanyuk et al. 2002)

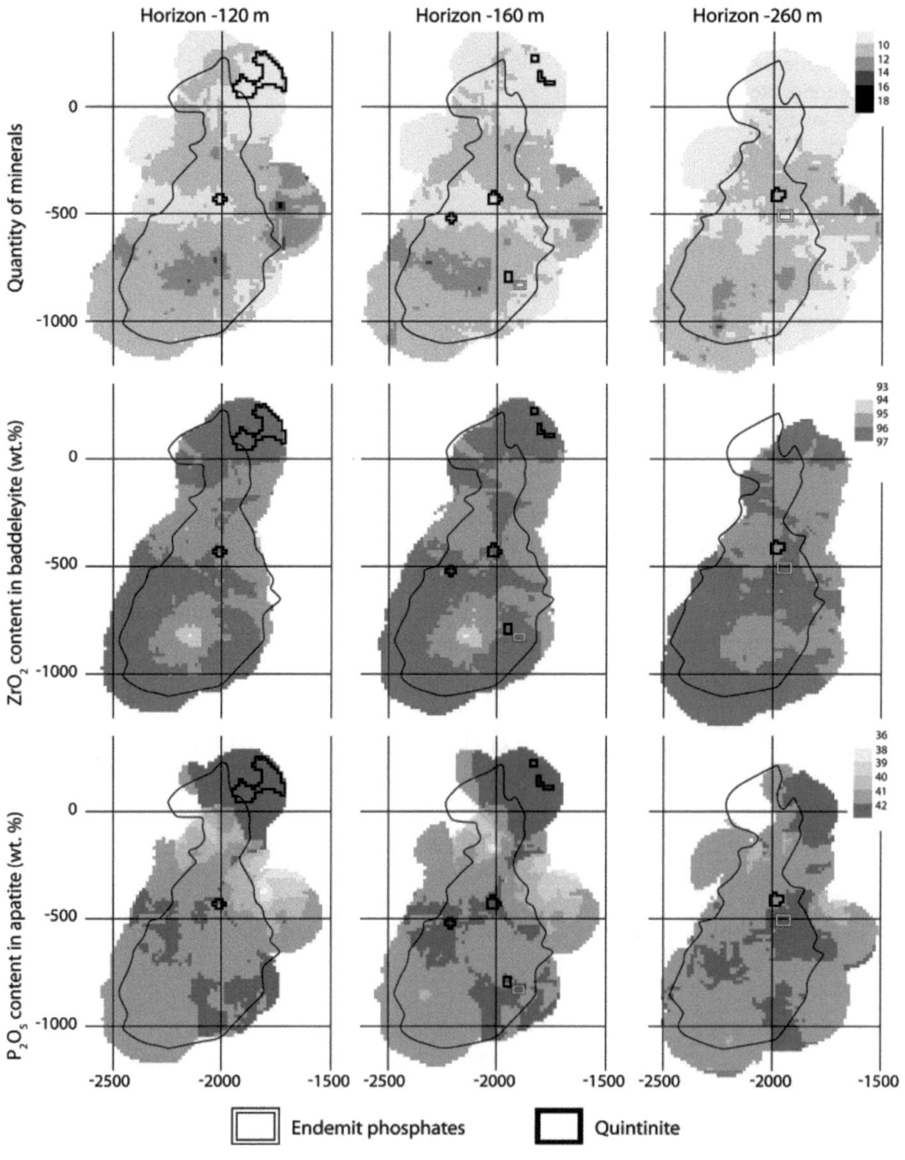

Fig. 8 Distribution of quantity of rock-forming and accessory minerals, ZrO_2 content in baddeleyite and P_2O_5 content in apatite within Kovdor Iron-Ore complex and localities of endemic phosphates and quintinite

Our study of the Iron-Ore Complex revealed that all above listed interesting minerals are again localized in areas with the simplest modal composition of rock and chemical composition of rock-forming minerals (Fig. 8). Mechanism of such reorganization can be comprehensively explained on the example of quintinite origin in magnetite-calcite ores.

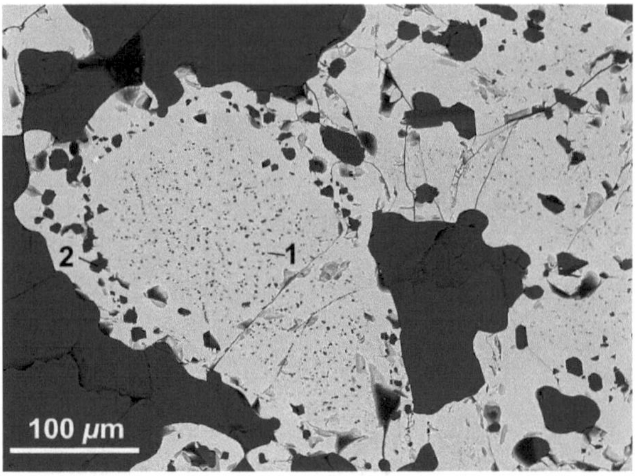

Fig. 9 Exsolution inclusions of spinel (**1**) and quintinite (**2**) in magnetite from calcite-apatite-forsterite-magnetite rock of the Kovdor Iron-Ore complex

Chemical composition of magnetite varies in a wide range because it is sufficiently enriched by Mg and Al (up to 3.5 wt.%) in comparatively high-temperature apatite-magnetite-forsterite rock and by Mg and Ti (up to 8.5 wt.%) in comparatively low-temperature magnetite-calcite rock. Exsolution of magnetite during the rock cooling gives numerous inclusions of spinel and/or ilmenite-geikielite (Fig. 9):

$$\underset{\text{Mg-Al-rich magnetite}}{4(Fe^{2+}{}_{0.75}Mg_{0.25})(Fe^{3+}{}_{1.50}Al_{0.50})O_4} = \underset{\text{magnetite}}{3Fe_3O_4} + \underset{\text{spinel}}{MgAl_2O_4},$$

$$\underset{\text{Mg-Ti-rich magnetite}}{2Fe^{3+}(Fe^{2+}Mg_{0.50}Ti_{0.50})O_4} = \underset{\text{magnetite}}{Fe_3O_4} + \underset{\text{ilmenite-geikielite}}{(MgFe)TiO_3}\ .$$

Primary quintinite is also a result of Mg-Al-rich magnetite self-cleaning (a low-temperature analog of spinel):

$$\underset{\text{Mg-Al-rich magnetite}}{8(Fe^{2+}{}_{0.5}Mg_{0.5})(Fe^{3+}{}_{1.75}Al_{0.25})O_4} + 8H_2O + H_2CO_3$$

$$= \underset{\text{magnetite}}{6Fe_3O_4} + \underset{\text{quintinite}}{Mg_4Al_2(CO_3)(OH)_{12}\bullet 3H_2O} + {}^{1}\!/_{2}O_2.$$

It forms small (up to 10 μm) rounded inclusions in marginal parts of magnetite crystals (Fig. 9), sometimes together with ilmenite-geikielite lamellae. Quintinite recrystallization forms large (up to 4 cm) well shaped crystals in voids of calcite-magnetite ore. In contrast to synthetic LDH grains, these crystals are accessible to single-crystal X-ray diffraction analysis, which allowed determining peculiarities of cation and anion ordering in their crystal structure (Krivovichev et al., this book).

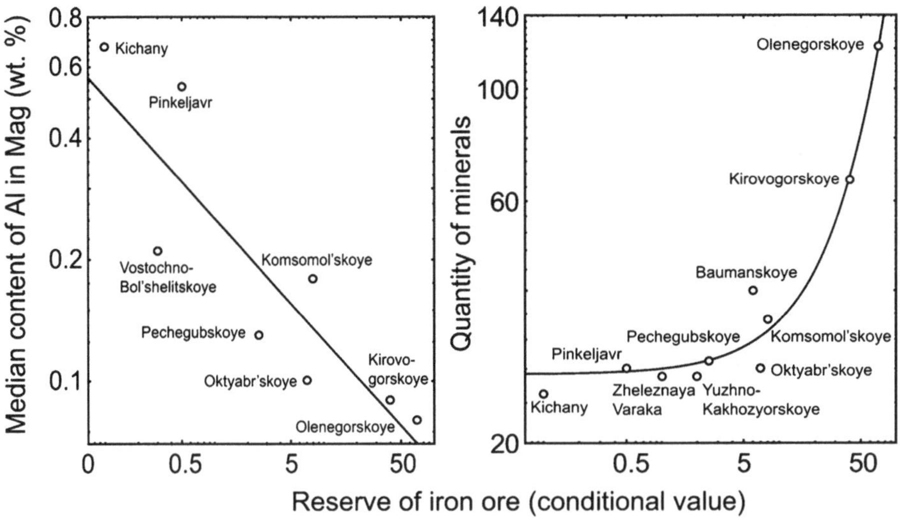

Fig. 10 Relation between size of BIF deposit, magnetite composition and quantity of minerals known in this deposit

5 Conclusion

Thus most of interesting for purposes of material science minerals will be localized in the largest object of separated geological SOC system (ore province, massif etc.). Within this object, those parts where composition of rock-forming minerals is closer to their ideal composition are mostly perspective. Usually, these are parts of deposits and ore bodies with the best ores (magnetite, apatite, etc.).

This rule works within geological SOC system of any nature (igneous, metamorphic or even sedimentary). For example, the largest deposit of the Kola Banded Iron-Formation, Olenegorskoye (Fig. 10), has the region best ores (with most primitive modal composition, highest magnetite content and lowest content of impurities in magnetite) and highest mineral diversity including two unknown phases (Ivanyuk et al. 2009).

Alkaline massifs are much more rich in rare minerals, and application of our rule to above described profile through the Khibiny massif (see Fig. 4) helped us to discover seven new minerals with unique properties (see Fig. 5): punkaruaivite (Yakovenchuk et al. 2010a, b; this book), chivruaiite (Men'shikov et al. 2006), ivanyukite-Na, ivanyukite-K and ivanyukite-Cu (Yakovenchuk et al. 2009; this book), strontiofluorite and polezhaevaite-(Ce) (Yakovenchuk et al. 2010a, b). All this allows us to develop a new technology of minerals discovering on the basis of the above rule.

Acknowledgements This work was partially supported by Russian Foundation for Basic Research (grant 10-05-00431) and JSC "Kovdorsky Mining and Dressing Plant".

References

Arzamastsev AA, Kaverina VA, Polezhaeva LI (1988) Daikovye porody Khibinskogo massiva i ego obramleniya (Dyke rocks of the Khibiny massif and its surrounding area). Kola Science Centre of AN SSSR Publishing Apatity (in Russian)

Bak P (1997) How nature works: the science of self-organized criticality. Oxford University Press, Oxford

Bayanova TB (2004) Age of reference geological complexes in the Kola region and duration of magmatic processes. Nauka St. Petersburg (in Russian)

Britvin SN (2008) Structural diversity of layered double hydroxides. In: Krivovichev S (ed) Minerals as advanced materials. Springer, Berlin/Heidelberg, pp 123–128

Ivanyuk GYu, Yakovenchuk VN, Pakhomovsky YaA (2002) Kovdor. Laplandia Minerals Apatity

Ivanyuk GYu, Goryainov PM, Pakhomovsky YaA, Konoplyova NG, Yakovenchuk VN, Bazai AV, Kalashnikov AO (2009) Self-organization of ore-bearing complexes. GEOS, Moscow (in Russian)

Khomyakov AP (1995) Mineralogy of hyperagpaitic alkaline rocks. Clarendon Press Oxford, Oxford

Men'shikov YuP, Krivovichev SV, YaA P, Yakovenchuk VN, Ivanyuk GYu, Mikhailova JA, Armbruster T, Selivanova EA (2006) Chivruaiite, Ca4(Ti, Nb)5[(Si6O17)2 (OH, O)5] •13–14H_2O, a new mineral from hydrothermal veins of Khibiny and Lovozero alkaline massifs. Am Miner 91:922–928

Ramsay W (1890) Petrographische Beschreibung der Gesteine des Lujavr-urt. Fennia 3:1–52

Ramsay W, Hackman V (1893) Das Nephelinsyenitgebiet auf der Halbinsel Kola. Fennia 11:1–225

Yakovenchuk VN, Ivanyuk GYu, Pakhomovsky YaA, Men'shikov YuP (2005) Khibiny. Laplandia Minerals Apatity

Yakovenchuk VN, Nikolaev AP, Selivanova EA, YaA P, Korchak JA, Spiridonova DV, Zalkind OA, Krivovichev SV (2009) Ivanyukite-Na-*T*, ivanyukite-Na-*C*, ivanyukite-K, and ivanyukite-Cu: new microporous titanosilicates from the Khibiny massif (Kola Peninsula, Russia) and crystal structure of ivanyukite-Na-*T*. Am Miner 94:1450–1458

Yakovenchuk VN, Ivanyuk GYu, Pakhomovsky YA, Selivanova EA, Men'shikov YuP, Korchak JA, Krivovichev SV, Spiridonova DV, Zalkind OA (2010a) Punkaruaivite, $LiTi_2$ $[Si_4O_{11}(OH)]$ $(OH)_2$•H_2O, a new mineral species from hydrothermal assemblages, Khibiny and Lovozero alkaline massifs, Kola peninsula, Russia. Can Miner 48:41–50

Yakovenchuk VN, Selivanova EA, Ivanyuk GYu, YaA P, Korchak JA, Nikolaev AP (2010b) Polezhaevaite-(Ce), $NaSrCeF_6$, a new mineral from the Khibiny massif (Kola Peninsula, Russia). Am Miner 95:1080–1083

Gas Release from Minerals

Klaus Heide

1 Introduction

Quite generally, volatiles are released from minerals during heat treatment in high vacuum conditions. Well known is the gas release during the decomposition of hydrates, hydroxides, carbonates, less so is the release of sulphur species by the decomposition of sulfates, sulfites and sulfides. Oxygen released during the thermal treatment of oxides with polyvalent cations. Most interest is the charge transfer between Fe^{2+} and Fe^{3+} in oxides and silicates and the formation of volatile oxygen. The thermal stability of halogenides in crystalline structures is strongly determined by the water presence. The decomposition of complex silicates and borates with halogen anions in the crystal structure is especially of interest.

Fluid inclusions are a second source of volatiles. The release is characterized by spiky increasing the partial pressure during the decrepitating of the inclusion. Beside water CH_4 or other hydro-carbons, CO_2, and noble gases are observed. It was shown by the gas release profiles that formal "volatile-free" minerals release traces of volatiles in relation to the genesis. The knowledge about the bonding and fixation of such traces in the crystal structure or on the crystal surface is small.

A special problem of analysis of natural samples is the separation of alteration processes from the primary genetic condition. The release of water, CO_2 or hydrocarbons above 800°C do not released from the decomposition of alteration products as hydrates, hydroxides or carbonates.

The paper will show that Gas Release Profiles (GRP) are a useful tool for the mineral characterization, the genetic interpretation and the quality control of natural and synthetic crystals and rocks.

K. Heide (✉)
Institut für Geowissenschaften, Friedrich-Schiller-Universität Jena, Burgweg 11,
Jena D-+49 – 07749, Germany
e-mail: ckh@uni-jena.de

S.V. Krivovichev (ed.), *Minerals as Advanced Materials II*,
DOI 10.1007/978-3-642-20018-2_3, © Springer-Verlag Berlin Heidelberg 2012

2 Analytical Technique

Degassing experiments with single mineral fragments were carried out using a high-vacuum-hot-extraction method with a quadrupol mass spectrometer for the detection of volatiles. With the **D**irectly coupled **E**volved **G**as **A**nalyzing **S**ystem (DEGAS) the volatile species were analyzed in multiple ion detection mode and correlated with the total pressure change in the sample chamber during heating. The volatile species are determined by the change in the partial pressure during the controlled heating rate of 10 K/min between room temperature and 1450°C. In contrast to degassing experiments using a Knudsen cell arrangement or scimmer and capillary systems, the DEGAS experiments occur under highly non-equilibrium conditions. This is a very good feature because reverse reaction between volatiles with each other and between the crystal or the melt are largely prevented. The distance between the sample in the crucible and the ion sources of mass spectrometer was minimized up to 15 cm. Interaction between the evolved volatile species in the molecular beam can be excluded under high vacuum conditions (Fig. 1).

The volatile species are characterized by the relation of mass number (m) and the electric charge (z) of ionized molecules or molecule fragments:

m/z *Assignment of possible species*	
m/z 1	$^{1}H^{+}$ as fragment from H_2O, H_2, and CH_x
m/z 2	$^{1}\mathbf{H_2}^{+}$ or as a fragment from H_2O and/or D^{+} as a product from D_2O
m/z 4	$^{4}\mathbf{He}^{+}$ and/or D_2^{+} as a fragment from D_2O
m/z 10	^{10}B
m/z 11	^{11}B
m/z 12	$^{12}C^{+}$ as fragment from CO and CO_2 or C_xH_y
m/z 13	$^{13}C^{+}$ as fragment from CO, CO_2 or C_xH_y and $^{12}CH^{+}$
m/z 14	$^{14}N^{+}$, $^{13}CH^{+}$ and $^{12}CH_2^{+}$
m/z 15	$^{12}\mathbf{CH_3}^{+}$, $^{13}CH_2^{+}$, NH_{+}
m/z 16	$^{16}O^{+}$ as fragment from H_2O, CO_2, and CO(?),$^{12}\mathbf{CH_4}^{+}$, $^{13}CH_3^{+}$, NH_2^{+}
m/z 17	$^{16}OH^{+}$ as fragment from $H_2{}^{16}O$, $\mathbf{NH_3}^{+}$**(?)**
m/z 18	$\mathbf{H_2{}^{16}O}^{+}$, NH_4^{+}, $^{16}O^{+}$
m/z 19	$^{18}OH^{+}$ as fragment from $H_2{}^{18}O$, $HD^{16}O^{+}$(?), $\mathbf{F}^{+}$
m/z 20	$H_2{}^{18}O^{+}$, $D_2{}^{16}O^{+}$(?), $\mathbf{HF}^{+}$
m/z 23	$^{23}\mathbf{Na}^{+}$
m/z 28	$\mathbf{N_2}^{+}$, $^{12}\mathbf{C}^{16}\mathbf{O}^{+}$ (primary and secondary), $^{12}C_2H_4^{+}$
m/z 32	$^{16}\mathbf{O_2}^{+}$ or $^{32}\mathbf{S}^{+}$ as fragments from CO, CO_2, H_2O, SO_2
m/z 34	$\mathbf{H_2{}^{32}S}^{+}$, $^{34}S^{+}$, $^{18}O^{16}O^{+}$
m/z 35	$^{35}\mathbf{Cl}^{+}$
m/z 36	$^{36}Ar^{+}$, $^{12}C_3^{+}$, $\mathbf{H^{35}Cl}^{+}$
m/z 37	$^{37}Cl+$, $^{12}C_3H^{+}$
m/z 40	$^{40}\mathbf{Ar}^{+}$, $^{12}C_3H_4^{+}$
m/z 44	$^{12}\mathbf{C}^{16}\mathbf{O_2}^{+}$
m/z 48	$^{12}C^{18}O_2^{+}$, $^{32}S^{16}O^{+}$
m/z 64	$^{32}S^{16}O_2^{+}$

(continued)

m/z 80	$^{32}S^{16}O_3^+$
m/z 81	$^{11}B_2{}^{16}O_3^+$
m/z 82	$Na^{11}B^{16}O_3^+$
m/z 84	AlF_3^+
m/z 85	SiF_3^+
m/z 104	SiF_4^+

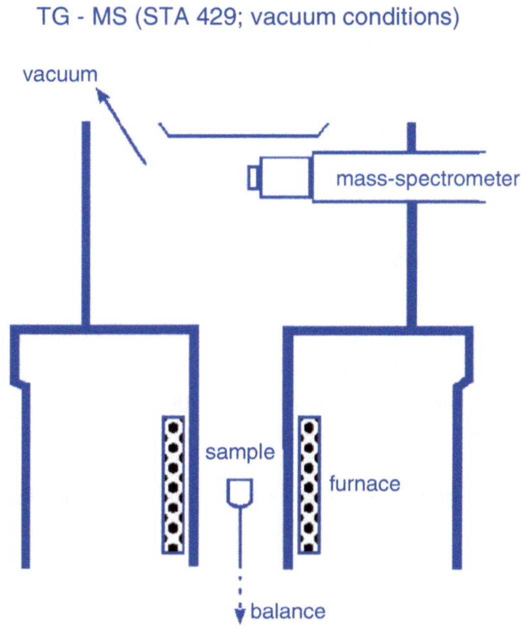

Fig. 1 Sketch of the DEGAS devices

Quantitative determinations are possible by calibration of the gas release profile with the decomposition of stoichiometric compounds. The calibration was possible for water, also in relation to spectroscopic and KFT determinations, for CO_2 by the decomposition of carbonates, for SO_2 by the decomposition of sulphates and for oxygen by vacuum reduction of haematite (Schöps et al. 2005).

In dependence of the total volatile content and the sample weight the minimum signals correspond to concentrations between 1 and 0.01 ppm.

The overall advantages of the newly developed method are (Heide et al. 2008):

- <100 mg of sample,
- no special sample preparation required,
- simultaneous qualitative and quantitative detection of different species,
- simultaneous detection of mass loss,
- linear regression over several units (ppm – wt.%),
- molecular masses between 1 and 200 detectable,
- time- and temperature resolved measurement.

3 Results

The gas-release-profiles (GRP) are determined by two processes:

- a continuous change of the degassing rate with a characteristic temperature maximum. These curves results by decomposition of solids or evaporation from solids or melts.
- A spiky change of partial pressure by bursting of fluid inclusions in solids or bubbles, formed in the melt during the heating.

Data of interest from the GRPs are the quality and quantities of released volatiles, the start and maximum temperatures of gas release, and temperature range of continuous or spiky gas release.

3.1 *Garnet*

As shown in Fig. 2, the GRP from a garnet of eclogite from Scandinavia is very complex and has indications for the decomposition of alteration products. The alteration is visible from the gas release maxima up to 300°C. The quantities of the alteration products are possible to determine by the weight loss and are in the order of 0.2–0.5 wt%.

Genetic significant are the gas release above 800°C. Remarkable are maxima of water releases by diffusion 1063°C resp. 928°C and of water release by a

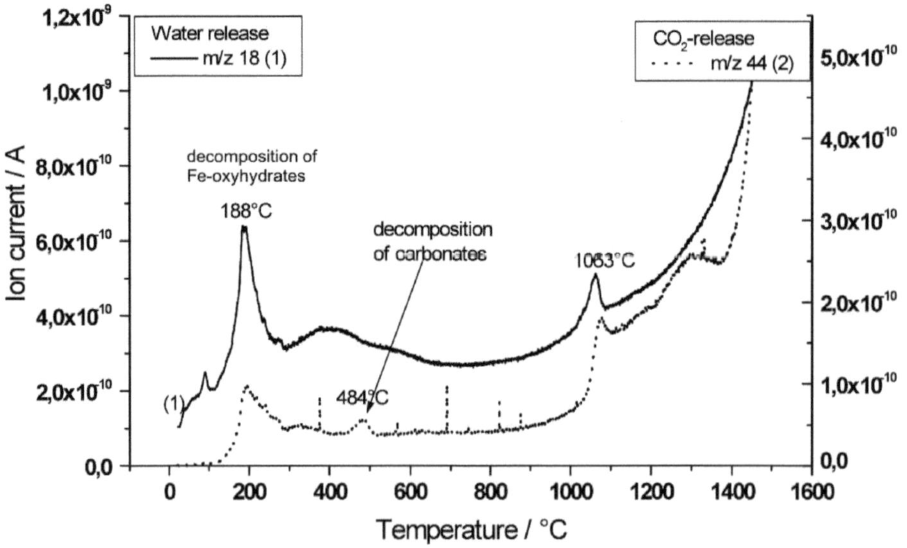

Fig. 2 GRP of a garnet from Scandinavian eclogite

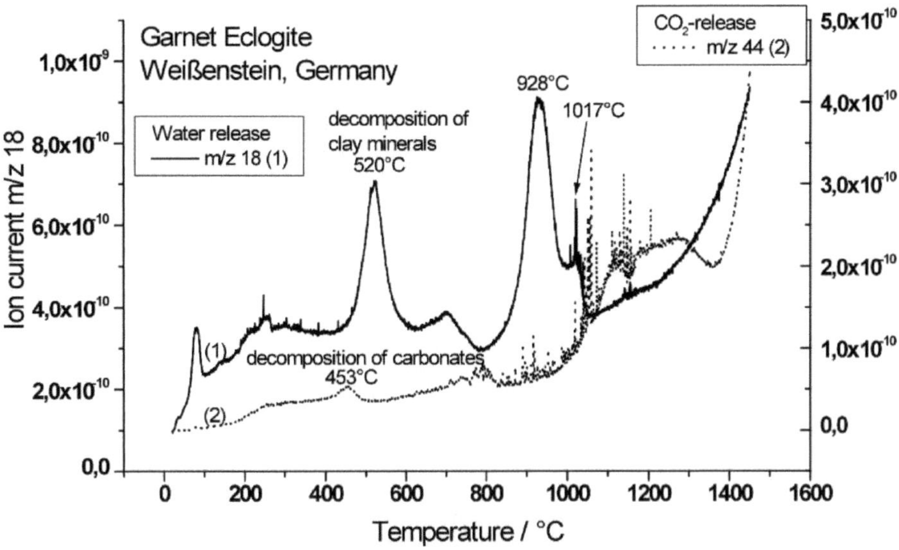

Fig. 3 Water and CO_2-release from eclogite garnet of Weißenstein, Germany

combination of diffusion and bubbling above 1000°C from eclogite garnets in different locations (Fig. 3). CO_2 was evolved in the same temperature range but by a different mechanism. More detailed analysis are necessary to explain these findings.

3.2 *Cordierite*

Cordierite is of high interest for the investigating of fluid-solid-interactions in geologic processes, because many of them are strongly affected by the nature of the fluid phase. The complex composition of fluids is of central importance. A few minerals can occlude various molecules or molecular species in micro/nano cavities, pores or channels. Thus, such minerals can be very useful for providing information on the nature of the fluid phase present during crystallization and/or entrapment.

Cordierite, ideal formula $(Mg,Fe)_2Al_4Si_5O_{18}\cdot(nH_2O,mCO_2\ldots)$, is a microporous silicate typically found in metamorphic rocks and sometimes in igneous rocks and it has also been reported in lunar rocks and meteorites. A number of molecular species can be occluded in its micropores as was shown by DEGAS experiments (Geiger et al. 2011). Significant for cordierites is a continuous gas release with a maximum of release rate in the range between 450°C and 1100°C (Fig. 4).

H_2O is found in concentration up to 3 wt% in many cordierites. In Fig. 5, H_2O degassing profiles for cordierites from different origin are shown and compared with the CO_2-release. They show different behavior and the maximum degassing

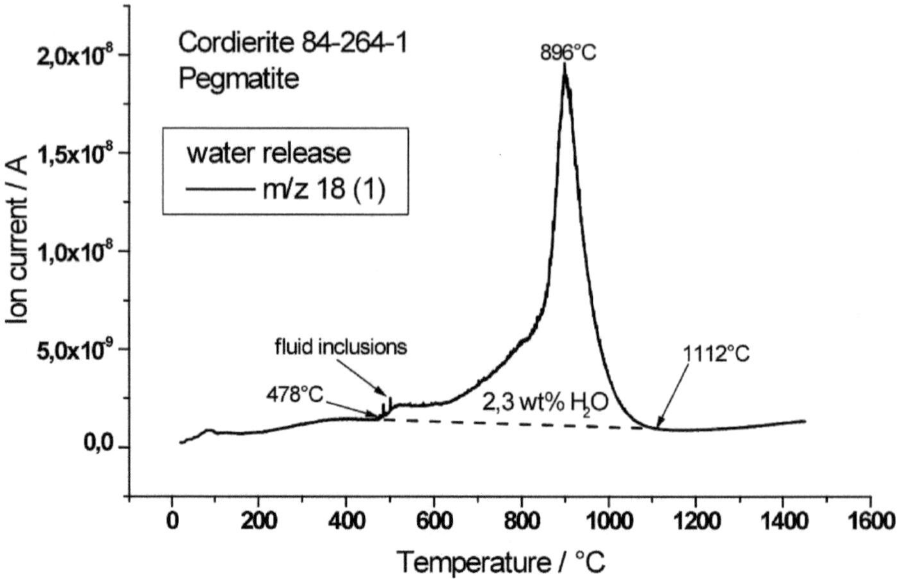

Fig. 4 Water release during the heat treatment of pegmatitic cordierite

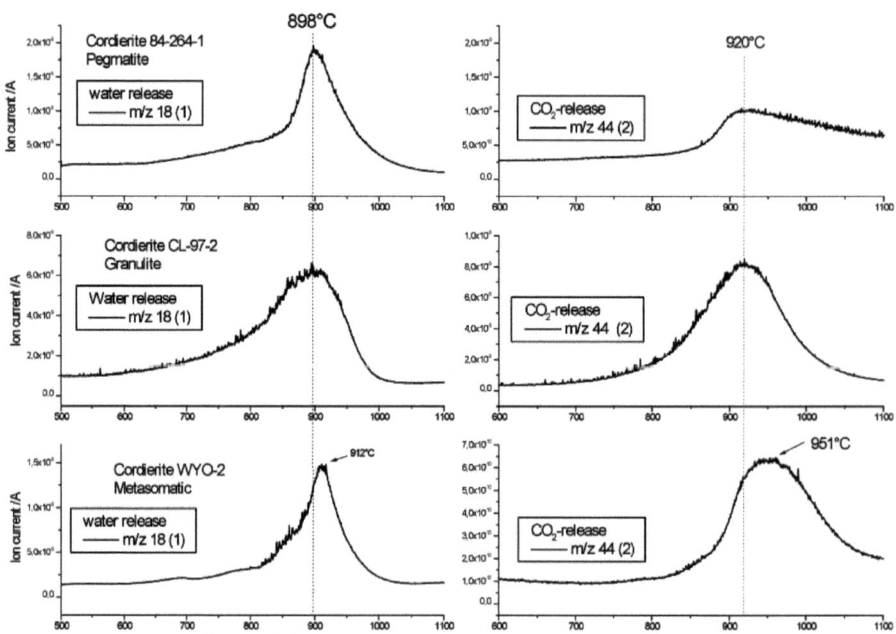

Fig. 5 H_2O and CO_2 content in cordierites from different genetic conditions

temperature varies both for the water (about 898 (pegmatite) and 912°C (muscovite schist)) and CO_2-release between 920°C and 950°C for the various cordierites.

The experimental observations suggest that the majority of the observed $^{12}C^+$, $^{13}C^+$, and CO^+ ion currents result from fragmentation of the CO_2 molecule. Owing to overlapping the different volatile species and fragments it is difficult to determine traces of CH_4 in the m/z16 signal. More sensitive to the methan detection is the m/z 15 (Heide et al. 2008). From the maxima of m/z 15 was concluded that only small amounts (<100 ppm) of primary CH_4 are occluded in the cordierite samples studied here. The high degassing temperature strongly suggests that the molecule is held in the channels and not on the surface or in some impurity phase. The data from metamorphic cordierite is characterized by a significant higher CO_2-content (Fig. 5).

3.3 *Haematite*

The high-temperature vacuum reduction of Fe^{3+}-compounds offers a very sensitive potential for the determination of the redox-state of solids and melts (Fig. 6).

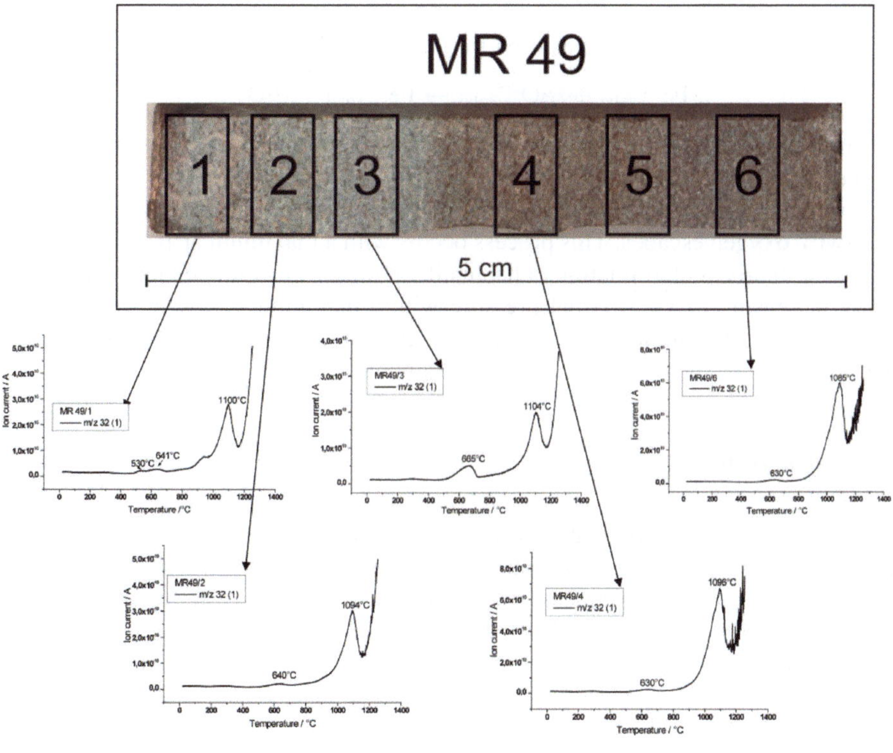

Fig. 6 Characterisation of bleaching phenomena in German red bed sediments by oxygen release (Hilse et al. 2010)

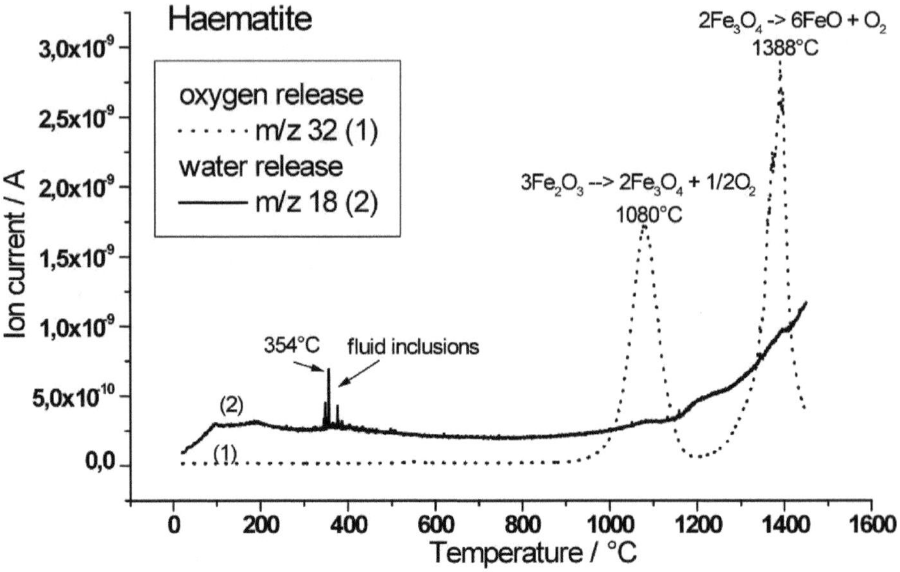

Fig. 7 Oxygen and water release during the thermal treatment of hematite in high vacuum

A typical redox reaction was used for the calibration:

$$Fe^{3+}{}_{solid} + {}^{1}\!/_{2}O^{2-}{}_{solid} \rightarrow Fe^{2+}{}_{solid} + {}^{1}\!/_{4}O_{2gas}$$

where the oxygen release is determined by the oxygen fugacity in the solids (rocks) or melt and in the surrounding atmosphere.

During the high temperature reduction of hematite to magnetite in vacuum 3.34 wt% oxygen escapes. This process occurs with a maximum at 1080°C (Fig. 7).

A quantitative determination of haematite in rocks is possible up to 30 ppm. The spiky increasing of H_2O-partial pressure with a maximum at 355°C could be an indication for the temperature of crystal growth. More systematic studies are necessary to use such data for genetic interpretation.

3.4 *Boracite*

The decomposition of boracites is determined by the bonding of chlorine in the borate structure and the thermal stability of the borate structure. As shown in Fig. 8, boron escaped in three characteristic processes with maxima at 1050°C, 1222°C and 1271°C. Chlorine escaped during the first step of boron release with a maximum at 1039°C (Fig. 9) mostly in the form of $Cl_{2,}$ but surprisingly in the form of small amounts of HCl. Since H_2O and CO_2 could also be detected (Fig. 10), the formation of HCl results most probably inter-crystalline thermal reaction between the mobile chlorine and the water traces. More structure analysis are necessary to explain the processes in detail.

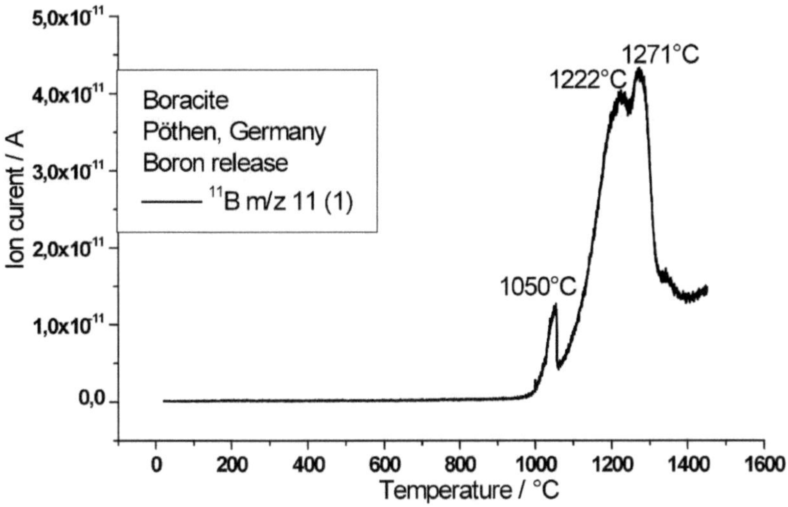

Fig. 8 Boron release during the heat treatment of boracite

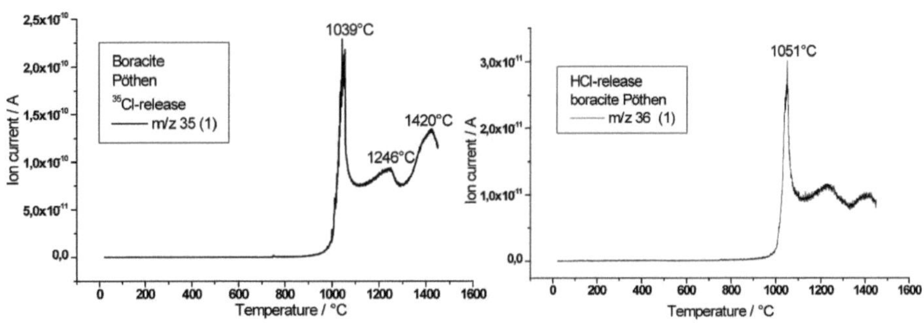

Fig. 9 Release of ^{35}Cl-species during the decomposition of boracite

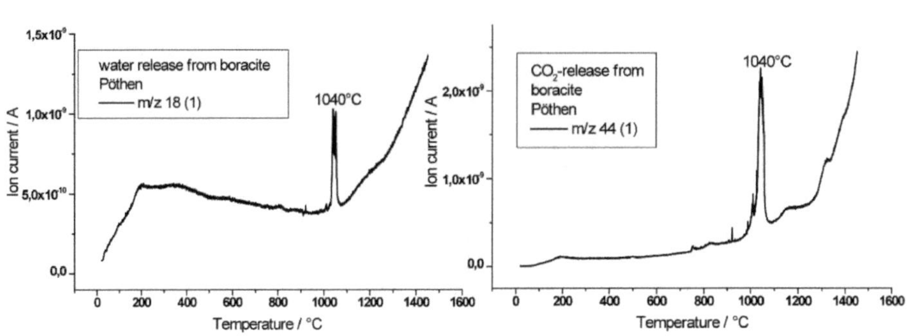

Fig. 10 Water and CO_2-release during the decomposition of boracite

3.5 *Charoite*

Charoite is an unusual mineral. It is probably due to a reaction of a limestone with a chemical unique intrusion. Its structure could only recently be determined (5). GRP experiments revealed that water release occurs in two separate temperature ranges (maxima 122°C and 409°C). In addition, fluorine release was observed above 1270°C (Figs. 11 and 12). Whereas these findings are in accordance with the

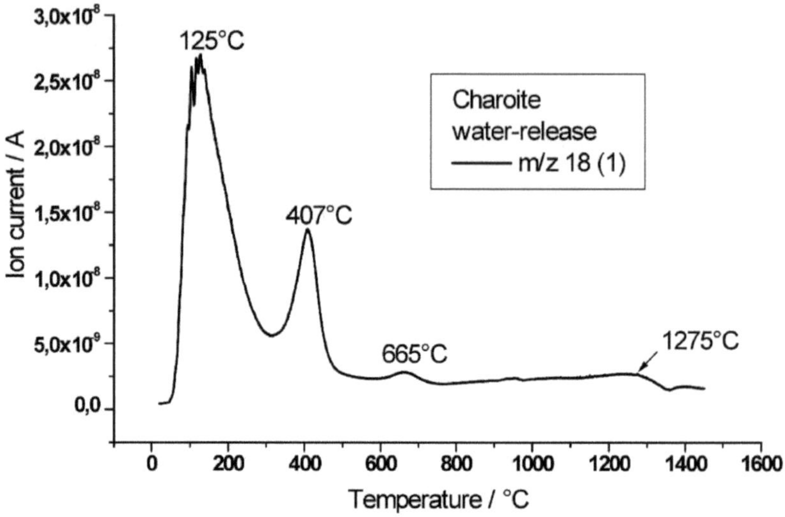

Fig. 11 Decomposition of charoite in high vacuum

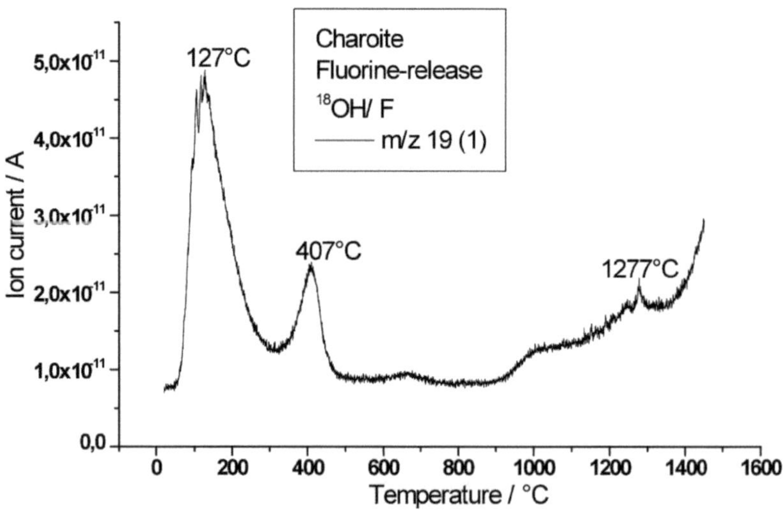

Fig. 12 Determination of the fluorine release from the GRP m/z 19

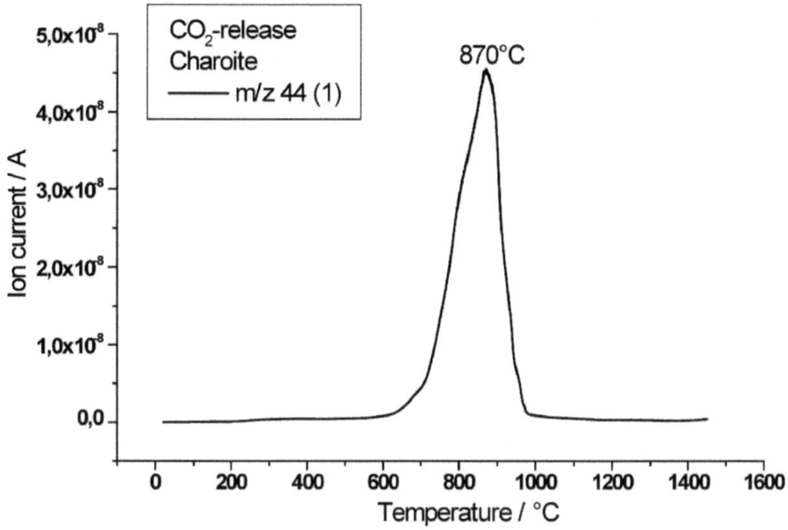

Fig. 13 CO_2 release during the heat treatment of charoite

formula as obtained from electron crystallography, the observed CO_2-release between 600°C and 1000°C is not. We believe that decomposition of carbonates from admixtures can be excluded. Possibly, our present GRP results might indicate that the composition of charoite is even more complex than given in the already quite complicated formula given above – at least for the sample used in our study. New experiments with new samples are planned (Fig. 13).

4 Conclusions

The main conclusions of this study are:

1. There is a gap between data about the major and trace element and the volatile concentration in crystals of the majority of minerals.
2. The release of volatiles allows the separation between alteration and pristine species in the crystalline solid.
3. The kinetic of gas release by diffusion and by bubbling allow the separation between fluid inclusions, melting processes and solid decomposition.
4. The qualitative composition and the maximum temperatures of the evolved gases are genetic indication.
5. The different bonding of volatile species in the crystalline structure is visible on the basis of degassing experiments.

Acknowledgements I would like to thank W. Depmeier for the inspiring and critical discussion. Financial support of the workshop "Minerals as Advanced Materials II" by the Deutsche Forschungsgemeinschaft under contract number DE 412/46-1 is gratefully acknowledged.

References

Heide K, Woermann E, Ulmer G (2008) Volatiles in pillows of the Mid-Ocean-Ridge-Basalt (MORB) and vitreous basaltic rims. Chem Erde 68:353–368

Hilse U, Goepel A, Pudlo D, Heide K, Gaupp R (2010) Characterization of bleaching phenomena in German red bed sediments by directly coupled evolved gas analysis (DEGAS). IMA Budapest 2010 ID Nr 482, MA97C10 p 701

Geiger CA, Rahmoun NS, Heide K (2011) Cordierite V: A high-temperature microporous silicate with occluded molecular species. Geochimica et Cosmochimica Acta (in press)

Schöps D, Schmidt CM, Heide K (2005) Simultaneous and quantitative DEGAS analysis of H_2O in silicate glasses. J Therm Anal Calor 80:749–752

The Principle of Duality in Isomorphism and Its Use in the Systematics of Minerals with Zeolite-Like Structures

Alexander P. Khomyakov

The modern mineralogical nomenclature is based on the principle that each new mineral species within a given structure type is defined by the dominance of a different chemical element at at least one structural site compared to the known minerals of that type (Nickel and Grice 1998). In realizing this principle, a major problem until recently was presented by zeolite-structured multicomponent minerals with variably occupied structural sites (e.g., eudialyte and labuntsovite groups) for which there were no reliable criteria for particular structure types. The key to solving this problem was provided the author's idea that these minerals are berthollides of the second kind, or non-stoichiometric compounds with a variable number and composition of atoms in structural sites. In the theory of isomorphism, they are usually regarded as isodimorphic solid solutions whose end-members belong to different structure types (Belov and Smirnova 1969; Filatov and Bubnova 1983). In the most general form, this idea can be formulated as a *principle of duality in isomorphism*, stating that *the end-members of continuous solid solutions can belong both to the same or to several different structure types* (Khomyakov 2008a).

According to this principle, the entire diversity of second-kind berthollides observed in nature can be described as combinations of a limited set of stoichiometrically complete structural and chemical end-members, for which the terms *structals* and *chemals*, respectively, are proposed (Khomyakov 2008b,c). Each structal corresponding to an experimentally determined structure type is characterized by an idealized structural formula with fully occupied sites, and each chemal is characterized by an idealized chemical formula with integer coefficients. The crystal chemical characteristics of each particular structurally characterized sample of a second-kind berthollide can be described by partial fractions of structals and mole fractions of chemals. Mineral species are defined by the combination of the structural and chemical end-members that dominate a solid solution. Chemals and structals that are subordinate in the composition and structure of a solid solution

A.P. Khomyakov (✉)
Institute of Mineralogy, Geochemistry and Crystal Chemistry of Rare Elements, Moscow, Russia
e-mail: noomin@mail.ru

S.V. Krivovichev (ed.), *Minerals as Advanced Materials II*,
DOI 10.1007/978-3-642-20018-2_4, © Springer-Verlag Berlin Heidelberg 2012

may be considered when defining chemical and structural varieties of mineral species. One structural end-member dominating different solutions can represent combinations with different chemals dominating the same solutions; as a consequence, one structure type can represent a series of several different mineral species.

A striking example of this type of berthollide is provided by eudialyte group minerals (EGM). In recent years, this group has expanded to more than 25 species (Khomyakov 2002, 2004, 2007). It consists of trigonal zircono- and titanosilicates with the general formula $A(1)_3A(2)_3A(3)_3A(4)_3A(5)_3B_{3-6}CD\{M_6Z_3[T_{24}O_{72}]\}$ $(O,OH)_{2-6}X_2$, in which braces enclose a heteropolyhedral framework. These include members with space groups $R\bar{3}m$, $R3m$ and $R3$, as well as members with a relatively low-order 12-layer structure ($a \sim 14.2$, $c \sim 30$ Å) and a high-order 24-layer structure ($a \sim 14.2$, $c \sim 60$ Å). The sites $A(1)$ to $A(5)$ are usually dominated by Na; in some minerals also by K, H_3O, Mn, Ca, Sr, REE; B = Fe^{2+}, Fe^{3+}, Mn, Zr, Na, K; C = Nb, W, Ti, Mn, Si; D = Si, Nb; M = Ca, Mn, Sr, REE, Y, Na; Z = Zr, Ti; T = Si; X = Cl, F, H_2O, CO_3. Extra-framework cations and anions are located in variable-sized cavities of eudialyte structures, in which single-crystal X-ray diffraction indicates there are not only high-occupancy structural sites, but also a large number of closely spaced low-occupancy sites. EGMs vary significantly both in the total number and in the composition of atoms occupying these sites.

Given these and other crystal chemical features of the eudialyte group, when assigning its specific members to distinct structure types, the author was guided by the above general principle and the related notion of chemals and structals as independent end-members of solid solutions with a variable number and composition of atoms in structural sites. On the basis of identifying the dominant structal in EGM structures, 24 approved members of this group were found to belong to 15 different structure types, named for minerals with precisely characterized structures. These include the alluaivite, aqualite, carbokentbrooksite, dualite, eudialyte $R\bar{3}m$, eudialyte $R3m$, golyshevite, ikranite, kentbrooksite, labyrinthite, mogovidite, oneillite, raslakite, rastsvetaevite, and voronkovite structure types.

The remaining minerals are assigned to the carbokentbrooksite (johnsenite-Ce and zirsilite-Ce) and kentbrooksite (andrianovite, feklichevite, ferrokentbrooksite, georgbarsanovite, khomyakovite, manganokhomyakovite, and taseqite) structure types as distinct members of appropriate isostructural series based on the dominant chemal. All the identified EGM structure types are homeotypic. Differing mainly in the geometric arrangement of atomic positions in the zeolite part of the structure, they are all closely related as they share the same topology of the heteropolyhedral framework $\{M_6Z_3[T_{24}O_{72}]\}$ and together make up a single eudialyte structure class, which is divided into two subclasses, with 12- or 24-layer structure motifs, and six families of structure types based on combinations of the two structure motifs with each of the three space groups identified in eudialytes.

EGMs contain about half the elements of the periodic table. Wide variations in their concentrations in most extra-framework sites and some framework sites coupled with a wide variety of EGM structure types make the eudialyte group truly unique in the number of possible mineral species. Hence, mineralogical and crystal chemical research on this group holds great promise for new mineral discoveries.

Equally promising for these discoveries is the labuntsovite group, which comprises highly variable (in composition and structure) zeolite-like titano- and niobosilicates with the general formula $(Na,K,Ca,Sr,Ba)_{2-6}[(Mg,Mn,Fe,Zn)_{0-1}(Ti,Nb)_4(O,OH)_4(Si_4O_{12})_2] \cdot nH_2O$ (Chukanov et al. 2002).

To conclude, we address the question of whether it is useful to introduce the terms *structals* and *chemals*, which we propose for the two fundamental types of end-members of solid solutions with a variable number and composition of atoms in structural sites. Initially, to define the specific nature of such solutions, we considered using the terms *structural end-member* and *chemical end-member*. However, this idea was abandoned because it was found that the term *structural end-member* is widely used in the Russian literature to refer to building blocks in the description of homologous series of minerals and its use in another sense would inevitably cause confusion. The introduction of the two short terms with a clear meaning is fully justified by the complexity of the isomorphism phenomena described by them. Based on the international Greek and Latin rooted words "structure" and "chemistry," these terms, we hope, will be accepted by the worldwide scientific community. The same can be said for the proposed *principle of duality in isomorphism*, reflecting the dual nature of the end-members of isodimorphic solid solutions.

References

Belov NV, Smirnova NL (1969) Basic concepts of isomorphism with reference to crystallochemistry. 2*nd* All-Union symposium on the isomorphism problem. Abstracts Moscow: 2–6 (in Russian)

Chukanov NV, Pekov IV, Khomyakov AP (2002) Recommended nomenclature for labuntsovite-group minerals. Eur J Miner 14:165–173

Filatov SK, Bubnova RS (1983) Isomorphism and adjacent crystallochemistry events (experiment of classification). Zap VMO 117(5):552–556 (in Russian)

Khomyakov AP (2002) The "Second Chapter" in the mineralogy and crystal chemistry of the eudialyte group. New ideas and conceptions in mineralogy Syktyvkar: IG Komi SC UB RAS: 108

Khomyakov AP (2004) Crystal chemical systematics of the eudialyte-group minerals. 32*nd* Int-l Geol Congr, Florence, Scientific sessions abstracts (part 1): 309–310

Khomyakov AP (2007) Crystal-chemical systematization of the eudialyte-group minerals. Third international symposium mineral diversity: Research and preservation. Working Papers. Sofia: Zemyata & Horata: 257–264 (in Russian with English abstract)

Khomyakov AP (2008a) The principle of duality in isomorphism and crystal chemical diversity of the mineral kingdom. Structure and diversity of mineral world. In: Proceedings international mineral seminar, Syktyvkar: IG Komi SC UB RAS: 13–14 (in Russian)

Khomyakov AP (2008b) Structals and chemals: Two fundamental end-member types of solid solutions with a varying number and composition of atoms in the structural sites. 33rd international geological congress, Oslo CD-ROM Abstract

Khomyakov AP (2008c) Eudialyte-group minerals as an example of solid solutions with a varying number and composition of atoms in the structural sites. 33rd international geological congress, Oslo CD-ROM Abstract

Nickel EH, Grice JD (1998) The IMA commission on new minerals and mineral names: procedures and guidelines on mineral nomenclature. Can Miner 36(3):913–927

“Ab-Initio” Structure Solution of Nano-Crystalline Minerals and Synthetic Materials by Automated Electron Tomography

Enrico Mugnaioli, Tatiana E. Gorelik, Andrew Stewart, and Ute Kolb

1 Introduction

Most of the newly discovered mineral phases, as well as many new synthesized industrial materials, appear only in the form of nano crystals, with a size not sufficient for single-crystal x-ray structure analysis. The development of techniques able to investigate the structure of nano crystalline materials is therefore one of the most important frontiers of crystallography. The most widespread technique providing relatively fast and well consolidated routes for structure analysis of bulk materials is x-ray powder diffraction (XRPD). Nevertheless, XRPD suffers from intrinsic 1-dimension reduction of information that greatly limits its applicability in presence of peak broadening and overlapping. Peak broadening is usually caused by very small crystallites, namely less than 50 nm. Overlapping of peaks is problematic mainly for intensity integration, but in case of polyphasic mixtures or significant amount of impurities it can be critical also for cell parameter determination and reflection indexing.

The main advantage of using electron radiation is the ability to collect structural information from areas of a few tens of nanometres, i.e. it is possible to collect single crystal information from crystals 2–3 orders of magnitude less than the smallest crystal accessible by single crystal x-ray, even using synchrotron radiation. This is possible because electrons have a stronger interaction with matter (Cowley et al. 2001) and electron beam can be easily deflected and focused in a probe of few nanometers.

High resolution transmission electron microscopy (HRTEM), the most established techniques for nano particle investigation, provides direct space structural information down to 0.8–0.5 Å resolution (Kisielowski et al. 2008a, b). Nevertheless, performing HRTEM requires a relatively high intensity illumination and a large electron dose on

E. Mugnaioli (✉) • T.E. Gorelik • A. Stewart • U. Kolb
Institut für Physikalische Chemie, Johannes Gutenberg-Universität,
Welderweg 11, Mainz 55128, Germany
e-mail: mugnaiol@uni-mainz.de

S.V. Krivovichev (ed.), *Minerals as Advanced Materials II*,
DOI 10.1007/978-3-642-20018-2_5, © Springer-Verlag Berlin Heidelberg 2012

the sample. Under these conditions, nearly all organic and most of inorganic materials suffer a fast deterioration due to beam damage. The necessity to orient the particle in order to record meaningful HRTEM images further increases the exposure time. This may lead to a modification of the crystalline structure or to complete amorphization or sublimation of the sample (Spence 2003; Reimer and Kohl 2008; Kolb et al. 2010).

On the contrary, electron diffraction can deliver structural information at a comparable resolution with significantly lower radiation on the sample. Classical drawbacks of electron diffraction are the presence of dynamic effects and the incompleteness of intensity data sets. Moreover beam damage could be still a major issue for many classes of materials, like organic or water-containing inorganic materials. *All these problems derive basically from the instrumentation and the method used for electron diffraction data collection.* In fact the conventional way of electron diffraction data collection is based on acquisition of oriented zones. Though, dynamic effects are maximized in oriented zones, and only a limited number of reflections are accessible (namely only reflections that belong to low-index zones). Additionally, a long time is needed for orienting the crystal, operation that implies time consuming and longer exposure of the sample to beam radiation.

The application of electron diffraction data for structure solution dates back to 1936 (Rigamonti 1936), but for a long time electron crystallography was considered of little use due to the presence of dynamic effects (Cowley 1956; Lipson and Cochran 1966). Nevertheless, excellent results were achieved from data sets collected by the electron diffraction camera developed by Pinsker and Vainshtein in Moskow during 1950 (Pinsker 1953; Vainshtein 1956, 1964). The technique developed by these authors, named oblique-textured electron diffraction (OTED), was able to solve a number of organic and inorganic phases, up to the localization of H atoms in the structure (see for example Zhukhlistov et al. 1997; Zhukhlistov and Zvyagin 1998). Nevertheless, sample preparation for OTED measurements is highly problematic because diffraction takes place on many crystals which need to be oriented approximately in the same way.

Pioneering structure investigations using electron diffraction data collected by a TEM (i.e. collected form single nano crystals) started only at the end of 1970' (Dorset and Hauptman 1976). In the following 30 years a growing interest for electron diffraction and crystallography has arose and both organic and inorganic structures have been extensively investigated (Dorset 1995). The leading idea of these works is that collecting data from very thin crystals (e.g. less than 50 Å), the dynamic effects are so reduced that a kinematic approximation (I proportional to F^2) is valid for structure solution (see for example Nicolopoulos et al. 1995; Voigt-Martin et al. 1995; Weirich et al. 1996; Wagner et al. 1999; Dorset et al. 2005; Dorset 2007).

In order to further reduce dynamic effects, the precession unit was invented by Vincent and Midgley (Vincent and Midgley 1994). This device precesses the beam around a conical path. Consequently, the sample is never fully oriented along a main axis and reflections are integrated by the precessing Ewald sphere. A number of structure solutions achieved using Precession Electron Diffraction (PED) data are reported in literature, even for relatively thick samples (see for example Weirich et al. 2006; Dorset et al. 2007; Gemmi et al. 2010).

Electron diffraction has been frequently used in combination with x-ray powder data. A recent example of structure solution of nano crystalline materials combining electron diffraction with *a priori* crystallochemical information and HRTEM and XRPD data is reported by Baerlocher et al. (2007). This method has been successful for the solution of some complicate zeolite structures, but until now it cannot be considered a routine path for structure analysis, as any working case needs a specific, different approach.

Therefore, there is still a need for routine methods allowing the acquisition of more complete and close-to-kinematic electron diffraction data. This demand recently led to the development of the Automated Diffraction Tomography (ADT) module (Kolb et al. 2007, 2008). ADT approach marks a radical change from the traditional way of electron diffraction data acquisition based on oriented zones. ADT acquisition consists of a homogenous sampling of the reciprocal space available in the tilt range of the microscope goniometer (usually $\pm 60°$), using a fixed tilt step of $1°$ or $0.5°$ and without orienting the crystal along a (main) crystallographic axis, principally avoiding oriented zones.

In order to perform long acquisitions even on beam sensitive samples, illumination and electron dose on the sample are significantly reduced (down to an electron dose rate of 5–15 e $Å^{-2}$ s^{-1}). The crystal position can be tracked by high angular annular dark field scanning-transmission electron microscopy (HAADF–STEM) and diffraction patterns are acquired in nano electron diffraction (NED) mode. With the use of a small condenser aperture (C2) of 10 μm it is possible to obtain a beam size of 50 nm with almost parallel illumination on the sample. Dedicated software allowing automation of data acquisition further reduces acquisition time and consequently beam damage (Kolb et al. 2007). Setting an exposure time of 1 s per diffraction pattern, it is possible to collect a full tilt series with a total exposure time of 2 min. Possibly beam damage can be further reduced using cooling conditions or slightly moving the beam around the crystal during the acquisition.

After the acquisition, a stack of 2-dimensional non-zonal diffraction patterns is stored. The tilt series is analyzed by the software package ADT-3D (Schömer et al. 2009). Knowing the tilt axis azimuth, the 3-dimensional reciprocal space can be reconstructed and used for direct visualization of disorder, twinning and polycrystallinity. Cell vectors (cell parameters+orientation matrix) are defined by automated routines based on clustering in difference vector space (Kolb et al. 2008). Finally, reflections are indexed and intensities integrated. Because all reflections inside the available tilt range are sampled, ADT intensity data sets have a significantly higher coverage of reciprocal space than those obtained by conventional electron diffraction zonal acquisition. Depending on the crystal family, completeness ranges from 60% for triclinic, to 70–90% for monoclinic, 90% for orthorhombic and 100% for cubic lattices.

Reflection intensities integrated by ADT are expected to be more kinematic than those measured in oriented patterns, as they are collected *off zone* and dynamic effects are therefore drastically reduced. Nevertheless a significant deviation from expected values is still present, mainly due to the excitation error, i.e. the distance the Ewald sphere cuts the reflection from the centre (Williams and Carter 1996). In order to perform a complete integration of reflection intensities, PED can be coupled with ADT (Mugnaioli et al. 2009). The resulting data sets proved to be of extremely high quality and in the last 2 years a number of complicate structures have been solved *ab initio* by direct methods with a fully kinematic assumption (Birkel et al. 2010; Kolb et al. 2010, 2011; Rozhdestvenskaya et al. 2010; Andrusenko et al. 2011; Denysenko et al. 2011).

In this chapter we show how ADT/PED data can be used for routine analysis of nano crystalline inorganic materials. In doing that, we selected two known structures and performed for them a complete *ab initio* structure analysis. The first example is a porous natural zeolite (natrolite – $Na_2Al_2Si_3O_{10}{\cdot}2H_2O$). Structure solution of this orthorhombic material is particularly challenging because this structure is non-centrosymmetric and contains Na^+ ions and water molecules in the pores. The second example is a layered monoclinic sodium titanate ($Na_2Ti_6O_{13}$) consisting of titania layers intercalated by Na^+ ions.

2 Experimental

For TEM and ADT investigations samples were dispersed in ethanol using ultrasonic bath and sprayed on a carbon-coated copper grid (Mugnaioli et al. 2009). TEM, STEM and ADT were carried out with a FEI TECNAI F30 S-TWIN transmission electron microscope working at 300 kV. TEM images and electron diffraction patterns were acquired with a CCD camera (14-bit GATAN 794MSC). STEM images were acquired by a FISCHIONE HAADF detector.

For structure analysis electron diffraction data were collected using ADT module (Kolb et al. 2007, 2008). During the acquisition crystal position was tracked in microprobe STEM mode and NED patterns were collected every 1° tilt inside the possible tilt range of the microscope goniometer (up to ±60°). Camera length was 380 mm, corresponding to a resolution of diffraction data of 0.75 Å. For high tilt experiments a FISCHIONE tomography holder was used. In order to improve reflection intensity integration, ADT was coupled with PED, performed using NanoMEGAS DigiStar unit (Vincent and Midgley 1994; Avilov et al. 2007). The precession angle was kept at 1.2°. The 3-dimensional electron diffraction data were processed using ADT-3D (Schömer et al. 2009) software package coupled with Matlab scripts written by ourselves (Kolb et al. 2008, 2009; Mugnaioli et al. 2009).

Ab initio structure solution was performed by direct methods implemented in the software SIR2008 (Burla et al. 2007). Least square refinement was performed by Shelx97 (Sheldrick 2008). For both *ab initio* structure solution and structure refinement, a full kinematic approach was used (I proportional to F^2) and reflection intensities data were used without any correction for absorption or geometry. Scattering factors for electrons come from Doyle and Turner (1968).

3 Results and Discussion

3.1 Porous Material: Natrolite ($Na_2Al_2Si_3O_{10}{\cdot}2H_2O$)

Among all micro-porous materials, zeolites are the most studied and widespread used for industrial applications. Most of synthetic zeolites cannot be grown in crystals large enough for single crystal x-ray analysis. On the other hand structure solution from powder x-ray diffraction data is not straightforward as many zeolites tend to crystallize as intergrown polyphasic assemblies (Thomas et al. 2001). Despite some remarkable results (Nicolopoulos et al. 1995; Wagner et al. 1999), zeolite structure solutions by electron diffraction data are hampered by dynamic effects and data incompleteness due to the preferential orientation of the crystals, often needle or platelet like (Dorset et al. 2005). Therefore, structure solution of zeolites is often limited to 2-dimensional solutions and only T-sites are correctly identified (Dorset et al. 2005, 2007; Gilmore et al. 2008).

A natrolite single crystal of 500 $\times$ 500 nm was selected for the ADT data acquisition (Fig. 1). For accurate cell vectors determination, one tilt series was collected without PED in order to measure accurate cell parameters (tilt range $\pm 30°$). Cell parameters determined automatically fit well with the expected ones (Capitelli and Derebe 2007) but the difference between **a** and **b** lengths is inside the experimental error, namely 2% (Table 1).

For intensity integration two tilt series with orthogonal tilt axes were collected using PED from the same crystal (tilt ranges $\pm 60°$ and $-60/+50°$, respectively). A third intensity data set was obtained merging the two data sets coming from the two tilts. As the two data collections come from the same crystal, a merging factor of 1 was used. Some important parameters of the first tilt series and of the merged data set are reported in Table 2. Both these two data sets were used for structure solution, giving very similar results.

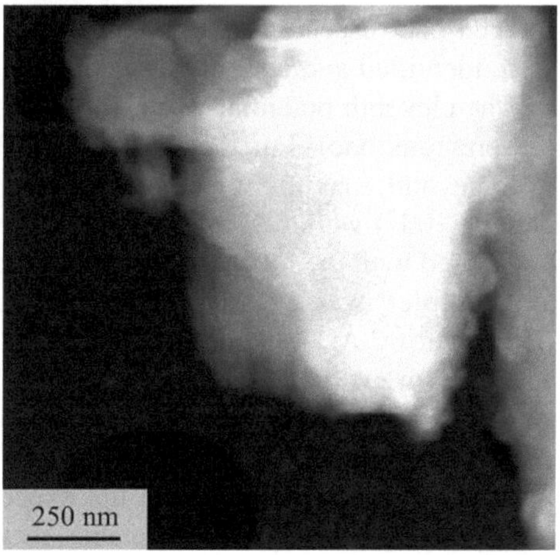

Fig. 1 A natrolite crystal similar to the one selected for ADT data collection

Table 1 Cell parameters determined experimentally from ADT data compared with cell parameters determined by Capitelli and Derebe (2007) by single-crystal x-ray

	Experimental ADT	Capitelli and Derebe 2007
a (Å)	18.25	18.29
b (Å)	18.50	18.64
c (Å)	6.60	6.59
α (°)	90.00	90
β (°)	90.61	90
γ (°)	90.23	90

Table 2 Important parameters of natrolite ADT/PED data collection. In the first column are reported the parameters of the intensity data set obtained from the first tilt series, in the second one the parameters of the complete data set obtained by merging the two tilt series

Tilt range	−60°/+60°	−60°/+60° and −60°/+50°
Total collected reflections	3,460	4,560
Independent reflections	719	743
Resolution, Å	0.75	0.75
Reflection coverage, %	96	99
R_{sym} (F), %	19.73	20.60
Overall U, $Å^2$	0.020	0.020
Final R (SIR2008), %	21.25	20.22
R1 (Shelx97), %	–	21.34

Any attempt of structure solution in tetragonal symmetry ended with a final residual R around 40–50% and did not converge to any crystallochemically meaningful solution. In orthorhombic symmetry, space group Fdd2 was uniquely defined by automatic extinction analysis performed by SIR2008. Selecting this symmetry, *ab initio* structure solution set converged nicely to the expected solution, with final R of 21.25% and 20.22% for the data set obtained from the first tilt series and the merged one respectively. In both cases all the ten non-hydrogen atomic positions were identified and correctly assigned by SIR2008 (Table 3).

The eleventh potential was significantly weaker (height of 50, less than half of the previous one). For the structure solution obtained with the merged data set, average and maximum deviations from the expected positions (Capitelli and Derebe 2007) were 0.100 and 0.183 Å respectively. The maximum deviation was associated with the intra-channel water molecule (Ow). Global thermal factor from Wilson plot was positive ($U = 0.020$ $Å^2$), as typical for structure solution performed by ADT intensity data sets, but not for zonal electron diffraction data sets (see for example Dorset et al. 2005).

Structural refinement was performed by Shelx97 without imposing any geometrical restraint or constrain and led to R1 of 21.34% (Fig. 2 and Table 4). Average and maximum deviations from the expected positions were both shorter. All the atoms had positive and reasonable thermal factors. Interestingly, the average

Table 3 Natrolite structure solution from SIR2008 using the merged data set, compared with the atomic positions found by Capitelli and Derebe (2007)

	SIR solution				Capitelli and Derebe 2007			
Atom	Height	x/a	y/b	z/c	x/a	y/b	z/c	Difference (Å)
Si1	220	0	0	0.420	0	0	0.411	0.059
Si2	220	0.098	0.037	0.043	0.097	0.039	0.038	0.049
Al1	240	−0.038	0.098	0.803	−0.037	0.094	0.796	0.094
Na1	170	−0.219	0.031	0.796	−0.221	0.031	0.794	0.038
O1	170	0.070	0.023	0.283	0.069	0.023	0.271	0.084
O2	160	−0.022	0.069	0.561	−0.023	0.068	0.545	0.108
O3	150	−0.068	0.183	0.792	−0.070	0.182	0.801	0.076
O4	140	−0.094	0.042	0.904	−0.098	0.035	0.911	0.161
O5	140	0.038	0.096	0.920	0.044	0.097	0.935	0.146
Ow	120	−0.313	0.060	1.029	−0.306	0.060	1.050	0.183

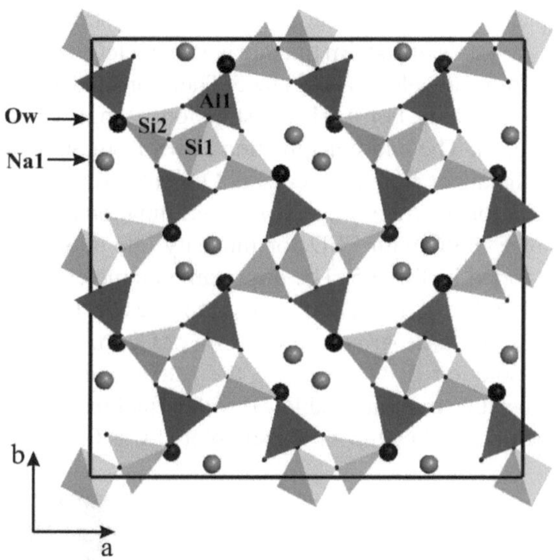

Fig. 2 Natrolite structure solved and refined by ADT intensity data in projection along [001]

Table 4 Natrolite structure refinement from Shelx97, compared with atomic positions found by Capitelli and Derebe (2007)

	Shelx97 refinement				
Atom	x/a	y/b	z/c	U_{iso} (Å^2)	Difference (Å)
Si1	0	0	0.420	0.026	0.058
Si2	0.095	0.038	0.043	0.025	0.044
Al1	−0.039	0.096	0.801	0.024	0.064
Na1	−0.218	0.030	0.786	0.035	0.072
O1	0.069	0.022	0.279	0.023	0.057
O2	−0.022	0.068	0.562	0.036	0.118
O3	−0.067	0.180	0.787	0.031	0.110
O4	−0.098	0.040	0.905	0.028	0.101
O5	0.038	0.096	0.924	0.034	0.129
Ow	−0.312	0.057	1.031	0.051	0.170

thermal factor for oxygen atoms was higher than the thermal factor associated with silicon and aluminium and the higher thermal factor was associated with H_2O molecule, which in fact is free to move inside the channel and may not have full occupancy.

3.2 *Layered Material: Sodium Titanate ($Na_2Ti_6O_{13}$)*

Sodium titanate ($Na_2Ti_6O_{13}$) crystals were found as impurities inside a commercial $Na_2Ti_3O_7$ powder (ABCR®, 95% purity). Sodium titanate can be used in various applications, such as ion-exchanging materials (e.g. to fix radioactive metal ions), humidity or gas sensors, starting material for the synthesis of other alkaline or earth alkaline titanates (e.g. lithium titanate used in lithium ion batteries).

The structure was known and was solved by Andersson and Wadsley (1962) in space group C2/m using x-ray single crystal data.

Crystals of sodium titanate have platelet shape, with preferential orientation along the more developed (100) face. Therefore it is not possible to sample reflections close to the a* axis. For ADT investigation, a single crystal of 300 × 800 nm was selected (Fig. 3). The entire acquisition arises from an area of 200 × 200 nm located close to the upper rim of the crystal. One tilt series in a range ±55° was acquired without PED for accurate cell vectors determination and another tilt series was acquired in a range ±55° with PED for intensity integration. Experimental cell parameters are reported in Table 5. The important parameters of the intensity data sets are listed in Table 6. No reflection extinctions beside C-centring was observed, limiting the possible space-groups to C2, C/m or C2/m.

Solution in C2/m using the full acquired resolution converged nicely to the expected structure (Table 7), with a final residual R of 36.98%. The overall thermal factor calculated from Wilson plot U was 0.012 Å^2. The structure was immediately

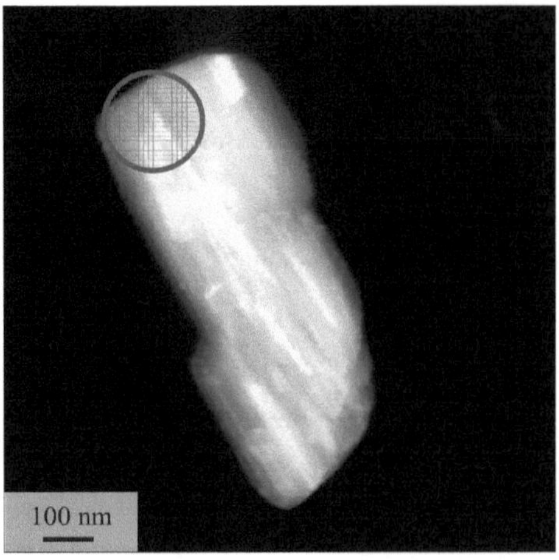

Fig. 3 The sodium titanate crystal selected for ADT data collection. The grey circle indicates the area of data collection

Table 5 Cell parameters determined experimentally from ADT data compared with the cell parameters determined by Andersson and Wadsley (1962) by single-crystal x-ray diffraction

	Experimental ADT	Andersson and Wadsley 1962
a (Å)	15.10	15.13
b (Å)	3.74	3.75
c (Å)	9.10	9.16
α (°)	90.32	90
β (°)	99.49	99.30
γ (°)	89.91	90

Table 6 Important parameters of sodium titanate ADT/PED data acquisition

Tilt range	−55°/+55°
Total collected reflections	1,497
Independent reflections	517
Resolution, Å	0.75
Reflection coverage, %	72
R_{sym} (F), %	13.07
Overall U, $Å^2$	0.012
Final R (SIR2008), %	36.98
R1 (ShelX97), %	28.72

Table 7 Sodium titanate structure solution from SIR2008, compared with atomic position found by Andersson and Wadsley (1962). g1 and g2 are two ghost positions

	SIR2008 solution					Andersson and Wadsley 1962			
Atom	Pos.	Height	x/a	y/b	z/c	x/a	y/b	z/c	Difference (Å)
Ti1	1	2,454	0.165	0	0.430	0.171	0	0.433	0.088
Ti2	2	2,314	0.115	0	0.099	0.114	0	0.090	0.089
Ti3	3	2,163	0.222	0	0.762	0.229	0	0.773	0.140
O1	4	1,403	0.295	0	0.567	0.282	0	0.574	0.207
O2	5	1,160	0.231	0	0.200	0.228	0	0.247	0.433
O3	6	1,102	0.126	0	0.598	0.124	0	0.617	0.177
O4	7	898	0.078	0	0.281	0.071	0	0.291	0.140
O5	8	831	0.370	0	0.898	0.358	0	0.884	0.222
O6	9	785	0.134	0	0.885	0.167	0	0.927	0.630
O7	10	779	0	0	0	0	0	0	0
g1	11	535	0.077	0	0.282	–	–	–	–
g2	12	494	0.744	0	0.107	–	–	–	–
Na1	13	491	0.457	0	0.237	0.454	0	0.251	0.134

recognisable. The first three potential peaks, significantly stronger than the following, corresponded nicely to the three expected independent Ti atoms of the structure. The following seven positions corresponded to the seven independent O atoms. Positions 11 and 12 were too close to other peaks to be taken into account, so they were considered as ghosts. Position 13 corresponded to the only independent Na atom

Table 8 Sodium titanate structure refinement from Shelx97, compared with atomic positions found by Andersson and Wadsley (1962)

	Shelx97 refinement				
Atom	x/a	y/b	z/c	U_{iso} (Å^2)	Difference (Å)
Ti1	0.164	0	0.437	0.021	0.171
Ti2	0.116	0	0.089	0.014	0.114
Ti3	0.225	0	0.767	0.016	0.229
O1	0.292	0	0.551	0.044	0.282
O2	0.238	0	0.249	0.031	0.228
O3	0.128	0	0.599	0.088	0.124
O4	0.078	0	0.253	0.054	0.071
O5	0.354	0	0.882	0.072	0.358
O6	0.153	0	0.905	0.039	0.167
O7	0	0	0	0.077	0
Na1	0.465	0	0.248	0.050	0.454

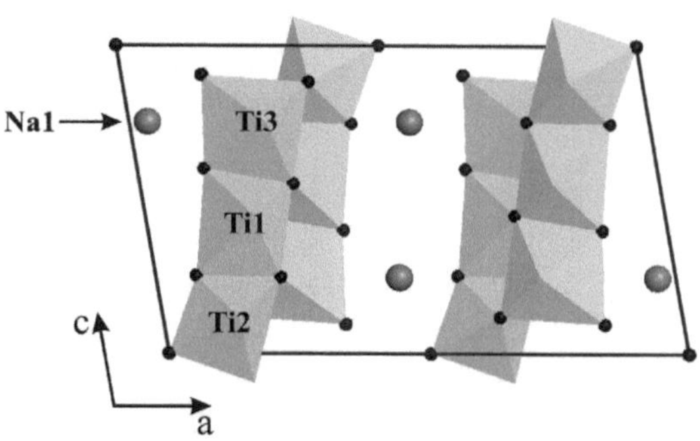

Fig. 4 Sodium titanate structure solved and refined by ADT intensity data, projection [010]

of the structure. Average and maximum deviations from expected positions (Andersson and Wadsley 1962) were 0.206 and 0.630 Å, respectively. Differences for oxygen atoms were significantly larger than for titanium atoms.

Structural refinement using Shelx97 (performed without any geometrical restraint or constrain) converged to a solution close to SIR2008 output (Table 8), with a final R1 of 28.72% (Fig. 4). Average and maximum deviations from expected positions were 0.152 and 0.454 Å, respectively. After refinement the maximum deviation from expected position belongs to interlayer sodium atom. All atoms have positive and reasonable thermal factors.

3.3 *About Completeness of Data*

Structure solution of natrolite gave a better result when compared with sodium titanate, both in terms of final residual and deviation from expected atom positions. This result is easily explicable considering the higher completeness of natrolite intensity data set (99% for natrolite vs. 72% for sodium titanate). The main reason of this difference is related with the different crystal family of the two materials.

Natrolite structure has an orthorhombic symmetry (reflection multiplicity 8), while sodium titanate is monoclinic (reflection multiplicity 4). Additionally, sodium titanate crystals have a strong preferential orientation, resulting in missing $00l$ reflections from intensity data set. On contrast, for natrolite all the three main directions are sampled.

For natrolite, merging two tilt series did not produce significant benefits in this specific case. Reflection completeness increased only of 3% (from 96% to 99%) and the two structure solutions were pretty comparable. Nevertheless, merging different tilt series can be of significant benefit in cases of lower symmetry and even a small percentage of reflections could be determinant for structure solution, especially if these reflections belong to a main direction. Finally it is worth to point out that the merging of two data sets was done using a merging factor of 1 and resulted in a very small increase of the R_{sym} (from 19.73% to 20.60%). Indeed the two tilt series come from the same crystal and was then reasonable to assume no change in thickness.

4 Conclusions

In this work the use of ADT technique for structure solution of nano crystalline materials starting from single nano crystal data is presented. Structure solution of medium structural complexity inorganic structure can be performed routinely even for materials considered beam sensitive for HRTEM or conventional zonal electron diffraction. Data collection by ADT module is significantly faster than conventional zonal electron diffraction acquisition. Even in presence of significant deviation from kinematic values, resulting in a final large residual (R = 20–30%), raw intensity data sets are suitable for the localization of all the non-H atoms of the structure, even oxygen in the vicinity of heavy atoms (or carbon in organic or hybrid organic-inorganic materials; Kolb et al. 2010; Denysenko et al. 2011). *Ab initio* structure solution can be performed with routines and software developed for x-ray crystallography. The whole procedure, from data acquisition to structure solution, is now close to the time scale of single crystal x-ray structure analysis.

Still, the strong deviation of reflection intensities from kinematical expected values, resulting in the high final residual of the solution, is problematic for structure refinement. Different kinds of deviation from kinematic values affect intensity data, as residual dynamic effects, lack of geometrical correction, a need for better recording systems, more accurate background modelling, more elaborated intensity determination (e.g. 3-dimensional shape fitting of reflections). Nevertheless the possibility of *ab initio* structure solution of complicate structures based on raw ADT data proves the high potential of this approach, and ample margins of improvement are achievable in the near future. Therefore, we firmly believe the ADT approach will open up new possibilities for nanomaterial investigation and engineering.

Acknowledgements The authors thank Iryna Andrusenko, Dominik Koll, Govanna Vezzalini and Rossella Arletti for providing the samples and for useful discussion. The work was supported by the Deutsche Forschungsgemeinschaft in the Sonderforschungsbereich 625. Financial support for the workshop "Minerals as Advanced Materials II" came from the Deutsche Forschungsgemeinschaft DE 412/46-1.

References

Andersson S, Wadsley AD (1962) The structures of $Na_2Ti_6O_{13}$ and $Rb_2Ti_6O_{13}$ and the alkali metal titanates. Acta Crystallogr 15:194–201

Andrusenko I, Mugnaioli E, Gorelik TE, Koll D, Panthöfer M, Tremel W, Kolb U (2011) Structure analysis of titanate nanorods by automated electron diffraction tomography. Acta Crystallogr B67:218–225

Avilov A, Kuligin K, Nicolopoulos S, Nickolskiy M, Boulahya K, Portillo J, Lepeshov G, Sobolev B, Collette JP, Martin N, Robins AC, Fischione P (2007) Precession technique and electron diffractometry as new tools for crystal structure analysis and chemical bonding determination. Ultramicroscopy 107:431–444

Baerlocher C, Gramm F, Massüger L, McCusker LB, He Z, Hovmöller S, Zou X (2007) Structure of the polycrystalline zeolite catalyst IM-5 solved by enhanced charge flipping. Science 315:1113–1116

Birkel CS, Mugnaioli E, Gorelik T, Kolb U, Panthöfer M, Tremel W (2010) Solution synthesis of a new thermoelectric $Zn_{1+x}Sb$ nanophase and its structure determination using automated electron diffraction tomography. J Am Chem Soc 132:9881–9889

Burla MC, Caliandro R, Camalli M, Carrozzini B, Cascarano GL, De Caro L, Giacovazzo C, Polidori G, Siliqi D, Spagna R (2007) *IL MILIONE*: a suite of computer programs for crystal structure solution of proteins. J Appl Crystallogr 40:609–613

Capitelli F, Derebe MG (2007) Single crystal X-ray diffraction study of a pure natrolite sample. J Chem Cristallogr 37:583–586

Cowley JM (1956) Electron-diffraction study of the structure of basic lead carbonate, $2PbCO_3{\cdot}Pb(OH)_2$. Acta Crystallogr 9:391–396

Cowley JM, Goodman P, Vainshtein BK, Zvyagin BB, Dorset DL (2001) Electron diffraction and electron microscopy in structure determination. In: Shmueli U (ed) International tables for crystallography, Volume B, reciprocal space, 2nd edn. Kluwer Academic, Dordrecht

Denysenko D, Grzywa M, Tonigold M, Streppel B, Krkljus I, Hirscher M, Mugnaioli E, Kolb U, Hanss J, Volkmer D (2011) Elucidating gating effects for hydrogen sorption in MFU-4-type triazolate-based metal organic frameworks featuring different pore sizes. Chem Eur J 17:1837–1848

Dorset DL (1995) Structural electron crystallography. Plenum Press, New York

Dorset DL (2007) Electron crystallography of organic materials. Ultramicroscopy 107:453–461

Dorset DL, Hauptman HA (1976) Direct phase determination for quasi-kinematical electron diffraction intensity data from organic microcrystals. Ultramicroscopy 1:195–201

Dorset DL, Roth WJ, Gilmore CJ (2005) Electron crystallography of zeolites – the MWW family as a test of direct 3D structure determination. Acta Crystallogr A61:516–527

Dorset DL, Gilmore CJ, Jorda JL, Nicolopoulos S (2007) Direct electron crystallographic determination of zeolite zonal structures. Ultramicroscopy 107:462–473

Doyle PA, Turner PS (1968) Relativistic Hartree-Fock X-ray and electron scattering factors. Acta Crystallogr A24:390–397

Gemmi M, Klein H, Rageau A, Strobel P, Le Cras F (2010) Structure solution of the new titanate $Li_4Ti_8Ni_3O_{21}$ using precession electron diffraction. Acta Crystallogr B66:60–68

Gilmore CJ, Dong W, Dorset DL (2008) Solving the crystal structures of zeolites using electron diffraction data. I. The use of potential-density histograms. Acta Crystallogr A64:284–294

Kisielowski C, Erni R, Freitag B (2008a) Object-defined resolution below 0.5 Å in transmission electron microscopy – recent advances on the TEAM 0.5 instrument. Microsc Microanal 14:78–79

Kisielowski C, Freitag B, Bischoff M, van Lin H, Lazar S, Knippels G, Tiemeijer P, van der Stam M, von Harrach S, Stekelenburg M, Haider M, Uhlemann S, Müller H, Hartel P, Kabius B, Miller D, Petrov I, Olson EA, Donchev T, Kenik EA, Lupini AR, Bentley J, Pennycook SJ, Anderson IM, Minor AM, Shmid AK, Duden T, Radmilovic V, Ramasse QM, Watanabe M, Erni R, Stach EA, Denes P, Dahmen U (2008b) Detection of single atoms and buried defects in three dimensions by aberration– corrected electron microscope with 0.5-Å information limit. Microsc Microanal 14:469–477

Kolb U, Gorelik T, Kübel C, Otten MT, Hubert D (2007) Towards automated diffraction tomography: Part I—data acquisition. Ultramicroscopy 107:507–513

Kolb U, Gorelik T, Otten MT (2008) Towards automated diffraction tomography. Part II—cell parameter determination. Ultramicroscopy 108:763–772

Kolb U, Gorelik T, Mugnaioli E (2009) Automated diffraction tomography combined with electron precession: a new tool for *ab initio* nanostructure analysis. In Moeck P, Hovmoeller S, Nicolopoulos S, Rouvimov S, Petrok V, Gateshki M, Fraundorf P (ed) Electron crystallography for materials research and quantitative characterization of nanostructured materials, Materials research society symposia proceedings 1184: GG01-05, Warrendale PA

Kolb U, Gorelik TE, Mugnaioli E, Stewart A (2010) Structural characterization of organics using manual and automated electron diffraction. Polym Rev 50:35–409

Kolb U, Mugnaioli E, Gorelik TE (2011) Automated electron diffraction tomography – a new tool for nano crystal structure analysis. Cryst Res Technol 46:542–554

Lipson H, Cochran W (1966) The determination of crystal structures, revised and enlarged edition. Cornell University Press, Ithaca.

Mugnaioli E, Gorelik T, Kolb U (2009) "Ab initio" structure solution from electron diffraction data obtained by a combination of automated diffraction tomography and precession technique. Ultramicroscopy 109:758–765

Nicolopoulos S, González-Calbet JM, Vallet-Regí M, Corma A, Corell C, Guil JM, Pérez-Pariente J (1995) Direct phasing in electron crystallography: Ab initio determination of a new MCM-22 zeolite structure. J Am Chem Soc 117:8947–8956

Pinsker ZG (1953) Electron diffraction. Butterworth, London

Reimer L, Kohl H (2008) Transmission electron microscopy, physics of image formation, 5th edn. Springer, New York

Rigamonti R (1936) La struttura della catena paraffinica studiata mediante i raggi di elettroni. Gazz Chim Ital 66:174–182

Rozhdestvenskaya I, Mugnaioli E, Czank M, Depmeier W, Kolb U, Reinholdt A, Weirich T (2010) The structure of charoite, $(K, Sr, Ba, Mn)_{15-16}(Ca, Na)_{32}[(Si_{70}(O, OH)_{180})](OH, F)_{4.0} * nH_2O$, solved by conventional and automated electron diffraction. Miner Mag 74:159–177

Schömer E, Heil U, Schlitt S, Kolb U, Gorelik TE, Mugnaioli E, Stewart A (2009) ADT-3D. A software package for ADT data visualizing and processing, Institute of Computer Science. Johannes Gutenberg University, Mainz. http://www.adt.chemie.uni-mainz.de

Sheldrick GM (2008) A short history of SHELX. Acta Crystallogr A64:112–122

Spence JCH (2003) High-resolution electron microscopy, 3rd edn. Oxford Univeristy Press, New York

Thomas JM, Terasaki O, Gai PL, Zhou W, Gonzalez-Calbet J (2001) Structural elucidation of microporous and mesoporous catalysts and molecular sieves by high-resolution electron microscopy. Acc Chem Res 34:583–594

Vainshtein BK (1956) Kinematic theory of intensities in electron diffraction patterns. Part 2. patterns from textures and polycrystalline aggregates. Sov Phys – Crystallogr 1:117–122

Vainshtein BK (1964) Structure analysis by electron diffraction. Pergamon Press, Oxford

Vincent R, Midgley PA (1994) Double conical beam-rocking system for measurement of integrated electron diffraction intensities. Ultramicroscopy 53:271–282

Voigt-Martin IG, Yan DH, Yakimansky A, Schollmeyer D, Gilmore CJ, Bricogne G (1995) Structure determination by electron crystallography using both maximum-entropy and simulation approaches. Acta Crystallogr A51:849–868

Wagner P, Terasaki O, Ritsch S, Nery JG, Zones SI, Davis ME, Hiraga K (1999) Electron diffraction structure solution of a nanocrystalline zeolite at atomic resolution. J Phys Chem B103:8245–8250

Weirich TE, Ramlau R, Simon A, Hovmöller S, Zou X (1996) A crystal structure determined with 0.02 Å accuracy by electron microscopy. Nature 382:144–146

Weirich TE, Portillo J, Cox G, Hibst H, Nicolopoulos S (2006) Ab initio determination of the framework structure of the heavy-metal oxide $Cs_xNb_{2.54}W_{2.46}O_{14}$ from 100 kV precession electron diffraction data. Ultramicroscopy 106:164–175

Williams DB, Carter CB (1996) Transmission electron microscopy. Plenum Press, New York

Zhukhlistov AP, Zvyagin BB (1998) Crystal structure of lizardite 1 T from electron diffraction data. Crystallogr Rep 43:950–955

Zhukhlistov AP, Avilov AS, Ferraris D, Zvyagin BB, Plotnikov VP (1997) Statistical distribution of hydrogen over three positions in the brucite $Mg(OH)_2$ structure from electron diffractometry data. Crystallogr Rep 42:774–777

Charoite, as an Example of a Structure with Natural Nanotubes

Ira Rozhdestvenskaya, Enrico Mugnaioli, Michael Czank, Wulf Depmeier, and Ute Kolb

Charoite from the Murun massif in Yakutiya, Russia (Vorob'ev 2008) was investigated using automated electron diffraction tomography (ADT) (Kolb et al. 2007, 2008; Mugnaioli et al. 2010) and precession electron diffraction (PED) (Mugnaioli et al. 2010, 2009), which allowed to determine the structure of charoite for the first time. The structure was solved *ab initio* in space group $P2_1/m$ by direct methods using a fully kinematic approach. The least squares refinements with 2878 reflections $F(hkl) > 4s_F$ converged to unweighted/weighted residuals R_1/wR_2 • 0.173/0.21 (Rozhdestvenskaya et al. 2010).

The structure of charoite with the crystal chemical formula $(K_{13.88}Sr_{1.0}Ba_{0.32}Mn_{0.36})_{S15.56}$ $(Ca_{25.64}Na_{6.36})_{S32}$ $[(Si_6(O_{11}(O,OH)_6)_2(Si_{12}(O_{18}(O,\ OH)_{12})_2(Si_{17}(O_{25}(O,OH)_{18})_2](OH,F)_{4.0}\cdot 3.18H_2O$, or, ideally, $(K,Sr,Ba,Mn)_{15-16}(Ca,Na)_{32}$ $[Si_{70}(O,OH)_{180}](OH,F)_{4.0}\cdot nH_2O$, can be visualized as being composed of three different silicate chains: a dreier double chain, $[Si_6O_{17}]^{10\bullet}$, a tubular loop-branched dreier triple chain, $[Si_{12}O_{30}]^{12\bullet}$ and a tubular hybrid dreier quadruple chain, $[Si_{17}O_{43}]^{18\bullet}$ (cf. Fig. 1). All chains run parallel to z axis and lie on the mirror plane. The chains are bonded by their apical oxygens to (Ca,Na)-octahedra bands,

I. Rozhdestvenskaya (✉)
Department of Crystallography, Saint Petersburg State University,
University emb. 7/9, St. Petersburg 199034, Russia
and
Department of Crystallography, Institute of Geowissenschaften,
Christian-Albrechts-University, Olshausenstrasse 40, D-24098 Kiel, Germany
e-mail: ivrozhdestvenska@mail.ru

E. Mugnaioli • U. Kolb
Institute of Physical Chemistry, Johannes Gutenberg-University, Welderweg 11,
D-55099 Mainz, Germany

M. Czank •
W. Depmeier
Department of Crystallography, Institute of Geowissenschaften,
Christian-Albrechts-University, Olshausenstrasse 40, D-24098 Kiel, Germany

S.V. Krivovichev (ed.), *Minerals as Advanced Materials II*,
DOI 10.1007/978-3-642-20018-2_6, © Springer-Verlag Berlin Heidelberg 2012

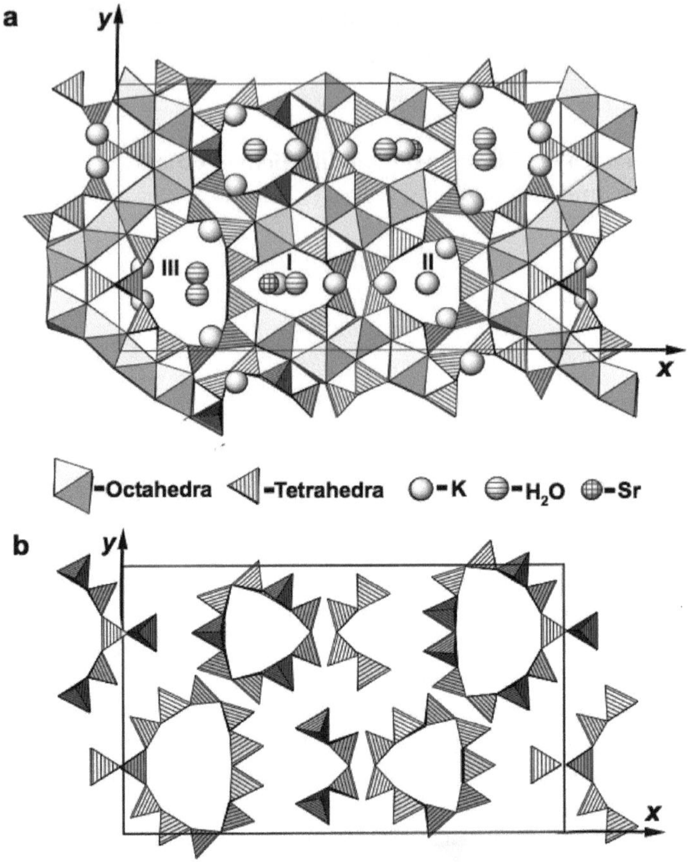

Fig. 1 The crystal structure of charoite-90. (**a**): in projection on plane (001) and (**b**): the arrangement of the three different silicate chains in projection on plane (001)

which are also parallel to *z*. The ions K^+, Ba^{2+}, Sr^{2+}, Mn^{2+} and H_2O molecules are located inside the tubes.

The columns of octahedra are joined by shared edges and build a block four octahedra wide. The apical oxygens link two such blocks to form a continuous zigzag sheet parallel to (100) (cf. Fig. 1a). Additional, isolated blocks four octahedra wide are present composed of a double column and two single columns joined by common apical oxygens. The octahedrally-coordinated sites in charoite are predominantly occupied by Ca. Only two octahedra have mixed Ca/Na occupancy.

The dreier double chain (I), $[Si_6O_{17}]^{10\bullet}$, is of the type as the chain found in the xonotlite structure (Kudoh and Takeuchi 1979). The dreier double chain forms a tubular unit together with two columns of Ca octahedra and four tetrahedra of the neighbouring tubular chain (cf. Fig. 2a, d). This tubular unit has a window of 8-membered ring (8MR) of terahedra in the plane parallel (100).

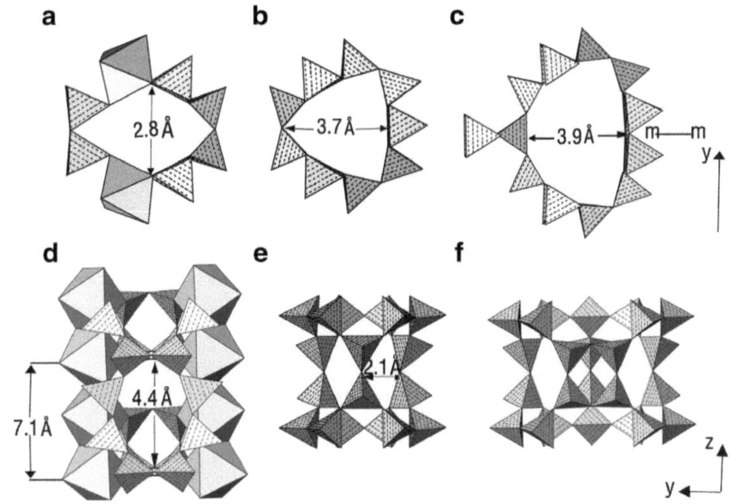

Fig. 2 The tubes of the charoite structure built up by the dreier double chain, the tubular dreier triple chain and the tubular hybrid dreier quadruple chain (in Fig. 3 these tubes will be assigned to I, II and III, respectively). (**a**), (**b**), (**c**): [001] projection. The smallest diameter of each tube is given. (**d**), (**e**), (**f**): [100] projection. Two typical values of the free diameters of the windows are indicated

The tubular dreier triple chain (II), $[Si_{12}O_{30}]^{12\bullet}$, is formed by three loop-branched dreier single chains, which are of the type as found in synthetic $Li_2Mg_2[Si_4O_{11}]$ (Czank and Bissert 1993). The tubular dreier triple chain of charoite has three-fold symmetry (cf. Fig. 2b, e) and differs from the chains of identical stoichiometry found in the structures of frankamenite (Rozhdestvenskaya et al. 1996), canasite (Rozhdestvenskaya et al. 1988) and miserite (Rozhdestvenskaya and Evdokimov 2006). There are three symmetrically equivalent 8MR windows present in this chain.

The tubular hybrid dreier quadruple chain (III), $[Si_{17}O_{43}]^{18\bullet}$, consists of two loop-branched dreier single chains, a dreier single chain and a dreier double chain. The dreier single chain is similar to that found in pectolite and the dreier double chain is of the type found in okenite (Merlino 1983). One tetrahedron of this large tubular chain is connected to four other tetrahedra, thus its connectivity is more typical for tetrahedra in framework structures (cf. Fig. 2c, f). This tubular chain has four 8MR windows of which two at a time are symmetrically equivalent.

Horizontal Si_2O_7 groups of each chain connect the separate bands of Ca-octahedra in such a way that apical oxygens of Si_2O_7 groups of neighboring chains are joined to different apices of vertical edges of octahedral (cf. Fig. 1). Therefore, adjacent chains are shifted relative to each other by half a translation along the *z* axis.

The free diameter of the windows as well as of the channels has been estimated from the distance between O atoms across the ring minus 2.7 Å (assuming the ionic radius of oxygen to be 1.35 Å). All the chains have windows of 8-membered rings. The free diameter of the windows along z axis is 4.4 Å and has values

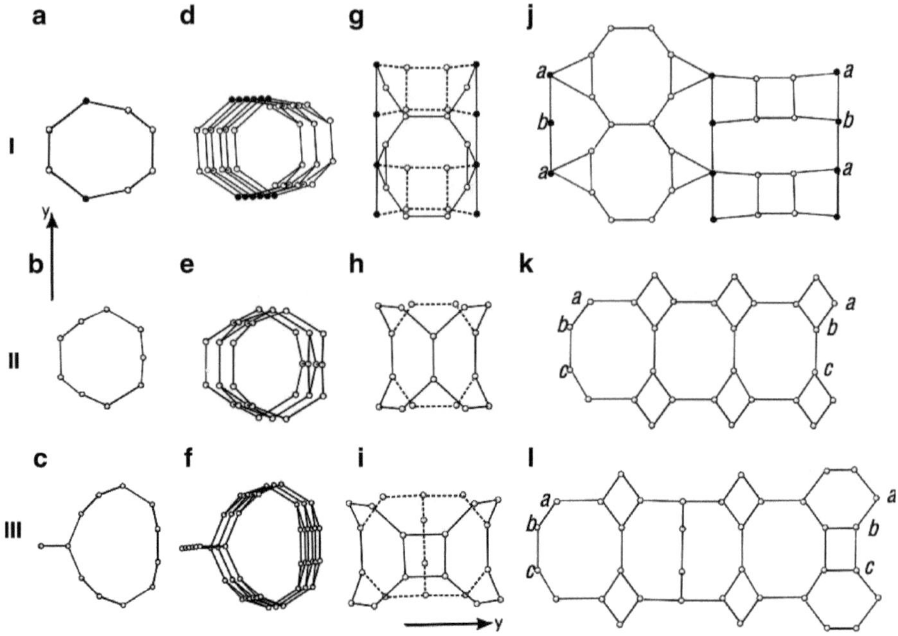

Fig. 3 The same tubes as in Fig. 2 represented only by the cations (black node – octahedron, white node – tetrahedron). (**a**), (**b**), (**c**): [001] projection of the tubes I, II and III, correspondingly. (**d**), (**e**), (**f**): the 3D nets of the tubes slightly out of the [001] projection by rotation on the b axis. (**g**), (**h**), (**i**): almost perpendicular view onto the tubes, for better visualization they are slightly out of the [100] projection by rotation on the b axis. (**j**), (**k**), (**l**): tapes of the unfolded tubular units I, II and III

perpendicular to z axis between 2.0 Å and 3.5 Å. The tubes in charoite exhibit free channels along the chains (cf. Figs. 2 and 3). In Fig. 2a the tube I is delimited by six terahedra and two octahedra, while the two others tubes (II and III) are defined by 9- and 11-membered rings of tetrahedra (Fig. 2b, c), respectively.

The smallest free diameters (effective pore widths) of the channels are given in Fig. 2a–c. The nodal representations (Krivovichev 2005) of the three different tubular units (now designated by I, II and III) are shown in Fig. 3: [001] projections (a, b, c); the 3D nets slightly out of the [001] projection by rotation on the b axis (d, e, f); [100] projections (g, h, i) and tapes of unfolded tubular units (j, k, l), respectively. All three tubular units are achiral as they have a mirror plane (Fig. 2). The channel of the large tubular chain $[Si_{17}O_{43}]^{18\bullet}$ has a structural protrusion. This big cavity has the largest free diameters of 6.9 and 6.5 Å. The cavity has four windows of 8MR of which two have free diameters of 2.2 • 4.4 Å and the other two of 3.5 • 4.4 Å, respectively.

The potassium atoms are located in the 8MR windows of the tubes. Another potassium atom and the H_2O molecule are located on the axis of the tubular loop-branched dreier triple chain at 1/2 of the translation along the *z* axes. The cavity within the large tubular chain $[Si_{17}O_{43}]^{18\bullet}$ is nearly empty.

The interior of the tube I associated with the bent dreier double chain has a higher occupancy. From the distances between of the peaks and the heights of the peaks on Fourier maps we infer that Sr^{2+}, H_2O molecules and the impurity cations Ba^{2+}, Mn^{2+}, K^+ are located inside this tube. The sites inside this tube are not fully occupied.

The term "nanotube" does not have a clear-cut definition. In general understandings it describes a nanoscale tube-like structure which can be found naturally in some minerals or be man-made from a variety of materials. The majority of known nanotubes are still carbon nanotubes, but various types and structures of non-carbon nanotubes, based on carbonitrides $B_x C_y N_z$, boron nitride BN, sulfides WS_2, MoS_2, selenides $NbSe_2$, halides $NiCl_2$, transition metal oxides SiO_2, TiO_2, MoO_3, V_2O_5 have also been synthesized (Pokropivny 2001). An overview of nanotubes and tubular-shaped materials in some mineral systems was given in (Krivovichev 2008). Nowadays SiO_2 nanotubes can be easily synthesized and have potential applications as nanoscale reactors (Ogihara et al. 2006). Despite the fact that up to now SiO_2 nanotubes are amorphous, crystalline SiO_2 nanotubes maybe synthesized in the future and the tubular chains of charoite may then serve as possible examples.

Acknowledgements The authors thank the Deutsche Forschungsgemeinschaft for financial support (DE 412/44-1).

References

Czank M, Bissert G (1993) The crystal structure of $Li_2Mg_2[Si_4O_{11}]$, a loop-branched dreier single chain silicate. Z Kristallogr 204:129–142

Kolb U, Gorelik T, Kübel C, Otten M, Hubert D (2007) Towards automated diffraction tomography: Part I – data acquisition. Ultramicroscopy 107:507–513

Kolb U, Gorelik T, Otten M (2008) Towards automated diffraction tomography. Part II – cell parameter determination. Ultramicroscopy 108:763–772

Krivovichev S (2005) Topology of microporous structures. In: Ferraris G, Merlino S (Eds) Rev Miner Geochem 57: 17–68

Krivovichev S (2008) Nanotubes in minerals and mineral-related systems. In: Krivovichev S (ed) Minerals as advanced materials I. Springer, Berlin/Heidelberg, pp 179–191

Kudoh Y, Takeuchi Y (1979) Polytypism in xonotlite: (I) structure of an A-1 polytype. Miner J 9:349–373

Merlino S (1983) Okenite, $Ca_{10}Si_{18}O_{46}$*$18H_2O$: the first example of a chain and sheet silicate. Amer Miner 68:614–622

Mugnaioli E, Gorelik T, Kolb U (2009) "Ab initio" structure solution from electron diffraction data obtained by a combination of automated diffraction tomography and precession technique. Ultramicroscopy 109:758

Mugnaioli E, Gorelik T, Stewart A, Kolb U (2010) "Ab initio" structure solution of nanocrystalline minerals and synthetic materials by automated electron tomography. In: Krivovichev S (ed) Minerals as advanced materials II. Springer, Berlin, this volume

Ogihara H, Takenaka S, Yamanaka I, Tanabe E, Genseki A, Otsuka K (2006) Synthesis of SiO_2 nanotubes and their application as nanoscale reactors. Chem Mater 18:996–1000

Pokropivny V (2001) Non-carbon nanotubes (Review). Part 2. Types and structure. Powder Metall Met Ceram 40:11–12, 582–594

Rozhdestvenskaya I, Evdokimov M (2006) Refinement of miserite crystal structure $(K_{1.29}\square_{0.21})$ $[Ca_{5.51}M_{0.49}{}^{3+}](Si_6(O,OH)_{15})(Si_2O_7)(F,OH)_2$ $0.25H_2O$, (M • Y, REE, Fe, Ti, Mn, Mg, Na) from the Dara-i-Pioz occurrence, Pamir, Tajikistan. Dokl Earth Sci 406:74–78

Rozhdestvenskaya I, Nikishova L, Bannova I, Lazebnik K (1988) Canasite: refinement and features of crystal structure, structural typomorphism. Miner Zh 10(4):31–41 (in Russian)

Rozhdestvenskaya I, Nikishova L, Lazebnik K (1996) The crystal structure of frankamenite. Miner Mag 60:897–905

Rozhdestvenskaya I, Mugnaioli E, Czank M, Depmeier W, Kolb U, Reinholdt A, Weirich T (2010) The structure of charoite, $(K,Sr,Ba,Mn)_{15\text{-}16}(Ca,Na)_{32}[(Si_{70}(O,OH)_{180})](OH,F)_{4.0}$ * nH_2O, solved by conventional and automated electron diffraction. Miner Mag 74:159–177

Vorob'ev E (2008) Charoite. Novosibirsk, Russia

Hydrothermal Alteration of Basalt by Seawater and Formation of Secondary Minerals – An Electron Microprobe Study

Christof Kusebauch, Astrid Holzheid, and C. Dieter Garbe-Schönberg

1 Introduction

Submarine hydrothermalism with extensive alteration of oceanic crust is a major process governing energy and mass fluxes from Earth's interior into the oceans. New estimates suggest that about 50% of hydrothermal crust alteration occurs along slow and ultra-slow spreading ridges in the deep sea (German et al. 2009). Hydrothermal systems in water depths >3,000 m can run at extreme temperatures under pressure (p) and temperature (T) conditions above the critical two-phase curve of seawater if enough heat from magma injections is provided. Exactly this type of submarine hydrothermalism has recently been discovered (Koschinsky et al. 2008): a high-pressure system in 3,000 m water depth at 5°S on the Mid Atlantic Ridge (MAR) characterized by extreme temperatures (>400°C) and high to very high concentrations of dissolved hydrogen, methane, transition metals (e.g., Cu, Fe), and unique rare earth element chemistry with LREE and HREE depletion and no Eu-anomaly (Schmidt et al. 2010). All processes during phase separation, ascent, and final venting of the hydrothermal fluids at the seafloor run entirely under p-T-conditions within the vapour-liquid two phase field above the critical seawater curve. New results from 2-D, three-phase models indicate that these high-pressure systems run very stable for long time periods and without the compositional fluctuations known from the East Pacific Rise (EPR) in ~2,500 m water depth (Coumou et al. 2008, 2009). Figure 1 illustrates p-T-positions of the MAR 5°S vent fluid samples from the high temperature hydrothermal fields Red Lion (RL), Sisters Peak (SP), and Turtle Pits (TP) together with representative chondrite-normalized REE + Y pattern of vent fluids of the three fields. Based on measured p-T-conditions vent site RL seems to run under p-T-conditions of the single-phase field of seawater, while sites SP and TP temporarily run under p-T-conditions of the vapour-liquid two phase field or at least very close to p-T-conditions of the

C. Kusebauch • A. Holzheid (✉) • C.D. Garbe-Schönberg
Institut für Geowissenschaften, Universität Kiel, Kiel 24098, Germany
e-mail: holzheid@min.uni-kiel.de

S.V. Krivovichev (ed.), *Minerals as Advanced Materials II*,
DOI 10.1007/978-3-642-20018-2_7,

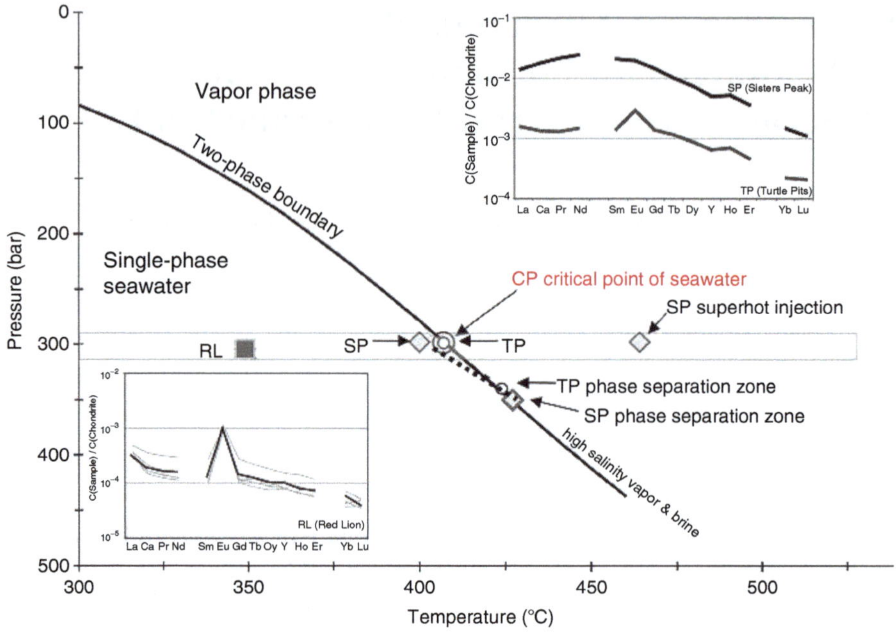

Fig. 1 REE pattern of MAR 5°S fluids and positions of the 5°S vent fluid samples (see text for more detail; the figure is modified after Koschinsky et al. 2008; the REE data are taken from Schmidt et al. 2010)

two-phase boundary. Differences in REE + Y pattern of the vent fluids of the three sites are obvious: RL fluids with temperatures of 350°C exhibit a continuous increase of chondrite-normalized REE concentrations towards La, and a positive Eu anomaly. These REE + Y patterns are similar to more 'typical' patterns known from many hydrothermal systems, i.e., systems that run under p-T-conditions of single phase fields (single-phase seawater or vapour phase, known from e.g., EPR vent sites). Contrary to RL, fluids from SP and TP with temperatures of 407°C and higher exhibit a variable REE + Y chemistry ranging from fluids with highly elevated REE concentrations, relative LREE and HREE depletion, and no Eu-anomaly (fluids from SP) to fluids with weakly elevated LREE abundance, HREE depletion, and a small positive Eu-anomaly (fluids from TP).

Our understanding of fluid-rock interaction, pressure-temperature regimes, and temporal variability of hydrothermal systems has been dominated by studies at the EPR in more shallow water depths <2,500 m (e.g., Jupp and Schultz 2000; Von Damm et al. 1995, 2003; Butterfield et al. 1997; Edmond et al. 1979). As a consequence, most hydrothermal experiments studying basalt-seawater interaction have been conducted mainly under p-T-conditions prevailing at the EPR (see Fig. 2 for compilation of p-T-conditions of literature experiments). Moreover, very little work has been done on trace element partitioning and composition of resulting fluids as well as on chemical alteration of basalt and chemistry of newly formed mineral assemblages. A systematic experimental study on basalt-seawater interaction

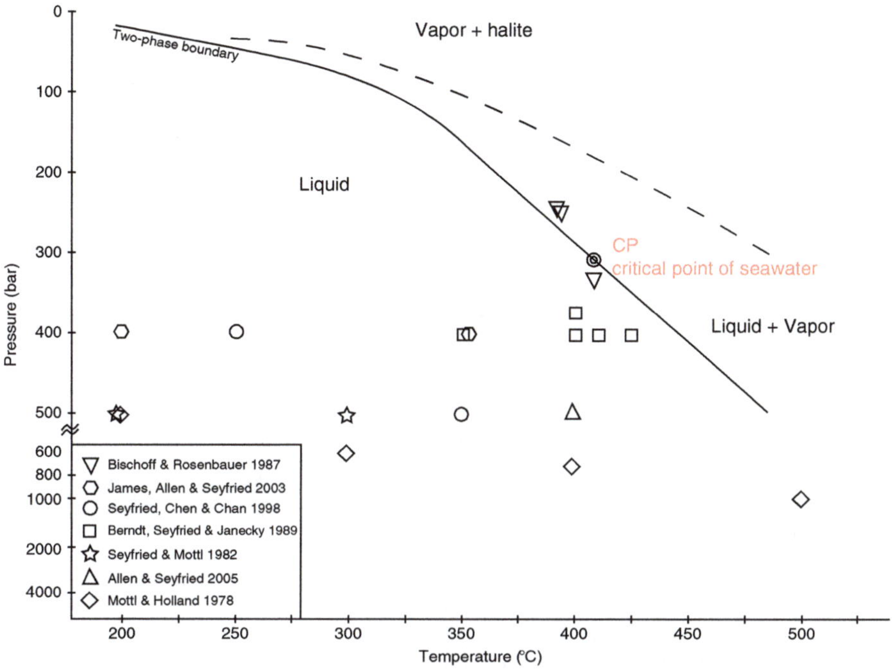

Fig. 2 Compilation of literature data on experiments at individual p-T-conditions

processes at p-T-conditions close to or within the vapour-liquid two phase field improving our understanding of hydrothermal systems in great water depths and at extreme temperatures, namely the origin of primary hydrothermal fluids, transport reactions, and element fluxes during ascent in such hydrothermal systems is missing so far. Moreover, the *in situ*-microanalysis of original but experimentally altered basaltic rock fragments, experimentally formed secondary mineral phases, and micro-volumes of experimentally produced hydrothermal fluids would furthermore allow derivation of element partitioning coefficients that are prerequisites for subsequent thermodynamic calculations. This systematic experimental approach has recently been performed in our laboratories. Our contribution to the proceedings of the second international meeting on 'Minerals as advanced materials' is part of that study, and this paper focuses on the hydrothermal alteration products, i.e., the altered basaltic rock fragments and the newly formed secondary mineral assemblages, and their major and minor element compositions.

2 Experimental and Analytical Methodology

All hydrothermal experiments are conducted using cold-seal pressure vessels. The experimental set-up and the starting material are illustrated in Fig. 3.

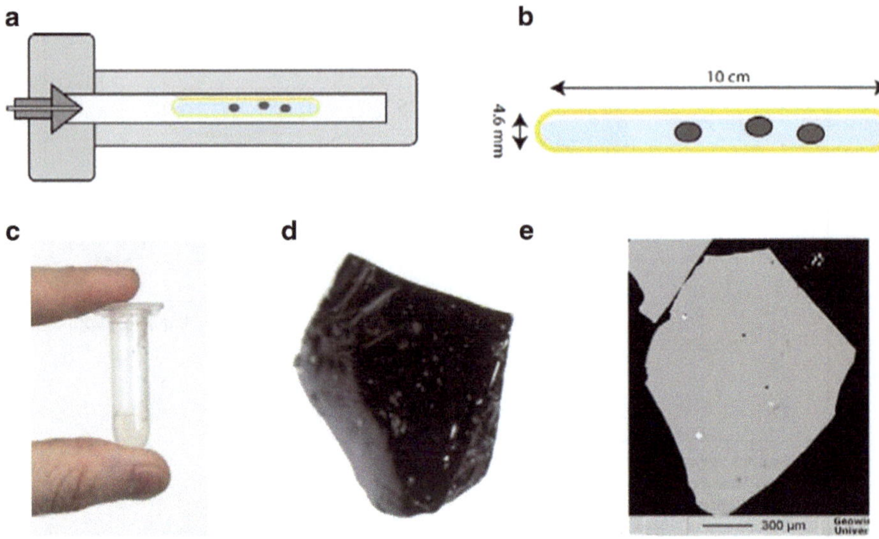

Fig. 3 Schematic experimental set-up and starting materials. (**a**): schematic cold-seal pressure vessel autoclave with the sample in the centre; (**b**): Au-capsule (smaller size) as container of the starting materials (basalt glass and seawater inside the Au-capsule); (**c**): bottom seawater sampled at MAR 5°S and used as starting fluid; (**d**) and (**e**): homogeneous unaltered mid-ocean ridge basalt glass sampled at MAR 5°S and used as solid starting material

The starting material consisted of original MORB (mid-ocean ridge basalt) glass fragments and Atlantic bottom seawater, all from the MAR 5°S, and was placed in Au capsules of two different sizes (length: 100 mm, OD/ ID: 5/4.6 mm or OD/ID: 7/6.6 mm) as containers. About ~50–75 mg basalt and ~400–600 mg seawater was placed in the Au capsules, resulting in internal pressures within the container of 400–1,500 bar at 400–500°C. The experimental conditions mimicked the p-T-conditions close to or within the vapour-liquid two phase field, i.e., at MAR 5°S. As above-mentioned the main emphasis of the present contribution is on the altered basaltic rock fragments and the secondary mineral assemblages. Representative aliquots of the experimental solid charges were therefore mounted in epoxy and polished for analysis by electron microscopy. The major and minor element chemistry of the alteration products were analyzed by electron microprobe (JEOL 8900RL, Institute of Geosciences, Kiel). Visual intuitive alteration and mineral growth phenomena were studied by optical and digital photomicrographs, the latter acquired by backscattered electron detection on the JEOL 8900RL electron microprobe.

In the following the visual intuitive alteration and mineral growth phenomena will be correlated with the experimental parameters run duration, temperature, pressure, and water-to-rock ratio of the starting material and with the changes in major and minor element chemistry. In general, the alteration phenomena are quite divers. Formation of veins is common. The veins criss-cross the glass fragments. Figure 4a is a detail of an optical photomicrograph of a representative sample that

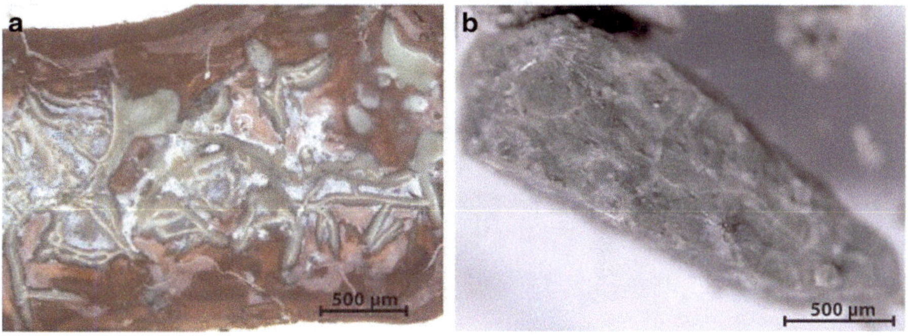

Fig. 4 Optical photomicrographs of a heavily altered glass fragment with a vein network (**a**) and idiomorphic plagioclase rosettes (**b**) grown at the emersion points of the veins

is pervaded by a vein network. The degree of alteration increases from the rim to the centre. The outer heavily altered parts do have the tendency to peel off (not shown in Fig. 4a). Most glass fragments are opaque and reddish-brown or green in colour, some fragments have still unaltered centres. Formation of secondary minerals is also common. Figure 4b shows as an example idiomorphic plagioclase radial arranged as spheres at the emersion point of the veins.

3 Alteration Phenomena of Basaltic Glass

3.1 Alteration of Glass Fragments as Function of Run Duration

Figure 5 compiles backscattered electron images of basalt-seawater interaction experiments at constant temperature, pressure, and water-to-rock ratio (500°C, 4,000 bar, ~ 10) but variable run durations (1, 3, 7, and 10 days). The experimental pressure and temperature conditions are within the single-phase seawater field. Areas of glass fragments of the shortest experiment (Fig. 5a) provide indications of beginning alteration as only a few veins exist in these glass regions (see glass fragment in the right lower corner of Fig. 5a). These regions are side by side to areas that are completely nerved by veins (centre part of Fig. 5a). All veins are chemical heterogeneous and not related to mineral chemistries. The degree of alteration increased at the 3 day experiment (Fig. 5b), but besides the more pronounced alteration no obvious difference exists compared to the 1 day experiment.

This is also the case for the 7 and 10 day experiments (Fig. 5c, d, respectively). As alteration progressed, larger regions of homogeneously altered glass are formed. The lack of significant changes in optical and chemical appearance of alteration as

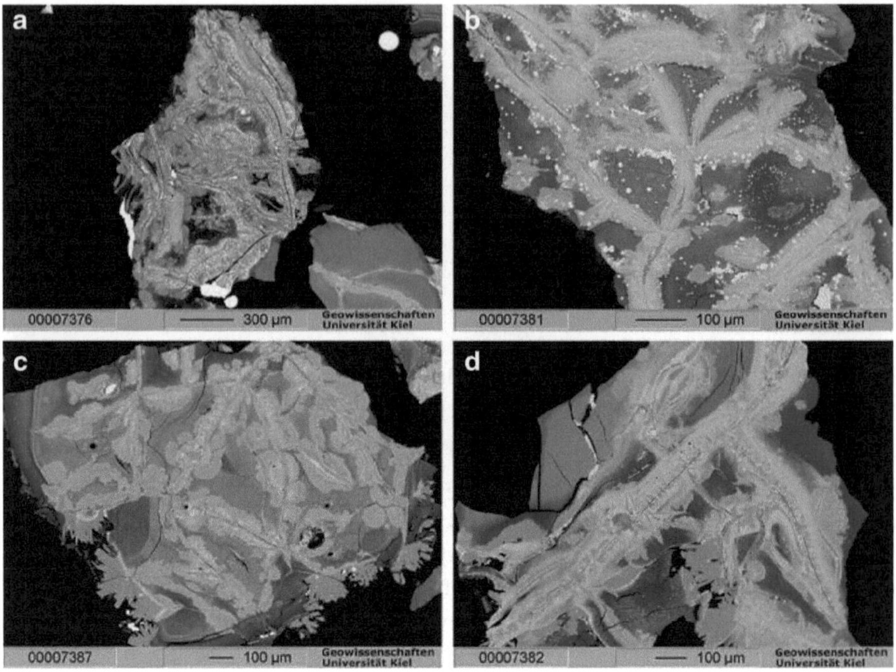

Fig. 5 Backscattered electron images of experiments at constant temperature, pressure, and water-to-rock ratio (500°C, 4,000 bar, ~10) but variable run durations (**a**: 1 day, **b**: 3 days, **c**: 7 days, **d**: 10 days)

function of run duration at constant temperature and pressure point to very fast formation of alteration textures in the basaltic glasses that coexist with seawater.

3.2 Alteration of Glass Fragments as Function of Temperature

Backscattered electron images of experiments at slightly various pressures (400 and 450 bar), constant run duration (10 days), and water-to-rock ratio (~10), but variable temperatures (450, 475, 500°C) are compared in Fig. 6. All these basalt-seawater interaction experiments were run at conditions of the vapour-liquid two phase field. The alteration textures are again similar. The only differences are the distribution and the progress of alteration within the glass fragments. The lowest temperature experiment (450°C, Fig. 6a) has still unaltered cores of the glass fragments. Chemical analyses of the altered regions in the lowest temperature experiment show Calcium loss. Regions with low Ca content are also observed in the intermediate temperature experiment (475°C, Fig. 6b). These low Ca content regions adjoin to regions composed of microcrystalline phases. The experiment at 500°C (Fig. 6c) is completely altered and the glass is totally replaced by heterogeneous microcrystalline phases.

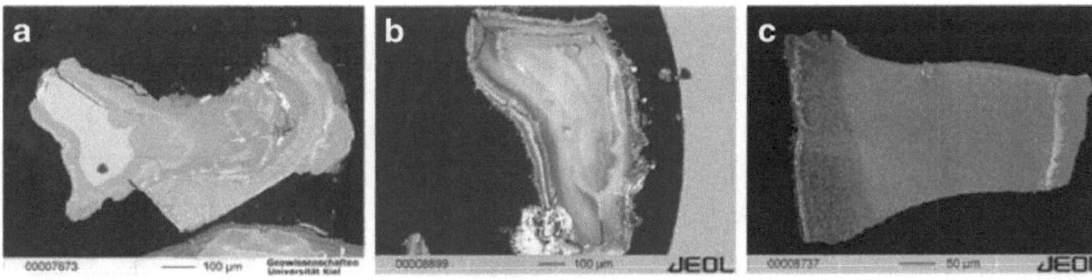

Fig. 6 Backscattered electron images of experiments at slightly various pressures, constant run duration and water-to-rock ratio (400 and 450 bar, 10 days, ~10) but variable temperatures (**a**: 450°C, **b**: 475°C, **c**: 500°C)

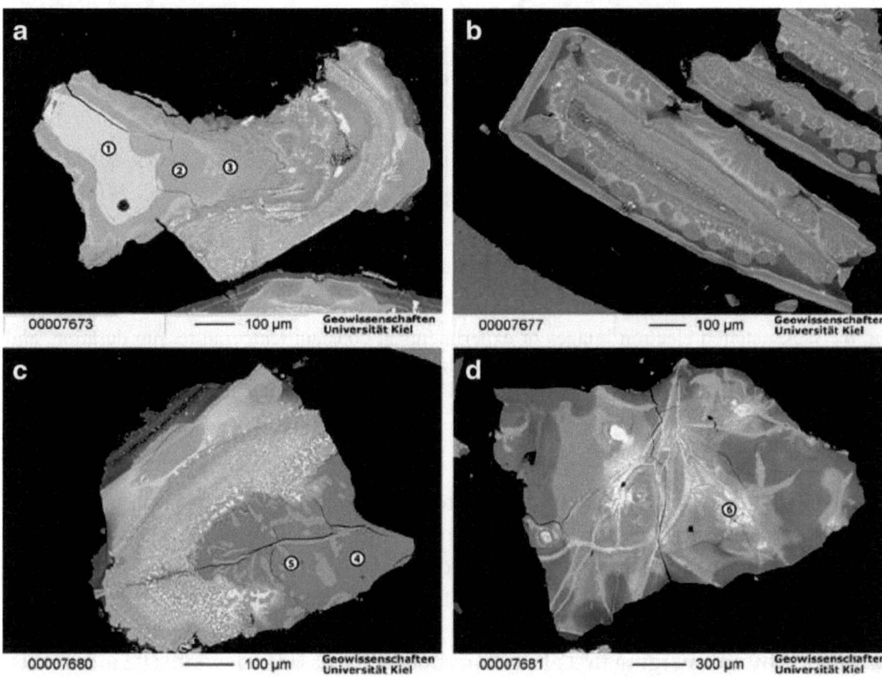

Fig. 7 Backscattered electron images of experiments at constant temperature, run duration, and water-to-rock ratio (450°C, 10 days, ~10) but variable pressures (**a**: 400 bar, **b**: 600 bar, **c**: 1,000 bar, **d**: 4,000 bar). Numbers within the images indicate spots of representative chemical analyses, that are listed in Table 1

3.3 *Alteration of Glass Fragments as Function of Pressure*

Figures. 7 and 8 display variations in alteration as function of pressure. All experiments in Fig. 7 were performed at a constant temperature of 450°C, these of Fig. 8 at 500°C. Both sets had a constant run duration (10 days) and water-to-rock

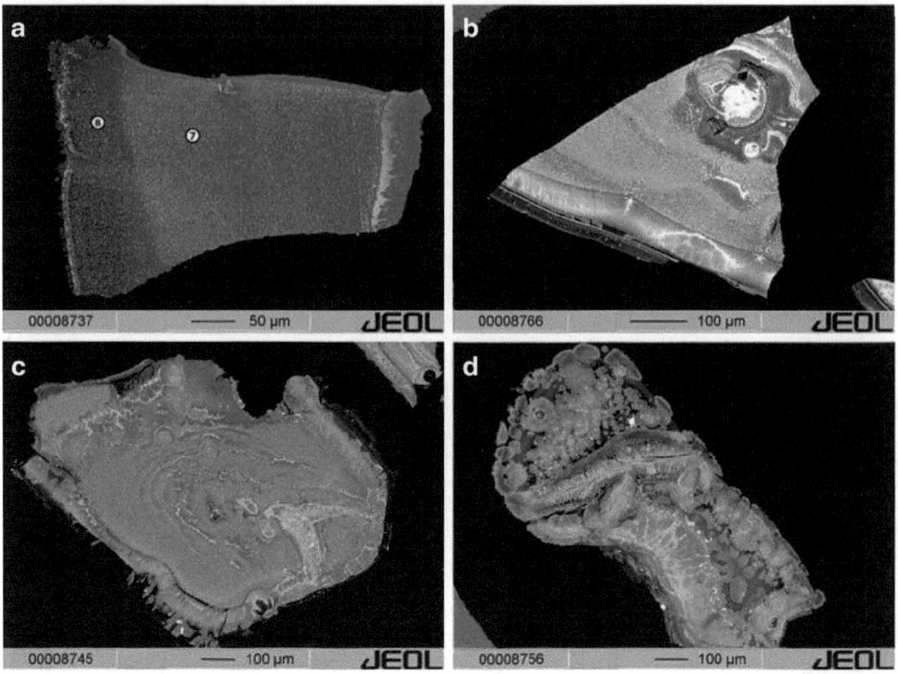

Fig. 8 Backscattered electron images of experiments at constant temperature, run duration, and water-to-rock ratio (500°C, 10 days, ~10) but variable pressures (**a**: 450 bar, **b**: 500 bar, **c**: 600 bar, **d**: 800 bar). Numbers within (**a**) indicate spots of representative chemical analyses, that are listed in Table 1

ratio (~10). The pressures varied from 400 to 4,000 bar in Fig. 7 experiments and from 450 to 800 bar in Fig. 8 experiments. The degree of alteration of the glass fragments increases with increasing pressure in both sets of experiments. At the 450°C experiments regions with unaltered basalt are still preserved in the experiment at the lowest pressure (400 bar, labeled #1 in Fig. 7a). These unaltered regions are surrounded by an emerging first alteration zone of about 100 μm width (#2 in Fig. 7a). This first zone is characterized by low CaO and SiO_2 contents (concentrations drop by 7–8 wt.% for CaO and 2–3 wt.% for SiO_2), but high water content (more than 10 wt.% – derived from the difference of 100 wt.% to the total of the analysis) compared to the unaltered basalt (#1 in Fig. 7a), see chemical analyses of #1 and #2 in Table 1. The first alteration zone borders on a narrow, only 20 μm wide zone (#3 in Fig. 7a). This second alteration zone is even more depleted in Ca and Si, but enriched in Fe, Mg, and Na compared to the unaltered regions, see chemical analysis of #3 in Table 1. The outermost region of the post-run charge at 400 bar exists of heterogeneous and very fine grained textures. No reasonable meaningful chemical analyses of these textures were possible due to particle sizes of less than 1 μm. At a pressure of 600 bar (Fig. 7b) the altered glass mainly consists of the above mentioned heterogeneous and very fine grained textures that appear in layers. No unaltered regions do exist anymore.

Table 1 Chemical composition (wt.%) of alteration products

Number in figures		1	2	3	4	5	6	7	8	9
Run #		22	22	22	24	24	26	43	43	36
P [bar]		400	400	400	1,000	1,000	4,000	450	450	400
T [°C]		450	450	450	450	450	450	500	500	450
Duration [days]		10	10	10	10	10	10	10	10	10
Water-to-rock		9.6	9.6	9.6	8.4	8.4	6.5	10.1	10.1	9.7
SiO_2	50.36	48.96	47.38	41.22	43.52	51.59	45.98	49.76	46.98	49.89
TiO_2	1.07	1.09	1.00	1.32	1.30	1.02	0.76	1.32	1.14	1.03
Al_2O_3	14.96	14.77	14.20	14.27	9.81	24.70	8.33	15.29	15.17	15.25
Cr_2O_3	0.04	0.06	0.01	0.08	0.07	0.06	0.03	0.07	0.05	0.03
FeO	9.16	9.25	8.21	14.75	9.89	1.52	13.09	7.77	5.70	9.32
MnO	0.18	0.17	0.22	0.16	0.13	0.05	0.52	0.09	0.05	0.16
MgO	7.90	8.03	7.89	10.96	18.65	1.68	7.58	9.16	7.41	8.18
CaO	11.81	12.01	3.45	1.83	3.23	10.70	18.92	8.50	8.02	7.58
Na_2O	2.58	2.99	2.36	3.00	1.62	4.50	0.53	3.52	3.63	4.18
K_2O	0.09	0.11	1.55	2.50	0.26	0.07	0.03	0.23	0.26	0.28
Total	98.15	97.44	86.26	90.09	88.48	95.88	95.76	95.71	88.41	95.89
Comments	Starting material (basalt)	Unaltered basalt	First alteration zone	Second alteration zone	Homogeneous area	First vein formation	Homogeneous whitish area	Heterogeneous area I	Heterogeneous area II	Microcrystalline phase

At a pressure of 1,000 bar (Fig. 7c) coarse grained (>1 μm) textured areas exist within the regions of very fine grained textures. Besides these heterogeneous regions up to 1 mm large homogeneous areas formed that appear greenish in color under reflected light. Chemical analyses of these homogeneous areas point to high water and Mg contents and low Si and Ca contents. A representative analysis is given as #4 in Table 1 (see Fig. 7c for location of analysis #4). Both, the water and Mg contents (>10 wt.% water; ~19 wt.% MgO) are high and the SiO_2 and CaO contents (43.5 wt.% SiO_2; 3.2 wt.% CaO) are low compared to the basaltic starting material (<2 wt.% water; 7.9 wt.% MgO; 50.4 wt.% SiO_2; 11.8 wt.% CaO). Within the homogeneous areas also first-time formation of veins (#5 in Fig. 7c) can be observed with high Al_2O_3 (22.6–24.7 wt.%), CaO (8.5–10.7 wt.%), and Na_2O (4.5–6 wt.%) contents and low MgO (1.7–3 wt.%) and FeO (1.5–2 wt.%) contents. Only the analysis with the highest Al, Ca, Na, and lowest Mg and Fe values is listed in Table 1. An increase of pressure to 4,000 bar causes disappearance of the dominant alteration phenomena at lower pressures, i.e., the heterogeneous and very fine grained textures vanished. Large-scale homogeneous areas formed at 4,000 bar (see #6 in Fig. 7d). These areas are greenish and whitish in color under reflected light and have high Fe and Ca contents and low Si, Al, and Na concentrations (see #6 in Table 1 as an example of the whitish area). The areas are pervaded by veins. The veins are identical in chemistry to the veins that formed at 1,000 bar (see #5 in Fig. 7c and Table 1). However, some veins at the 4,000 bar experiments are zoned and host flamy formed phases with very high concentrations of CaO (up to 23 wt.%). The experiments at 500 C (Fig. 8) show similar alteration phenomena as function of pressure as the 450°C experiments did. Heterogeneous and very fine grained textures formed at lower pressures (450 and 500 bar – Figs. 8a, b, respectively), indicating the presence of microcrystalline phases. Representative areas of these heterogeneous regions are highlighted as #7 and #8 in Fig. 8a. Chemical analyses of these spots are given in Table 1. The absence of unaltered basalt is most likely related to the higher temperature and therefore to faster alteration reactions compared to the 450°C experiments. Increasing pressure (600 and 800 bar – Fig. 8c, d, respectively) causes formation of vein networks similar in chemistry to these of the experiments at 450°C.

3.4 *Alteration of Glass Fragments as Function of the Water-to-Rock Ratio*

The influence of the water-to-rock ratio on the alteration of MORB glass was determined at constant temperature, pressure, and run duration (450°C, 400 bar, 10 days). Following the terminology of Seyfried and Mottl (1982) rock-dominated experiments with water-to-rock ratios of 10, 5, 2.5, and 1, and one experiment at fluid-dominated conditions with a water-to-rock ratio of 44 were performed. Backscattered electron images of the rock-dominated experiments are compared in Fig. 9a–d. All rock-dominated experiments show similar alteration phenomena.

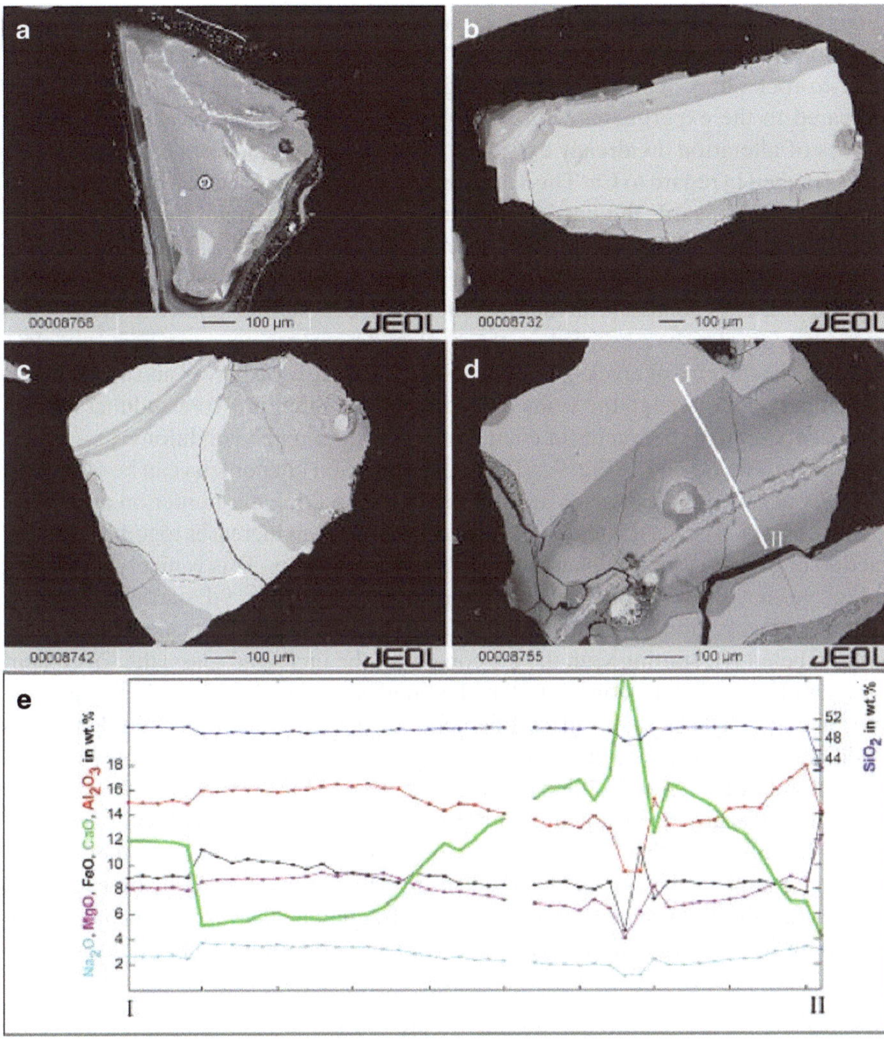

Fig. 9 Backscattered electron images of experiments at constant temperature, run duration, and pressure (450°C, 10 days, 400 bar) but variable water-to-rock ratio (**a**: 10, **b**: 5, **c**: 2.5, **d**: 1). The number within (**a**) indicates the spot of a representative chemical analysis, that is listed in Table 1. (**e**) Illustrates the change in chemistry along the profile I–II in (**d**)

Basaltic glasses that reacted with fluids in experiments with a water-to-rock ratio of 10 preserved small patches of unaltered MORB glass in the center. The still unaltered regions border on glassy areas that are leached in regard to Ca. These areas are surrounded by heterogeneous and very fine grained regions that consist of microcrystalline phases (#9 in Fig. 9a). The chemical compositions of the microcrystalline phases (#9 of Table 1) are similar in chemistry to the heterogeneous and very fine grained regions of the other experiments (see e.g., #7 of Table 1 and Fig. 8a, that belongs to an experiment at 500°C, 450 bar and a water-to-rock ratio of 10). Similar alteration phenomena can be observed in the post-run charges of experiments with water-to-rock ratios of 5 (Fig. 9b). Although the unaltered region is significant larger compared to the experiments with water-to-rock ratios of 10, areas with different degrees of alteration do already exist: the unaltered region borders on a glassy area that is leached in regard to Ca. This Ca poor area merges into an area with progressive increase of Ca towards the outside margin of the glass fragment that was in direct contact with the fluid during the experiment. The increase of Ca is accompanied by an increasing frequency of occurrence of microcrystalline phases. A simultaneous increase of the Ca content with the fraction of microcrystalline phases towards veins or cracks can also be observed in these glass fragments as some post-run glass fragments have veins or cracks. The veins are sporadically dispersed and are different in chemistry compared to the veins that occurred at higher pressure and that formed vein networks. In experiments at even more fluid-dominated conditions, i.e., water-to-rock ratios of 2.5 and 1 (Fig. 9c, d), similar alteration phenomena can be observed: the cores of the glass fragments are still unaltered and directly border on areas with low Ca content. The Ca content in the altered regions than increases together with the fraction of microcrystalline phases towards veins or cracks. This is nicely illustrated in Fig. 9e that shows profile I-II through the image of Fig. 9d starting within an unaltered glass region (labeled I) via a vein or crack and ending in an altered region again (labeled II): starting with the Ca content of the unaltered glass, the Ca content significantly drops at the contact to the altered glass and increases again towards the vein or crack. The post-run glass fragments of the experiment at fluid-dominated conditions (water-to-rock ratio = 44) are homogeneously altered with only a small remnant area of microcrystalline phases (Fig. 10a). Macroscopically, the post-run glass fragments are lucent and more or less colorless. This is in stark contrast to the starting basaltic glass that is dark brown to black in color and to all other altered post-run glass fragments that are opaque reddish-brown or green. Iron-rich phases formed as layers at the outside margin of the glass fragments. The iron-rich phases are oscillating with heavily altered glass layers. Figure 10b shows a profile through the

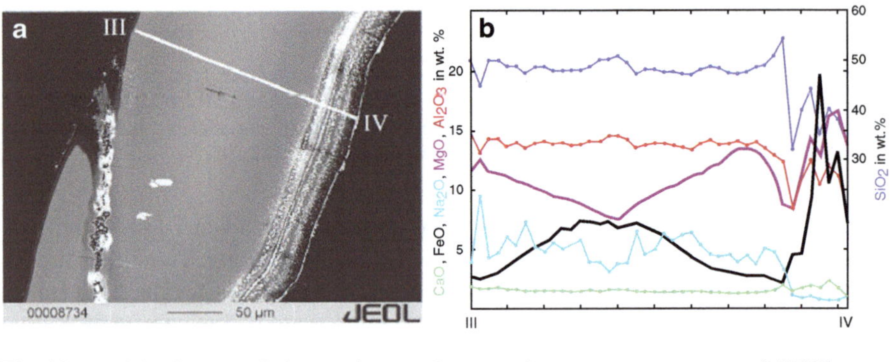

Fig. 10 (**a**) A backscattered electron image of an experiment at a temperature of 450°C, a run duration of 10 days, a pressure of 400 bar, and a water-to-rock ratio of 44. (**b**) Illustrates the chemistry changes along the profile III–IV shown in (**a**)

image of Fig. 10a starting within a homogeneously altered glass region (labeled III) and ending at the outside margin of the glass fragments where Fe-rich phases alternate with heavily altered glass layers (labeled IV). The homogeneously altered glass is relatively enriched in FeO (7 wt.%) and depleted in MgO (7.5 wt.%), while the heavily altered glass at the rim is depleted in FeO (2.65 wt.%) and enriched in MgO (12 wt.%), resulting in V- and U-shaped distribution patterns of Mg and Fe, respectively. Fe-Mg exchange reactions in minerals usually result in complementary Fe-Mg distributions as Fe^{2+} diadochically replaces Mg^{2+} in the crystal structure. Not shown in the profile is the significant content of water that is present in the homogeneously altered glass region as well as in the heavily altered glass layers.

The observed V- and U-shaped distribution patterns of Mg and Fe might therefore be attributed to leaching processes within the amorphous glass structure and not to a solid state solution-precipitation reaction as the controlling process. This hypothesis is strengthened by the observation of Mg increase in the altered glass towards the outside margin of the glass fragment that was in direct contact with the fluid during the experiment. As secondary minerals are absent at the outside margin of the glass, the Mg increase points to incorporation of Mg from the fluid into the amorphous glass structure without formation and therefore involvement of secondary minerals at the outer rim of the glass fragment.

4 Formation of Secondary Minerals

The most common newly formed minerals are plagioclase, clinopyroxene, amphibole, anhydrite, and iron ore phases. Plagioclase is found in all post-run charges in varying quantities and sizes. The plagioclase crystals appear either as columnar crystals (about 40 μm wide and up to 120 μm long columns) or spherically grown with diameters of 100–250 μm. Fig. 11a, b show examples of both appearances. The columnar crystals are observed in all post-run charges and crystallize on the surface

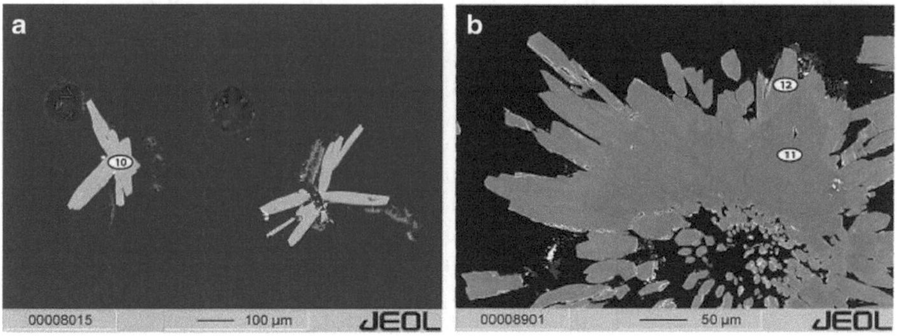

Fig. 11 Backscattered electron images of newly grown plagioclase crystals. (**a**): columnar crystals; (**b**): spherically grown and chemically zoned crystals. The numbers within the images indicate spots of representative chemical analyses listed in Table 2

of altered glass fragments. The spheres of plagioclase are only observed in post-run charges of experiments at 500°C and 450 bar and preferentially crystallize at the emersion point of veins which criss-cross the glass fragments. Both, columnar and spherical plagioclases, are intermediate to albite-rich in composition with albite contents between 55 and 80%. The plagioclases are homogeneous in composition. Analysis #10 of Table 2 represents a chemical analysis of a homogeneous columnar plagioclase (see Fig. 11a for location). Only in some experiments at 500°C and 450 bar zoned plagioclase crystals with albite-rich cores ($Ab_{70–81}$) and less albite-containing rims ($Ab_{50–60}$) exist, see chemical analyses of the spherical plagioclase of Fig. 11b (analysis #11: core; analysis #12: rim). Pyroxenes are mainly observed in post-run charges of experiments at relatively high temperatures and low pressures (500°C, 450–500 bar) as well as in experiments with low water-to-rock ratios. In experiments at higher pressures pyroxenes as secondary minerals are totally absent. The pyroxenes appear either as bottle-green, columnar pyroxenes (Fig. 12a,b) or as radial arranged acicular pyroxenes (Fig. 12c,d). Pyroxenes of both groups are enclosed by amphibole crystals. The radial arranged acicular pyroxenes are either grouped to larger spheres or they formed felts of pyroxenes (Fig. 12e). Small plates of pyroxenes (Fig. 12f) also exist in some of the post-run charges. All pyroxenes are clinopyroxenes. The diopside content ranges from Di_{35} to Di_{60}, with most pyroxenes having a diopside content of slightly less than Di_{50}. The bottle-green pyroxenes have low Al and high Si contents (Al_2O_3: <1.5 wt.%; SiO_2: ~50 wt.%) and formed only at experimental conditions of 500°C and 450 bar.

Table 2 Chemical composition of secondary minerals (wt.%; b.d. = below detection limit)

Number in figures	10	11	12	13	14	15
Run #	32	43	43	44	44	55
P [bar]	600	450	450	450	450	400
T [°C]	500	500	500	500	500	450
Duration [days]	10	10	10	10	10	10
Water-to-rock	9.8	10.1	10.1	9.9	9.9	44
SiO_2	61.65	60.85	54.72	51.03	54.17	45.00
TiO_2	b.d.	b.d.	b.d.	b.d.	b.d.	b.d.
Al_2O_3	24.93	24.46	28.68	0.18	1.26	8.12
Cr_2O_3	0.00	0.05	0.00	0.00	0.00	0.00
FeO	0.28	0.16	0.03	16.93	12.22	3.03
MnO	b.d.	b.d.	b.d.	0.36	0.16	0.04
MgO	b.d.	0.03	0.02	7.65	15.69	22.55
CaO	5.83	6.11	10.61	23.54	13.29	1.74
Na_2O	7.19	8.05	5.45	0.17	0.29	0.25
K_2O	0.03	0.03	b.d.	b.d.	0.02	0.02
Total	99.91	99.76	99.50	99.87	97.09	80.75
Comments	Columnar plagioclase	Plagioclase (core)	Plagioclase (rim)	Clinopyroxene	Amphibole	Smectite

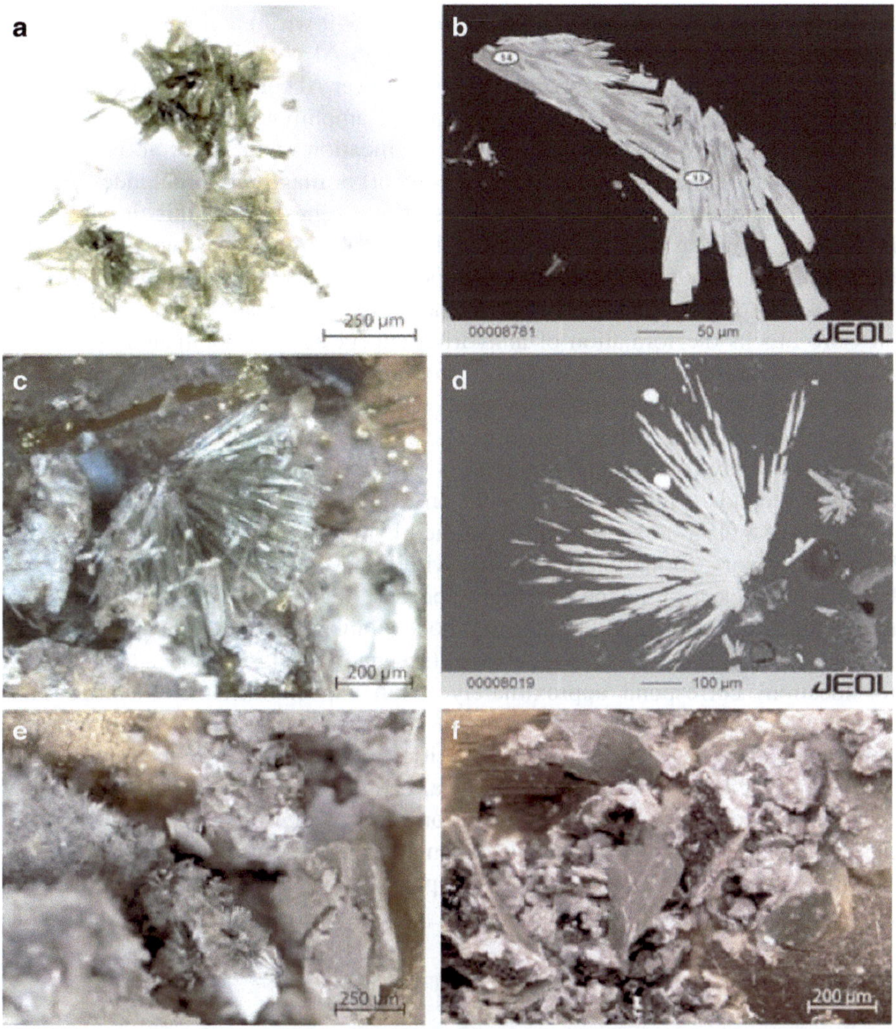

Fig. 12 Optical photomicrographs and backscattered electron images of secondary minerals. (**a**): bottle-green, columnar pyroxene (with low Al contents); (**b**): BSE image of bottle-green, columnar pyroxene (centre, #13) enclosed by amphibole (#14); (**c**): radial arranged acicular pyroxene (with high Al contents); (**d**): BSE image of radial arranged acicular pyroxene enclosed by amphibole; (**e**): felt of pyroxenes; f: small plates of pyroxenes. Chemical analysis of #13 and #14 are listed in Table 2

Radial arranged acicular pyroxenes have higher Al_2O_3 concentrations (2–7 wt.%) and a SiO_2 content that is always less than 50 wt.%.

The radial arranged acicular pyroxenes are only observed in post-run charges of experiments with experimental conditions either of 500°C and 450 bar or of 450°C and 400 bar, and with water-to-rock ratios of 1–5. Amphiboles exclusively formed as outer covers of pyroxenes. In analogy to the pyroxenes, the amphiboles can also be divided into 2 groups: amphiboles with high (5–9 wt.%) and with lower Al_2O_3 content (<1.5 wt.%). Amphiboles with e.g., high Al content are associated with pyroxenes with high Al content and these with lower Al content are associated with pyroxenes with lower Al content. In general, the amphiboles are higher in Mg than the corresponding pyroxenes. Using the classification of Leake et al. (1997), the amphiboles are either actinolites (low Al content) or magnesiohornblende (high Al content). Representative chemical analyses of pyroxenes and amphiboles are listed in Table 2 (analysis #13: pyroxene; analysis #14: amphibole; see Fig. 12b for location). Anhydrite is found in some post-run charges in various quantities. It seems that higher pressure supports the formation of anhydrite, while lower pressure blocks the formation of anhydrite, except at fluid-dominated conditions as anhydrite formed even under lower pressure conditions at high water-to-rock ratios. Anhydrite appears usually as small white plates with different thicknesses. An example is given in Fig. 13a (upper left corner of the backscattered electron image). The clay-like mineral smectite is described in the literature as a further common secondary mineral (e.g., Mottl and Holland 1978; Hajash and Chandler 1981). Hajash (1984) experimentally studied hydrothermal alteration processes of basalt and reports formation of smectite with honeycomb structures. Fig. 13b is a backscattered electron image of a highly porous silicate. The highly porous structure in the 2-D section might build a 3-D honeycomb structure. As the high Mg, Al, and water contents and the low Ca content (see analysis #15 in Table 2 and Fig. 13b for location) correspond approximately to an expected chemical composition of smectite, the mineral in Fig. 13b might be smectite. However, a clear proof of existence cannot be provided as the quantities were too small to perform a X-ray diffractometer analysis. In addition to the already macroscopical visible secondary minerals, phases of only a few μm in size formed as well. Iron ore phases are among these tiny phases. They were observed as small spherules within the altered glass at high pressure (4,000 bar) or as individual phases next to the altered glass at lower pressure (450 bar and 500°C).

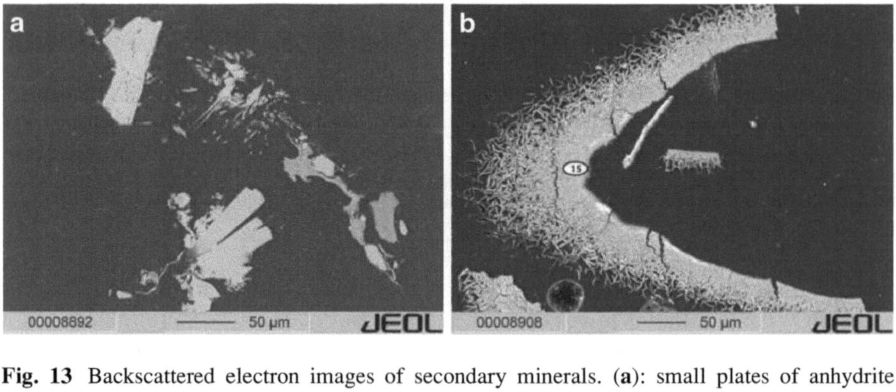

Fig. 13 Backscattered electron images of secondary minerals. (**a**): small plates of anhydrite (*upper left corner*) next to columnar plagioclase crystals (*lower centre part of the image*); (**b**): highly porous silicates with high Mg and Al and low Ca contents (possibly smectite). Chemical analyses of #15 is listed in Table 2

5 Concluding Remarks

A systematic experimental study on basalt-seawater interaction was performed to mimic processes of submarine hydrothermal systems that run at extreme temperatures and in great water depths, e.g., at the venting site 5° S on the Mid Atlantic Ridge (MAR) . This part of the study focuses on the hydrothermal alteration of basalt, newly formed, i.e., secondary, mineral assemblages, and their major and minor element compositions. Various alteration phenomena are observed in all post-run charges. The degree of alteration of basalt depends on parameters like run duration, temperature, pressure and water-to-rock ratio. With increasing degree of alteration of basalt not only the bulk chemistry of the glass but also the former amorphous glass structure changes and gets replaced by heterogeneous microcrystalline phases. The observed newly-formed minerals in the post-run charges, namely plagioclase, clinopyroxene, amphibole, anhydrite, iron ore phases, and probable smectite, are typical secondary minerals as described in the review article by German and Von Damm (2003). The knowledge of the alteration phenomena of the basalt and the newly formed secondary minerals helps to better understand the natural alteration processes of oceanic crust. In combination with the derived water-rock partitioning behaviour of transition metals and REE the hydrothermal alteration of basalt and consequential formation of secondary minerals, including the REE pattern of newly formed minerals, will add to our knowledge of basalt-seawater interaction processes and will improve our understanding of transport reactions and element fluxes during ascent in hydrothermal systems that run at extreme temperatures and great water depths. Preliminary results on the partitioning behaviour of transition metals between coexisting basaltic glass and seawater led to the conclusion, that the experimental post-run fluids are 'primary' fluids directly produced by interaction between seawater and rock, i.e., the 'unusual' pattern of MAR 5°S fluids might represent a fluid originating close to the reaction zone of the entire hydrothermal system (Kusebauch et al. 2010; Schmidt et al. 2010).

Acknowledgements The authors are grateful to P. Kluge for laboratory assistance and to A. Fehler for excellent preparation of polished sections of very tiny samples. The work was partly supported by grants from the priority program 1144 "From Mantle to Ocean: Energy-, Material- and Life-cycles at Spreading Axes" of the German Science Foundation (DFG) to DGS and further DFG grants to AH.

References

Allen DE, Seyfried WE (2005) REE controls in ultramafic hosted MOR hydrothermal systems: an experimental study at elevated temperature and pressure. Geochim Cosmochim Acta 69:675–683

Berndt ME, Seyfried WE, Janecky DR (1989) Plagioclase and epidote buffering of cation ratios in mid-ocean ridge hydrothermal fluids: experimental results in and near the 'supercritical' region. Geochim Cosmochim Acta 53:2283–2300

Bischoff JL, Rosenbauer RJ (1987) Phase separation in seafloor geothermal systems: an experimental study of the effects on metal transport. Amer J Sci 287:953–978

Butterfield DA, Jonasson IR, Massoth GJ, Feely RA, Roe KK, Embley RE, Holden JF, McDuff RE, Lilley MD, Delaney JR (1997) Seafloor eruptions and evolution of hydrothermal fluid chemistry. Phil Trans R Soc Lond Ser A Math Phys Eng Sci 355:369–386

Coumou D, Driesner T, Heinrich CA (2008) The influence of phase separation on the flow patterns of mid-ocean ridge hydrothermal systems. Geochim Cosmochim Acta 72:A184 (Goldschmidt Conference Abstracts)

Coumou D, Driesner T, Weis P, Heinrich CA (2009) Phase separation, brine formation, and salinity variation at Black Smoker hydrothermal systems. J Geophys Res 114:B03212

Edmond JM, Measures C, McDuff RE, Chan LH, Collier R, Grant B, Gordon LI, Corliss JB (1979) Ridge crest hydrothermal activity and the balances of major and minor elements in the ocean – Galapagos data. Earth Planet Sci Lett 46:1–18

German CR, Von Damm KL (2003) Hydrothermal processes. In: Holland HD, Turekian KK (eds) Treatise on geochemistry, vol 6. Elsevier – Pergamon, Oxford, pp 181–222

German CR, Thurnherr AM, Knoery J, Charlou JL, Jean-Baptiste P, Edmonds HN (2009) Export fluxes from submarine venting to the ocean: A synthesis of results from the rainbow hydrothermal field, 36°N MAR. Geochim Cosmochim Acta 73:A428 (Goldschmidt Conference. Abstracts)

Hajash A (1984) Rare-earth element abundances and distribution patterns in hydrothermally altered basalts – experimental results. Contrib Miner Petrol 85:409–412

Hajash A, Chandler GW (1981) An experimental investigation of high-temperature interactions between seawater and rhyolite, andesite, basalt and peridotite. Contrib Miner Petrol 78:240–254

James RH, Allen DE, Seyfried WE (2003) An experimental study of alteration of oceanic crust and terrigenous sediments at moderate temperatures (51 to 350°C): Insights as to chemical processes in near-shore ridge-flank hydrothermal systems. Geochim Cosmochim Acta 67:681–691

Jupp T, Schultz A (2000) A thermodynamic explanation for black smoker temperatures. Nature 403:880–883

Koschinsky A, Garbe-Schönberg D, Sander S, Schmidt K, Gennerich HH, Strauss H (2008) Hydrothermal venting at pressure-temperature conditions above the critical point of seawater, 5°S on the Mid-Atlantic Ridge. Geology 36:615–618

Kusebauch C, Garbe-Schönberg D, Holzheid A (2010) Water-rock-interaction in submarine hydrothermal systems and their influence on fluid chemistry: an experimental approach. Annual meeting of the German Mineralogical Society, Münster, 20.-22.09.2010, Abstract # S11-T08 (CD-ROM)

Leake BE, Woolley AR, Arps CES, Birch WD, Gilbert MC, Grice JD, Hawthorne FC, Kato A, Kisch HJ, Krivovichev VG, Linthout K, Laird J, Mandarino JA, Maresch WV, Nickel EH, Rock NMS, Schumacher JC, Smith DC, Stephenson NCN, Ungaretti L, Whittaker EJW, Guo YZ (1997) Nomenclature of amphiboles: Report of the subcommittee on amphiboles of the International Mineralogical Association, Commission on New Minerals and Mineral Names. Can Miner 35:219–246

Mottl MJ, Holland HD (1978) Chemical exchange during hydrothermal alteration of basalt by seawater.1. Experimental results for major and minor components of seawater. Geochim Cosmochim Acta 42:1103–1115

Schmidt K, Garbe-Schönberg D, Bau M, Koschinsky A (2010) Rare earth element distribution in >400 degree C hot hydrothermal fluids from 5 degrees S, MAR: The role of anhydrite in controlling highly variable distribution patterns. Geochim Cosmochim Acta 74:4058–4077

Seyfried WE, Mottl MJ (1982) Hydrothermal alteration of basalt by seawater under seawater-dominated conditions. Geochim Cosmochim Acta 46:985–1002

Seyfried WE, Chen X, Chan LH (1998) Trace element mobility and lithium isotope exchange during hydrothermal alteration of seafloor weathered basalt: an experimental study at 350 degrees C, 500 bars. Geochim Cosmochim Acta 62:949–960

Von Damm KL, Oosting SE, Kozlowski R, Buttermore LG, Colodner DC, Edmonds HN, Edmond JM, Grebmeier JM (1995) Evolution of East Pacific Rise hydrothermal vent fluids following a volcanic eruption. Nature 375:47–50

Von Damm KL, Lilley MD, Shanks WC, Brockington M, Bray AM, O'Grady KM, Olson E, Graham A, Proskurowski G, Party SS (2003) Extraordinary phase separation and segregation in vent fluids from the southern East Pacific Rise. Earth Planet Sci Lett 206:365–378

Sorbents from Mineral Raw Materials

Anatoly I. Nikolaev, Lidiya G. Gerasimova, and Marina V. Maslova

In 2000, the share of Russia in the world mineral output amounted to 8.3%, to which a significant contributor was the Murmansk Region producing Cu, Ni, Co, Al, P, REE, Zr, Ti, Ta, Nb, Fe ores and other minerals. The Europe-biggest mining companies "Apatit" and "JSC Kola Mining & Metallurgical Company", and some other leading enterprises ("KovdorGOK", "Kovdorsluyda", "Olkon", "Lovozerskaya Mining Co."), are producing up to 100% phosphoric raw materials in Russia, 10–12% iron ores, 18–20% Ni, Cu and Co, up to 100% Zr, 80–90% mica, about 35% ceramic raw materials, 70–80% Nb- and Ta-bearing concentrates. During the twentieth century 200 claims and licenses were granted to the effect that about 40% of the explored regional mineral resources were involved into exploitation.

Mineral processing at mining, dressing and chemical plants is known to be accompanied by the formation of solid and liquid wastes of both natural and synthetic origin. Stored in tailing dumps, theses wastes pose a serious environmental hazard. At the same time, mineral components of the tailings can either be turned into natural geochemical barriers, preventing the dissolved toxic elements from escaping into the environment, or converted into valuable engineering materials (Brylyakov et al. 2009).

This issue has been recently addressed at Kola Science Centre by developing a set of innovation technologies whereby reject materials from technogenic deposits are converted into fertilizers, tanning and cleaning agents, pigments, and concretes. Moreover, the research has been focused on joint processing of natural products with electronic and metallurgical wastes (Gerasimova and Nikolaev 2007). Table 1 provides data on chemical composition of major mineral tailings accumulated at enterprises operating in Kola Peninsula and North Kareliya.

A.I. Nikolaev (✉) • L.G. Gerasimova • M.V. Maslova
Institute of Chemistry and Technology of Rare Elements and Mineral Raw Materials,
Kola Science Centre, Russian Academy of Science, Fersman St. 14,
Apatity 184200, Murmansk region, Russia
e-mail: nikol_ai@chemy.kolasc.net.ru

S.V. Krivovichev (ed.), *Minerals as Advanced Materials II*,
DOI 10.1007/978-3-642-20018-2_8, © Springer-Verlag Berlin Heidelberg 2012

One of the promising trends in mineral recovery is their application as adsorbents (Gerasimova and Maslova 2009). Table 2 demonstrates the data on the removal capacity of some of the magnesium-silicate minerals of the Kola Peninsula suggesting their applicability as natural adsorbents of toxic elements (Rakaev et al. 2004). The removal capacity was maximal for calcinated sungulite, in which the O-H···O bond between the layers extends as a result of thermal expansion and structural deformation (Kremenetskaya. et al. 2010).

Since the structure of sungulite (like that of talc) is laminated, fine grinding can turn it into filler for rubber compounds (Gerasimova and Nikolaev 2009).

Table 3 presents the relationship between the surface area and removal capacity of titanite and apatite – minerals from apatite-nepheline ore concentration (Fedorov et al. 2003). The surface area was increased through mechanical activation in a vibratory mill. The result of increasing amorphization of the titanite mineral

Table 1 Principal raw materials for sorbent production

Mineral	Deposit	Composition
Apatite	Khibiny	$Ca_5[PO_4]_3(OH,F)$
Nepheline	Khibiny	$K,Na_3[AlSiO_4]_4$
Titanite	Khibiny	$CaSiTiO_5$
Olivine	Kovdor and Khabozero	$(Mg,Fe)_2[SiO_4]$
Sungulite	Kovdor and Khabozero	$Mg_3Si_2O_5(OH)_4$
Plagioclase	Loukhi	$Na[Al\ Si_3O_8]$
Mica-group minerals:		
Muscovite	Chupa and Kovdor	$KAl[AlSi_3O_{10}](OH)_2$
Phlogopite		$KMg[AlSi_3O_{10}](OH,F)_2$
Biotite		$K(Mg, Fe)_3[AlSi_3O_{10}](OH, F)_2$

Table 2 Direct utilization of mineral concentrates as adsorbents (the solutions contained about 20 mg/l of metals)

Cations	Sungulite	Sungulite[a]	Olivine
Fe(III)	99.5	>99.5	99.2
Fe(II)	93.2	>99.5	81.4
Cu(II)	89.6	>99.5	64.9
Ni(II)	62.7	>99.5	46.4
Co(II)	58.3	98.5	34.2
Zn(II)	78.0	99.2	60.0
Al(III)	68.0	97.5	46.5

[a]calcinated at 600°C

Table 3 Surface area (S) and ion removal capacity (A) for some metals of titanite and apatite

		A, meq/g			
Minerals	S, m^2/g	Cs^+	Sr^{2+}	Co^{2+}	Ni^{2+}
Titanite (fraction up to 200 μm)	1.02	0.05	0.09	0.18	0.24
Titanite (fraction up to 20 μm)	3.80	0.43	0.67	0.80	0.95
Apatite (fraction up to 100 μm)	3.17	0.17	015	0.75	0.90
Apatite (fraction up to 10 μm)	27.90	0.46	0.85	0.94	1.05

particles was boosted removal capacity due to the growth of excess negative charge on the mineral particle surface. By destroying the apatite grains via grinding, the surface is ionized, increasing the calcium ion-exchange capacity accordingly (Gerasimova et al. 2005).

The research also included the removal capacity of the anortosite mineral. The metal concentration in solutions purified was, g/l: Fe – 0.401, Ni – 0.516, Co – 0.577, Cu – 0.464. The material constitution of the sample is given in Table 4.

With diminishing of the mineral particles size, the sorption of non-ferrous metal cations increases (Table 5). Apparently, the diffusion of coarser iron and copper cations into the defective surface layer of the particles is hindered, so that their removal capacity is lower than that of nickel and cobalt cations.

The spent mineral sorbent can be recovered by high-temperature processing. By this procedure, the adsorbed components are transferred to the insoluble state by forming compounds of the spinel type, colouring the product in different shades. Recovered sorbents are used as functional fillers for building and paint-and-varnish materials (Gerasimova and Shchukina 2009).

The adsorption of the oxotitanium (IV) cation on the flaky mica particles has been investigated. It has been shown that increasing the adsorbate acidity reduces adsorption, whereas increasing temperature results in a reverse process, namely, desorption at 50°C, and the titanium (IV) hydrolysis at 75°C (Fig. 1).

Table 4 Chemical composition of anorthosite

Component	Al_2O_3	SiO_2	P_2O_5	Cl	$K_2O + Na_2O$	CaO	TiO_2	MnO_2	Fe_2O_3
mas. %	25.20	54.56	0.17	0.2	0.49 + 10.1	1.85	0.22	0.04	0.62

Table 5 The effect of grain size of anortosite on its removal capacity

	Removal capacity with respect to metal cations (mg/g)			
Grain-size composition	Fe^{3+}	Ni^{2+}	Co^{2+}	Cu^{2+}
<100 μm	0	1.0	0	0.4
>100 μm	0.7	1.8	1.6	2.2
>40 μm	2.6	2.6	4.7	2.7

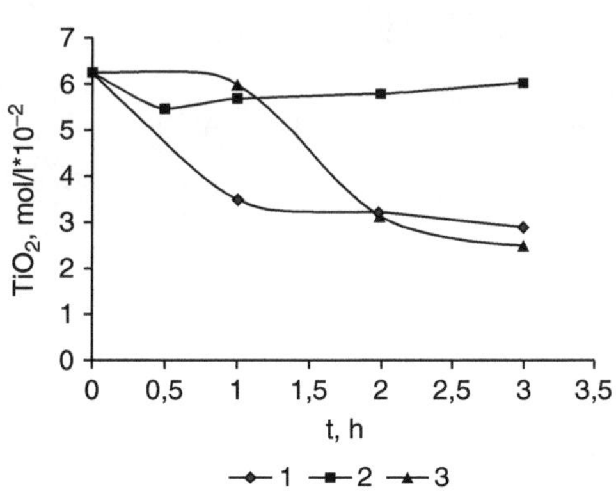

Fig. 1 Change of titanium(IV) content in the presence of mica flakes at the temperature of 1–25°C, 2–50°C, 3–75°C. $C_{H2SO4free} = 0.15$ mol/l

Smaller flake sizes uptake greater substance quantities. These findings were used when developing a technology of pearl pigments, where the principal operation consists in adsorbing on the mica particle surface of the negatively charged oxotitanium (IV) cation. After being treated under conditions described above, the mica particles develop a surface charge facilitating the physico-chemical processes, whereby fine hydroxide layers are formed on them (Fig. 2).

We have proposed an original method for depositing hydroxide coatings, based on electrodialysis of titanium salt solutions (iron, cobalt, etc.), in the presence of mica. The process occurs at a low current density at electrodes 0.12–0.15 A/cm^2 with a pH controlled within 6–6.5 in a reaction chamber. The hydroxide nano-coating formed on mica particles is consolidated by heat treatment to remove water and form the oxide shell crystal structure. The electrochemical method underlying the mica preparation and used for the nano-coating deposition onto mica particles, makes controllable both the process proper and consumer properties of the target product (Gerasimova et al. 1991, 1995).

The possibility of producing from titanite mineral of pigment fillers for building and paint-and-varnish materials, and plastic has been investigated (Fedorov et al. 2001; Gerasimova and Bubnov 2006; Gerasimova et al. 2003).

By studying the material containing 92–94% of titanite mineral ($CaSiTiO_5$) ground in ball, impact-centrifugal and vibrating mills it has been discovered that the resulting powders differ in grain size and, correspondingly, in optic properties (the latter being determined by an X-Rtte SP-62 spectrophotometer (see Table 6).

Another result of dispersing is changed surface layer morphology due to ionization and amorphization (Fig. 3).

Fig. 2 SEM-image of feed phlogopite particles (*left*) and same after splitting and grinding (*right*)

Table 6 Variations in powder optic properties

Grain size, mcm	Up to 2.5	Up to 6.6	Up to 7.5	Up to 10.0	10–30
Whiteness, %	87.0	84.2	83.6	81.5	79.0
Yellowness, %	12.4	13.5	14.9	15.8	16.3

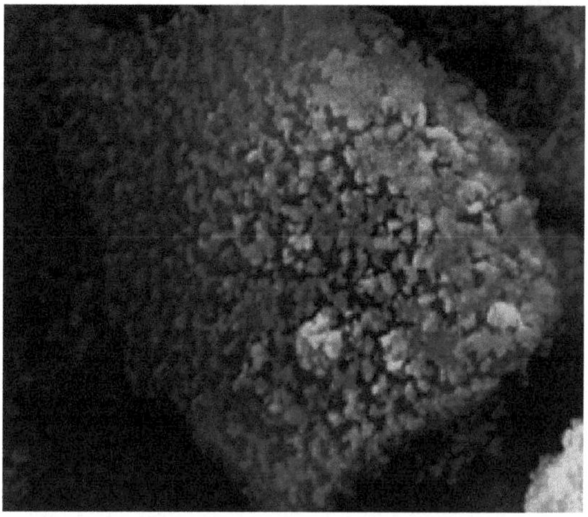

Fig. 3 SEM-image of filler particles with a size less than 3 μm

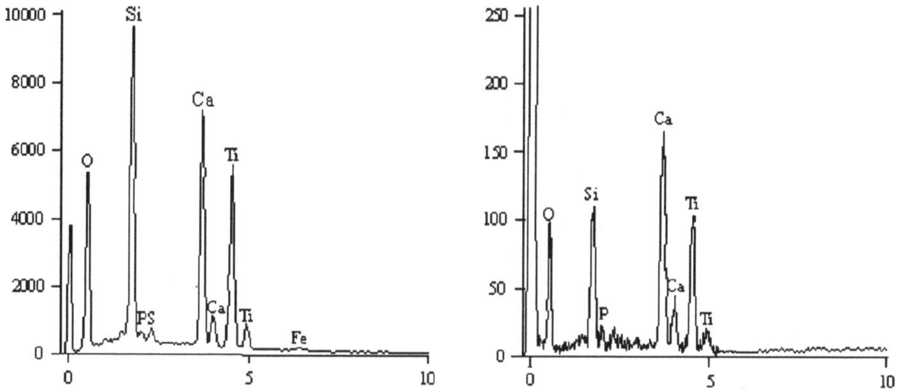

Fig. 4 Energy-disperse analysis of the surface of titanite particles ground in a ball mill (*left*) and in a vibrating mill (*right*)

The surface area of the powder obtained by attrition (in a ball mill) has the highest index (1.92 m^2/g) due to prevailing of a mesoporous structure on the particle surface (the t-plot). As displayed in the desorption isotherm, the average pore diameter is 21.7 nm. The pore system on the particles dispersed in a vibrating mill (the crushing-grating method) is mostly represented by wide mesopores. The powder obtained by splitting in an impact-centrifugal mill has lower surface area indices (1.42 m^2/g) and a smaller pore volume due to prevailing micropores. All diffractograms are practically identical. The energy variance analysis has shown that the greatest transformations on the surface of the three samples compared are created by the attrition method. The intensity of Si response is over 9000, that of Ca – 7000, whereas after crushing these indices diminish to 110 and 170, respectively. This is indicative of a high degree of amorphization in titanite particles (Fig. 4).

The findings presented here suggest the possibility of turning minerals and technogenic mineral wastes of the Kola Peninsula into engineering products such as sorbents, pigments and various fillers. The research has involved physical and chemical properties of both starting and dispersed minerals and the effect of their surface properties on adsorption ability of the materials. It has been suggested that spent mineral sorbents can be utilized in the production of pigments and fillers of a broad colour range.

References

Brylyakov YuE, Nikolaev AI, Gerasimova LG (2009) Outlooks for technologies processing the concentrates of apatite-nepheline ores complex processing. Min J 9:62–65 (in Russian)

Fedorov SG, Nikolaev AI, Brylyakov YuE, Gerasimova LG, Vasilieva NYa (2003) Chemical processing of mineral concentrates of the Kola Peninsula. Apatity (in Russian)

Fedorov SG, Brylyakov YuE, Gerasimova LG (2001) A method for the production of an atmosphere-resistant pigment from sphene concentrate. RF Patent No 2177016

Gerasimova LG, Bubnov AV (2006) Mineral pigment containing titanium dioxide. Build Mater 7:32–33 (in Russian)

Gerasimova LG, Maslova MV (2009) Ion-exchange materials for cleaning wastes from radionuclides and non-ferrous heavy metals. J Water Purif Prep Suppl 4:76–79 (in Russian)

Gerasimova LG, Nikolaev AI (2007) Recovery of solid wastes in the production of pigments and other inorganic materials. Ecol Aspects Ind Prod 2:34–43 (in Russian)

Gerasimova LG, Nikolaev AI (2009) Magnesium-containing fillers for rubber compounds. CaoutchVulcanized Rubber 6:3–5 (in Russian)

Gerasimova LG, Shchukina ES (2009) Producing pigment fillers from alumosilicate mineral wastes. Build Mater 6:76–78 (in Russian)

Gerasimova LG, Zhdanova NM, Okhrimenko RF (1991) A method for the production of a pigment from sphene concentrate. RF Patent No 1640894

Gerasimova LG, Zhdanova NM, Okhrimenko RF (1995) The production of nacreous pigment by electrodialysis. Paint Lacq Mater Appl 6:8–10 (in Russian)

Gerasimova LG, Lazareva IV, Maslova MV, Okhrimenko RF (2003) Producing a mineral pigment from sphene concentrate. Paint Lacq Mater Appl 2–3:34–36 (in Russian)

Gerasimova LG, Maslova MV, Lazareva IV, Matveev VA, Alekseev AI (2005) Sphene concentrate in sorbent production. Ore Benefication 4:31–33 (in Russian)

Kremenetskaya IP, Belyaevsky AT, Vasilieva TN (2010) Amorphization of serpentine minerals in magnesial-silicate reagent technology used for heavy-metal immobilization. Chem Sustain Dev 1:41–49 (in Russian)

Rakaev AI, Nikolaev AI, Kukushkina AN, Alekseeva SA, Varyukhina IM (2004) Comprehensive utilization of non-ore olivinite sources of the Khabozero deposit. Ore Benefication 2:40–44 (in Russian)

Natural Double Layered Hydroxides: Structure, Chemistry, and Information Storage Capacity

Sergey V. Krivovichev, Victor N. Yakovenchuk, and Elena S. Zhitova

1 Introduction

Layered double hydroxides (LDHs) constitute an important group of materials with many applications ranging from catalysis and absorption to carriers for drug delivery, DNA intercalation and carbon dioxide sequestration (Rives 2001; Duan and Evans 2006). The structures of LDHs are based upon double brucite-like hydroxide layers $[M_n^{2+}M_m^{3+}(OH)_{2(m+n)}]^{m+}$, where $M^{2+} = Mg^{2+}$, Fe^{2+}, Mn^{2+}, Zn^{2+}, etc.; $M^{3+} = Al^{3+}$, Fe^{3+}, Cr^{3+}, Mn^{3+}, etc. The positive charge of the layer is compensated by interlayer species that may consist of anions (CO_3^{2-}, Cl^-, SO_4^{2-}, etc.) or both anions and cations (Na^+, Ca^{2+}, Sr^{2+}, etc.). Structural features of LDHs such as cation ordering, charge distribution and polytypism have an immediate influence upon their properties and have been under extensive experimental and theoretical investigations recently. In particular, Mg-Al cation order is important for catalytic activity of MgAl LDHs correlated with the numbers of Al^{3+} sites at the closest distance from an Al^{3+} site (Kim et al. 2003). Different distribution of Al in a Mg hydroxide matrix also results in different charge distribution in the interlayer, which is critically important for intercalation reactions.

Another area of interest in LDHs is the theory of the origin of life on Earth. Bernal (1951, 1967) first proposed that minerals played a crucial importance in the origin of life and, along this line, Cairns-Smith (1982) suggested that clay mineral

S.V. Krivovichev (✉)
Nanomaterials Research Center, Kola Science Center, the Russian Academy of Sciences, 14 Fersman Street, 184209 Moscow, Russia
and
Department of Crystallography, Faculty of Geology, St. Petersburg State University, University Emb. 7/9, 199034 St. Petersburg, Russia
e-mail: skrivovi@mail.ru

V.N. Yakovenchuk • E.S. Zhitova
Department of Crystallography, Faculty of Geology, St. Petersburg State University, University Emb. 7/9, 199034 St. Petersburg, Russia

S.V. Krivovichev (ed.), *Minerals as Advanced Materials II*,
DOI 10.1007/978-3-642-20018-2_9, © Springer-Verlag Berlin Heidelberg 2012

complexes with organic molecules could have constituted the first organisms, capable of self-assemblage, replication and mutation, and that carbon based life may have gradually evolved from such an organic–inorganic hybrid state. In his concept identified as a ‘genetic takeover’, Cairns-Smith proposed that the mechanisms of replication, information storage and transfer were first adapted by crystals of clay minerals, where bits of information were ciphered in crystal defects. Due to excellent intercalation properties of clays, they accommodated in their interlayers variety of biologically active molecules that took over replication mechanisms from mineral crystals and transferred them to biological systems.

Kuma et al. (1989) and Arrhenius (2003) proposed that, instead of clays, LDHs may serve as matrices for the origin of life in prebiotic systems. In particular, he noted that LDHs easily form as as hydration products at ocean weathering of basalts, at serpentinization, etc. Of special interest is the fougerite or green rust, a mixed-valent Fe^{2+}–Fe^{3+} LDH, that could have a widespread occurence in the Archean ocean. In development of these ideas, Greenwell and Coveney (2006) proposed mechanism of prebiotic information storage and transfer in LDH matrices on the basis of replication and conservation of M^{3+}–M^{2+} cation arrangements. According to their proposal (Fig. 1), pattern of M^{3+} sites in one layer may be transferred through the crystal by controlled location of interlayer anions (CO_3^{2-}, SO_4^{2-}). Information is stored in LDH by the arrangement of trivalent cation sites. The presence of this site is denoted ‘1’ and its absence as ‘0’. In the model shown in Fig. 1, the sequence of the sites provides a string of binary data. When entering into

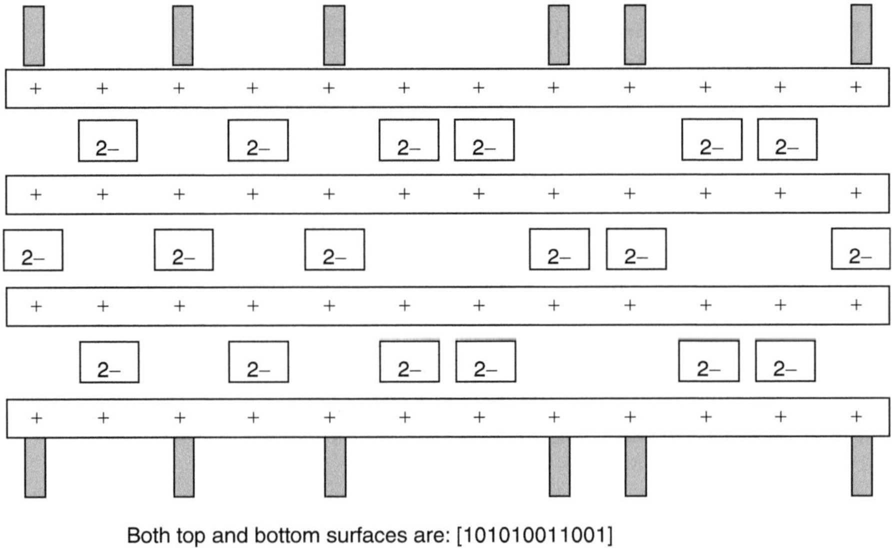

Fig. 1 Schematic to show information encoded at the external surfaces of LDH polytypes with a three-layer repeat structure. The '+' signs indicate positions of trivalent metal sites, '2-' in boxes are divalent anions. See text for details (Reproduced with permission from Greenwell and Coveney (2006))

interlayer space, divalent anions will position into sites located around the positive charge ('1'). Thus the arrangement of the positive sites in the next layer will be exactly complementary to that of the previous layer. In the absence of stacking faults, defects and impurities, the alternating sequences of bit strings will propagate along the direction perpendicular to the layers. When LDH crystal gets too large and cleaves, it will replicate itself either by splitting or by epitaxial growth of another crystal.

If the proposed mechanism is realized in natural systems, it can be exploited by humans under laboratory conditions for creation of new subnanometer matrices for information storage and transfer. Thus the purpose of this contribution is to investigate possibility of this mechanism on the basis of current knowledge on natural LDHs. First, we briefly summarize basic chemical and structural factors affecting chemical and structural diversity of LDH minerals. Then we shall discuss the reliability of cation ordering in the information transfer and storage on the basis of our recent experimental studies in this field.

2 Natural LDHs: Structure and Chemistry

There are at least 38 minerals that can be considered as direct analogs of LDH materials with octahedrally coordinated metal cations (Table 1).

The main factors affecting their chemical diversity are:

1. Chemical nature of the M^{2+} and M^{3+} cations;
2. The M^{2+}:M^{3+} ratio with the ratios 2:1 and 3:1 being the most common, which is the result of most preferable cation ordering patterns (Fig. 2); in minerals with M^{2+}:M^{3+} = 2:1, the M^{2+}–M^{3+} ordering was experimentally confirmed in LDH sulfates (motukoreaite, shigaite, and nikisherite), LDH antimonates (zincalstibite and cualstibite) and LDH carbonates (quintinites, see below); it is also noteworthy that Mg-Al/Fe ordering has been observed in wermlandite, the only mineral with the ratio M^{2+}:M^{3+} = 7:2 (Rius and Allmann 1984);
3. Chemical nature of interlayer anions and cations (if any): there are minerals that contain Na^+ or Ca^{2+} cations in the interlayer space.

Another important aspect of the crystal chemistry of LDHs is polytypism.

Polytype diversity of LDHs has been considered in details by Bookin and Drits (1993), Bookin et al. (1993), and Drits and Bookin (2001), who elaborated a structural nomenclature for the polytypes observed in this group, which is summarized below.

In the plane perpendicular to the direction of layer stacking, cations and anions may occupy three distinct sites: **A**, **B** and **C** (similar to the sites of spheres in closest packings). The upper (**A**, **B**, **C**) and lower (**a**, **b**, **c**) case symbols are reserved for positions of hydroxyls and cations, respectively. For instance, if hydroxyl anions of the layer are in the **A** and **C** sites, the cations occupy the **b** sites, and the layer has

Table 1 Natural double layered hydroxides and their crystal chemical parameters

$M^{2+}:M^{3+}$	Mineral name	Chemical formula	Space group	Unit-cell parameters	Ref
LDHs with monovalent interlayer anions					
3:1	Iowaite	$[Mg_3Fe^{3+}(OH)_8][Cl(H_2O)_2]$	*R*-3*m*	3.118–24.113	1
3:1	Woodallite	$[Mg_3Cr^{3+}(OH)_8][Cl(H_2O)_2]$	*R*-3*m*	3.103–24.111	2
3:1	Meixnerite	$[Mg_3Al(OH)_8][(OH)(H_2O)_2]$	*R*-3*m*	3.046–22.93	3
	Droninoite	$[Ni_3Fe^{3+}(OH)_8][Cl(H_2O)_2]$	*R*-3*m*	6.206–46.184	4
2:1	Chlormagaluminite	$[Mg_4Al_2(OH)_{12}][Cl_2(H_2O)_2]$	*P*6$_3$/*mcm*	5.29–15.46	5
?	Jamborite	$[(Ni^{2+},Ni^{3+},Fe)(OH)_2]$ $[OH, S,H_2O]$	hex	3.07–23.3	6
LDH carbonates					
3:1	Pyroaurite	$[Mg_6Fe_2{}^{3+}(OH)_{16}]$ $[(CO_3)(H_2O)_4]$	*R*-3*m*	3.111–23.504	7
	Sjögrenite	$[Mg_6Fe_2{}^{3+}(OH)_{16}]$ $[(CO_3)(H_2O)_4]$	*P*6$_3$/*mmc*	3.113–15.61	8
	Hydrotalcite	$[Mg_6Al_2(OH)_{16}]$ $[(CO_3)(H_2O)_4]$	*R*-3*m*	3.054–22.81	9
	Manasseite	$[Mg_6Al_2(OH)_{16}]$ $[(CO_3)(H_2O)_4]$	n.d.	6.12–15.34	7
	Stichtite	$[Mg_6Cr_2{}^{3+}(OH)_{16}]$ $[(CO_3)(H_2O)_4]$	*R*-3*m*	3.096–23.507	10
	Barbertonite	$[Mg_6Cr_2{}^{3+}(OH)_{16}]$ $[(CO_3)(H_2O)_4]$	*P*6$_3$/*mmc*	3.097–15.619	10
	Desautelsite	$[Mg_6Mn_2{}^{3+}(OH)_{16}]$ $[(CO_3)(H_2O)_4]$	*R*-3*m*	3.114–23.39	11
	Reevesite	$[Ni_6Fe_2{}^{3+}(OH)_{16}]$ $[(CO_3)(H_2O)_3]$	*R*-3*m*	3.085–23.355	12
	Takovite	$[Ni_6Al_2(OH)_{16}]$ $[(CO_3)(H_2O)_3]$	*R*-3*m*	3.025–22.595	13
2:1	Comblainite	$[Mg_4Co_2{}^{3+}(OH)_{12}]$ $[(CO_3)(H_2O)_4]$?	*R*-3*m*	3.038–22.79	14
	Caresite-3*T*	$[Fe_4{}^{2+}Al_2(OH)_{12}]$ $[(CO_3)(H_2O)_3]$	*P*6$_3$22	10.985–15.10	15
	Quintinite-2*H*	$[Mg_4Al_2(OH)_{12}]$ $[(CO_3)(H_2O)_3]$	*P*-62*m*	5.283–15.150	16
	Quintinite-3*T*	$[Mg_4Al_2(OH)_{12}]$ $[(CO_3)(H_2O)_3]$	*P*3$_1$12 ?	10.558–22.71	15
	Charmarite-2*H*	$[Mn_4{}^{2+}Al_2(OH)_{12}]$ $[(CO_3)(H_2O)_3]$	*P*6$_3$22	10.985–15.10	15
	Charmarite-3*T*	$[Mn_4{}^{2+}Al_2(OH)_{12}]$ $[(CO_3)(H_2O)_3]$	*P*3$_1$12 ?	10.985–22.63	15
2:1	Karchevskyite	$[Mg_{18}Al_9(OH)_{54}][Sr_2(CO_3,PO_4)_9(H_2O,H_3O)_{11}]$	trigonal	16.055–25.66	32
	Zaccagnaite	$[Zn_4Al_2(OH)_{12}]$ $[(CO_3)(H_2O)_3]$	*P*6$_3$/*mmc*	3.073–15.114	17
5:1	Coalingite	$[Mg_{10}Fe_2{}^{3+}(OH)_{24}]$ $[(CO_3)(H_2O)_2]$	*R*-3*m*	3.12–37.4	18
6:1	Brugnatellite	$[Mg_6Fe^{3+}(OH)_{14}]$ $[(CO_3)_{0.5}(H_2O)_4]$?	n.d.	5.47–15.97	19

(continued)

Table 1 (continued)

M^{2+}:M^{3+}	Mineral name	Chemical formula	Space group	Unit-cell parameters	Ref
LDH sulfates					
5:4	Carrboydite	$[Ni_5Al_4(OH)_{18}]$ $[(SO_4)_2(H_2O)_{10}]$	n.d.	9.14–10.34	20
5:3	Glaucocerinite	$[Zn_{10}Al_6(OH)_{32}]$ $[(SO_4)_3(H_2O)_{18}]$	n.d.	3.070–32.65	21
3:1	Honessite	$[Ni_6Fe_2^{3+}(OH)_{16}]$ $[(SO_4)(H_2O)_4]$	n.d.	3.083–26.71	22
	Hydrohonessite	$[Ni_6Fe_2^{3+}(OH)_{16}]$ $[(SO_4)(H_2O)_7]$	n.d.	3.087–33.4	22
2:1	Motukoreaite	$[Mg_6Al_3(OH)_{18}]$ $[Na_{0.6}(SO_4,CO_3)_2(H_2O)_{12}]$	*R*-3*m*	9.172–33.51	23
11:3	Mountkeithite	$[Mg_{11}Fe_3^{3+}(OH)_{28}]$ $[(SO_4)_{1.5}(H_2O)_{11}]$?	n.d.	10.698–22.545	24
2:1	Shigaite	$[Mn_6Al_3(OH)_{18}]$ $[Na(SO_4)_2(H_2O)_{12}]$	*R*-3	9.512–33.074	25
	Nikisherite	$[Fe_6^{2+}Al_3(OH)_{18}]$ $[Na(SO_4)_2(H_2O)_{12}]$	*R*-3	9.347–33.000	26
7:2	Wermlandite	$[Mg_7Al_2(OH)_{18}]$ $[Ca(SO_4)_2(H_2O)_{12}]$	*P*-3*c*1	9.303–22.57	27
2:1	Woodwardite	$[Cu_4Al_2(OH)_{12}]$ $[(SO_4)(H_2O)_{2–4}]$	*R*-3*m*	5.306–26.77	28
5:4	Hydrowoodwardite	$[Cu_5Al_4(OH)_{18}]$ $[(SO_4)_2(H_2O)_n]$	*R*-3*m*	3.070–31.9	29
1:1 ?	Zincowoodwardite	$[Zn_2Al_2(OH)_8]$ $[(SO_4)(H_2O)_n]$	*P*-3	3.036–8.85	30
5:3	Natroglaucocerinite	$[Zn_{20}Al_{12}(OH)_{64}]$ $[Na_6(SO_4)_9(H_2O)_{36}]$	n.d.	n.d.	31
LDH antimonites					
2:1	Zincalstibite	$[Zn_2Al(OH)_6][Sb(OH)_6]$	*P*-3	5.321–9.786	33
	Cualstibite	$[Cu_2Al(OH)_6][Sb(OH)_6]$	*P*-3	9.150–9.745	33
LDH with variable interlayer content					
?	Fougerite	$[Fe_{1-x}^{2+}Fe_x^{3+}Mg_y(OH)_{2+2y}]$ $[x/nA^{n-}\cdot m(H_2O)]$ with $x/(1+y) = 0.25–0.33$ and $m \leq (1-x+y)$	variable	variable	34

n.d. = not determined

References: (1) Braithwaite et al. 1994; (2) Grguric et al. 2001; (3) Koritnig and Suesse 1975; (4) Chukanov et al. 2009; (5) Kashaev et al. 1982; (6) Morandi and Dalrio 1973; (7) Taylor 1973; (8) Ingram and Taylor 1967; (9) Allmann and Jepsen 1969; (10) Mills et al. 2011; (11) Dunn et al. 1979; (12) Song and Moon 1998; (13) Bish and Brindley 1977; (14) Piret and Deliens 1980; (15) Chao, Gault, 1997; (16) Arakcheeva et al. 1996; (17) Merlino and Orlandi 2001; (18) Pastor-Rodriguez and Taylor 1971; (19) Fenoglio 1938; (20) Nickel and Clarke 1976; (21) Raade et al. 1985; (22) Bish and Livingstone 1981; (23) Rius and Plana 1986; (24) Hudson and Bussell (1981); (25) Cooper and Hawthorne 1996; (26) Huminicki and Hawthorne 2003; (27) Rius and Allmann 1984; (28) Nickel 1976; (29) Witzke 1999; (30) Witzke and Raade 2000; (31) mineral approved in 1995, but description has not yet been published; (32) Britvin et al. 2008; (33) Bonaccorsi et al. 2007; (34) Trolard et al. 2007

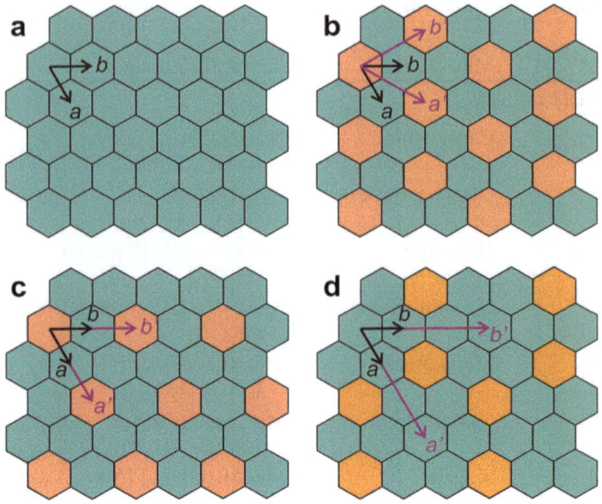

Fig. 2 2D M^{2+}-M^{3+} cation superstructures in natural LDHs (M^{2+} and M^{3+} sites are shown as *green* and *orange* hexagons, respectively): (**a**) ideal brucite-like layer; (**b**) layer with M^{2+}:M^{3+} = 2:1; (**c**) layer with M^{2+}:M^{3+} = 3:1; (**d**) layer with M^{2+}:M^{3+} = 7:2

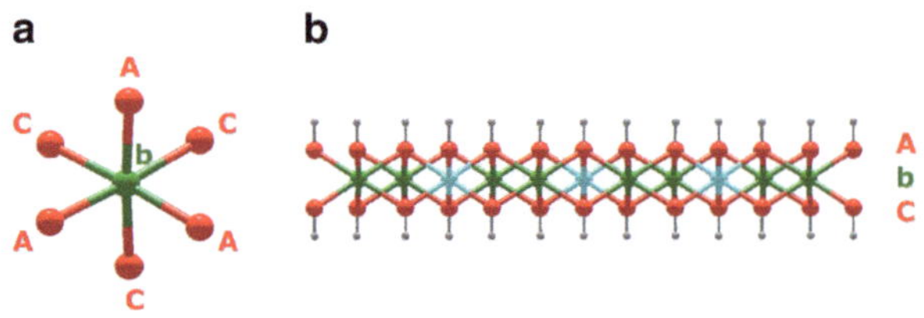

Fig. 3 Projection of metal-centered octahedron with identification of cation position as b and anion positions as A and C (**a**) and projection of double hydroxide layer parallel to its extension (**b**)

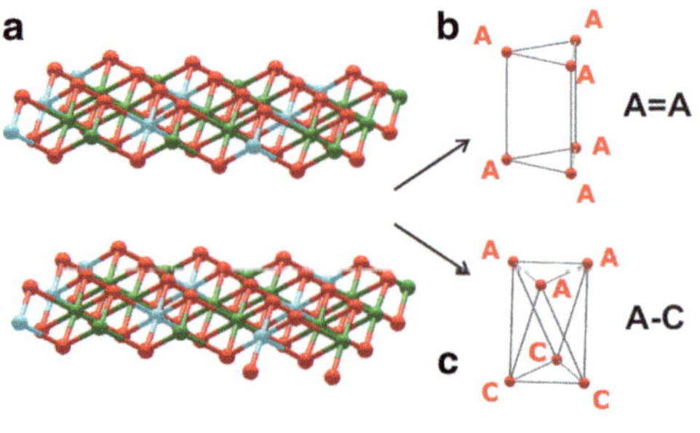

Fig. 4 The interlayer space in the structures of LDHs (**a**) may consist either from trigonal prisms (**b**): P-type interlayer or octahedra (**c**): O-type interlayer

a structural formula **AbC** (Fig. 3). Since positions of the cations within the layer are uniquely determined by the hydroxyl positions, they can be omitted.

Stacking of two layers may result in formation of two different types of interlayers (Fig. 4). In the case when upper hydroxide sheet of the lower layer and lower hydroxide sheet of the upper layer have the same notations (e.g., **A** and **A**), interlayer may be represented as consisting of trigonal prisms (Fig. 4b). This type of interlayer is referred to as a P-type and denoted with an equal sign (=) (e.g., ...**A** = **A**...). In the case when upper hydroxide sheet of the lower layer and lower hydroxide sheet of the upper layer have different notations (e.g., **A** and **B**), interlayer may be represented as consisting of elongated octahedra (Fig. 4c). This type of interlayer is referred to as a O-type and denoted with a dash (–) (e.g., ...**A–B**...).

Using this simple and elegant scheme of notations, Bookin and Drits (1993) derived all two- and three-layer LDH polytypes and six-layer rhombohedral polytypes and calculated their powder X-ray diffraction patterns. For instance, there are exactly three two-layer polytypes, ...**AC** = **CA** = **AC**..., ...**AC-AB-AC**..., and ...**AC-BA** = **AC**..., denoted as $2H_1$, $2H_2$, and $2H_3$, respectively. Out of these three polytypes, $2H_1$ polytype appears to be the most common in minerals (it is obviously one observed for manasseite). In this polytype, all cations occupy the **b** sites and all interlayers are of the P-type. Among three-layer polytypes, the most common is the $3R_1$ polytype that has the structure ...**AC** = **CB** = **BA** = **AC**..., i.e. all its interlayers are of the P-type. On the basis of the nomenclature proposed by Bookin and Drits (1993), Bookin *et al.* (1993) investigated experimentally studied natural LDHs and reported occurrence of two- and three-layer polytypes $2H_1$ and $3R_1$ in CO_3-bearing minerals. In contrast, in sulfate-bearing LDHs, the situation is more complex: both one- and three-layer polytypes are observed with both P- and O-type interlayers. The nomenclature was further illustrated and exemplified by Drits and Bookin (2001). However, despite its exhaustive and rigorous character in derivation of the polytypes, this nomenclature does not take into account such important structural feature of LDHs as an ordering of M^{2+} and M^{3+} cations within the double hydroxide layers.

3 Cation Ordering in Natural Mg_2Al LDHs: Recent Results

All the structural studies on synthetic LDHs deal with powder samples that prevent elucidation of such fine details of structure architecture as formation of superstructures due to cation ordering.

Detection of the cation ordering is especially problematic in synthetic powder samples and, in particular, in synthetic quintinites (Mg_2Al-CO_3 LDHs). Richardson and Braterman (2007) investigated short-range order in Mg-Al LDHs with Mg: Al = 2:1 and 3:1 by infrared (IR) spectroscopy in the region between 400 and 250 cm^{-1}. They studied both fresh and aged (for 24 h) LDHs and found out that aged Mg_2Al specimen shows a sharp 447 cm^{-1}band, which is diagnostic of lattice ordering and which is absent in the IR spectra of fresh Mg_2Al and both fresh and aged Mg_3Al samples. On this basis, Richardson and Braterman (2007) concluded that a disorderly as-formed Mg_2Al material through a solution-precipitation process transforms into material with regular Mg-Al order. The driving force for the ordering is to avoid direct contacts of $Al(OH)_6$ octahedra, which, in Mg_2Al LDHs, is possible only through formation of a regular honeycomb superstructure. In contrast, in Mg_3Al LDHs, Al-Al avoidance may be achieved in a disordered fashion, i.e. without formation of a periodic superstructure.

Sideris et al. (2008) investigated Mg-Al ordering in Mg-Al carbonate LDHs with the Mg:Al ratio of ca. 5:1, 4:1 and 2:1 by means of combined ^{1}H magic angle spinning and ^{25}Mg triple-quantum magic angle spinning nuclear magnetic resonance (NMR) spectroscopy and found that the Mg^{2+} and Al^{3+} are not randomly distributed in the metal hydroxide sheets and that in Mg_2Al LDH they are ordered in a honeycomb arrangement.

As to our knowledge, the only evidence of formation of 2D cation superstructures in LDHs with $M^{2+}:M^{3+} = 3:1$ comes from nanoscale imaging of surfaces of hydrotalcite crystals with ordering pattern shown in Fig. 2c. It is worthy to note that, though other superstructures can also be theoretically constructed, the one shown in Fig. 2c has the most uniform distribution of trivalent cations over octahedral layer.

When compared to the monocation $[M(OH)_2]$ octahedral layer (Fig. 2a), cation ordering results in formation of 2D supercells related to the a_{br} parameter (~3.1–3.2 Å) of the brucite-like layer as following: (1) for the hexagonal layer with $M^{2+}:M^{3+} = 2:1$ (Fig. 2b): $a' = b' = 3^{1/2}a_{br}$ (~5.2–5.4 Å); (2) for the hexagonal layer with $M^{2+}:M^{3+} = 2:1$ (Fig. 2c): $a' = b' = 2a_{br}$ (~ 6.3–6.4 Å); (3) for the hexagonal layer with $M^{2+}:M^{3+} = 7:2$ (Fig. 2d): $a' = b' = 3a_{br}$ (~9.3 Å).

In contrast to synthetic materials, natural LDHs are known to form single crystals accessible to single-crystal X-ray diffraction analysis, which allowed to identify basic features of their crystal chemistry and to demonstrate pecularities of cation and anion ordering. Figure 5 show crystals of quintinite (Mg_2Al-CO_3 LDHs) varieties from Kovdor alkaline massifs, Kola peninsula, Russia (Krivovichev et al. 2010a, b; Zhitova et al. 2010). In this mineral deposit, LDH-group minerals form at the late stages of hydrothermal activity as a result of secondary hydrothermal alteration of spinel crystals. Chemical composition of the samples studied by the wave-length dispersion spectrometry using a Cameca MS-46 electron microprobe and infrared spectroscopy provided the same (within standard errors) chemical formula, $[Mg_4Al_2(OH)_{12}](CO_3)(H_2O)_3$. Single-crystal X-ray diffraction study revealed that the diffraction pattern of the samples quintinite-$2H$-$3c$ and quintinite-$1M$ was characterized by the presence of strong and sharp Bragg reflections and weakly discrete diffuse-like lines. Whereas sharp Bragg reflections originate from basic layer stacking of metal hydroxide layers, weak reflections are indicative of formation of 3-D cation superlattices due to the Mg-Al ordering. For instance, diffuse streaks in diffraction pattern of quintinite-$2H$-$3c$ are extended along **c*** and centered at $h-k \neq 3n$ relative to a supercell indexing (R-cell, $a = 5.2745(7)$, $c = 45.36(1)$ Å). Indexing of sharp Bragg reflections only resulted in a small subcell with parameters $a = 3.045$, $c = 15.12$ Å, which are approximately in agreement with unit-cell parameters of the $2H_1$ polytype of Mg-Al LDHs. In the large supercell, indices of the sharp Bragg reflections correspond to conditions $h-k = 3n$ and $l = 3n$ (Fig. 6a, b). Similar situation is observed also for quintinite-$1M$.

Fig. 5 Crystals of quintinite (natural Mg_2Al-CO_3 LDH) polytypes from hydrothermal veins of the Kovdor alkaline massif, Kola peninsula, Russia: (**a**) quintinite-2*H*-3*c*, (**b**) quintinite-1*M*, (**c, d**) quintinite-2*H*-1*c*

The unit-cell parameters obtained for the two samples are given in Table 2. It is noteworthy that, in quintinite-2*H*-3*c*, Mg-Al ordering results in formation of threefold superstructure relative to the usual hexagonal 2*H* polytype, whereas, in quintinite-1*M*, cation ordering and superlattice formation results in dramatic reduction of symmetry: from rhombohedral (as in 3*R* polytype) to monoclinic (and thus the sample should be qualified as a 1*M* polytype).

The structures of all studied LDH crystals consists of metal hydroxide layers, $[Mg_2Al(OH)_6]^+$, and disordered interlayer (Fig. 7a). According to LDH polytype nomenclature (see above), the layer stacking in quintinite-2*H*-3*c* can be described as $\ldots\mathbf{AC} = \mathbf{CA} = \mathbf{AC}\ldots$, with **A** and **C** being positions of hydroxide ions and **b** position of cations (in the same manner as in closest packing of equal spheres). The sequence of layers within the unit cell can be described as $\ldots\mathbf{AC} = \mathbf{CA} = \mathbf{AC} = \mathbf{CA} = \mathbf{AC} = \mathbf{CA}\ldots$, since the unit cell contains exactly six double hydroxide layers.

Thus, in terms of layer stacking sequence, the structure clearly has a pseudo-period with *c* parameter of about 15.12 Å, i.e. three times smaller than one observed experimentally. The reason for tripling the *c* parameter is Mg-Al ordering in the $[Mg_2Al(OH)_6]^+$ layer. There are three symmetry-independent octahedral cation

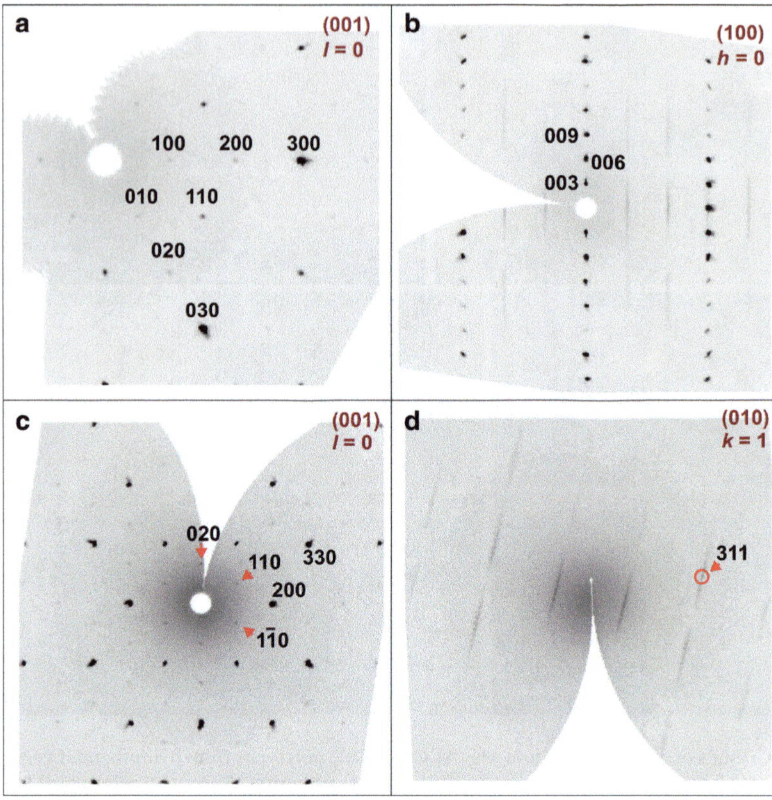

Fig. 6 Reconstructed sections of reciprocal diffraction space showing sharp and strong Bragg reflections and weak diffuse-like superstructure reflections for crystal quintinite-2*H*-3*c* (**a** and **b**) and quintinite-1*M* (**c** and **d**). Superstructure reflections are indicated by *red arrows*

Table 2 Crystallographic data for natural Mg_2Al-CO_3 LDHs (quintinites)

Sample	Quintinite-2*H*-3*c*	Quintinite-1*M*
Symmetry	Rhombohedral	Monoclinic
Space group	*R*32	*C*2/*m*
u [Å]	5.2745	5.2662
b [Å]	–	9.114
c [Å]	45.364	7.766
β [°]	–	103.17
V [Å^3]	1093.0	362.9

sites in the structure of quintinite-2*H*-3*c*. Since the site-scattering factors of Mg^{2+} and Al^{3+} cations are nearly identical, the only way to distinguish between Mg and Al sites is to analyse distribution of the M-O bond lengths. The structure refinement indicates one M site with the M bond lengths in the range of 1.936–1.940 Å

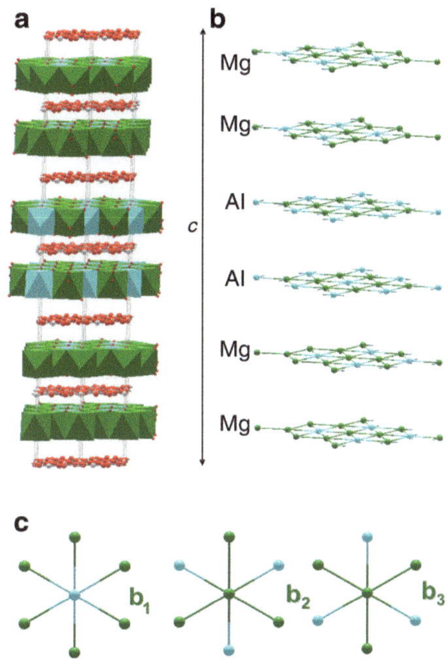

Fig. 7 Crystal structure of quintinite-2*H*-3*c* (**a**) and stacking of Mg_2Al cation arrays with the $[Mg_4Al_2]$ repeat sequence (**b**) and three different positions of the arrays (**c**)

(assigned to Al) and two M sites with the M-O bond lengths of 2.042–2.045 and 2.070–2.077 Å (assigned to Mg). According to the Bookin and Drits (1993) nomenclature, in 2*H* polytype, all anions are either in **A** or **C** positions, whereas all cations are in the **b** positions. However, the sequence of the **b** positions if seen along the *c* axis is occupied by Mg and Al cations differently. This sequence can be written as [MgMgMgMgAlAl] or $[Mg_4Al_2]$, taking into account that the content given in the square brackets corresponds to the *c* parameter repeat. Considering possible relative positions of the 2-D Mg-Al cation array, one may distinguish exactly three different Mg_2Al arrays related to each other by either *a* or *b* translations (Fig. 7b). These arrays may be indicated as $\mathbf{b_1}$, $\mathbf{b_2}$ and $\mathbf{b_3}$ (since all cations are in the b positions). Therefore the full description of the layer sequence (i.e. description that takes into account cation ordering) can be written as $\ldots = \mathbf{Ab_1C} = \mathbf{Cb_1A} = \mathbf{Ab_2C} = \mathbf{Cb_2A} = \mathbf{Ab_3C} = \mathbf{Cb_3A} = \ldots$

The layer stacking in quintinite-1*M* (Fig. 8a) can be described as … = **AB** = **BC** = **CA** = …, i.e. it corresponds to a rhombohedral polytype structure typical for most synthetic Mg-Al LDHs. However, because of the cation ordering, the situation becomes more complex and can be deciphered from the analysis of relative position of the 2-D Mg_2Al cation arrays (Fig. 8b–e). First, it is obvious that cations in quintinite-1*M* are located in all possible sites, **a**, **b**, and **c**, so that the full description of the layer sequence should be written as … = **AcB** = **BaC** = **CbA** = … In the case of complete Mg-Al disorder, structure with this sequence would have a rhombohedral symmetry (space group *R*-3*m*), but cation ordering results in symmetry reduction and formation of superstructure. As can be seen from

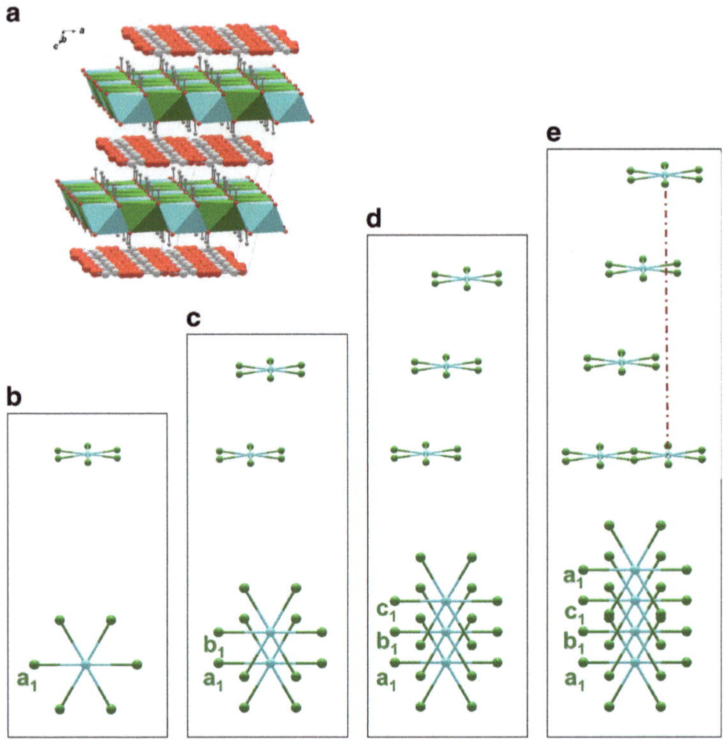

Fig. 8 Crystal structure of quintinite-1*M* (**a**) and schemes of stacking of Mg_2Al cation arrays along the direction perpendicular to the layers (**b–e**)

Fig. 8b–d, Al cations in adjacent Mg_2Al arrays are located in one plane, which is a mirror plane *m* in the space group *C*2/*m*. This results in disappearance of the threefold symmetry axis perpendicular to the layers and transition from rhombohedral to monoclinic symmetry. In order to distinguish between **a**, **b** and **c** positions occupied by Mg and Al cations in the Mg_2Al array, we identify them as $\mathbf{a_1}$, $\mathbf{a_2}$, $\mathbf{a_3}$, etc. Thus, the complete layer stacking sequence can be described as $\ldots = \mathbf{Ac_1B} = \mathbf{Ba_1C} = \mathbf{Cb_1A} = \ldots$ It is of theoretical interest that the sequence $\ldots = \mathbf{Ac_1B} = \mathbf{Ba_1C} = \mathbf{Cb_2A} = \ldots$ possesses a trigonal symmetry with Al cations in adjacent layers segregated along a 3_1 screw axis, though this situation has not yet been observed in LDHs. According to the traditional nomenclature of polytypes, sample 2 should be called quintinite-1*M*, since it contains exactly one layer per monoclinic unit cell.

The advantage of single-crystal diffraction is the possibility to analyse electron density distribution in certain areas of a structure. Figure 9a shows electron density distribution map in quintinite-1*M* at the interlayer level of $z = 0.25$. It can be seen that electron density maxima corresponding to the O atoms of carbonate groups are associated into almost continuous toroidal regions, which makes the refinement

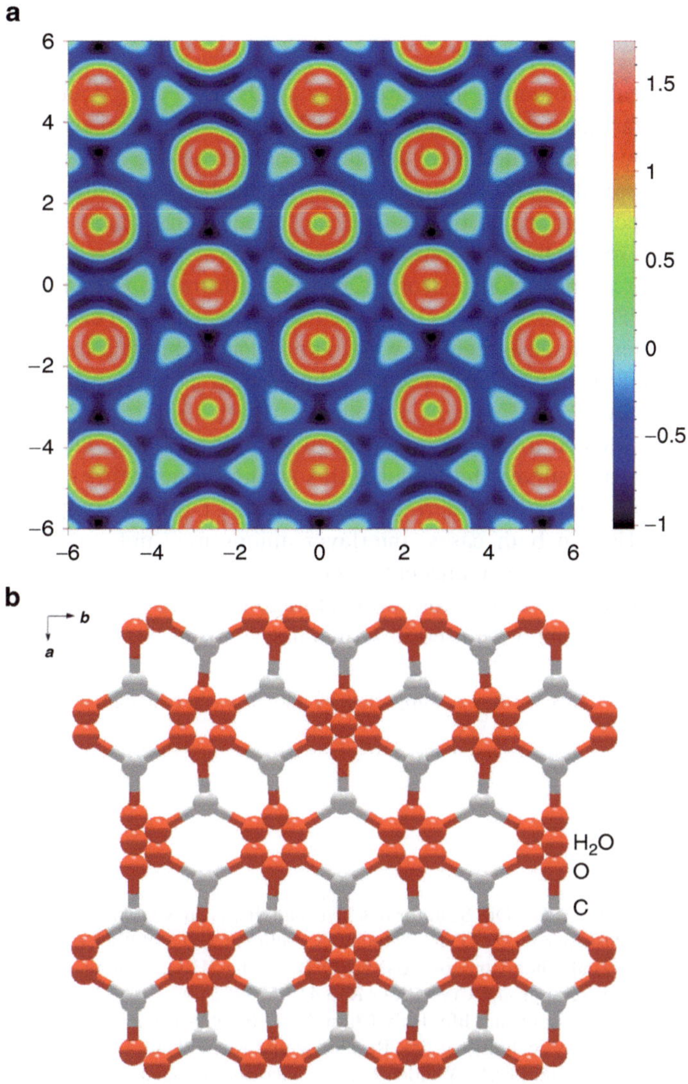

Fig. 9 Electron density Fourier map of interlayer in quintinite-1*M* (**a**) and model of disordered interlayer arrangement of carbonate anions and water molecules (**b**)

procedure a difficult task. Since there are no indications of any dynamic disorder effects in the structure (e.g., rotation of carbonate triangles), we suppose that the observed continuous character of the tori is due to positional disorder. Figure 9b shows an approximate identification of the electron density peaks as proposed by the refinement.

4 Discussion

The results on cation ordering in natural LDHs with $M^{2+} = Mg^{2+}$ and $M^{3+} = Al^{3+}$, which are by far the most common natural LDHs (we have observed that many samples previously identified as hydrotalcite and manasseite are in fact quintinite polytypes), demonstrate that cation ordering in double hydroxide layers is very conservative and is governed by the tendency to minimize repulsion between the M^{3+} sites. As a result, the arrangement of di- and trivalent cations is controlled by electrostatic forces that would strongly restricts information storage capacity of LDH matrices. The tendency of LDH to periodic arrangement of cations revealed in the recent studies is in disagreement with their role as prebiotic information storage materials (recall the definition of life by Schroedinger as '*aperiodic* crystal').

In the Greenwell and Coveney's model of information transfer in LDHs, the crucial role is devoted to the arrangement of interlayer anions. Our experimental results and literature survey indicate that this arrangement is either disordered and not controlled by the cation ordering (in carbonate LDHs) or completely ordered (in sulfate LDHs). In both cases, interlayer anions may not serve as agents in information transfer between adjacent layers.

Thus the role of LDHs as information storage matrices in prebiotic systems is rather doubtful and cannot be used in artificial systems as well.

Acknowledgements This work was supported by the Russian Foundation of Basic research (grant 10-05-00431).

References

Allmann R, Jepsen HP (1969) Die Struktur des Hydrotalkits. N Jb Mineral Mh 1969:544–551

Arakcheeva AV, Pushcharovskii DYu, Atencio D, Lubman GU (1996) Crystal structure and Comparative crystal chemistry of $Al_2Mg_4(OH)_{12}(CO_3)\cdot 3H_2O$, a new mineral from the hydrotalcite-manasseite group. Crystallogr Rep 41:972–981

Arrhenius GO (2003) Crystals and life. Helv Chim Acta 86:1569–1586

Bernal JD (1951) The physicalbasis of life. Routledge and Kegan Paul, London

Bernal JD (1967) The origin of life. Weidenfeld and Nicolson, London

Bish DL, Brindley GW (1977) A reinvestigation of takovite, a nickel aluminum hydroxy-carbonate of the pyroaurite group. Am Mineral 62:458–464

Bish DL, Livingstone A (1981) The crystal chemistry and paragenesis of honessite and hydrohonessite: the sulphate analogues of reevesite. Mineral Mag 44:339–343

Bonaccorsi E, Merlino S, Orlandi P (2007) Zincalstibite, a new mineral, and cualstibite: crystal chemical and structural relationships. Am Mineral 92:198–203

Bookin AS, Drits VA (1993) Polytype diversity of the hydrotalcite-like minerals. I. Possible polytypes and their diffraction patterns. Clays Clay Miner 41:551–557

Bookin AS, Cherkashin VI, Drits VA (1993) Polytype diversity of the hydrotalcite-group minerals. II. Determination of the polytypes of experimentally studied varieties. Clays Clay Miner 41:558–564

Braithwaite RSW, Dunn PJ, Pritchard RG, Paar WH (1994) Iowaite, a re-investigation. Mineral Mag 58:79–85

Britvin SN, Chukanov NV, Bekenova GK, Yagovkina MA, Antonov AV, Bogdanova AN, Krasnova NI (2008) Karchevskyite,[Mg18Al9(OH)54][Sr2(CO3, PO4)9(H2O, H3O)11], a new mineral species of the layered double hydroxide family. Geol Ore Deposits 50:556–564
Cairns-Smith AG (1982) Genetic takeover and the mineral origins of life. Cambridge University Press, Cambridge
Chao GY, Gault RA (1997) Quintinite-2H, quintinite-3T, charmarite-2H, charmarite-3T and caresite-3T, a new group of carbonate minerals related to the hydrotalcite/manasseite group. Can Mineral 35:1541–1549
Chukanov NV, Pekov IV, Levitskaya LA, Zadov AE (2009) Droninoite, $Ni_3Fe_{3+}Cl(OH)_8 \cdot 2H_2O$, a new hydrotalcite-group mineral species from the weathered Dronino meteorite. Geol Ore Deposits 51:767–773
Cooper MA, Hawthorne FC (1996) The crystal structure of shigaite, $[AlMn_2{}^{2+}(OH)_6]_3(SO_4)_2$ $Na(H_2O)_6(H_2O)_6$, a hydrotalcite-group mineral. Can Mineral 34:91–97
Drits VA, Bookin AS (2001) Crystal structure and X-ray identification of layered double hydroxides. In: Rives V (ed) Layered double hydroxides: present and future. Nova Scientific Publishers, New York, pp 39–92
Duan X, Evans DG (eds) (2006) Layered double hydroxides, vol 119, Structure and bonding. Springer, Berlin
Dunn PJ, Peacor DR, Palmer TD (1979) Desautelsite, a new mineral of the pyroaurite [hydrotalcite] group. Am Mineral 64:127–130
Fenoglio M (1938) Ricerche sulla brugnatellite. Period Mineral 9:1–13
Greenwell HC, Coveney PV (2006) Layered double hydroxide minerals as possible prebiotic information storage and transfer compounds. Orig Life Evol Biosph 36:13–37
Grguric BA, Madsen IC, Pring A (2001) Woodallite, a new chromium analogue of iowaite from the Mount Keith nickel deposit, Western Australia. Mineral Mag 65:427–435
Hudson DR, Bussell M (1981) Mountkeithite, a new pyroaurite-related mineral with an expanded interlayer containing exchangeable $MgSO_4$. Mineral Mag 44:345–350
Huminicki DMC, Hawthorne FC (2003) The crystal structure of nikischerite, $NaFeAl_3(SO_4)_2(OH)_{18}(H_2O)_{12}$, a mineral of the shigaite group. Can Mineral 41:79–82
Ingram L, Taylor HFW (1967) The crystal structures of sjögrenite and pyroaurite. Mineral Mag 36:465–479
Kashaev AA, Feoktistov GD, Petrova SV (1982) Chlormagaluminite, $(Mg, Fe^{2+})_4Al_2(OH)_{12}(Cl_2, CO_3)_2{\bullet}2H_2O$ – a new mineral of the manasseite-sjoegrenite group. Zap Vses Miner Obshch 11:121–127 (in Russian)
Kim D, Huang C, Lee H, Han I, Kang S, Kwon S, Lee J, Han Y, Kim H (2003) Hydrotalcite-type catalysts for narrow-range oxyethylation of 1-dodecanol using ethyleneoxide. Appl Catal A: Gen 249:229–240
Koritnig S, Suesse P (1975) Meixnerite, $Mg_6Al_2(OH)_{18}$ $\bullet 4H_2O$, ein neues magnesium-aluminium-hydroxid-mineral. Tscherm Miner Petrogr Mitt 22:79–87
Krivovichev SV, Yakovenchuk VN, Zhitova ES, Zolotarev AA, Pakhomovsky YA, Ivanyuk GY (2010a) Crystal chemistry of natural layered double hydroxides. 1. Quintinite-*2H–3c* from the Kovdor alkaline massif, Kola peninsula, Russia. Mineral Mag 74:821–832
Krivovichev SV, Yakovenchuk VN, Zhitova ES, Zolotarev AA, Pakhomovsky YA, Ivanyuk GYu (2010b) Crystal chemistry of natural layered double hydroxides. 2. Quintinite-*1M*: First evidence of a monoclinic polytype in M^{2+}-M^{3+} layered double hydroxides. Mineral Mag 74:833–840
Kuma K, Paplawsky W, Gedulin B, Arrhenius G (1989) Mixed-valence hydroxides as bioorganic host minerals. Orig Life Evol Biosph 19:573–602
Merlino S, Orlandi P (2001) Carraraite and zaccagnaite, two new minerals from the Carrara marble quarries: their chemical compositions, physical properties, and structural features. Am Mineral 86:1293–1301
Mills SJ, Whitfield PS, Wilson SA, Woodhouse JN, Dipple GM, Raudsepp M, Francis CA (2011) The crystal structure of stichtite, re-examination of barbertonite, and the nature of polytypism in MgCr hydrotalcites. Amer Mineral 96:179–187

Morandi N, Dalrio G (1973) Jamborite: a new nickel hydroxide mineral from the Northern Apennines, Italy. Am Mineral 58:835–839
Nickel E (1976) New data on woodwardite. Mineral Mag 43:644–647
Nickel EH, Clarke RM (1976) Carrboydite, a hydrated sulfate of nickel and aluminum: a new mineral from Western Australia. Amer Mineral 61:366–372
Pastor-Rodriguez J, Taylor HFW (1971) Crystal structure of coalingite. Mineral Mag 38:286–294
Piret P, Deliens M (1980) La comblainite, $((Ni_x{}^{2+}, Co_{1-x}{}^{3+})(OH)_2)(CO_3)_{(1-x)/2}.yH_2O$, nouveau mineral du groupe de la pyroaurite. Bull Minéral 103(1):113–117
Raade G, Elliott CJ, Din VK (1985) New data on glaucocerinite. Mineral Mag 49:583–590
Richardson MC, Braterman PS (2007) Infrared spectra of oriented and nonoriented layered double hydroxides in the range from 4000 to 250 cm^{-1}, with evidence for regular short-range order in a synthetic magnesium-aluminum LDH with Mg: Al = 2: 1 but not with Mg: Al = 3: 1. J Phys Chem C111:4209–4215
Rius J, Allmann R (1984) The superstructure of the double layer mineral wermlandite $[Mg_7(Al_{0.57}Fe_{0.43}{}^{3+})_2(OH)_{18}]^{2+}[(Ca_{0.6}Mg_{0.4})(SO_4)_2(H_2O)_{12}]^{2-}$. Z Kristallogr 168:133–144
Rius J, Plana F (1986) Contribution to the superstructure resolution of the double layer mineral motukoreaite. N Jahrb Mineral Monatsh 1986:263–272
Rives V (ed) (2001) Layered double hydroxides: present and future. Nova Science Publishers, New York
Sideris PJ, Nielsen UG, Gan ZH, Grey CP (2008) Mg/Al ordering in layered double hydroxides revealed by multinuclear NMR spectroscopy. Science 321:113–117
Song Y, Moon HS (1998) Additional data on reevesite and its Co-analogue, as a new member of the hydrotalcite group. Clay Miner 33:285–296
Taylor HFW (1973) Crystal structures of some double hydroxide minerals. Mineral Mag 39:377–389
Trolard F, Bourrie F, Abdelmoula M, Refait P, Feder F (2007) Fougerite, a new mineral of the pyroaurite-iowaite group; description and crystal structure. Clays Clay Miner 55:323–334
Witzke T (1999) Hydrowoodwardite, a new mineral of the hydrotalcite group from Koenigswalde near Annaberg, Saxony/Germany and other localities. N Jahrb Mineral Monatsh 1999:75–86
Witzke T, Raade G (2000) Zincowoodwardite, $[Zn_{1-x}Al_x(OH)_2][(SO_4)_{x/2}(H_2O)_n]$, a new mineral of the hydrotalcite group. N Jahrb Mineral Monatsh 2000:455–465
Zhitova ES, Yakovenchuk VN, Krivovichev SV, Zolotarev AA, Pakhomovsky YA, Ivanyuk GY (2010) Crystal chemistry of natural layered double hydroxides. 3. The crystal structure of Mg, Al-disordered quintinite-2*H*. Mineral Mag 74:841–848

Fixation of Chromate in Layered Double Hydroxides of the TCAH Type and Some Complex Application Mixtures

Herbert Pöllmann and Jürgen Göske

1 Introduction

High chromate concentrations can be released from ferrochromium industry, painting and lacquer paints, tanning and impregnation and also from high temperature processes like cement production (occupational disease "Cement- or Bricklayer-eczema"). A treatment for fixation of chromate can be the ion exchange by LDH-phases (Auer 1992; Auer et al. 1990, 1991; Göske and Pöllmann 1996, 1999a, b; Göske 1999; Göske et al. 1997; Pöllmann and Auer 2010).

Some of these compounds among the group of layered calcium aluminate hydrates are important during the process of cement hydration and also as deterioration reaction products like the influence of deicing salts or carbonation reactions. Their chemical composition can be described by the following general formula: $3CaO{\cdot}Al_2O_3{\cdot}CaX{\cdot}nH_2O$ and $3CaO{\cdot}Al_2O_3{\cdot}CaY_2{\cdot}nH_2O$ with $X = SO_4^{2-}$, CO_3^{2-}, CrO_4^{2-}, ... and $Y = OH^-$, Cl^-, NO_3^-, ... Structurally they consist of positively charged layers of $[Ca_2Al(OH)_6]^+$ and negatively charged interlayers of $[OH{\cdot}nH_2O]^-$.

Detailed structural studies were done with Monosulfate ($3CaO{\cdot}Al_2O_3{\cdot}CaSO_4{\cdot}12H_2O$) and Monochloride ($3CaO{\cdot}Al_2O_3{\cdot}CaCl_2{\cdot}12H_2O$) and Monocarbonate ($3CaO{\cdot}Al_2O_3{\cdot}CaCO_3{\cdot}11H_2O$) (Allmann 1970; Terzis et al. 1987; Renaudin 1998).

TetraCalciumAluminateHydrate (TCAH) with the formula $C_3A{\cdot}Ca(OH)_2{\cdot}nH_2O$ is especially suitable as a storage mineral, because of its layered structure and capability to incorporate different cations and anions. It is widely used to incorporate efficiently for example heavy metals, because the OH-group in the interlayers can be exchanged for 1^-- or 2^--anions.

H. Pöllmann • J. Göske (✉)
Department of Mineralogy/Geochemistry, Martin-Luther-University Halle (Saale), Von Seckendorffplatz 3, 06120 Halle, Germany

ZWL Lauf, Hartdstr. 39b, Lauf, Germany
e-mail: herbert.poellmann@geo.uni-halle.de; juergen.goeske@gmx.de

S.V. Krivovichev (ed.), *Minerals as Advanced Materials II*,
DOI 10.1007/978-3-642-20018-2_10,

The family of layered double hydroxides (LDHs) posesses different possibilities to incorporate different cations and anions a structural schemata is given in figure 1.

In the overall formula $[Me_2Me(OH)_6]^{n+}[anion{\cdot}k{\cdot}H_2O]^{n-}$ the following replacement variations in the structure are possible:

1. Diadochic substitution of metals(2+)
2. Diadochic substitution of metals(3+)
3. Diadochic substitution of interlayer anions
4. Incorporation on H_2O-places
5. Additional incorporation in interlayer
6. Sorption on charged surfaces of layered structures

Some selected examples of crystal chemistry possibilities of lamellar Ca-Al-hydroxysalts with the general formula $[A_4B_2(OH)_{12}][(X)_n{\cdot}nH_2O]$ (A = Ca, B = Al) are given in Table 1.

Storage minerals can be used or added for the immobilization of inorganic and organic substances in contaminated water, gas and soil (Pöllmann 1993, 1994, 2010; Göske et al. 1996, 2003; Pöllmann et al. 1996; Auer and Pöllmann 1994). The composition of the harmful substance to be immobilized depends on the structure and the kind of the compounds. The immobilization of toxic substances

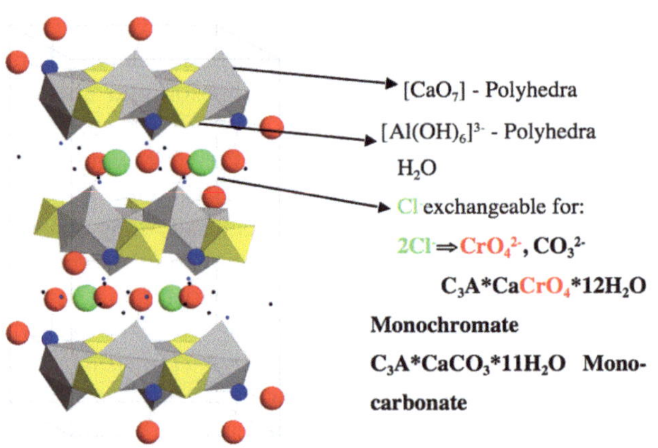

Fig. 1 Part of structure of $C_3A.CaCl_2.10H_2O$ (Terzis et al. 1987)

Table 1 Selected examples of crystal chemical possibilities for lamellar Ca-Al-hydroxysalts with the general formula $[A_4B_2(OH)_{12}][(X)_n{\cdot}nH_2O]$

A	B	X
Fe^{2+} Co^{2+} Ni^{2+} Ca^{2+} Cd^{2+} Zn^{2+} Mg^{2+} Mn^{2+} Cu^{2+}	Co^{3+} Fe^{3+} Al^{3+} Ga^{3+} Cr^{3+}	SO_4^{2-} CrO_4^{2-} SeO_4^{2-} SeO_3^{2-} $B(OH)_4^-$ SiO_4^{4-} OH^- Cl^- SO_3^{2-} NO_2 NO_3^- IO_3^- ClO_3^- BrO_3^- MnO_4^- I^- Br^- (C_6H_5)-(CH_2)n-COO^- (CH_3)-$(CH_2)_n$-COO^- OOC-$(CH_2)_n$-COO^{2-}

is based on the crystallo-chemical fixation and physical aggregation due to the crystallization of metal-metal-hydroxysalts. These compounds offer a great possibility for diadochism of cations and anions. The fixation of non-polar and polar molecules is guaranteed by additional bonding mechanisms (hydrogen bond, residual bond, etc.) and crystallo-chemical replacement.

It is well known that chromates are capable to accelerate cement hydration. Chromate is the reason for the so called "Cement- or Bricklayer-eczema". Formation of these lamellar calcium-aluminate hydrates (in connection to hydration of calcium aluminates) can proceed under variable conditions:

1. Formation in concrete due to harmful substances in the environment.
2. Formation due to additives controlling cement-hydration.
3. Formation under controlled conditions as storage minerals

$$C_3A + CaSO_4.2H_2O + aq \rightarrow C_3A.CaSO_4.H_{12}$$
$$CA + 2CaO + CaSO_4.2H_2O + aq \rightarrow C_3A.CaSO_4.H_{12}$$

The following syntheses methods for layered double hydroxide minerals can be applied:

1. Titration-method
2. Slow precipitation
3. Fast precipitation
4. Precipitation at constant pH-values
5. Synthesis by hydolysis of alkoxydes
6. Acid synthesis from metal oxydes
7. Hydrothermal synthesis
8. High temperature decomposition reaction including rehydration
9. Interlayer exchange reaction

2 Experimental

The starting materials for TCAH production were CaO and $C_{12}A_7$ in stochiometric mixtures (well known mixture for chromate immobilisation using chromate-TCAH solid solutions (Auer 1992) with different additions of other materials).

The immobilization strategy was followed by using stochiometric amounts of CaO and $C_{12}A_7$ with the addition of different other "materials" in the amounts of 1:0.5, 1:1 and 0.5:1 and Na_2CrO_4_solutions with chromate concentrations of 0.1-, 0.05- and 0.01-M concentrations. The stochiometric ratio of CaO:Al_2O_3 was 2:1 following the ratio of TCAH (W/F-ratio was set to 15/1, derived from the amount of CaO + $C_{12}A_7$). The concentration of 0.1-molar chromate is equivalent to the starting chromate amount, leading theoretically to 100% of monochromate.

The different trials were performed parallel and filtrated after 1, 5, 10, 24 and 72 h. The filtrates were analyzed for their chromate contents, the precipitates were investigated for their phase composition at 100% r.h. in a climate chamber.

The following different test mixtures for chromate fixation (selected chromate solutions: 0.1-, 0.05- and 0.01-M) from aqueous solutions with different starting mixtures were investigated. Chromate concentrations were determined by ICP and UV–VIS-techniques. The X-ray experiments were performed using a Bruker D5000 diffractometer equipped with a special wet cell sample holder. SEM investigations were performed using a KONTRON scanning electron microscope with a cryo stage (Göske et al. 1999).

The following list provides an overview of the different mixtures applied for chromate fixation.

1. Experiment: Technical mineral mixtures ($Ca(OH)_2$, $Al(OH)_3$ and $CaCO_3$) with chromate solution
2. Experiment: Technical mineral mixtures ($Ca(OH)_2$, $Al(OH)_3$ and $CaCO_3$), treated for 5 h at about 80°C with chromate solution
3. Experiment: CaO + Al-Brine with chromate solution
4. Experiment: CaO + Al-Brine, after 1 day with chromate solution
5. Experiment: (CaO + H_2O) + Al-brine with chromate solution
6. Experiment: (CaO + H_2O) + Al-brine, after 1 day with chromate solution
7. Experiment: $C_{12}A_7$ + CaO with chromate solution
8. Experiment: $C_{12}A_7$ + CaO, after 1 day with chromate solution
9. Experiment: $C_{12}A_7$ + CaO, after 1 day decanted, with chromate solution
10. Experiment: $C_{12}A_7$ + CaO, after 1 day filtrated, with chromate solution
11. Experiment: ($C_{12}A_7$ + CaO) : BKA = 0.5 : 1, with chromate solution
12. Experiment: ($C_{12}A_7$ + CaO) : BKA = 1 : 1, with chromate solution
13. Experiment: ($C_{12}A_7$ + CaO) : BKA = 1 : 0.5, with chromate solution
14. Experiment: ($C_{12}A_7$ + CaO) : $Al(OH)_3$ = 0.5 : 1, with chromate solution
15. Experiment: ($C_{12}A_7$ + CaO) : $Al(OH)_3$ = 1 : 1, with chromate solution
16. Experiment: ($C_{12}A_7$ + CaO) : $Al(OH)_3$ = 1 : 0.5, with chromate solution
17. Experiment: ($C_{12}A_7$ + CaO) : technical faujasite = 0.5 : 1, with chromate solution
18. Experiment: ($C_{12}A_7$ + CaO) : technical faujasite = 1 : 1, with chromate solution
19. Experiment: ($C_{12}A_7$ + CaO) : technical faujasite = 1 : 0.5, with chromate solution
20. Experiment: ($C_{12}A_7$ + CaO) : clinoptilolite = 0.5 : 1, with chromate solution
21. Experiment: ($C_{12}A_7$ + CaO) : clinoptilolite = 1 : 1, with chromate solution
22. Experiment: ($C_{12}A_7$ + CaO) : clinoptilolite = 1 : 0.5, with chromate solution
23. Experiment: ($C_{12}A_7$ + CaO) : synthetic zeolite type A = 0.5 : 1, with chromate solution
24. Experiment: ($C_{12}A_7$ + CaO) : synthetic zeolite type A = 1 : 1, with chromate solution
25. Experiment: ($C_{12}A_7$ + CaO) : synthetic zeolite type A = 1 : 0.5, with chromate solution
26. Experiment: ($C_{12}A_7$ + CaO) : alkaline activated bentonite = 0.5 : 1, with chromate solution

27. Experiment: ($C_{12}A_7$ + CaO) : alkaline activated bentonite = 1 : 1, with chromate solution
28. Experiment: ($C_{12}A_7$ + CaO) : alkaline activated bentonite = 1 : 0.5, with chromate solution
29. Experiment: ($C_{12}A_7$ + CaO) : highly active adsorption material based on bentonite = 0.5 : 1, with chromate solution
30. Experiment: ($C_{12}A_7$ + CaO) : highly active adsorption material based on bentonite = 1 : 1, with chromate solution
31. Experiment: ($C_{12}A_7$ + CaO) : highly active adsorption material based on bentonite = 1 : 0.5, with chromate solution
32. Experiment: ($C_{12}A_7$ + CaO) : technical flocculation material = 0.5 : 1, with Chromate solution
33. Experiment: ($C_{12}A_7$ + CaO) : technical flocculation material = 1 : 1, with chromate solution
34. Experiment: ($C_{12}A_7$ + CaO) : technical flocculation material = 1 : 0.5, with chromate solution
35. Experiment: ($C_{12}A_7$ + CaO) : alkaline activated bentonite based on montmorillonite = 0.5 : 1, with chromate solution
36. Experiment: ($C_{12}A_7$ + CaO) : alkaline activated bentonite based on montmorillonite = 1 : 1, with chromate solution
37. Experiment: ($C_{12}A_7$ + CaO) : alkaline activated bentonite based on montmorillonite = 1 : 0.5, with chromate solution
38. Experiment: ($C_{12}A_7$ + CaO) : organic modified bentonite = 0.5 : 1, with chromate solution
39. Experiment: ($C_{12}A_7$ + CaO) : organic modified bentonite = 1 : 1, with chromate solution
40. Experiment: ($C_{12}A_7$ + CaO) : organic modified bentonite = 1 : 0.5, with chromate solution
41. Experiment: ($C_{12}A_7$ + CaO) : technical kaolinite = 0.5 : 1, with chromate solution
42. Experiment: ($C_{12}A_7$ + CaO) : technical kaolinite = 1 : 1, with chromate solution
43. Experiment: ($C_{12}A_7$ + CaO) : technical kaolinite = 1 : 0.5, with chromate solution
44. Experiment: ($C_{12}A_7$ + CaO) : natural kaolinite = 0.5 : 1, with chromate solution
45. Experiment: ($C_{12}A_7$ + CaO) : natural kaolinite = 1 : 1, with chromate solution
46. Experiment: ($C_{12}A_7$ + CaO) : natural kaolinite = 1 : 0.5, with chromate solution

3 Results

After mixing of the different mixtures a filtration occurred after definite time cuts and the solid was separated from the solutions. The determined chromate concentrations in solution are summarized in Figs. 2–5.

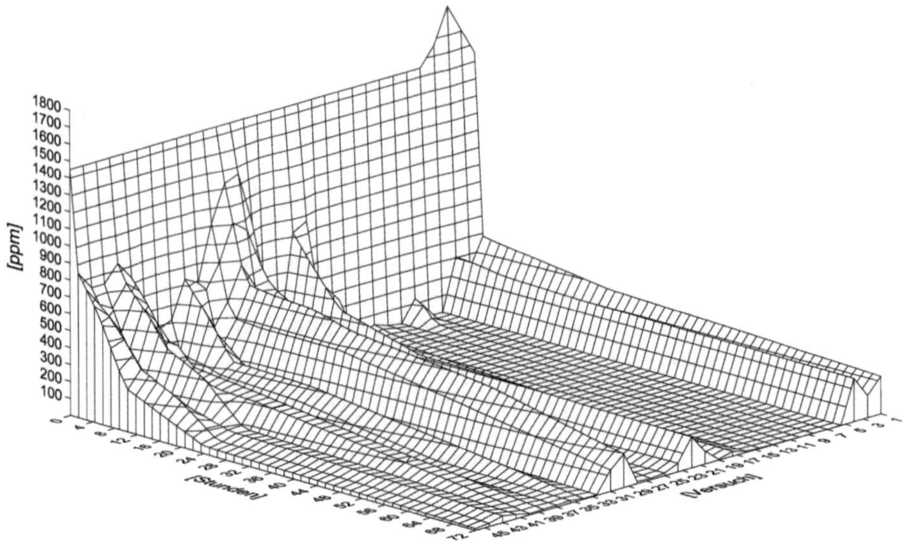

Fig. 2 Fixation of 0.1 M chromate solution by using different mixtures with additives

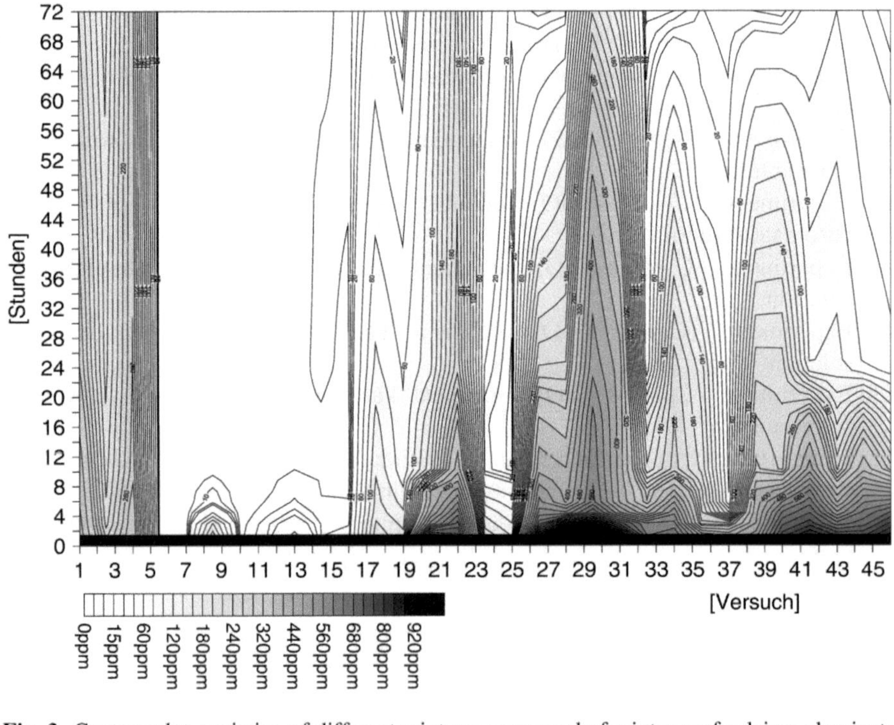

Fig. 3 Contour plot: variation of different mixtures composed of mixtures of calcium aluminate hydrate, bentonite, zeolites, brown coal fly ash, aluminium hydroxide, kaolinite, and the varying chromate bonding capacity as a function of time, 0.1 M chromate solution

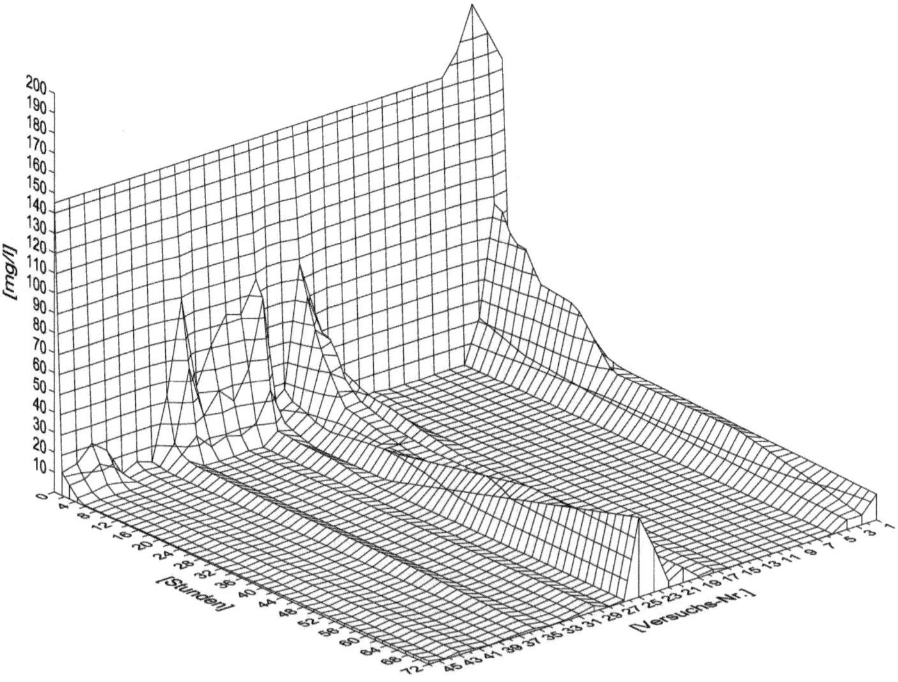

Fig. 4 Fixation of 0.01 M chromate solution by using different mixtures with additives

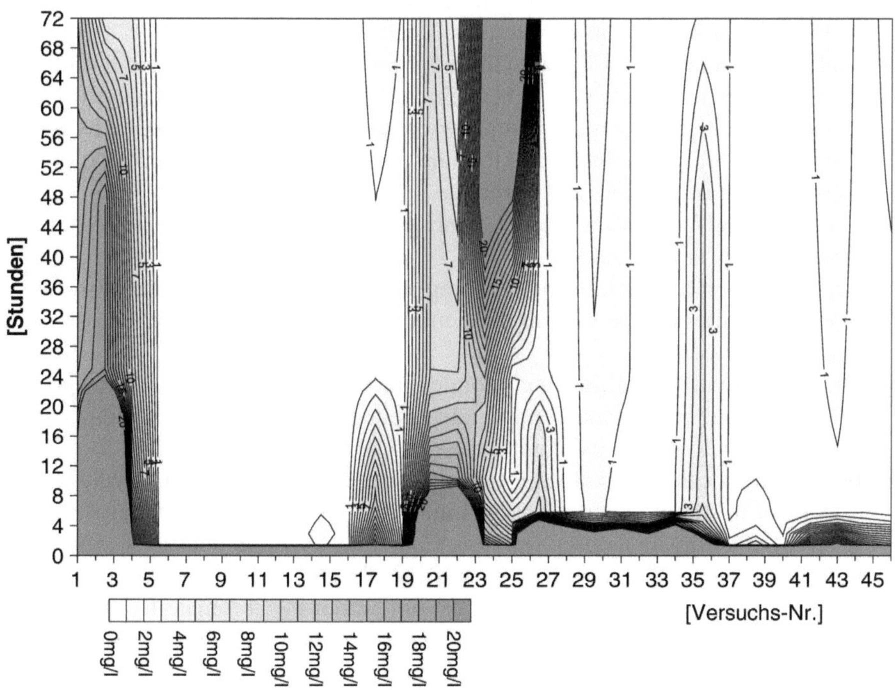

Fig. 5 Variation of different mixtures composed of mixtures of calcium aluminate hydrate, bentonite, zeolites, brown coal fly ash, aluminium hydroxide, kaolinite, and the varying chromate bonding capacity as a function of time, 0.01 M chromate solution

The fixation of chromate starting with initial 0.1 M concentrations in different reaction mixtures (compositions 1–46, see above) with increasing times is shown in Figs. 2 and 3.

The quality of immobilization can be clearly seen in Fig. 2, showing several mixtures with complete fixation of chromate in the interlayer of TCAH. Other mixtures also can be recognized, which are not usable for complete immobilization.

The fixation of chromate starting with initial 0.01 M concentrations in different reaction mixtures (compositions 1–46) with increasing times, up to 72 h is shown in Figs. 4 and 5. Due to the lower concentrations an increased variety of immobilization mixtures can be used and fix completely the total chromate from solution in the TCAH-solids (Göske and Pöllmann 1999a, b).

Summarizing the overall results on the different fixation possibilities of chromate the following statements can be made:

- For an optimized syntheses of storage minerals a complete and detailed knowledge of industrial raw materials is necessary to get best improved results.
- Characterisation and pre tests are necessary to get knowledge on optimal additions and improve capability of reaction mixtures.
- Quality control and continuous quality of educts is essential.
- Technical raw materials as mineral mixtures composed from $Ca(OH)_2$, $Al(OH)_3$ and $CaCO_3$ are easily usable and can be used for *in situ fixation of* chromate.
- Aluminium containing waste materials as $NaAl(OH)_4$ can be used in stochiometric mixtures with free lime CaO fort he syntheses of TCAH. The knowledge on Al-contents of solutions is necessary.
- Additional anions like sulphate (from brown coal fly ash) interfere with chromate fixation and can lead to other mineral formations like sulphate-ettringite, Monosulfate or other lamellar sulphate containing calcium aluminate hydrates. The predictability is reduced due to more complex reactions and competition of incorporation of sulphate and chromate.
- For chromate fixation the use of Gibbsite as additional pH-buffer system in reactive walls is favourable.
- Beside fixation by crystallochemical replacement in LDH additional adsorptive working zeolites can be used in these mixtures to increase possibilities of multi barrier systems.
- Sodium activated zeolites incorporate calcium ions (sodium exchange for calcium) resulting in some portlandite reduction. For later reaction stages, zeolites provide slowly calcium ions.
- For excellent calcium exchange capacity a low Si/Al-ratio, respectively a high al-content in structure is necessary (Linde A > clinoptilolite > faujasite).
- Kinetics of chromate fixation (mixture CaO + $C_{12}A_7$) is increasingly influenced with increasing calcium exchange capacity of added zeolite types.
- Alkaline activated bentonites do almost not hinder the chromate fixation.
- Also sodium activated bentonites can be used in these mixtures.
- The use of technical and natural china clays (Quartz content ca. 12%) does not influence the anion fixation up to 0.5 M ratios negatively.

- The additional offer of sodium activated bentonites, zeolites and china clays some enhanced sorption capacity is offered in the multi barrier systems with LDH´s as main component.
- Resorption processes in lamellar mineral reservoirs ($CO_3^{2-} \leftrightarrow CrO_4^{2-}$, $2OH^- \leftrightarrow CrO_4^{2-}$) under real application conditions cannot be excluded totally.
- The modelling of the different processes and immobilization products must be critically characterized and tested and verified by practical tests.
- The calculations and optimizations were tested and verified by practical tests and a confidence area of 90% of theoretical and practical results seems to be obtainable.

The determination of the different phases in the insoluble residues by X-ray diffraction gives indication of the formation of different chromate containing phases with compositions in the solid solution range of $3CaO.Al_2O_3.(x)Ca(OH)_2.(y)CaCrO_4.(z)CvaCO_3.nH_2O$.

Depending on the concentrations of chromate different amounts of chromate is incorporated in TCAH giving rise to solid solution series between MCr and TCAH (Fig. 6).

The examination of the storage minerals and reaction products in the residues by SEM show always arrangements of typical platelets of chromate-containing Afm-phases (Figs. 7 and 8).

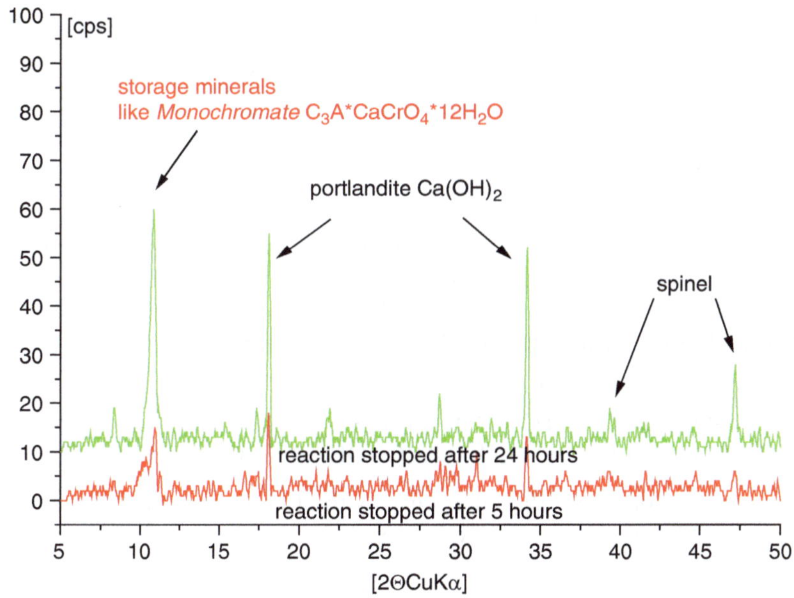

Fig. 6 XRD diagrams of residues after different times of reaction showing AFm-phases and relevant other phases of the used mixtures (Stöber et al. 1997)

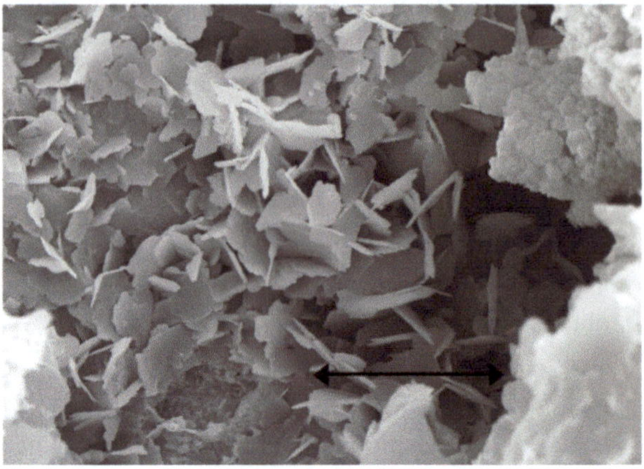

Fig. 7 Scanning electron microscopy (SEM) image of a residue, reaction stopped after 24 h, scale bar = 10 μm

Fig. 8 Cryo-SEM image, reaction time: 10 h/r.h, scale bar = 15 μm, type TCAH solid solution with partial incorporation of chromate

Following the reaction by using cryo-SEM shows, that very typical metal-metal-hydroxi salts form a 3-dimensional framework of platelets. Depending on the amount of chromate in the solid a more or less intensive yellow colour of the solid is obtained. The following different phases were described (Auer et al. 1991):

1. $3CaO.Al_2O_3.CaCrO_4.12H_2O$
2. $3CaO.Al_2O_3.1/2CaCrO_4 1/2Ca(OH)_2.12H_2O$
3. $3CaO.Al_2O_3.(0–0.17)CaCrO_4(1–0.83)Ca(OH)_2.12H_2O$ ($0 < x < 0.17$)

By using solutions from Bag filter dusts with of chromium (VI) concentrations of originally 70 mg/l, turned to decrease to less than 0.03 mg/l after conversion of TCAH into a lamellar storage mineral like monochromate solid solutions. Suitable procedures can be transferred from laboratory scale to the scale of technical industrial processing. TCAH can be synthesized through stochiometric compositions of various educts.

4 Discussion

It could be shown clearly that the use of lamellar calcium aluminate hydroxi salts for uptake of chromate is very useful and easy. To perform different industrial applicable mixtures even additions of other industrial raw materials, like china clay, bentonite, zeolites, Al-hydroxide and fly ashes produce interesting mixtures with variable uptake of chromate. The best, out of different mixtures, can reduce chromate in solution drastically. The overall reaction schemata can be followed as given in Fig. 9.

Depending on the used mixtures of lamellar metal metal hydroxysalts and accompanying materials and concentrations of chromate to be fixed, the final concentrations in solution are as low as <0.1 mg/l. These mixtures are favourable for a primary ion exchange in calcium aluminate hydroxi salts and fixation of chromate from solutions.

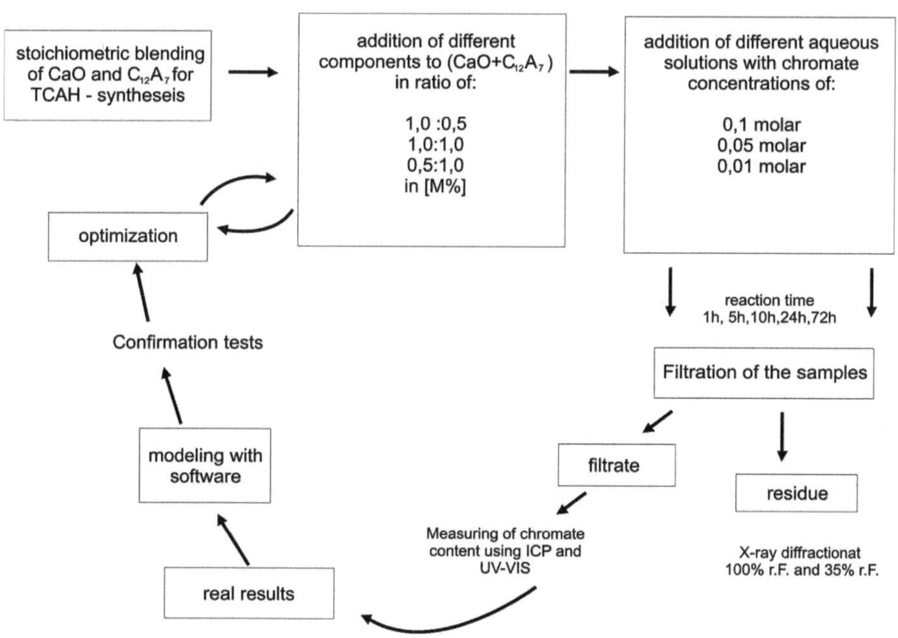

Fig. 9 Reaction schemata of chromate immobilisation

References

Allmann R (1970) $[Me(II)_{1-x}Me(III)_x(OH)_2]^{x+}$, Doppelschichtstrukturen mit brucitähnlichen Schichtionen. Chimia 24:99–108

Auer S (1992) Bindung umweltrelevanter Ionen in Ettringit und in Schichtstrukturen vom Typus TCAH. Dissertation, Universität Erlangen-Nürnberg

Auer S, Pöllmann H (1994) Synthesis and characterization of lamellar cadmium aluminium hydroxide salts with SO_4^{2-}, CO_3^{2-}, Cl^- and NO_3^-. J Solid State Chem 108:1–10

Auer S, Pöllmann H, Kuzel HJ (1990) Chromathaltige calciumaluminat-hydroxisalze. Eur J Mineral 2:7

Auer S, Pöllmann H, Kuzel HJ (1991) Kristallographie CrO_4^{2-}-haltiger Schichtstrukturen vom Typus TCAH unter besonderer Berücksichtigung von SO_4^{2-}, Cl^-, OH^- und CO_3^{2-}. Z Kristallogr 188:8

Göske J (1999) Die mineralogisch-chemische Barriere – Fixierung der Anionen SO_4^{2-}, Cl^-, NO_3^- and CrO_4^{2-} in Speichermineralen, unter dem Aspekt verschiedener Rohstoffmischungen, Abmischungen and Modellkonzentrationen. Hall Jb Geowiss, Bh 7, 134 S

Göske J, Pöllmann H (1996) Fixation of Cr (VI) in mineral reservoirs. Abstracts ICAM 15th general meeting, 2–5 Juni 1996, Warschau/Polen. S. 211

Göske J, Pöllmann H (1999a) Speicherminerale vom Typ TCAH – Einsatz in einem Multibarrierensystem. Abstr 4 Symp Techn Angew Mineral, KTM, Freiberg

Göske J, Pöllmann H (1999b) Immobilisierung von Chromat in TCAH, Optimierung des Prozesses mit Hilfe der statistischen Versuchsplanung. Eur J Mineral 11:86

Göske J, Prieto G, Pöllmann H (1996) Fixation of harmful substances in lamellar metal-metal-hydroxisalts. In: 2nd international symposium on environmental geochemistry in tropical countries, Cartagena/Colombia, p 50

Göske J, Stöber S, Pöllmann H (1997) Application of the TTK camera in studies of fixation of CrO_4^{2-} in calcium aluminate hydrates, type specimen TCAH. Paar Physica Application Note A31/A04-A

Göske J, Czurratis P, Risch A, Pöllmann H, Ecker M (1999) Use of EDX Detectors to study leaching behaviour of stabilised industrial waste materials. Microsc Anal 57:17–18

Göske J, Pöllmann H, Auer S (2003) Fixation of chromate in lamellar calcium aluminate hydrates (AFM-Phases) and its recycling process. ICCC, Durban

Pöllmann H (1993) Immobilisation von Schadstoffen durch Speichermineralbildung. Das Konzept der Inneren Barriere. Fachtagung Innsbruck-Igls Dez 1993, S. 51–59

Pöllmann H (1994) Immobile Fixierung von Schadstoffen in Speichermineralen. Geowissenschaften und Umwelt. Springer, Heidelberg, pp S 331–S 340

Pöllmann H (2010) Mineralization of Industrial wastes. Shaker-Vlg, Aachen / Germany, 440 p

Pöllmann H, Auer S (2010) Investigations of Cr^{6+}-containing AFM-phases in the system CaO-Al_2O_3-CrO_4^{2-}-H_2O. In preparation

Pöllmann H, Gebhard G., Göske J (1996) Immobilization of pollutants by mineral storage. Abstracts ICAM 15th general meeting, 2–5 Juni 1996, Warschau / Polen. S. 216

Renaudin G (1998) Christallochemie des aluminates calciques hydrates. Diss Nancy III pp 99–266

Stöber S, Göske J, Pöllmann H (1997) TTK low-temperature camera, investigations on lamellar calcium aluminate hydrates. Paar Physica Application Note A31/A03-A

Terzis A, Filippakis S, Kuzel HJ, Burzlaff H (1987) The crystal structure of $Ca_2Al(OH)_6Cl \cdot 2H_2O$. Z Kristallogr 181:29–34

Crystal Chemistry of Lamellar Calcium Aluminate Sulfonate Hydrates: Fixation of Aromatic Sulfonic Acid Anions

Stefan Stöber and Herbert Pöllmann

1 Introduction

In ordinary portland cements the calcium aluminate and calcium aluminate ferrate phase react with the sulphate phase (gypsum or different calcium sulphate hydrates) to ettringite (Aft-phases) and are transferred with additional C_3A, to lamellar calcium aluminate ferrate hydrates (Afm-phases). The interaction of admixtures with cement pastes does not only manipulate the properties of the cement paste, the crystal chemistry of the hydrated phases were influenced, too (Pöllmann 1989a; Kuzel 1969).

Lamellar calcium aluminate hydrates are built up of sequences of so called positively charged mainlayers $[Ca_2Al(OH)_6]^+$ and negatively charged interlayers $[X{\cdot}nH_2O]^-$ (Renaudin et al. 1999; Allmann 1968, 1977; Terzis et al. 1987). In the interlayer, water molecules and inorganic or organic single or double charged anions build up a framework. Under the influence of temperature, relative humidity and ion pressure, water molecules and interlayer anions can be easily exchanged. In order to identify newly formed lamellar calcium aluminate hydrate phases in cement pastes, the properties of these phases and the stability concerning temperature and relative humidity must be identified (Kuzel 1969).

2 Experimental

2.1 Synthesis

The synthesis of lamellar calcium aluminate sulfonate hydrates comprise the following process steps. In a first step, $CaCO_3$ was sintered at 1,000°C for 1 h in

S. Stöber (✉) • H. Pöllmann
Institute for Geosciences Mineralogy/Geochemistry, Martin-Luther University Halle, Von Seckendorff - Platz 3, 06120 Halle (Saale), Germany
e-mail: stefan.stoeber@geo.uni-halle.de

S.V. Krivovichev (ed.), *Minerals as Advanced Materials II*,
DOI 10.1007/978-3-642-20018-2_11,

order to obtain pure CaO. CA ($CaO{\cdot}Al_2O_3$) was prepared by homogenising CaO and Al_2O_3 in a laboratory mill for 5 min and sintered at 1,250°C for 2 days with several meantime millings.

In a second step calcium salts of particularly substituted benzenesulfonic acids were prepared by the reaction of reagent grade $CaCO_3$ and different benzenesulfonic acids shown in Tables 1 and 2 (Stöber and Pöllmann 1999). Stoichiometric concentrations of the benzenesulfonic acids were added drop by drop to a strongly diluted $CaCO_3$ slurry which was stirred until the solutions became translucent. After storing the solutions in a heating cabinet at 30°C for 1 week, transparent compounds were obtained. The salts were checked by X-ray powder diffraction (XRPD), the crystal water concentration was determined by thermal gravimetry (TG). The compounds were stored in a refrigerator in order to prevent dehydration.

In a final step, carbonate free water was prepared by cooking deionized water for half an hour with is essential to prevent the crystallisation of lamellar phases like $C_3A{\cdot}0.5Ca(OH)_2{\cdot}0.5CaCO_3{\cdot}11.5H_2O$ and $C_3A{\cdot}CaCO_3{\cdot}11H_2O$. Because the synthesis of any Aluminous ferrous monophase (Afm-phase) is sensitive to CO_2, present as CO_3^{2-} - ions in the paste, the whole synthesis was carried out under the exclusion of CO_2. The assembly of raw materials (CA, calcium salts of particularly substituted benzenesulfonic acids, CaO and H_2O) and filtration of the obtained precipitate after the end of the reaction was done in a glove box under N_2-atmosphere. Furthermore the samples were stored in a box filled with soda lime in order to prevent carbonation during the reaction time.

Table 1 Different applied benzenesulfonic acids

Molecular formula	Trivial name	IUPAC name
$C_7H_7SO_3H$	p-Toluenesulfonic acid	4-Methylbenzenesulfonic acid
$C_8H_9SO_3H$	Xylenesulfonic acid	2,4-Dimethylbenzenesulfonic acid
$C_9H_{11}SO_3H$	Mesitylenesulfonic acid	2,4,6-Trimethylbenzenesulfonic acid

Table 2 Synthesis of different calcium benzenesulfonate hydrates under excess of water

Decarbonation process
$CaCO_3 + 2C_7H_7SO_3H + x\ H_2O \rightarrow Ca^{2+} + 2C_7H_7SO_3^- + xH_2O + CO_2$
$CaCO_3 + 2C_8H_9SO_3H + xH_2O \rightarrow Ca^2 + 2C_7H_7SO_3^- + xH_2O + CO_2$
$CaCO_3 + 2C_9H_{11}SO_3H + xH_2O \rightarrow Ca^2 + 2C_9H_{11}SO_3\text{-} + xH_2O + CO_2$
Evaporation and crystallisation of different calcium benzenesulfonate hydrates
$Ca^{2+} + 2C_7H_7SO_3^- + xH_2O \rightarrow Ca(C_7H_7SO_3)_2{\cdot}2H_2O + xH_2O$
$Ca^{2+} + 2C_7H_7SO_3^- + xH_2O \rightarrow Ca(C_8H_9SO_3)_2{\cdot}4H_2O + xH_2O$
$Ca^{2+} + 2C_9H_{11}SO_3^- + xH_2O\ Ca(C_9H_{11}SO_3)_2{\cdot}4H_2O + xH_2O$

Table 3 Synthesis of different Afm - phases by paste reaction

Eq. 1	$CA + 2CaO + Ca(C_7H_7SO_3)_2 + xH_2O \rightarrow C_3A{\cdot}Ca(C_7H_7SO_3)_2{\cdot}nH_2O + xH_2O$
Eq. 2	$CA + 2CaO + Ca(C_8H_9SO_3)_2 + xH_2O \rightarrow C_3A{\cdot}Ca(C_8H_9SO_3)_2{\cdot}nH_2O + xH_2O$
Eq. 3	$CA + 2CaO + Ca(C_9H_{11}SO_3)_2 + xH_2O \rightarrow C_3A{\cdot}Ca(C_9H_{11}SO_3)_2{\cdot}nH_2O + xH_2O$

Stoichiometric concentrations of CA, CaO and the calcium salts of particularly substituted benzenesulfonic acids were filled into polyethylene bottles according to the molar concentrations shown in Table 3. The raw materials were mixed with deionized and decarbonised water at a water – solid ratio (w/s) of 10. The samples were stored on a shaking tool in order to obtain better homogenization. The paste reacted for 4 months at a temperature of 25°C in order to prevent the formation of hydrogrossulare C_3AH_6. Furthermore the samples were stored in a box filled with soda lime in order to prevent carbonation during the reaction time.

Binary systems with the general formula $C_3A{\cdot}Ca(X)_2{\cdot}nH_2O - C_3A{\cdot}Ca(Y)_2{\cdot}nH_2O$ with $X = C_7H_7SO_3$, $C_8H_9SO_3$ and $C_9H_{11}SO_3$; $Y = OH$ were investigated. The reactants were mixed in accordance to the equation $C_3A{\cdot}(1-x)Ca(X)_2{\cdot}xCa(Y)_2{\cdot}n\,H_2O$ $0 \leq x \leq 1$ and treated during the synthesis according to the synthesis of pure Afm-phases.

2.2 *Instruments*

X-ray powder diffraction data were collected with a D 5000 diffractometer (Siemens) equipped with the low temperature camera attachment TDK 450 (Anton Paar) at 100, 94, 85, 75, 53% relative humidity (r.h) and under a nitrogen atmosphere. The stable compounds at 35% r.h. were investigated by means of a D 5000 diffractometer (Siemens) equipped with a flat sample holder. During data collection, the samples were exposed to the ambient atmosphere without any protection. The samples were dried in a box under a CO_2-free atmosphere over saturated salt solutions for 3 weeks. Particle size, habit and surface properties of different $[Ca_2Al(OH)_6]^+[X{\cdot}nH_2O]^-$ (X = sulfonic anion) crystals were studied by SEM (scanning electron microscope) of JEOL in combination with the EDX (electron diffraction X-ray system) for semi quantitative analysis. The Ca and Al concentration of the precipitates were measured by ICP-OES. C and S concentrations were analysed quantitatively by direct element analysis (TOC). The H_2O concentration of the main layer $[Ca_2Al(OH)_6]^+$ and inter layer $[X{\cdot}nH_2O]^-$ were determined by Karl – Fischer Titration and by thermal analysis (TG/DTG-DTA) of SEIKO INSTRUMENTS. The dehydration reactions were determined by differential scanning calorimetry (DSC) on a SEIKO DSC 220. The Fourier transformed IR-spectrometer BRUKER was used to determine the spectrum in the area of 4,000–400 cm^{-1} using KBr in order to detect contamination of CO_2.

3 Results

3.1 Calcium Aluminate p-Toluenesulfonate Hydrate $[Ca_2Al(OH)_6]^+[C_7H_7SO_3{\cdot}nH_2O]^-$

After a reaction time of 4 months $[Ca_2Al(OH)_6]^+$ $[C_7H_7SO_3{\cdot}nH_2O]^-$ crystallized as thin platy hexagonal crystals with a diameter of 10 μm and a thickness of approximately 1 μm in a mother solution at a measured pH – value of 12–13. Calcium aluminate p-toluenesulfonate hydrate was synthesized purely, which was proofed by PXRD. After removing the precipitate from the polyethylen bottles PXRD analysis were conducted at 100% r.h. in order to obtain data for indexing the powder pattern and refining the lattice parameters. Silicon was added to the pastes for the correction of sample height errors. After drying the sample under an inert atmosphere, PXRD-data of this compound was collected on a D5000 diffractometer at 35% r.h. The chemical formula of the precipitate dried at 35% r.h. is shown in Table 4.

The unit-cell parameters were refined by least-squares methods at 100% and 35% r.h. Calcium aluminate p-toluenesulfonate hydrate crystallises in the hexagonal crystal system (Stöber and Pöllmann 1999). In the primitive centred unit cell a single sequence of "mainlayer – interlayer – layer" perpendicular c_o exists. After drying at 35% r.h. the layer distances c´ at 100% r.h. compared to the layer distances c´ at 35% r.h. shows minor differences (Table 5). It was not possible to determine the lattice parameter a_o at 100% r.h. because of a strong texture was due the alignment of the platy shaped crystals in the sample holder. Although a special side-loading sample carrier was used the preferred orientation could not be reduced sufficiently, thus no h0l reflections were detected in the powder pattern.

The dehydration of $[Ca_2Al(OH)_6]^+[C_7H_7SO_3{\cdot}5H_2O]^-$ in combination with the layer distance variation is illustrated in Table 6. With the rise in temperature $[Ca_2Al(OH)_6]^+[C_7H_7SO_3{\cdot}5H_2O]^-$ three losses in weight at 32°C, 58°C and at 128°C were detected in the temperature – loss of weight plot of the thermogravimetric analysis (TGA) (Fig. 1). In combination with a Karl–Fischer–Titrator equipped with a furnace the compound was heat treated at the corresponding three temperatures

Table 4 Chemical composition of $[Ca_2Al(OH)_6]^+[C_7H_7SO_3{\cdot}5H_2O]^-$ at 35% r.h. (in wt. %)

Chemical formula	CaO	Al_2O_3	H_2O	$C_7H_7SO_3^-$
$[Ca_2Al(OH)_6]^+[C_7H_7SO_3{\cdot}5H_2O]^-$	22.2	10.9	29.2	36.9

Table 5 Lattice parameters of $[Ca_2Al(OH)_6]^+[C_7H_7SO_3{\cdot}nH_2O]^-$ at 100% r.h. and 35% r.h.

r.h. [%]	a_o [nm]	b_o [nm]	c_o [nm]	c´ [nm]	α	β	γ
100	–	–	1.7177	1.7177	90.0	90.0	120
35	0.5779	0.5779	1.7202	1.7202	90.0	90.0	120

Table 6 Dehydration process of $[Ca_2Al(OH)_6]^+[C_7H_7SO_3{\cdot}nH_2O]^-$

T [°C]	Mass loss [%] (in air)	$[(C_7H_7SO_3)_2{\cdot}nH_2O]^{2-}$	c′ [nm]	H_2O [mole]
25	–	$[(C_7H_7SO_3)_2{\cdot}9H_2O]^{2-}$	1.7202	15
32	6.8	$[(C_7H_7SO_3)_2{\cdot}5.5H_2O]^{2-}$	1.6223	11.5
58	11.3	$[(C_7H_7SO_3)_2{\cdot}3H_2O]^{2-}$	1.4973	9
128	17.6	$[(C_7H_7SO_3)_2]^{2-}$	–	6
234	26.2	–	–	3

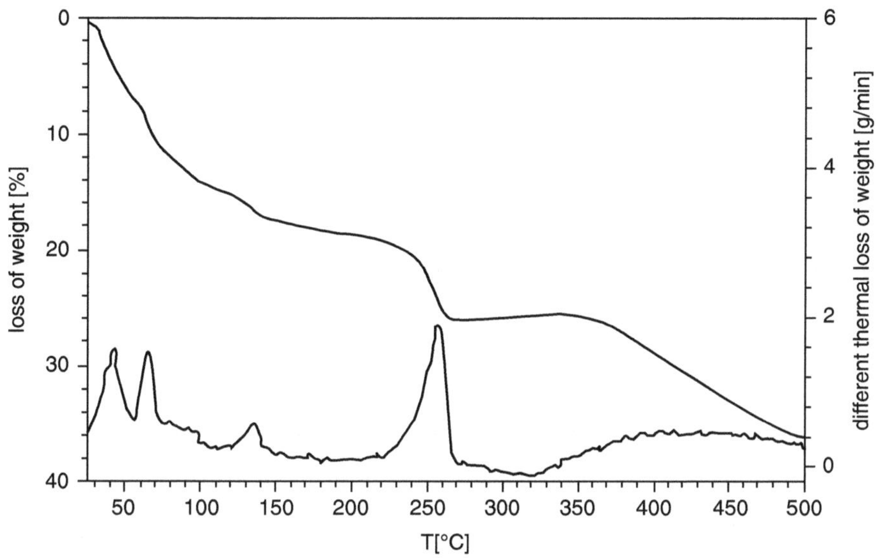

Fig. 1 Thermal analysis of $C_3A{\cdot}Ca(C_7H_7SO_3)_2{\cdot}15H_2O$

(32°C, 58°C and at 128°C) and the gas phase titrated. All three weight losses match hydration processes with water weight losses of 3.5, 2.5 and finally 5 mole H_2O. The dehydration of $[C_7H_7SO_3{\cdot}5H_2O]^-$ was finished at 128°C. The layer distance c′ decreased from the initial value at 1.702 nm to 1.493 nm at 58°C. Further c′ values could not be determined by lattice parameter refinement, because the structure became X-ray amorphous at 128°C due to two dimensional stacking faults and degradation of the layer modules $[Ca_2Al(OH)_6]^+$ and $[C_7H_7SO_3{\cdot}nH_2O]^-$ in the crystal structure. A further weight loss at approximately 235°C is equal to the dehydration step of the $[Ca_2Al(OH)_6]^+$-layer. In combination with PXRD at non ambient temperatures the water content of each hydration stage was determined by thermal analysis (Fig. 2). The dehydration of the $[Ca_2Al(OH)_6]^+$-layer is partly overlapped by the pyrolysis of the p-toluenesulfonic group at 350°C (Fig. 2).

In cement pastes not only pure salts with the chemical composition $[Ca_2Al(OH)_6]^+$ and $[C_7H_7SO_3{\cdot}nH_2O]^-$ occurs, because sulfate, chloride or hydroxide ions are present. As a first step the formation and stability of Calcium aluminate

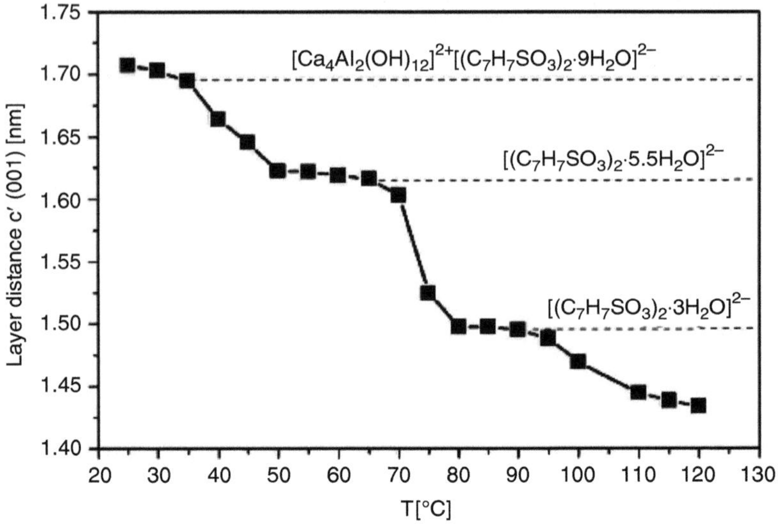

Fig. 2 Variation of the interlayer distance c´ of $C_3A{\cdot}Ca(C_7H_7SO_3)_2{\cdot}15H_2O$

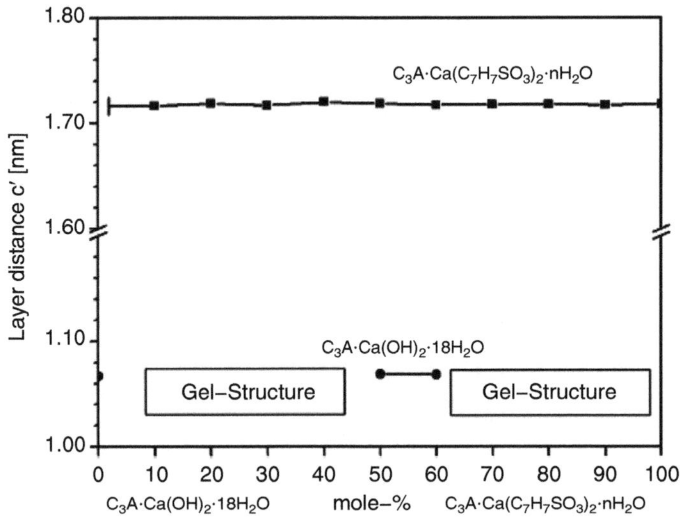

Fig. 3 Interlayer distance c´ in the system $C_3A{\cdot}Ca(OH)_2{\cdot}18H_2O$–$C_3A{\cdot}Ca(C_7H_7SO_3)_2{\cdot}nH_2O$ at 100% r.h.

p-toluenesulfonate hydroxy hydrate solid solutions were investigated. In the system $C_3A{\cdot}Ca(C_7H_7SO_3)_2{\cdot}nH_2O$–$C_3A{\cdot}Ca(OH)_2{\cdot}nH_2O$ (Fig. 3) the phase assemblages were analysed by PXRD. The evaluation of the patterns yielded in the range of 10–90 mole-% $C_3A{\cdot}Ca(OH)_2{\cdot}nH_2O$ the coexistence of two different Afm-phases. One compound was identified as the pure $[Ca_2Al(OH)_6]^+[C_7H_7SO_3{\cdot}nH_2O]^-$ salt because the refined layer distance c´ at 1.7177 nm is equivalent to the previously described Calcium aluminate p-toluenesulfonate hydrate synthesized as a pure phase. The other Afm-phase was identified as Calcium aluminate hydroxide hydrate in samples containing 50 and 60 mole-% $C_3A{\cdot}Ca(OH)_2{\cdot}nH_2O$. In all other precipitates gel phases were also detected in equilibrium with calcium aluminate p-toluenesulfonate hydrate. After drying and analysing the samples at 35% r.h. the phase assemblage in the system differed slightly form the phase assemblage at 100% r.h. just the gel phases recrystallized to Calcium aluminium hydroxy hydrate and Calcium aluminium p-toluenesulfonate hydrate.

3.2 Calcium Aluminate Xylenesulfonate Hydrate $[Ca_2Al(OH)_6]^+[C_8H_9SO_3{\cdot}nH_2O]^-$

After a reaction time of 3 months the precipitate of pure calcium aluminium xylenesulfonate hydrate $[Ca_2Al(OH)_6]^+[C_8H_9SO_3{\cdot}nH_2O]^-$ was filtered and dried to 35% r.h. SEM-analysis gave the proof that the precipitate consists of elongated platy crystals with an average diameter of 5–10 μm (Fig. 4). The chemical analysis of the precipitates dried to 35% r.h. is shown in Table 7. On the basis of the analysed concentrations the chemical formula of calcium aluminate xylenesulfonate hydrate was determined as $[Ca_2Al(OH)_6]^+[C_8H_9SO_3{\cdot}4H_2O]^-$.

Fig. 4 SEM image of crystals of $C_3A{\cdot}Ca(C_8H_9SO_3)_2{\cdot}14H_2O$

Table 7 Chemical composition $C_3A{\cdot}Ca(C_8H_9SO_3)_2{\cdot}14H_2O$ (in wt. %)

Chemistry	CaO	Al_2O_3	$C_8H_9SO_3$	H_2O	Total
Theoretical	22.3	10.9	39.8	27.0	100.0
Measured	21.8	10.9	40.2	27.0	99.9

At 100% and 35% r.h. PXRD measurements were performed with comparable instrumental settings used for the data collection of $[Ca_2Al(OH)_6]^+[C_7H_7SO_3{\cdot}nH_2O]^-$. Unlike the indexing and lattice parameter refinement by least-squares methods of $[Ca_2Al(OH)_6]^+[C_7H_7SO_3{\cdot}nH_2O]^-$, the crystal system of $C_3A{\cdot}Ca(C_8H_9SO_3)_2{\cdot}nH_2O$ was identified as trigonal with an *R*-centered unit cell at both relative humidities. The crystal structure of this compound is comparable with the structure of $3CaO{\cdot}Al_2O_3{\cdot}CaSO_4{\cdot}12H_2O$ (Allmann 1968).

The structure with space group *R*-3 contains three sequences of $[Ca_2Al(OH)_6]^+$ and $[1/2\ SO_4{\cdot}3H_2O]^-$ modules which are aligned perpendicular [001] as one translation period. The refinement of the metric parameters at both humidities transformed on a hexagonal base can be seen in Table 8.

Further least squares refinements of the metric parameters at 94, 86, 76 and 55% r.h. (Table 9) demonstrated that the layer distances keep their distance at 1.6298 nm, even the compound is exposed to different humidities for 3 months. Thermogravimetric analysis in combination with Karl–Fischer titration experiments gave the proof that the decrease of the relative humidity between 94% and 35% has no influence on the water concentration in $[C_8H_9SO_3 \cdot nH_2O]^-$. However, the majority of the investigated Afm-phases (Kuzel 1969; Pöllmann and Stöber 1997; Pöllmann 1989b) loose interlayer water during the dehydration process between 100% and 35% r.h.

With the rise of temperature $[Ca_2Al(OH)_6]^+[C_8H_9SO_3{\cdot}4H_2O]^-$ dehydrates immediately to $[C_8H_9SO_3{\cdot}2.25\ H_2O]^-$ at 34°C. The TGA graphs showed a weight loss of 5.7 w.-% which is equal to 1.75 mole H_2O. Further dehydration reactions take place at 66°C, 86°C and 161°C. The water concentration decreased to 2, 1.25 and respectively 0 mole H_2O. DSC-analysis support the results obtained by TGA analysis and demonstrated that all dehydration reactions between 34°C and 161°C belong to endothermal reactions.

The complete destruction of the compound was detected at 260°C. Three mole water were dehydrated. In Fig. 5, the dehydration process can not be

Table 8 Lattice parameters of $C_3A{\cdot}Ca(C_8H_9SO_3)_2{\cdot}\ nH_2O$

Synthesis	a_o [nm]	c_o [nm]	c´ [nm]
Paste 100% r.h.	–	9.7789	1.6298
Paste 35% r.h.	0,5797	9,7986	1.6331

Table 9 Lattice parameters of $C_3A{\cdot}Ca(C_8H_9SO_3)_2{\cdot}\ nH_2O$ at different relative humidities

Synthesis	a_o [nm]	c_o [nm]	c´ [nm]
Paste at 94% r.h.	–	9.7788	1.6298
Paste at 86% r.h.	–	9.7788	1.6298
Paste at 76% r.h.	–	9.7788	1.6298
Paste at 55% r.h.	–	9.7788	1.6298

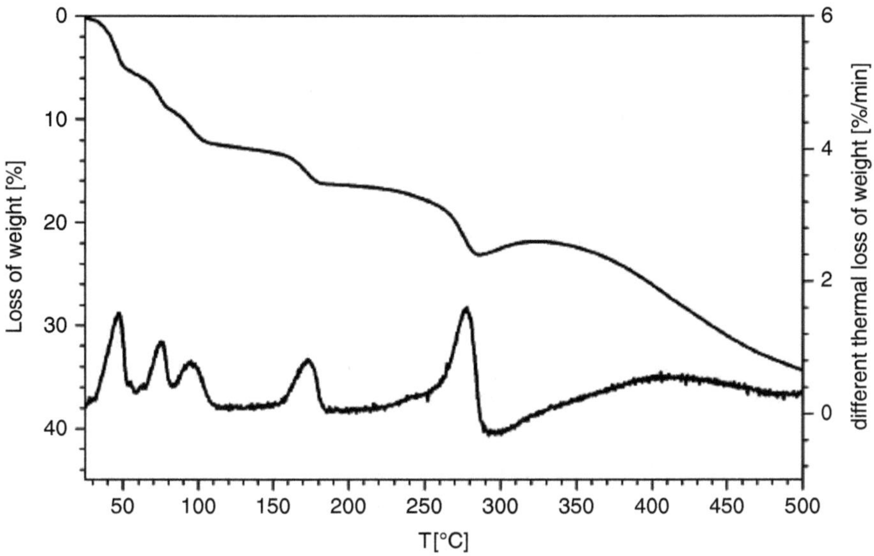

Fig. 5 TGA curve of $[Ca_4Al_2(OH)_{12}]^{2+}$ $[(C_8H_9SO_3)_2{\cdot}8H_2O]^{2-}$

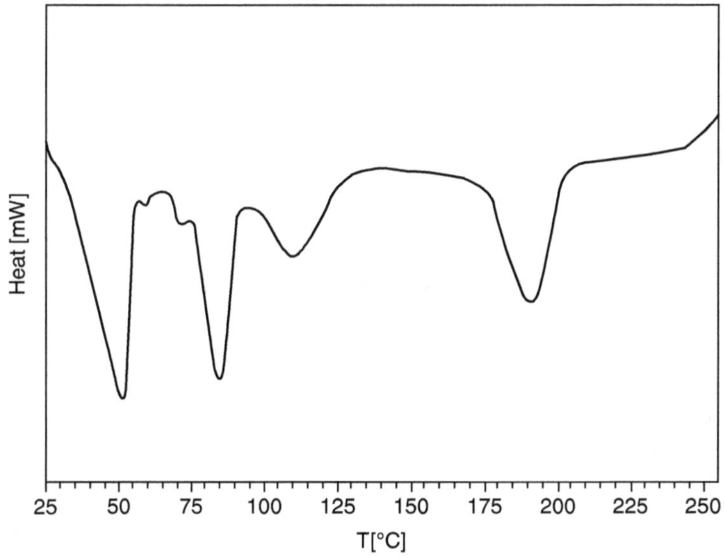

Fig. 6 DSC curve of $C_3A{\cdot}Ca(C_8H_9SO_3)_2{\cdot}14H_2O$ at 35% r.h.

determined in detail, because the TGA/DTA graph showed a slight increase of the weight in the range of 300–350°C again. This is due the oxidation of the SO_3-group. In the DSC-graph (Fig. 6) the endothermic peak of the dehydration reaction $[Ca_2Al(OH)_6]^+$ is overlapped by the heat generation of the exothermal organic pyrolysis.

According to the measurements with DSC and TGA/DTA, the layer distance c´ is direct proportional to the dehydration processes (Fig. 7). Each shrinkage step of the layer distance c is a result of the dehydration reactions detected by DSC. In compliance with the temperature driven shrinkage of the layer distance c´ of compound $[Ca_2Al(OH)_6]^+[C_7H_7SO_3{\cdot}nH_2O]^-$ the lattice parameters at evolved temperature could not be refined because the structure becomes amorphous. The values of the layer distances c´determined on the basis of the lattice parameter c are listed in Table 10 until 160°C.

The binary system $[Ca_2Al(OH)_6]^+[C_8H_9SO_3{\cdot}nH_2O]^-$ –$[Ca_2Al(OH)_6]^+[OH{\cdot}nH_2O]^-$ (Fig. 8) was investigated in the range of 0–100 mol% $C_8H_9SO_3^-$ with $C_3A{\cdot}xCa(OH)_2{\cdot}(1-x)Ca(C_8H_9SO_3)_2{\cdot}nH_2O$ $0 \le x \le 100$, ($x = 0.1$) In all samples with the initial chemical compositions $C_3A{\cdot}xCa(OH)_2{\cdot}(1-x)Ca(C_8H_9SO_3)_2{\cdot}nH_2O$ $0 \le x \le 90$ $[Ca_2Al(OH)_6]^+[C_8H_9SO_3{\cdot}nH_2O]^-$ coexists with an amorphous gel phase. Neither an intermediate phase nor the formation of solid solutions were detected. At 35% r.h. the gel phase was transformed into Calcium aluminate hydroxy hydrate and Calcium aluminate xylenesulfonate hydrate. The phase assemblages with initial chemical compositions $C_3A{\cdot}xCa(OH)_2{\cdot}(1-x)Ca(C_8H_9SO_3)_2{\cdot}nH_2O$ $0 \le x \le 90$ were identified as a mixture of $[Ca_2Al(OH)_6]^+[C_8H_9SO_3{\cdot}4H_2O]^-$ and $[Ca_2Al(OH)_6]^+[OH{\cdot}3H_2O]^-$.

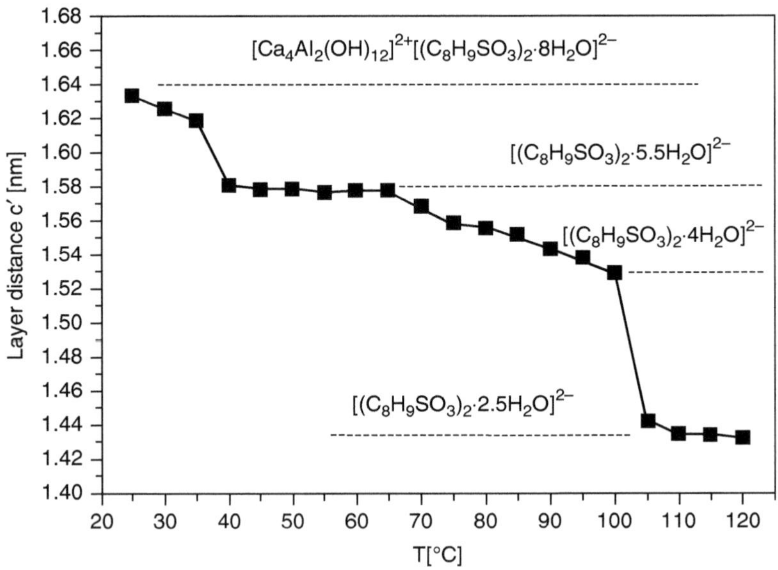

Fig. 7 Decrease of the layer distance shrinkage c´ as a function of the temperature

Table 10 Overview on the dehydration of $C_3A{\cdot}Ca(C_8H_9SO_3)_2{\cdot}14H_2O$ at 35% r.h.

T [°C]	Weight loss [%]	Layer distance c′ [nm]	$[(C_8H_9SO_3)_2{\cdot}nH_2O]^{2-}$	H_2O [mole]
25	–	1.633	$[(C_8H_9SO_3)_2{\cdot}8H_2O]^{2-}$	14
34	5.7	1.577	$[(C_8H_9SO_3)_2{\cdot}5.5H_2O]^{2-}$	11.5
66	8.8	1.565	$[(C_8H_9SO_3)_2{\cdot}4H_2O]^{2-}$	10
86	12.7	1.434	$[(C_8H_9SO_3)_2{\cdot}2.5H_2O]^{2-}$	8.5
161	16.0	–	$[(C_8H_9SO_3)_2{\cdot}0H_2O]^{2-}$	6
259	23.1	–	–	1.5

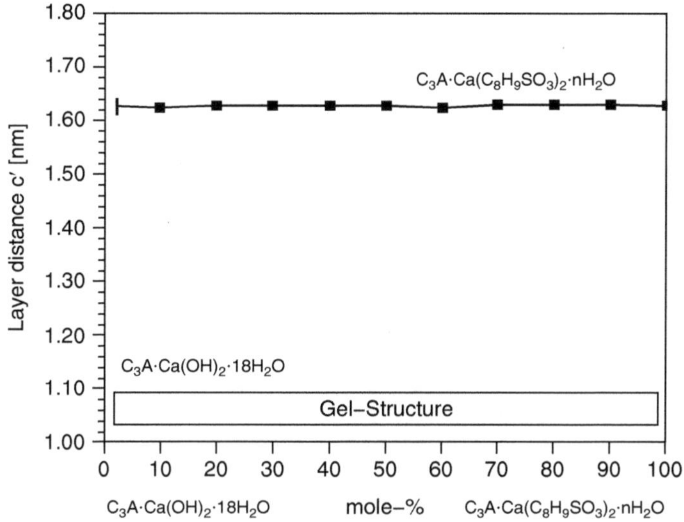

Fig. 8 Layer distances c′ in the system $C_3A{\cdot}Ca(OH)_2{\cdot}18H_2O$–$C_3A{\cdot}Ca(C_8H_9SO_3)_2{\cdot}$ nH_2O 100% r.h.

3.3 *Calcium Aluminate Mesitylenesulfonate Hydrate $[Ca_2Al(OH)_6]^+[C_9H_{11}SO_3{\cdot}nH_2O]^-$*

After a reaction time of 3 months platy, idiomorphic crystals were observed in the precipitate investigated by SEM. Immediate PXRD analysis were conducted at 100% r.h., just after filtration of the precipitate in the glove-box. PXRD gave the proof that Calcium aluminium mesitylenesulfonate hydrate $[Ca_2Al(OH)_6]^+$ $[C_9H_{11}SO_3{\cdot}nH_2O]^-$ was synthesised without additional minor phases. After drying the salt from 100% to 35% r.h. the chemical composition of Calcium aluminium mesitylenesulfonate hydrate was determined for the chemical composition $[Ca_2Al(OH)_6]^+$ $[C_9H_{11}SO_3{\cdot}4H_2O]^-$ 35% r.h. (Table 11).

Additional X-ray measurements at 35% r.h. were carried out and the lattice constants were refined (Table 12). $[Ca_2Al(OH)_6]^+[C_9H_{11}SO_3{\cdot}nH_2O]^-$

Table 11 Chemical analysis of $C_3A{\cdot}Ca(C_9H_{11}SO_3)_2{\cdot}15H_2O$ at 35% r.h. (wt. %)

Chemistry	CaO	Al_2O_3	$C_9H_{11}SO_3$	H_2O	Total
Theoretical	21.2	10.4	40.8	27.6	100.0
Measured	21.6	10.5	41.0	27.6	100.7

Table 12 Lattice parameters of $C_3A{\cdot}Ca(C_9H_{11}SO_3)_2{\cdot}15H_2O$

Relative humidity (%)	a_o [nm]	b_o [nm]	c_o [nm]	c´ [nm]	β [°]
100	0.93555	1.1453	1.7555	1.7554	90.548
35	0.9783	1.1369	1.8038	1.7544	103.5

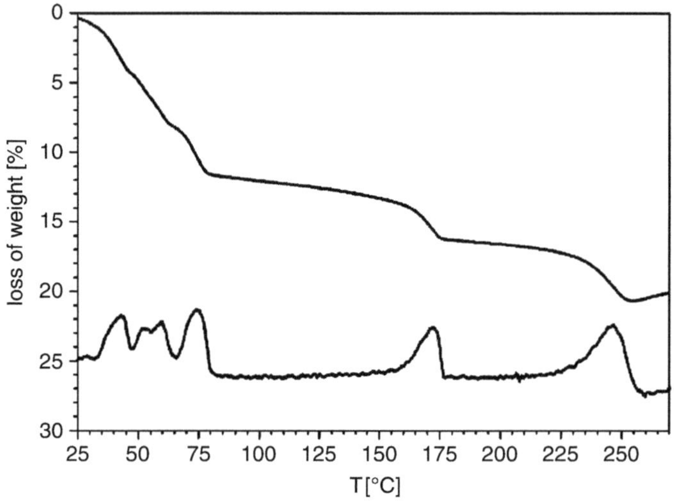

Fig. 9 Thermal gravimetry of $C_3A{\cdot}Ca(C_9H_{11}SO_3)_2{\cdot}15H_2O$

crystallizes in opposition to $[Ca_2Al(OH)_6]^+[C_7H_7SO_3{\cdot}nH_2O]^-$ and $[Ca_2Al(OH)_6]^+[C_8H_9SO_3{\cdot}nH_2O]^-$ in a monoclinic primitive cell at 100 and 35% r.h. In c-direction one sequence of $[Ca_2Al(OH)_6]^+[C_9H_{11}SO_3{\cdot}nH_2O]^-$ per unit cell was determined. A crystal structure for monoclinic Afm-phases has been determined by Terzis et al. (1987). $3CaO{\cdot}Al_2O_3{\cdot}CaCl_2{\cdot}10H_2O$ (Friedels salt) with space group *C2/c*, which undergoes a phase transformation from monoclinic to trigonal symmetry with space group *R*-3*c*.

The thermal stability of the salt was investigated by TGA (Fig. 9), DSC (Fig. 10), Karl–Fischer titration and additional non-ambient PXRD measurements (Fig. 11). It was proofed that, $[(Ca_2Al(OH)_6)]^{2+}[(C_9H_{11}SO_3)_2{\cdot}9H_2O]^{2-}$ has different hydration levels. In the temperature range of 35–68°C 7 mole H_2O were removed from the inter layer $[(C_9H_{11}SO_3)_2{\cdot}9H_2O]^{2-}$. DSC measurements followed the dehydration process from the initial level $[(C_9H_{11}SO_3)_2{\cdot}9H_2O]^{2-}$, $[(C_9H_{11}SO_3)_2{\cdot}6.5H_2O]^{2-}$, $[(C_9H_{11}SO_3)_2{\cdot}4.5H_2O]^{2-}$ to $[(C_9H_{11}SO_3)_2{\cdot}2H_2O]^{2-}$. Four endothermal reactions, with slight overlapping were identified, whereas the final and stronger bonded water molecules in the interlayer dehydrated at 160°C (Table 13).

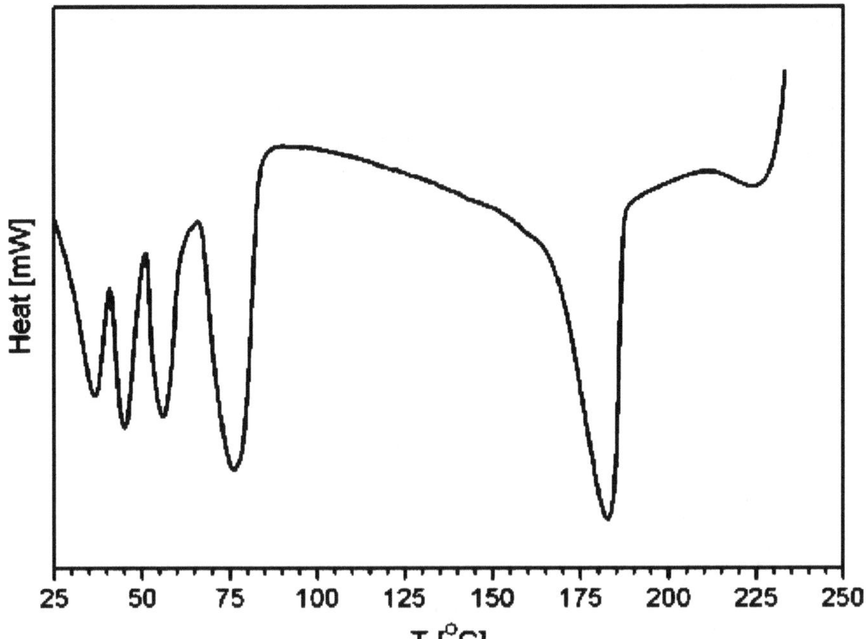

Fig. 10 DSC analysis of $C_3A{\cdot}Ca(C_9H_{11}SO_3)_2{\cdot}15H_2O$

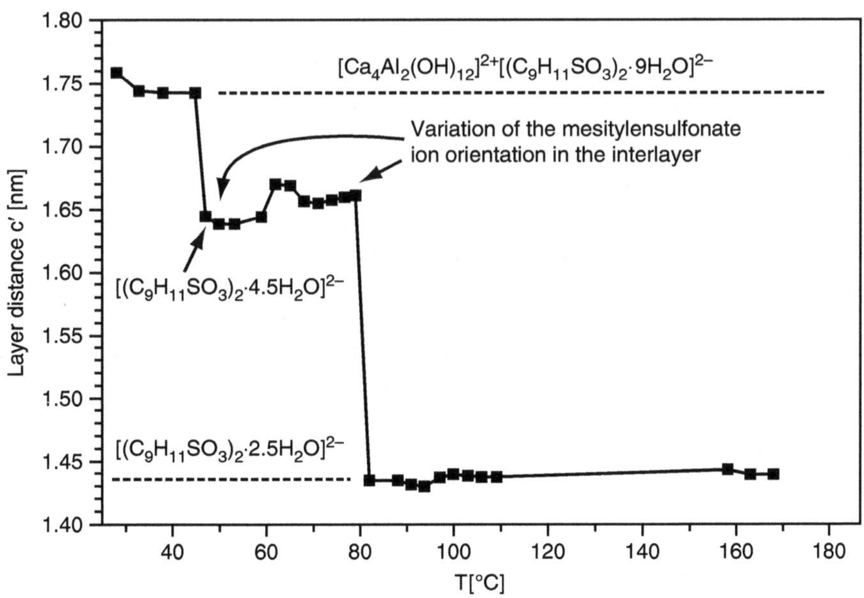

Fig. 11 Variation of the layer distance of $C_3A{\cdot}Ca(C_9H_{11}SO_3)_2{\cdot}15H_2O$

Table 13 Variation of the c´ parameter and water concentration as a result of the weight loss of phase $C_3A{\cdot}Ca(C_9H_{11}SO_3)_2{\cdot}15H_2O$

T [°C]	Weight loss [%]	c´ [nm]	$[(C_9H_{11}SO_3)_2{\cdot}nH_2O]^{2-}$	H_2O [mole]
25	–	1.7544	$[(C_9H_{11}SO_3)_2{\cdot}9H_2O]^{2-}$	15
35	4.4	–	$[(C_9H_{11}SO_3)_2{\cdot}6.5H_2O]^{2-}$	12.5
47/54	8.3	1.6382	$[(C_9H_{11}SO_3)_2{\cdot}4.5H_2O]^{2-}$	10
68	12.9	1.4352	$[(C_9H_{11}SO_3)_2{\cdot}2H_2O]^{2-}$	8
160	16.6	–	$[(C_9H_{11}SO_3)_2{\cdot}0H_2O]^{2-}$	6
227	20.7	–	–	4

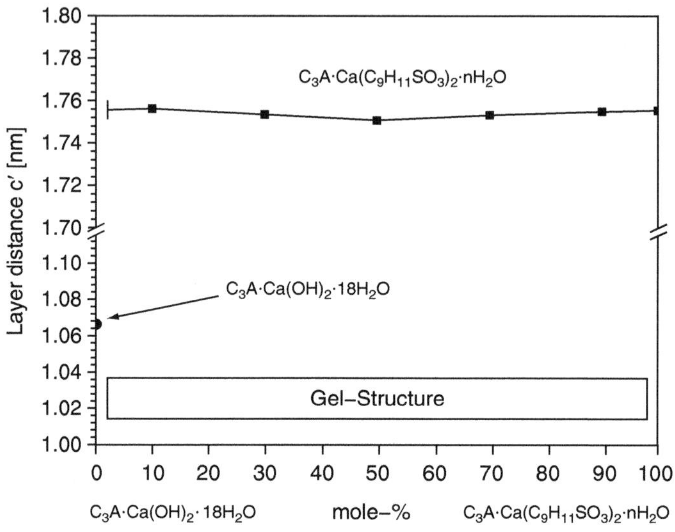

Fig. 12 Layer distances c´ in the system $C_3A{\cdot}Ca(OH)_2{\cdot}18H_2O$–$C_3A{\cdot}Ca(C_9H_{11}SO_3)_2{\cdot}nH_2O$ 100% r.h.

The variation of the interlayer shrinks at 50°C from 1.7544 to 1.6382 nm. Instead of a decreased interlayer dimension, the c´-value increased because of a variable orientation of the mesitylenesulfonic acid ion in the inter layer. The dehydration reaction $[(C_9H_{11}SO_3)_2{\cdot}6.5H_2O]^{2-}$ –> $[(C_9H_{11}SO_3)_2{\cdot}2H_2O]^{2-}$ at 68°C resulted in a strong decrease of c´ from 1.6382 to 1.4352 nm. The variation of c ´as a function of the temperature was finally refined at 170°C, because the thermal process reduced the crystallinity until the structure became amorphous. At least, the 6 mole H_2O of the main layer $[(Ca_2Al(OH)_6)]^{2+}$ were removed from the structure at 227°C in good agreement with experimental findings of the TGA of $C_3A{\cdot}Ca(C_8H_9SO_3)_2{\cdot}14H_2O$.

In the system $[Ca_2Al(OH)_6]^+[C_9H_{11}SO_3{\cdot}nH_2O]^-$ –$[Ca_2Al(OH)_6]^+[OH{\cdot}nH_2O]^-$ (Fig. 12) the stable phases were $[Ca_2Al(OH)_6]^+[C_9H_{11}SO_3{\cdot}nH_2O]^-$ and some gel phases at 100% r.h. and at 35% $[Ca_2Al(OH)_6]^+[C_9H_{11}SO_3{\cdot}4.5H_2O]^-$ and $[Ca_2Al(OH)_6]^+[OH{\cdot}3H_2O]^-$ under equilibrium. No solid solutions or intermediate phases were detected. On the basis of the general formula $C_3A{\cdot}xCa(C_7H_7SO_3)_2(1-x)$

$Ca(OH)_2 \cdot nH_2O$ $0 \leq x \leq 1$ with x = [0.1, 0.2, 0.3,...,1], the concentration of the crystallized lamellar Afm-phase can be determined. Gel formation was determined in the binary systems at 100% r.h. stable together with calcium aluminate mesitylenesulfonate hydrate. As a result of the drying the gel phase recrystallised as calcium aluminate hydroxide hydrate and calcium aluminate mesitylenesulfonate hydrate.

4 Discussion

Three different salts $[Ca_2Al(OH)_6]^+$ $[C_7H_7SO_3 \cdot 4{,}5H_2O]^-$, $[Ca_2Al(OH)_6]^+$ $[C_8H_9SO_3 \cdot 4H_2O]^-$, and $[Ca_2Al(OH)_6]^+$ $[C_9H_{11}SO_3 \cdot 4{,}5H_2O]^-$ were synthesised applying equilibrium paste reactions. The formation of minor phases like portlandite or ettringite were not identified by PXRD and TGA/DTA. Metric parameters were determined at different relative humidities. Due to the fixation of different substituted benzenesulfonic acid anions (toluenesulfonic-, xylenesulfonic- and mesitylenesulfonic ions) in the lamellar structure, the phases crystallizes hexagonal, trigonal and monoclinic symmetry with a one, or six sequence $[Ca_2Al(OH)_6]^+$ $[X \cdot nH_2O]^-$ X = $C_7H_7SO_3^-$, $C_8H_9SO_3^-$ and $C_9H_{11}SO_3-$ per unit cell in c_o direction. Although the sample preparation was improved, applying side-loading techniques, the texture effect based on the preferred orientation of the platy crystals, parallel (00l) caused a lack of reflections indexed with different miller indices. Therefore the lattice constants, especially at high relative humidities were not refined. The thermal resistance of the salts were tested by TGA/DTA measurements. For each salt the different dehydration reactions were identified. It could be clearly pointed out, that water molecules fixed in the interlayer were dehydrated from the structure between 30°C and 200°C. All three lamellar Calcium Aluminate Hydrates lost their interlayer water not continuously, but at certain temperature specific levels. As a result of interlayer dehydration reactions the layer distance variation was identified by PXRD at non ambient temperatures. The variation of the layer distance of calcium aluminate xylenesulfonate hydrate and calcium aluminium toluenesulfonate hydrate is directly related with the loss of water of the interlayer. However, the mesitylenesulfonate ion in the interlayer of calcium aluminium mesitylenesulfonate hydrate structure shows a non uniform orientation at non ambient temperatures.

In all investigated binary systems containing different benzenesulfonic acid anion concentrations the phases assemblages were built up of two different calcium aluminate hydrates, one containing benzenesulfonic ions and the other hydroxide ions in the interlayer at 35% r.h. At 100% r.h. Low concentrations of benzenesulfonic ions caused the formations of amorphous gel structures, which recrystallised with water removal to calcium aluminate hydrates, containing benzenesulfonic ions or hydroxide ions. Furthermore superstructures or solid solution were not detected in any system.

References

Allmann R (1968) Die Doppelschichtstruktur der plättchenförmigen Calcium – Aluminium – Hydroxisalze am Beispiel des $3CaO{\cdot}Al_2O_3{\cdot}CaSO_4{\cdot}12H_2O$. N Jb Min Mh 140:144

Allmann R (1977) Refinement of the hybrid layer structure $Ca_2Al(OH)_6^+$ $1/2SO_4$ $3H_2O$. N Jb Min Mh 3:136–144

Kuzel HJ (1969) Beitrag zur Kristallchemie der Calciumaluminathydrate. Habilitationsschrift, Frankfurt am Main

Pöllmann H (1989) Study of the hydration mechanisms and formation of new hydrates applying organic additives to the aluminate phase of cement. In: Proceedings of 11th conference on cement Micr., New Orleans, Louisiana, pp 324–343

Pöllmann H (1989b) Mineralogisch-Kristallographische Untersuchungen an Hydratationsprodukten der Aluminatphase hydraulischer Bindemittel. Habilitationsschrift/Universittät Erlangen-Nürnberg, Erlangen

Pöllmann H, Stöber S (1997) Hydration characteristics and new hydrates using organic additives (Carboxylates and Sulphonates). In: Proceedings of the 10th international congress on chemical cement, Goteborg, 3iii032

Renaudin G, Francois M, Evrad O (1999) Order and disorder in the lamellar hydrated tetracalcium compound monocarboaluminate compound. Cem Concr Res 29:63–69

Stöber S, Pöllmann H (1999) Synthesis of a lamellar calcium aluminate hydrate (Afm phase) containing benzenesulfonic acid ions. Cem Concr Res 29:1841–1845

Terzis A, Fillipakis S, Kuzel HJ, Burzlaff H (1987) The crystal structure of $Ca_2Al(OH)_6Cl{\cdot}2H_2O$. Z Kristallogr 189:29–34

Use of Layered Double Hydroxides (LDH) of the Hydrotalcite Group as Reservoir Minerals for Nitrate in Soils – Examination of the Chemical and Mechanical Stability

T. Witzke, L. Torres-Dorante, F. Bullerjahn, and H. Pöllmann

1 Introduction

The leaching of nitrate resulting from both fertilizer application in agricultural production and also from natural transformation of nitrogen from organic compounds and/or ammonium into the nitrate form is an important way of nitrogen losses in crop production particularly during autumn and winter. Since nitrate as an anion is not adsorbed to the soil particles, it remains mobile in the soil. An efficient way to reduce nitrate leaching would be to control its movement in the soil by adsorbing it from the soil solution in the same way as clay minerals do it for cations like potassium and ammonium. The addition of an anion exchanging material to the soil could reduce this nitrate loss in the soil and the nitrate load in groundwater. The material should have a high selectivity and a high storage capacity for nitrate, and should be stable and active in the soil for several growing periods. The suitability of layered double hydroxides (LDH), known for their anion exchange properties, as a reservoir mineral for nitrate was studied (Bull 2001; Torres-Dorante 2007; Torres-Dorante et al. 2008; Witzke and Torres-Dorante 2009). LDH with the general composition $[Mg_{1-X}Al_X(OH)_2][(A)_X\ (H_2O)_n]$ ($A = NO_3$, Cl) and different Mg:Al ratios were used for the experiments.

T. Witzke (✉) • F. Bullerjahn • H. Pöllmann
Department of Mineralogy / Geochemistry, Martin-Luther-University Halle (Saale), Von Seckendorff-Platz 3, 06120 Halle (Saale), Germany
e-mail: thomas.witzke@geo.uni-halle.de

L. Torres-Dorante
Research Centre Hanninghof, YARA International ASA, Hanninghof 35, 48249 Dülmen, Germany

S.V. Krivovichev (ed.), *Minerals as Advanced Materials II*,
DOI 10.1007/978-3-642-20018-2_12, © Springer-Verlag Berlin Heidelberg 2012

2 Layered Double Hydroxides of the Hydrotalcite Group

All members of this group are built of two different, alternating layers, a M^{2+}- M^{3+}-hydroxide layer of the brucite type and with full occupation of the cation positions, and an anion-water layer, also called interlayer. The M^{2+}:M^{3+} ratio is variable within wide ranges. The positive layer charge is compensated by the interlayer anions. Bonding between the layers is very weak and allows a good anion exchange. The interlayer water can be removed reversible.

The general formula of hydrotalcite group LDH is

$[M^{2+}{}_{1-X}M^{3+}{}_X(OH)_2][(A^{Y-})_{X/Y}(H_2O)_n]$ with
$M^{2+} = Mg^{2+}, Mn^{2+}, Fe^{2+}, Ni^{2+}, Cu^{2+}, Zn^{2+}, Co^{2+}$
$M^{3+} = Al^{3+}, Cr^{3+}, Mn^{3+}, Fe^{3+}, Co^{3+}, Ni^{3+}, Ga^{3+}, In^{3+}, Y^{3+}$
$A = OH^-, F^-, Cl^-, Br^-, I^-, NO_3^-, IO_3^-, CO_3^{2-}, SO_3^{2-}, SeO_3^{2-}, ClO_4^-, MnO_4^-, SO_4^{2-}, CrO_4^{2-}, MoO_4^{2-}, WO_4^{2-}, S_2O_3^{2-}, C_2O_4^{2-}, Si_2O_5^{2-}, Cr_2O_7^{2-}, V_2O_7^{4-}, B_3O_4(OH)_2^-, CNO^-, Fe(CN)_6^{3-}, Fe(CN)_6^{4-}, V_4O_{12}^{4-}, Mo_7O_{24}^{6-}, W_7O_{24}^{6-}, V_{10}O_{28}^{6-}, CoW_{12}O_{40}^{5-}, H_2W_{12}O_{40}^{6-}, SiW_9V_3O_{40}^{7-}, SiW_{11}O_{39}^{8-}$ and other anorganic anions, $HCOO^-$ (formate), CH_3COO^- (acetate) and other carbon acids, $C_6H_4(COO)_2^{2-}$ (terephthalate), $H_{1+2n}C_nSO_2O^-$ (aliphatic sulfonates), vitamins, amino acids and several other organic anions.

In a few compounds, additional medium to large sized interlayer cations (plus charge compensating anions) are present, e.g. Na^+, Ca^{2+}, Sr^{2+}, Mg^{2+}, Ni^{2+}, $(UO_2)^{2+}$.

All members of the group are trigonal or hexagonal with a strong subcell containing just one hydroxide layer and one interlayer. The subcell is derived from the brucite cell and shows a lattice parameter a' ~3.0–3.1 Å, the parameter c' depends on the interlayer content (Fig. 1).

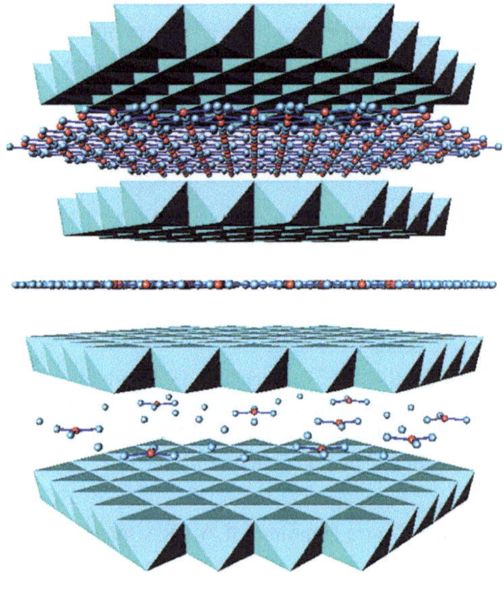

Fig. 1 Structure model of a Mg-Al nitrate LDH with Mg: Al = 5:1. The uppermost interlayer show all possible atom positions, lowermost the real occupation

Usually, an increasing content of trivalent cations in the hydroxide layer, and accordingly a higher anion content for charge compensation, reduce the basal spacing slightly due to a shortening of the hydrogen bonds (Mascolo and Marino 1980). The nitrate LDH shows contrary to other members of the hydrotalcite group, a strong increase of the basal spacing with higher Al^{3+} content: from 8.07 Å for Mg: Al = 6:1 to 9.03 Å for Mg:Al = 2:1 with a transition zone with a steep increase around the 3:1 ratio (Witzke and Torres-Dorante 2010). This behavior is explained by different models. In LDH with low Al content the nitrate anions are orientated parallel to the hydroxide layer as known for the carbonate in hydrotalcite (Allmann and Jepsen 1969). According to Kruissink et al. (1981) or Del Arco et al. (2000), in the material with high Al content and high basal spacing the nitrate is aligned tilted or perpendicular to the hydroxide layer. This model requires a high number of interlayer positions, but all with a very low occupancy to fit to the trigonal symmetry. From FTIR spectroscopic analyses, Xu and Zeng (2001) favour another model. In nitrate LDH with high basal spacing the nitrate is always aligned parallel to the hydroxide layer, but bounded via H bonds only to one layer. Unresolved are the water positions, especially in the transition from low to high basal spacing.

The different behavior of the nitrate LDH has a strong influence on the anion exchange properties and the stability of the material.

3 Formation and Stability of Natural and Synthetic LDH of the Hydrotalcite Group

Natural layered double hydroxides are generally formed under ambient or slightly elevated temperatures, under ambient pressures and under slightly acidic to alkaline conditions. Several members of the group can be formed during metasomatic alteration of mafic und ultramafic rocks at temperatures up to 200°C. Motukoreaite, a Mg-Al carbonate-sulphate LDH is known as submarine alteration product of volcanic rocks and glasses (Bryner et al. 1991). Several minerals like honessite were found in the oxidation zone of sulphide deposits, especially in nickel and copper-zinc deposits (Nickel and Wildman 1981). Hydrowoodwardite (Cu-Al sulphate LDH) is known as recent precipitation product from heavy metal bearing mine waters (Witzke 1999) at pH 6.5 or higher. Fougerite, a Fe^{2+}-Fe^{3+}-OH-LDH and member of the "Green Rust" family (Trolard et al. 2007), was observed in hydromorphic soils (gleysol) under reducing and neutral to alkaline conditions. Exposed to air, fougerite transform rapidly to lepidocrocite or goethite.

LDH compounds of the hydrotalcite group were observed also in anthopogene influenced environments. The formation of "Green Rust" is well known as an intermediate stage of iron or steel under reducing conditions, sometimes influenced by iron-reducing bacteria (Jorand et al. 2007). Heavy metal bearing LDH were found as precipitation products on slag dumps, a fly ash deposit and in the soil below a Ni raw material deposit (Witzke et al. 1995). Stoffyn et al. (1977) reported the neoformation of hydrotalcite in marine sediments due to pollutional inputs of $NaHCO_3$- and NaOH-enriched waters from a bauxite factory. The pore water in the zone of hydrotalcite formation showed pH values between 8.2 and 8.86, whereas above and below a pH value of 6.5 was measured.

References on the stability of LDH at low pH values are rare. In precipitation experiments of Zn-Al/Fe/Cr LDH using spring water highly loaded with Zn and other heavy metals from a fly ash/slag dump, the formation of LDH starts at pH 6 (Schreck and Witzke 1999). The removal of heavy metals at this low pH was very ineffective, but it indicates the stability field of the LDH. At values around pH 8 and higher a quantitative precipitation of heavy metals in LDH structures was observed.

Staß (without year, probably 1990) reported in a study on the medicament Talcid® (Bayer AG), composed mainly of hydrotalcite and used as an antacid, that the LDH buffer in an acidic environment (0.1 M HCl) to pH 4, even at an surplus of acid and with different ratios of Talcid® (respectively hydrotalcite) to acid.

Slight differences in the formation enthalpies and other thermodynamic properties of Mg-Al carbonate LDH were observed by Allada et al. (2005a) for material with different Mg:Al ratio. The solubility of carbonate, nitrate, chloride or iodide LDH in water was studied by Allada et al. (2005b). From thermodynamical predictions they conclude a much lower solubility of carbonate LDH in water compared to the nitrate because of the different bonding strength of the interlayer species. Study of the dissolution of the LDH with Mg:Al = 2.70–2.85 and different anions showed an enthalpy of solution for all material in the same order (ΔH_{sol} in KJ/mol = −78.67 for the carbonate and −93.72 for the nitrate LDH). It should be noted that the nitrate LDH used in the study belongs to the transition zone between the low and the high basal spacing which shows generally a low crystallinity and high structural disorder (Witzke and Torres-Dorante 2010).

4 Dissolution of LDH Under Soil Conditions

As most soils have acidic to neutral soil solutions, a certain dissolution of the LDH is expected as a result of the proton attack on the hydroxide layer according to the reaction (for a Mg-Al nitrate LDH with Mg:Al = 5:1)

$$[Mg_{0.833}Al_{0.167}(OH)_2][(NO_3)_{0.167}(H_2O)_{0.5}] + 2\,H^+ \rightarrow 0.833\,Mg^{2+} + 0.167\,Al^{3+} + 0.167\,NO_3^- + 2.5\,H_2O$$

Several factors may have an influence on the stability of the LDH in soil: the properties of the soil, e.g. the soil type, soil pH, the amount of soil solution and the diffusion rate in the soil, the chemical composition of the LDH, e.g. Mg:Al ratio, other cations and present anions, crystallographical properties like the crystal size and crystallinity of the material, and technological properties like granule diameter and shape, mechanical hardness, or coating.

5 Experimental

The LDH (Fig. 2) was synthesized by the co-precipitation method using Mg nitrate and Al nitrate solution (or the chlorides respectively) according to the stoichiometry of the material at pH 10. The pH value was controlled by titration with NaOH in the laboratory experiments and with KOH in an up-scaled synthesis in a pilot plant.

For the herein described stability experiments, the synthesized Mg-Al nitrate and chloride and additional carbonate LDH samples were used. The nitrate LDH has the composition $[Mg_{0.83}Al_{0.17}(OH)_2]$ $[(NO_3)_{0.12}$ $(CO_3)_{0.03}$ $(H_2O)_{0.5}]$ (Mg:Al = 5:1), and the lattice parameter a = 3.089, c = 24.33 and c' = 8.110 Å with an average crystal thickness of 0.023 μm. For the chloride LDH was found an empirical formula $[Mg_{0.82}Al_{0.18}(OH)_2]$ $[Cl_{0.18}$ $(H_2O)_{0.5}]$, and the lattice parameter a = 3.086, c = 24.46 and c' = 8.153 Å. As a synthetic Mg-Al carbonate LDH (= hydrotalcite) a product of the company Süd-Chemie (München) was used. The material has Mg:Al = 2:1, a X-ray diffraction analysis yielded the lattice parameter a = 3.051, c = 22.91 and c' = 7.637 Å, the average crystal thickness is 0.018 μm.

The dissolution of the material was studied in laboratory experiments using powdered LDH (12 mg) and 0.0001 M HCl solution (500 ml, pH 4) at room temperature under constant stirring. The dissolution of the LDH was measured by the increasing pH value according to the consumption of acid in the unbuffered system or by the Mg release in the solution.

For the soil experiments, three different soils with different pH values were used, a loamy sand soil (pH 5.5), a sandy clay loam (6.7), and a silt loam (7.6). An amount of 55 mg granulated Mg-Al chloride LDH was applied to 26 g of each soil. The soil moisture was kept at 65% of the total water holding capacity. The LDH dissolution was measured by the Mg release. The trials run for 12 month (Torres-Dorante 2007).

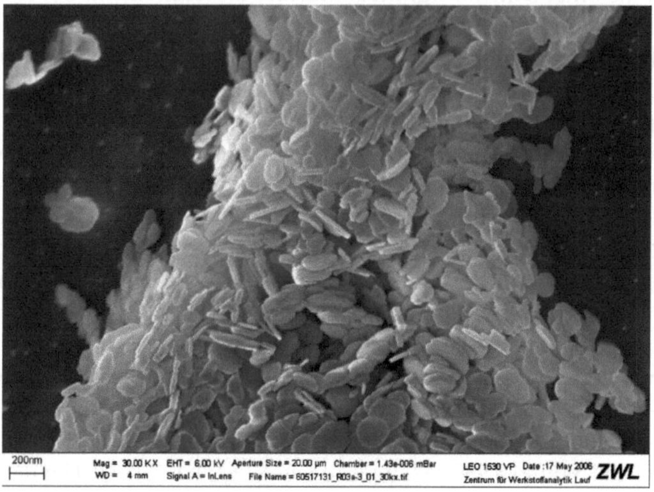

Fig. 2 Mg-Al chloride LDH. SEM micrograph J. Göske, ZWL Lauf

For the experiments irregularly shaped granules of 3–4 mm average diameter were used. The investigation of the optimal granule size was done also with smaller granules. For a few experiments cylindrical shaped granules were used.

6 Influence of Soil Parameters on the LDH Stability

The stability of the LDH was studied in long-time experiments in soil under cropping and non-cropping conditions. All the experiments showed, that a large amount of the applied LDH is stable in different soils and under different conditions. The LDH granules could be removed even after 500 day trials and four growing periods from the soil, without visible changes in the granule size or shape. The X-ray diffraction diagrams of LDH granules separated from the soil after the experiments show no differences in the peaks, their shape and sharpness compared to the diagrams of the starting material. An exposure of the LDH to the soil didn't change the crystallinity or the structural disorder of the material. Only a reduction of the basal spacing c' from ~8.1 Å to ~8.0 Å due to the exchange of nitrate or chloride interlayer anions for carbonate was found. Cropping or non-cropping conditions had no influence on the properties of the LDH after the trials. Nevertheless, in the laboratory and greenhouse experiments the dissolution of a certain amount of LDH, the release of Mg and Al and a change of the soil pH were observed.

The experiments with three different soil types and pH values described in the experimental setup chapter, showed the main dissolution effect within the first month. The amount of dissolution was 14.5, 9.2 and 6.6 wt.% of the initial added LDH in the soils with a pH of 5.5, 6.7 and 7.6, respectively. Only for the loamy sand soil (pH 5.5) the dissolution increased further within 6 month to the maximum value of 18.4%, in the other soils no significant further dissolution was observed (Fig. 3). As final soil pH values of 6.4, 7.1 and 7.7 were measured (Torres-Dorante 2007).

During the experiments in the soil, the LDH granules show a remarkable increasing mechanical stability. X-ray diffraction and chemical analyses of LDH granules separated from the soil show the presence of calcium carbonate phases (aragonite, calcite) and a silica phase in the granules. Aragonite was present in all analyzed granules, and in several samples additional calcite was observed. The quantity of calcium carbonate never exceeds 5% of the granule weight and shows no relation to the soil type, initial pH or cropping/non-cropping conditions. The first aragonite or calcite could be detected after 1 week in the soil. Separate X-ray analyses of core and rim samples showed that the calcium carbonates precipitate in the granule core, but not at the surface.

The precipitation of aragonite or calcite is influenced by several factors like pH value, over-saturation, CO_2 partial pressure, other anions and cations (especially Cl and Mg) in the solution, organic compounds or biological activity (Alexandersson 1972; Burton and Walter 1987; Munnecke and Westphal 2004). The increase of the pH value and the Mg release by slight dissolution of the LDH and a higher chloride

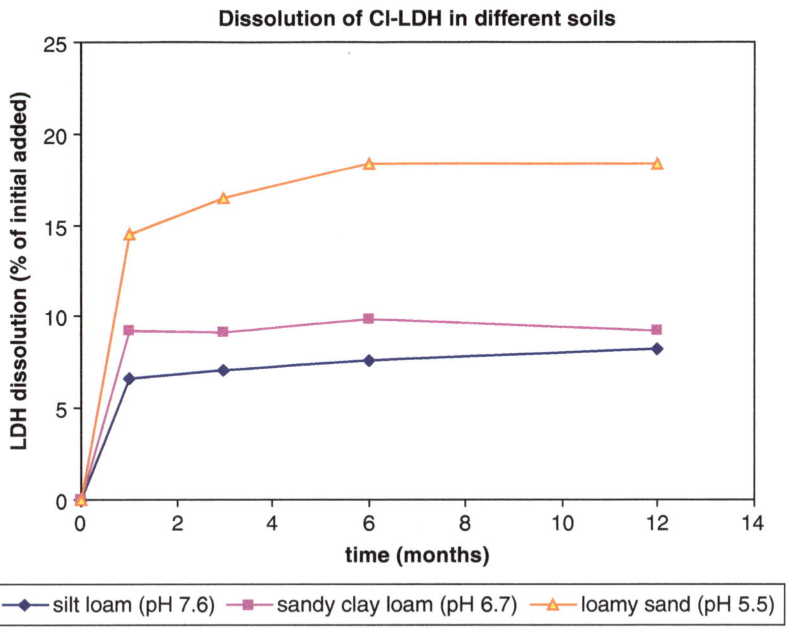

Fig. 3 Dissolution of Mg-Al chloride LDH granules in different soils with different initial pH values under non-cropping conditions

content due to the anion exchange (if a chloride LDH as starting material is used) prefer a precipitation of aragonite instead of calcite in the granules. Examination of granule cross-sections with SEM showed a relatively homogenous distributed silica phase. As no silica phase could be found in the X-ray diagrams, the material is very probably amorphous. Both, the calcium carbonate phases and the silica increase the mechanical stability of the granule, but do have no negative effect on the nitrate exchange properties of the LDH.

Scanning electron micrographs of the surface of the LDH granules from the soil experiments show no dissolution pits or other dissolution effects. The material is intensely overgrown by small roots (Fig. 4).

7 Influence of the LDH Chemistry on the Stability Under Soil Conditions

Whereas the amount of total dissolved LDH depends on the soil pH and the diffusion in the soil, the dissolution rate can be delayed by several factors.

One of these factors is the chemistry of the LDH. The positive effect of the incorporation of Zn in the hydroxide layer on the crystallinity and stability is known, e.g. for solid solutions of Cu/Zn-Al sulphate LDH (Witzke et al. 1997). A slower dissolution was also observed for Co-bearing nitrate LDH compared to

Fig. 4 Mg-Al chloride LDH after a long-time trial in soil, overgrown with small roots. SEM micrograph

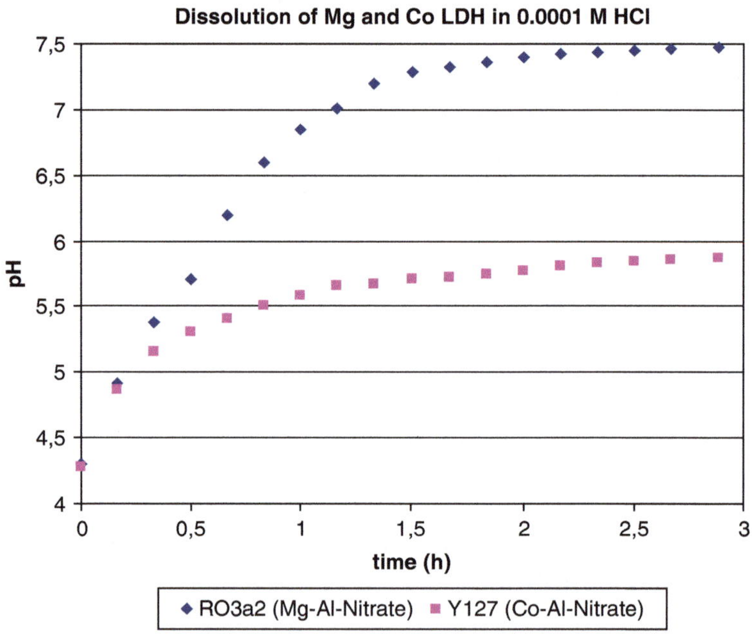

Fig. 5 Dissolution of Mg-Al nitrate and Co-Al nitrate LDH similar crystallinity, measured by the change of the pH value

the Mg nitrate LDH with the same crystallinity and the same M^{2+}-M^{3+} ratio (Fig. 5). The application of Zn or Co bearing LDH as a nitrate reservoir mineral is not recommended, as no additional heavy metals should be yielded into the soil, but it can be used to demonstrate the influence of different cations in the hydroxide layer on the dissolution rate.

The partial or complete replacement of Mg by Fe^{2+} changes the chemistry towards the "Green Rust" phases, which are known from soils, but unstable under oxidizing conditions. Another way of iron incorporation into the LDH is the replacement of Al by Fe^{3+}. This has an influence on the dissolution rate (Bullerjahn 2008). LDH with M^{2+}:M^{3+} = 5:1 ratio (M^{2+} = Mg) shows a faster dissolution compared to the pure Mg-Al LDH if 40% or more of the Al is replaced by Fe^{3+}. But if only 20% of the Al is replaced, the dissolution and the increase of the pH value is slower than for the iron-free LDH. A low amount of trivalent iron could therefore improve the properties of the LDH used in the soil.

The influence of different M^{2+}:M^{3+} ratios was not studied here. As shown by Allada et al (2005a), the expected influence is small and from anion exchange experiments only LDH with Mg:Al = 5:1 was suitable as a reservoir mineral for nitrate in soil.

The dissolution of LDH with similar crystallinity and different anions (chloride, nitrate, and carbonate) shows no significant differences in the amount of dissolved LDH over longer periods. Slight differenced were observed on a short-time scale, the nitrate LDH showed a slower dissolution (Fig. 6). Under soil conditions, a chloride or nitrate LDH is not stable and turns within one to three days by anion exchange with the soil solution into a form with mixed anions. If a chloride LDH is used as a starting material, the chloride is nearly completely lost and replaced by carbonate and nitrate. Starting with a nitrate LDH, a certain quantity of the nitrate will be replaced by carbonate in a short time.

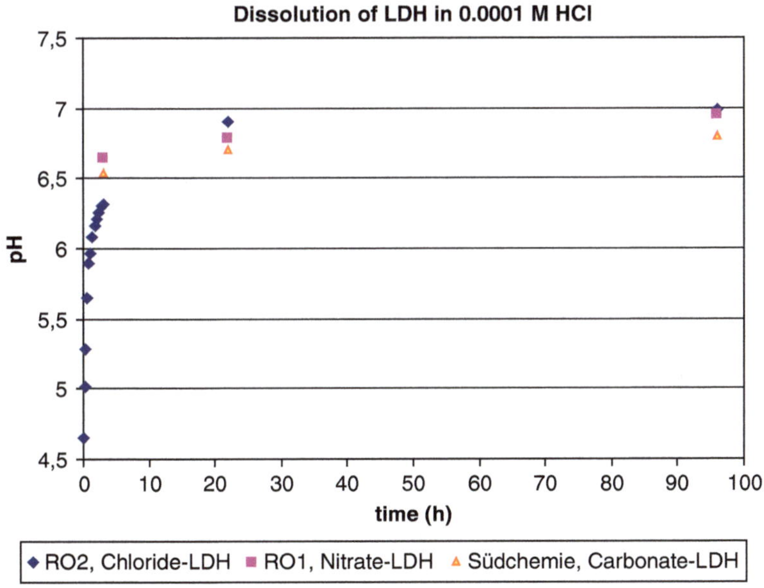

Fig. 6 Dissolution of chloride, nitrate, and carbonate LDH with similar crystallinity, measured by the change of the pH value

8 Influence of the LDH Crystallinity on the Stability Under Soil Conditions

As the lattice energy of a crystal is lower with an increasing disorder of the structure, an influence of the crystallinity on the dissolution rate is expected. Material with a higher degree of structural disorder and with a lower crystal size should reach the dissolution equilibrium faster. The crystallinity can be estimated from the X-ray diffraction data from 00 l peak broadening, presence and shape of the h0l/0kl band and the 110 and 113 peaks. For these experiments a Mg-Al chloride LDH with an average crystal thickness of 0.013 μm (sample RO3a2) was used, compared to the same sample after thermal treatment (50 °C for 24 h and 80 °C for 24 h, sample no. RO3a7). The latter shows an average crystal thickness of 0.028 μm. The crystal diameter of around 1 μm was not significantly influenced by the thermal treatment.

As expected, the material with the higher crystallinity shows a much slower dissolution (Fig. 7). For the application in the soil, only material with a high crystallinity should be used to avoid rapid dissolution effects.

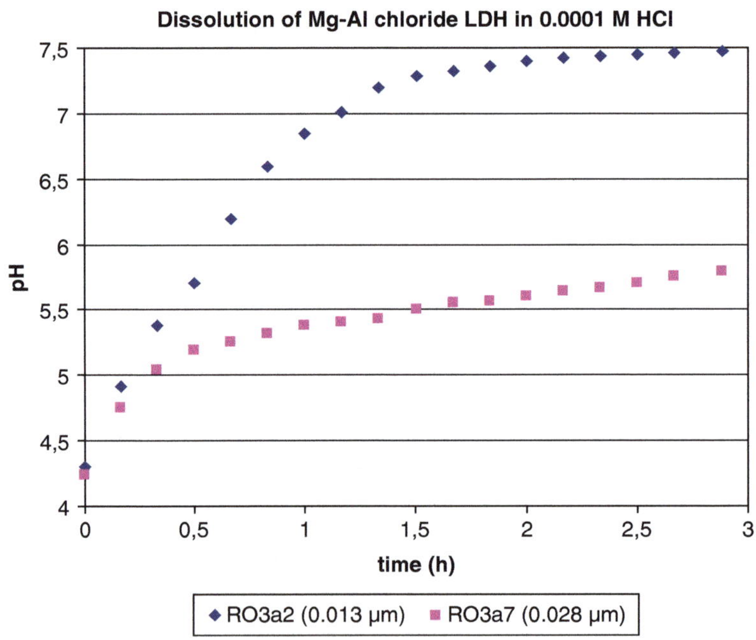

Fig. 7 Dissolution of Mg-Al chloride LDH with same composition but different average crystal thickness, measured by the change of the pH value

9 Influence of Technological Properties on the Stability of the LDH in the Soil

For the application in the soil, the LDH material has to be granulated. The use of powder leads to a rapid dissolution (Bull 2001). Experiments with Mg-Al chloride LDH (Mg:Al = 5:1) in a loamy sand soil (initial soil pH 5.5) with granule diameters of 1–2.5 and 3–4 mm showed a slightly higher Mg release due to dissolution of the smaller granules within the first 200 days, but no significant differences between the two samples after 500 days (Fig. 8).

Beside the granule diameter and shape, the particle density (respectively the porosity) has an influence on the dissolution rate. To reduce the porosity and increase the particle density, different compaction pressures were used for the production of the granules (Mg-Al chloride LDH, Mg:Al = 5:1). An influence of the compaction pressure was only observed for the soil with pH = 5.5, for soils with higher pH values no significant differences in the behaviour could be observed (Torres-Dorante 2007) (Table 1).

Lower pore volume reduce the amount of soil solution in the granules. A pH value "comfortable" for the LDH can be reached inside and close to the granules with less dissolved LDH. In soils with higher pH values the quantity of dissolved LDH is generally smaller, the pore volume compared to the granule surface has not such a significant influence here.

Coating on the LDH granules could reduce the dissolution and Mg release. As coating material a Ca-Al nitrate or carbonate LDH (hydrocalumite group, Tetra Calcium Aluminate Hydrate, TCAH), structurally closely related to the hydrotalcite

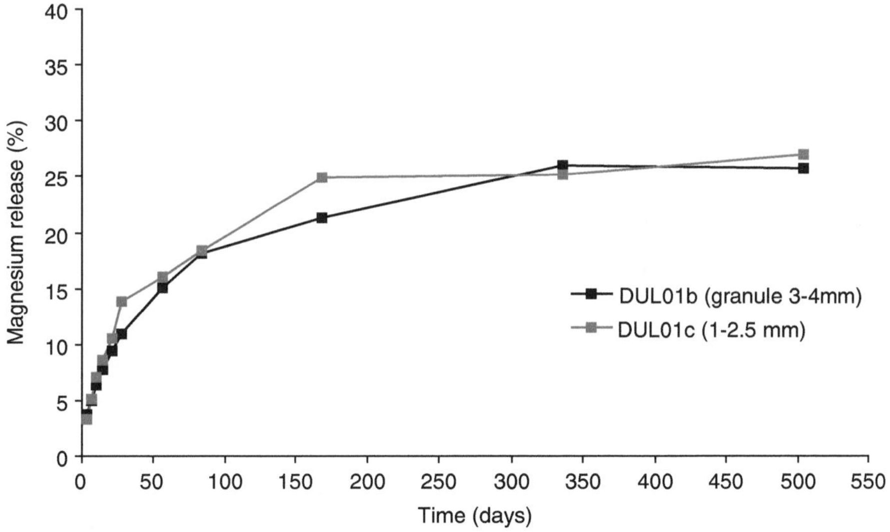

Fig. 8 Release of Mg due to dissolution of the Mg-Al chloride LDH (of total Mg in LDH) depending on the granule size, long-time experiment in loamy sand soil (pH 5.5)

Table 1 Dissolution of the LDH according to the compaction pressure

Compaction (bar)	Particle density (g/cm^3)	Initial pH	pH after 12 months	LDH dissolution, %
170	1.0	5.5	6.6	23.0
300	1.07	5.5	6.6	21.7
430	1.17	5.5	6.4	16.9

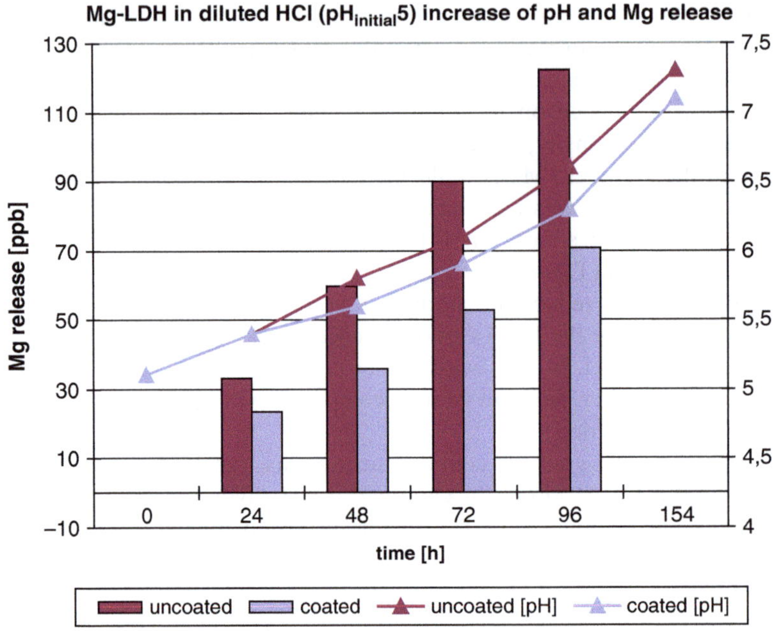

Fig. 9 Mg release and pH increase of coated and uncoated material in diluted HCl

group, was used. Different approaches were studied: (1) rolling of LDH granules in a TCAH suspension, (2) precipitation of TCAH from a solution on the granules, (3) precipitation (topotactic growing) of TCAH on LDH crystals (Bullerjahn 2008; Bullerjahn et al. 2009). The best results were obtained with the first approach. By rolling of the LDH granules in TCAH suspension a homogenous coating was obtained. Drying of the coating leads sometimes to small cracks. In the coating were formed small TCAH crystals of equal size and a rather low porosity. As the coating material is structurally and chemically closely related to the granule material, the pH value should show no large differences between coated and uncoated in samples in the dissolution experiments. In fact, only a slightly slower increase of the pH vale was observed for the coated material (Fig. 9).

The dissolution of a certain amount of TCAH produces a "comfortable" pH environment for the LDH and reduces the dissolution for around 40% after 96 h. Furthermore, the release of Ca from the TCAH supports the formation of calcium carbonate in the granules. In the second approach, the precipitation of TCAH at

Fig. 10 TCAH crystals grown on LDH granules, SEM micrograph

the granule surface, the growing of TCAH crystals was observed, but wide ranges in the crystal size prevent an effective coating of the granules (Fig. 10). The inhomogenous texture of the coating allows no protection of the LDH from dissolution. The topotactic growing of TCAH on the Mg-Al nitrate LDH was not successful, despite the strong structural relation. Only isolated or groups of TCAH crystals were observed.

10 Discussion

A Mg-Al chloride or nitrate LDH with Mg:Al ratio of 5:1 works as reservoir mineral for nitrate in soil. The material can absorb nitrate from the soil solution by anion exchange and can release the nitrate over several cycles (Torres-Dorante 2007). A part of the anion exchange capacity is lost due to dissolution of a certain amount of LDH, especially at low soil pH values and at the first days to weeks in the soil. The dissolution leads to the release of Mg and Al and to an increasing pH of the soil solution. Highest pH values were found close to the granules. At pH values above 7 the LDH becomes stable. Even after long time in the soil, the LDH granules can be separated and the material shows no crystallographical changes except a lower basal spacing due to anion exchange.

It is impossible to avoid the dissolution of a certain quantity of the LDH, but an optimized material can delay the release of Mg and Al, the increase of the pH and the reduction of the anion exchange capacity. The material used as a reservoir mineral for nitrate in soil must have a high crystallinity, and of the total trivalent cations up to 20% of Al can be replaced by Fe. Rapid dissolution can be avoided by an ideal granule size of ca. 3–4 mm diameter, a spherical granule form and the reduction of porosity by compaction. A suitable coating material is TCAH, which can significantly reduce the LDH dissolution and support the increase of the mechanical stability by calcium carbonate precipitation in the granules.

Acknowledgements The authors thank the financial support of YARA International ASA and the state Mecklenburg-Vorpommern to the research projects.

References

Alexandersson T (1972) Intragranular growth of marine aragonite and Mg-calcite: evidence of precipitation from supersaturated seawater. J Sediment Res 42:441–460

Allada RK, Navrotsky A, Boerio-Gates J (2005a) Thermochemistry of hydrotalcite-like phases in the $MgO-Al_2O_3-CO_2-H_2O$ system: a determination of enthalpy, entropy, and free energy. Am Mineral 90:329–335

Allada RK, Pless JD, Nenoff T, Navrotsky A (2005b) Thermochemistry of Hydrotalcite-like phases intercalated with CO_3^{2-}, NO_3^-, Cl^-, I^-, and ReO_4^-. Chem Mater 17:2455–2459

Allmann R, Jepsen HP (1969) Die Struktur des Hydrotalkits. N Jb Min Mh 544–551

Bryner V, Rodgers KA, Courtney SF, Postl W (1991) Motukoreaite from Brown's Island, New Zealand, and Stradner Kogel, Austria: a scanning electron microscopic study. N Jb Miner Abh 163:291–304

Bull C (2001) Development and evaluation of Layered Double Hydroxides (LDHs) for nitrate exchange in soil. Ph.D. thesis, University of Hannover, Hannover

Bullerjahn F (2008) Untersuchungen zur chemischen und mechanischen Stabilisierung eines Mg-Al-LDHs. M.Sc. thesis, Naturwissenschaftliche Fakultät III der Martin-Luther-Universität Halle-Wittenberg

Bullerjahn F, Witzke T, Pöllmann H, Torres-Dorante LO (2009) Untersuchungen zur Verbesserung der chemischen Stabilität eines Bodenadditivs auf LDH-Basis durch Coating. Hallesches Jahrbuch für Geowissenschaften 31:34

Burton EA, Walter LA (1987) Relative precipitation rates of aragonite and Mg calcite from seawater: temperature or carbonate ion control? Geology 15:111–114

Del Arco M, Gutierrez S, Martin C, Rives V, Rocha J (2000) Effect of the Mg:Al ratio on borate (or silicate)/nitrate exchange in hydrotalcite. J Solid State Chem 151:272–280

Jorand F, Zegeye A, Landry F, Ruby C (2007) Biogenesis and biocycling of FeII-FeIII hydroxysalt green rust. Geophys Res Abstr 9:04912

Kruissink EC, van Reijen LL, Ross JRH (1981) Coprecipitated nickel-alumina catalysts for methanation at high temperature. J Chem Soc Faraday Trans 1 77:649–663

Mascolo G, Marino O (1980) A new synthesis and characterization of magnesium-aluminium hydroxides. Mineral Mag 43:619–621

Munnecke A, Westphal H (2004) Shallow-water aragonite recorded in bundles of limestone-marl alterations – the Upper Jurassic of SW Germany. Sediment Geol 164:191–202

Nickel EH, Wildman JE (1981) Hydrohonessite – a new hydrated Ni-Fe hydroxy-sulphate mineral: its relationship to honessite, carrboydite, and minerals of the pyroaurite group. Mineral Mag 44:333–337

Schreck P, Witzke T (1999) Immobilisierung von Schwermetallen im Haldensickerwasser durch natürliche Mineralfällung. Umwelt 2000. Schriftenreihe der Deutschen Geologischen Gesellschaft 9:125

Staß H (without year, probably 1990) Hydrotalcite, physicochemical characteristics of a layer-lattice antacid. In: Bayer AG (ed) Advances in medicine, pocketbook for clinicians and general practitioners. Urban & Vogel GmbH, Munich

Stoffyn M, Dodge CH, Mackenzie FT (1977) Neoformation of hydrotalcite due to industrial inputs in marine sediments. Am Mineral 62:1173–1179

Torres-Dorante LO (2007) Evaluation of Layered Double Hydroxide (LDH) mineral as a long-term nitrate exchanger in soil. Cuvillier Verlag, Göttingen

Torres-Dorante LO, Lammel J, Kuhlmann H, Witzke T, Olfs HW (2008) Capacity, selectivity, and reversibility for nitrate exchange of a layered double-hydroxide (LDH) mineral in simulated soil solutions and in soil. J Plant Nutr Soil Sci 171:777–784

Trolard F, Bourrie G, Abdelmoula M, Refait P, Feder F (2007) Fougerite, a new mineral of the pyroaurite-iowaite group: description and crystal structure. Clays Clay Miner 55:323–334

Witzke T (1999) Hydrowoodwardite, a new mineral of the hydrotalcite group from Königswalde near Annaberg, Saxony/ Germany and other localities. N Jb Mineral Mh 75–86

Witzke T, Torres-Dorante L (2009) Untersuchungen zum Einsatz von LDH's der Hydrotalcit-Gruppe als Zwischenspeicher für Nitrat im Boden. Hallesches Jahrbuch für Geowissenschaften 31:265

Witzke T, Torres-Dorante L (2010) Layered Double Hydroxides (LDH) of the hydrotalcite group as reservoir minerals for nitrate in soil. III International Workshop on Layered Materials, Bochum, 14–15 May, Book of Abstracts, p 105

Witzke T, Pöllmann H, Tschapek B (1995) Formation of takovite in a nickel-contaminated soil. Ber D Min Ges Beihefte 1 Eur J Min 7:272

Witzke T, Göske J, Pöllmann H (1997) Mischkristalle von Cu- und Zn-haltigen Verbindungen mit Hydrotalcit-Struktur. Ber D Min Ges, Beih 1, Eur J Min 9:392

Xu ZP, Zeng HC (2001) Abrupt structural transformation in hydrotalcite-like compounds $Mg_{1-x}Al_x(OH)_2(NO_3)_x \cdot n\ H_2O$ as a continuous function of nitrate anions. J Phys Chem B105:1743–1749

Nanocrystalline Layered Titanates Synthesized by the Fluoride Route: Perspective Matrices for Removal of Environmental Pollutants

Sergey N. Britvin, Yulia I. Korneyko, Vladimir M. Garbuzov, Boris E. Burakov, Elena E. Pavlova, Oleg I. Siidra, A. Lotnyk, L. Kienle, Sergey V. Krivovichev, and Wulf Depmeier

Layered oxide compounds with structures based upon combinations of $[TiO_6]$ and $[NbO_6]$ octahedral building blocks attract considerable scientific and industrial interest due to their structural diversity and wide field for technological applications. The group of layered titanates is by far the best studied among this family of layered oxides (Bavykin et al. 2006; Doong and Kao 2008). In general, the crystal structures of layered titanates can be derived from the crystal structure of lepidocrocite, orthorhombic γ-FeOOH (England et al. 1983). The basic structural units of lepidocrocite and of the simplest representatives of layered titanates can be described as infinite double layers (slabs) composed of edge-sharing $[MO_6]$ octahedra where M denotes Fe^{3+} (in lepidocrocite) or Ti^{4+} (in titanates) (Fig. 1). In the case of titanates, the occupancy of the titanium site may be less than 1. Aside from infinite double layers of edge-shared octahedra, layered titanate structures may also be constructed from corrugated layers composed of three-, four- or five-membered $[TiO_6]$ "packs" connected to each another by corners of $[TiO_6]$ octahedra (Table 1). As an example of a corrugated layer that of tetratitanates $[Ti_4O_9]$ is shown in Fig. 2. The octahedral layers of layered titanates are negatively charged, and can thus act as host structures for positively charged species. In lepidocrocite, the charge balance is achieved by partial protonation of oxygen atoms. In the layered titanates, charge compensation is maintained by intercalation

S.N. Britvin (✉) • O.I. Siidra • S.V. Krivovichev
Department of Crystallography, Geological Faculty, St. Petersburg State University, Universitetskaya Nab. 7/9, 199034 St. Petersburg, Russia
e-mail: sbritvin@gmail.com

Y.I. Korneyko • V.M. Garbuzov • B.E. Burakov • E.E. Pavlova
V.G. Khlopin Radium Institute, 28, 2-nd Murinsky Ave., 194021 St. Petersburg, Russia

A. Lotnyk • L. Kienle
Institute for Material Science, Synthesis and Real Structure, University Kiel, Kaiserstr. 2, 24143 Kiel, Germany

W. Depmeier
Institute for Geosciences, University Kiel, Olshausenstr. 40, 24118 Kiel, Germany

S.V. Krivovichev (ed.), *Minerals as Advanced Materials II*,
DOI 10.1007/978-3-642-20018-2_13, © Springer-Verlag Berlin Heidelberg 2012

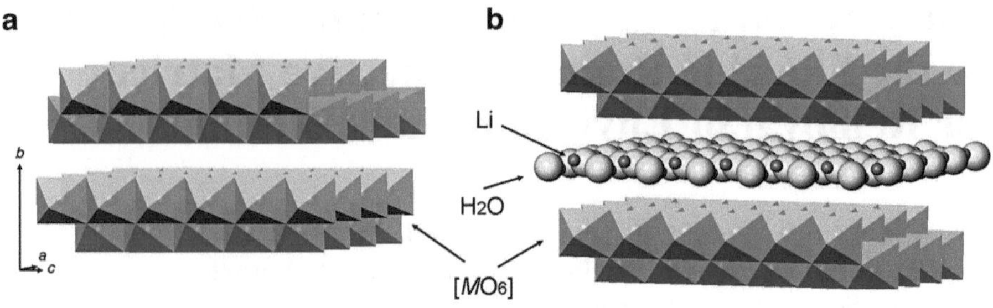

Fig. 1 Crystal structure of lepidocrocite, γ-FeOOH (Ewing 1935) (**a**) and orthorhombic layered titanate, $Li_{0.7}Ti_{2-x}O_4{\cdot}H_2O$ where $x = 0.7/4$ (**b**) (solved and refined based on data of Sasaki et al. 1995)

Table 1 Some typical representatives of layered titanates

Layer unit	Hydration stage	Representative compound	Reference
$[Ti_{2-x}O_4]^{4x-}$	0	$Cs_xTi_{2-x/4}O_4$	Grey et al. 1987
	1	$Cs_{0.48}H_{0.22}Ti_{1.825}O_4{\cdot}0.5H_2O$	Sasaki et al. 1995
$[Ti_3O_7]^{2-}$	0	$Na_2Ti_3O_7$	Andersson and Wadsley 1961
$[Ti_4O_9]^{2-}$	0	$Tl_2Ti_4O_9$	Verbaere and Tournoux 1973
	1	$K_{0.5}H_{1.5}Ti_4O_9{\cdot}0.6H_2O$	Sasaki et al. 1989
	2	$Cs_{1.5}H_{0.5}Ti_4O_9{\cdot}2H_2O$	Sasaki et al. 1989
	3	$Na_{0.98}H_{1.02}Ti_4O_9{\cdot}3.3H_2O$	Sasaki et al. 1985
$[Ti_5O_{11}]^{2-}$	0	$Cs_2Ti_5O_{11}$	Kwiatkowska et al. 1987
	1	$Cs_2Ti_5O_{11}{\cdot}H_2O$	Kwiatkowska et al. 1987

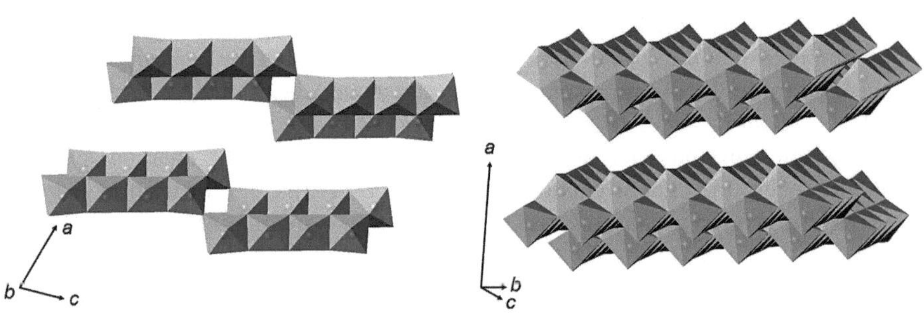

Fig. 2 Corrugated $[Ti_4O_9]$ layers in the structure of $Tl_2Ti_4O_9$ (Verbaere and Tournoux 1973)

of different cations into the interlayer space of the host structures. Typical intercalating cations are alkali metals (Li^+, Na^+, K^+, Rb^+, Cs^+) or hydronium (H_3O^+). Depending on the nature of intercalating cations as well as on the environmental conditions (i.e., temperature, relative humidity), one or more layers of H_2O molecules can be introduced into the interlayers. By selecting the appropriate cation and hydration stage, flexible tuning of interlayer space can be achieved without disruption of the structure (Table 1). In spite of numerous publications focusing on the synthesis and properties of layered titanates, the structures of many of them are not yet completely understood. Among those remarkable examples are the easily synthesized sodium layered titanates. Different synthesis conditions yield sodium titanates with different crystal morphology (nanoleafs, nanotubes, etc.), but with essentially the same bulk composition and identical powder and electron diffraction patterns. It is still a matter of debate to which structure type those sodium titanates should be ascribed. Among the publications one can find those claiming that sodium titanates belong to the lepidocrocite-type structures (Ma et al. 2003; Peng et al. 2008), or to a new structure type which can be described by the formula $Na_{2-x}H_xTi_2O_4(OH)_2$ (Yang et al. 2003), or that they are really trititanate $Na_{2-x}H_xTi_3O_7{\cdot}nH_2O$ (Morgado et al. 2007). The diversity of opinions can easily be explained by the enormous experimental difficulties encountered when studying these compounds. The main problems are their poor crystallinity, along with deformation of the crystals even on the nanoscale level and, possibly, stacking faults resulting in structural disorder between the adjacent layers.

During recent studies, the authors of the present paper have developed a new soft method for the synthesis of nanocrystalline niobates, tantalates, titanates and even germanates (Britvin et al. 2010a,b; 2011). The new method has been called "the fluoride route", by analogy with the well-known approach to the preparation of zeolitic frameworks. In essence, the "fluoride route" is based on alkaline hydrolysis of stable fluorometalate ions, like $[TaF_7]^{2-}$, $[NbOF_5]^{2-}$, $[TiF_6]^{2-}$ or $[GeF_6]^{2-}$, under ambient conditions, followed by short-time boiling of the precipitated compounds. Due to the extreme stability of these fluoro complexes, the precipitation of hydroxide compounds of Ta, Nb, Ti or Ge is shifted from the typically observed acidic medium (pH 1–1.5) to alkaline conditions (pH 8–9). This shift results in complete rebuilding of the total synthetic environment yielding a variety of complex oxide frameworks and layered structures instead of formation of amorphous hydroxides usually observed under acidic conditions. It is an essential advantage of the fluoride approach that it allows the one-step preparation of nanocrystalline titanates, niobates, tantalates and germanates with any arbitrary charge-balancing cation, either inorganic or organic one, provided that the used alkali is soluble in aqueous fluoride systems. Depending on the synthesis conditions, frameworks, layered structures or salts of Lindqvist isopolyanions ($[Nb_6O_{19}]^{8-}$ and $[Ta_6O_{19}]^{8-}$) can be obtained. For instance, a diversity of nanocrystalline pyrochlores, which conventionally are synthesized by solid state calcination, could be prepared by precipitation and subsequent boiling (Britvin et al. 2010a). As a consequence of the extreme "softness" of the method, exotic unstable cations, like hydrazinium $(N_2H_5)^+$, could be easily incorporated into the resultant compounds, as in case of zeolite hydrazinium germanate (Britvin et al. 2010b) and layered hydrazinium titanate LHT-9 (Britvin et al. 2011).

We have found that the fluoride method is especially suitable for the synthesis of different nanocrystalline titanates (Table 2). As an example, sodium layered titanates can be synthesized in one stage, by mixing of solution of H_2TiF_6 with concentrated NaOH solution followed by gentle boiling of obtained precipitate in the parent solution. Synthesized sodium titanate has atomic Ti/Na ratio = 2.

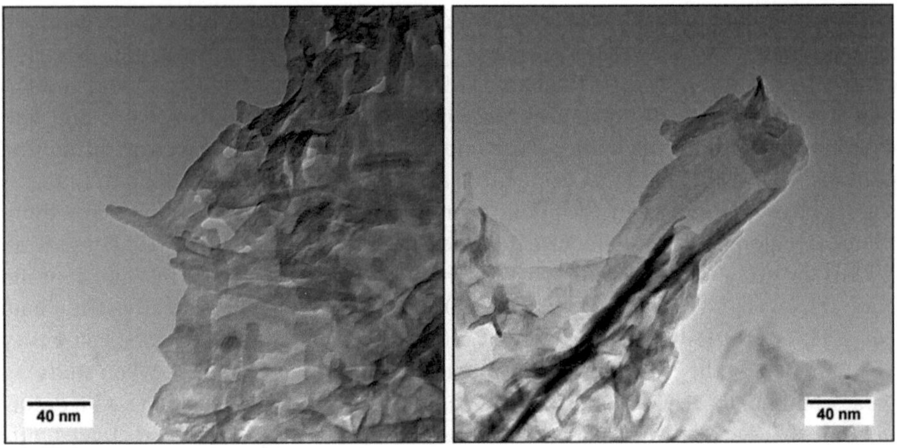

Fig. 3 Crystal morphology of layered sodium titanate

Table 2 Nanocrystalline titanates synthesized via fluoride route starting from H_2TiF_6

Base	Synthesized titanate	Structure type
NaOH	$NaTi_2O_4(OH)$?	Pseudo-lepidocrocite
KOH	$K_2Ti_6O_{13}$	Jeppeite ($K_2Ti_6O_{13}$)
RbOH	$Rb_2Ti_6O_{13}$	Jeppeite ($K_2Ti_6O_{13}$)
CsOH	$Cs_xTi_{2-x/4}O_4$	Pseudo-lepidocrocite

Table 3 Affinity of nanocrystalline sodium titanate toward some of environmental pollutants[a]

Cation	C_i	C_f	K_D
Ba^{2+}	42.94	1.38	7551
Cd^{2+}	57.48	0.67	21262
Pb^{2+}	53.90	0.02	641417
Sr^{2+}	48.63	4.66	2358

[a]Experimental details are given in the text

It crystallizes in form of fine leaflets or nanotubes, depending on variations of reaction conditions (Fig. 3). Preliminary experiments revealed that synthesized titanate possesses high adsorption capacity toward to ions of several environmental toxicants (Table 3). The adsorption experiments were performed by immersion of sodium titanate (100 mg batch) into 20 mL of solution containing ions of the elements listed in the table (in form of nitrate salts) and holding the suspension with periodical shaking for 8 h. The concentration of the ions was measured in aliquots of the supernatant before and after exchange procedure. The distribution coefficients were calculated according to the equation:

$$K_D = \frac{V(C_f - C_i)}{mC_f}$$

where V is a volume of supernatant solution (mL), m – weight of used adsorbent, C_i and C_f – initial and final concentrations of ion in supernatant solution, μg/mL. Noteworthy extremely high affinity of nanocrystalline sodium titanate toward to Pb^{2+}: here we can see nearly complete removal of lead from supernatant solution. The nature of this phenomenon requires further investigation, as well as extended experiments with other representatives of synthesized titanates.

Additionally, we suggest to consider layered titanates as prospective sorbents for dangerous radionuclides such as ^{137}Cs, ^{90}Sr, ^{99}Tc and transuranium elements (Pu, Np, Am, Cm). After radionuclide sorption, the sorbents might be converted (through sintering or hot pressing) into new stable crystalline phases, which are chemically resistant and stable under environmental conditions. Preliminary experiments have demonstrated that dispersed powders of layered titanates interact actively with aqueous solutions of Tc^{7+} in a broad range of pH, technetium activity and various chemical composition of solution. This process is accompanied with reducing Tc from (7+) to (4+) valence state.

References

Andersson S, Wadsley AD (1961) The crystal structure of $Na_2Ti_3O_7$. Acta Cryst 14:1245–1249

Bavykin DV, Friedrich JM, Walsh FC (2006) Protonated titanates and TiO_2 nanostructured materials: synthesis, properties, and applications. Adv Mater 18:2807–2824

Britvin SN, Siidra OI, Lotnyk A, Krivovichev SV, Depmeier W (2010a) Niobate and tantalate pyrochlores: soft synthesis by the fluoride route. Eur J Inorg Chem 2010:1082–1088

Britvin SN, Spiridonova DV, Siidra OI, Lotnyk A, Kienle L, Krivovichev SV, Depmeier W (2010b) Synthesis, structure and properties of hydrazinium germanate pharmacosiderite, $(N_2H_5)_3Ge_7O_{15}(OH)\cdot 2.5H_2O$. Micropor Mesopor Mat 131:282–288

Britvin SN, Lotnyk A, Kienle L, Krivovichev SV, Depmeier W (2011) Layered hydrazinium titanate: advanced reductive adsorbent and chemical toolkit for design of titanium dioxide nanomaterials. J Am Chem Soc 133:9516–9525

Doong R, Kao I (2008) Fabrication and characterization of nanostructured titanate materials by the hydrothermal treatment method. Recent Pat Nanotech 2:84–102

England WA, Birkett JE, Goodenough JB, Wiseman PJ (1983) Ion exchange in the $Cs_x[Ti_{2-x/2}Mg_{x/2}]O_4$ structure. J Solid State Chem 49:300–308

Ewing FJ (1935) The crystal structure of lepidocrocite. J Chem Phys 3:420–424

Grey IE, Li C, Madsen IC, Watts JA (1987) The stability and structure of $Cs_x(Ti_{2-x/4})O_4$ $0.61 < x < 0.65$. J Solid State Chem 66:7–19

Kwiatkowska J, Grey IE, Madsen IC, Bursill LA (1987) An X-ray and neutron diffraction study of $Cs_2Ti_5O_{11}$ and $Cs_2Ti_5O_{11}(X_2O)$, X = H, D. Acta Cryst B43:258–265

Ma R, Bando Y, Sasaki T (2003) Nanotubes of lepidocrocite titanates. Chem Phys Lett 380:577–582

Morgado E Jr, de Abreu MAS, Moure GT, Marinkovic BA, Jardim PM, Araujo AS (2007) Characterization of nanostructured titanates obtained by alkali treatment of TiO_2-anatases with distinct crystal sizes. Chem Mater 19:665–676

Peng C-W, Richard-Plouet M, Ke T-Y, Lee C-Y, Chiu H-T, Marhic C, Puzenat E, Lemoigno F, Brohan L (2008) Chimie douce route to sodium hydroxo titanate nanowires with modulated structure and conversion to highly photoactive titanium dioxides. Chem Mater 20:7228–7236

Sasaki T, Watanabe M, Komatsu Yu, Fujiki Y (1985) Na^+/H^+ ion-exchange process on layered hydrous titanium dioxide. Bull Chem Soc Jpn 58:3500–3505

Sasaki T, Komatsu Y, Fujiki Y (1989) Rb^+ and Cs^+ incorporation mechanism and hydrate structures of layered hydrous titanium dioxide. Inorg Chem 28:2776–2779
Sasaki T, Watanabe M, Michiue Y, Komatsu Y, Izumi F, Takenouchi S (1995) Preparation and acid-base properties of a protonated titanate with the lepidocrocite-like layer structure. Chem Mater 7:1001–1007
Verbaere A, Tournoux M (1973) Contribution a l'etude des titanates de metaux monovalents. Etude structurale de $Tl_2Ti_4O_9$. Bull Soc Chim Fr 4:1237–1241
Yang J, Jin Z, Wang X, Li W, Zhang J, Zhang S, Guo X, Zhang Z (2003) Study on composition, structure and formation process of nanotube $Na_2Ti_2O_4(OH)_2$. J Chem Soc Dalton Trans: 3898–3901

Minerals as Materials – Silicate Sheets Based on Mixed Rings as Modules to Build Heteropolyhedral Microporous Frameworks

Marcella Cadoni and Giovanni Ferraris

1 Introduction

In general, a material is classified as porous if its crystal structure is crossed by channels with a free aperture (effective channel width = *ecw*) at least 3.2 Å wide, such that the nitrogen molecule, N_2, can pass through. In oxysalts, the *ecw* is calculated by subtracting the ionic diameter of O^{2-} (2.8 Å) to the O. . .O distances across the channel (McCusker et al. 2003).

For the purpose of finding new microporous crystalline materials for technological applications, frameworks built on more than one type of coordination polyhedra (*i.e. heteropolyhedral frameworks*) offer a range of crystal-chemical opportunities by far wider than zeolites (Rocha and Lin 2005). In fact, the higher coordination polyhedra (*e.g.* octahedra) that occur in the heteropolyhedral frameworks can host a variety of cations. Heteropolyhedral microporous structures are well represented in the mineral realm and often were the incentive for the synthesis of new compounds suitable for technological applications (*e.g.* ion exchange and luminescence related to the presence of lanthanides in the framework). Several heteropolyhedral groups of minerals possess channels and cavities and potentially can act as technological porous materials (Ferraris and Merlino 2005): labuntsovites, eudialytes, tobermorites, palysepioles, rhodesite-type structures, pharmacosiderites, sodalites, tunnel oxides, tobermorite-type phases, cancrinites, chrysotile and apatites.

With reference to the type of porous crystal structures discussed in this article, we recall that Krivovichev (2009) has defined a *ring symbol* as a string $p_1^{r1}p_2^{r2}\ldots p_n^{rn}$, where $p_1, p_2\ldots p_n$ are numbers of nodes in a ring and $r_1, r_2\ldots r_n$ are relative numbers of the corresponding rings in a graph. If the nodes represent TO_4

M. Cadoni • G. Ferraris (✉)
Dipartimento di Scienze Mineralogiche e Petrologiche and Nanostructured Interfaces and Surfaces (NIS) Centre of Excellence, Università di Torino,
Via Valperga Caluso, 35, I-10125 Torino, Italy
e-mail: giovanni.ferraris@unito.it

S.V. Krivovichev (ed.), *Minerals as Advanced Materials II*,
DOI 10.1007/978-3-642-20018-2_14,

tetrahedra, one obtains two-dimensional sheets that occur in (alumino)silicates and not necessarily are strictly planar. Some geometrical aspects of these sheets have been analysed (Liebau 1985; Haile and Wuensch 1997) emphasising the condensation of wollastonite-type chains to form xonotlite-type double chains that contain eight-rings. A detailed nomenclature to classify the tetrahedral sheets, in particular naming tertiary, quaternary and non-tertiary those tetrahedra sharing three, four and less than three vertices, has been introduced (Liebau 1985). The most common $p_1^{r1}p_2^{r2}\ldots p_n^{rn}$ silicate sheet is 6^1 that occurs in layered silicates like micas; this sheet is generated by the periodic repetition of a pyroxene-type chain and all tetrahedra point to a same direction. In other cases mixed up- (u) and down-pointing (d) tetrahedra allow the formation of geometrical isomers. If only tertiary tetrahedra occur, as in micas and in the rare earth elements (REE) compounds described in this paper, $T/O = 2/5$; the presence of non-tertiary tetrahedra implies $T/O < 2/5$. Finally, quaternary tetrahedra require $T/O > 2/5$ and the sheet must be either branched or doubled to form microporous layers, as in the rhodesite series (Cadoni and Ferraris 2009, 2010). The sheets can act as building modules either of *s.s.* layered structures, where the strong bonds are confined within layers that are interconnected by weak bonds (typically phyllosilicates), or of heteropolyhedral frameworks. The latter are obtained when the silicate sheets are interconnected by strong bonds *via* five-, six-, or higher-coordinated polyhedra and often are crossed by channels.

The microporous structural characteristics of rhodesite-type structures – based on 4^18^1sheets – synthetic heteropolyhedral silicates with formula $A_3REESi_6O_{15}\cdot 2.25H_2O$ (A = Na, K, H_3O; REE = Ce, La, Eu) – based on $4^15^16^18^2$ sheets – and of polymorphically related phases (including sazhinite) – based on 5^28^1, $4^16^18^1$ and $4^15^16^28^2$ sheets – will be discussed.

2 Synthesis

Crystals of synthetic compounds suitable for single-crystal X-ray diffraction were obtained as products of hydrothermal syntheses, at about 500 K and autogenous pressure, using 25 ml Teflon-lined stainless-steel autoclaves as exemplified in the following; for details see Cadoni et al. (2010) and Cadoni and Ferraris (2009, 2010).

TR09 (Table 1) with some amorphous material was obtained from a gel of composition $1SiO_2$: $0.6Na_2O$: 0.12Sr: $40H_2O$, autoclaved for 10 d. The gel was prepared from a strong alkaline solution containing NaOH, NaCl and fumed SiO_2 to which an aqueous solution of $SrCl_2$ was added drop wise and thoroughly stirred.

For the $A_3REESi_6O_{15}\cdot 2.25H_2O$ compounds (Table 2) an alkaline solution was made by mixing fumed SiO_2 in a NaOH solution; REE salts were then added and the solution thoroughly stirred. Gels with composition $0.8\text{-}xNa_2O{:}xK_2O{:}1.0SiO_2{:}0.05REE_2O_3{:}55H_2O$ were autoclaved for 8 d.

Table 1 Members of the rhodesite mero-plesiotype series

Name	Chemical formula	a, b, c(Å), β (°)	Space group
Seidite-(Ce)[a]	$Na_4(Ce,Sr)_2\{Ti(OH)_2[Si_8O_{18}]\}(O,OH,F)_4 \cdot 5H_2O$	24.61, 7.23, 14.53, 94.6	*C2/c*
Rhodesite[b]	$KCa_2[Si_8O_{18}(OH)] \cdot 6H_2O$	23.416, 6.555, 7.050	*Pmam*
Macdonaldite[c]	$BaCa_4[Si_8O_{18}(OH)]_2 \cdot 10H_2O$	14.081, 13.109, 23.560	*Cmcm*
Delhayelite[d]	$K_4Na_2Ca_2[AlSi_7O_{19}]_2F_2Cl$	24.579, 7.057, 6.581	*Pmmn*
Hydrodelhayelite[e]	$KCa_2[AlSi_7O_{17}(OH)_2] \cdot 3H_2O$	6.648, 23.846, 7.073	$Pn2_1m$
Monteregianite-(Y)[f]	$KNa_2Y[Si_8O_{19}] \cdot 5H_2O$	9.512, 23.956, 9.617, 93.85	$P2_1/n$
Fivegite[g]	$K_4Ca_2[AlSi_7O_{17}(O_{2-x}OH_x)][(H_2O)_{2-x}OH_x]Cl$	24.345, 7.037, 6.540	$Pm2_1n$
AV-5[h]	$KNa_2Ce[Si_8O_{19}] \cdot 5H_2O$	9.579, 24.053, 9.695, 93.654	$P2_1/n$
Eu-AV-9[i]	$KNa_2Eu[Si_8O_{19}] \cdot 5H_2O$	23.973, 14.040, 6.566, 90.35	*C2/m*
Tb-AV-9[i]	$KNa_2Tb[Si_8O_{19}] \cdot 5H_2O$	23.945, 14.019, 6.554, 90.288	*C2/m*
Er-AV-9[j]	$KNa_2Er[Si_8O_{19}] \cdot 5H_2O$	23.951, 14.013, 6.550, 90.288	*C2/m*
Nd-AV-9[k]	$KNa_2Nd[Si_8O_{19}] \cdot 5H_2O$	–	–
Gd-AV-9[k]	$KNa_2Nd[Si_8O_{19}] \cdot 5H_2O$	–	–
TR03[l]	$KNaCa_2[Si_8O_{19}] \cdot 5H_2O$	6.585, 23.776, 7.025	$Pn2_1m$
TR04[l]	$KNa_3Sr[Si_8O_{19}] \cdot 4.3H_2O$	6.5699, 23.7225, 7.0225, 91.81	$P2_1/m$
TR09[m]	$Sr_2Na[Si_8O_{19}] \cdot 4H_2O$	22.7681, 6.9352, 13.5789, 2.58	*C2/c*
TR10[m]	$SrNa_4[Si_8O_{19}] \cdot 4H_2O$	22.4102, 7.0292, 13.3140, 2.54	*P2/c*

[a] Ferraris et al. (2003)
[b] Hesse et al. (1992)
[c] Cannillo et al. (1968)
[d] Cannillo et al. (1970), Pekov et al. (2009)
[e] Ragimov et al. (1980)
[f] Ghose et al. (1987)
[g] Pekov et al. (2010)
[h] Rocha et al. (2000)
[i] Ananias et al. (2001)
[j] Ananias et al. (2004)
[k] Rocha et al. (2004) without crystal data
[l] Cadoni and Ferraris (2009)
[m] Cadoni and Ferraris (2010)

3 Rhodesite Series

Minerals and synthetic phases (Table 1) belonging to the rhodesite mero-plesiotype series (Ferraris and Gula 2005) are well-known modular (Ferraris et al. 2008) microporous silicate phases based on a heteropolyhedral framework (Ferraris and Merlino 2005). The crystal structures of this series are based on a double silicate layer that is formed via corner sharing of two apophyllite-type tetrahedral sheets. The layer is crossed by eight-membered channels along two different directions: the *ecw* of the channels is in the range 3.5–3.8 Å along [001] and about 2.7 × 4.3 Å along [010] (Fig. 1). The silicate module alternates with different types of "octahedral" (*O*) layer where Na, K, Ca, Ba, Sr and lanthanides are typical cations.

Table 2 Chemical and crystal data of compounds with formula A_3REE[Si_6O_{15}]·$n$$H_2O$

A_3	REE	n	Unit cell parameters (Å)	Space group	Si_6O_{15} sheet
Na_3[a]	Ce	2	7.50, 15.62, 7.35	*Pmm*2	$4^16^18^1$
Na_3[b]	La	2	7.415, 15.515, 7.164	*Pmm*2	$4^16^18^1$
Na_3[c]	Nd	2	7.385, 30.831, 7.117	*Cmm*2	$4^15^16^18^1$
$Na_{2.74}K_{0.26}$[d]	Ce	2.25	7.413, 30.965, 7.176	*Cmm*2	$4^15^16^18^1$
$Na_{2.72}K_{0.28}$[d]	La	2.25	7.422, 31.039, 7.196	*Cmm*2	$4^15^16^18^1$
Na_3[d]	La	2.25	7.413, 30.965, 7.176	*Cmm*2	$4^15^16^18^1$
$Na_{2.74}(H_3O)_{0.26}$[d]	$La_{0.68}Eu_{0.32}$	2.25	7.391, 30.873, 7.127	*Cmm*2	$4^15^16^18^1$
$Na_{2.4}$[e]	Ce	2	7.309, 14.971, 7.135	*Pman*	5^28^1
K_3[f]	Nd	2	7.279, 16.008, 15.004	*Pmcb*	$4^16^18^1$

[a] Sazhinite-(Ce); Shumyatskaya et al. (1980)
[b] Sazhinite-(La); Cámara et al. (2006)
[c] Haile et al. (1997)
[d] Cadoni et al. (2010)
[e] Jeong et al. (2007); to make easier the comparison with sazhinite and the *Cmm*2 synthetic compounds, the *b* and *c* parameters given in the original paper are exchanged with consequent modification of the space group symbol from *Pmna* to *Pman*
[f] Haile and Wuensch (2000); the *a*, *b* and *c* parameters given in the original paper are renamed, in the order, *b*, *c* and *a* with consequent modification of the space group symbol from *Pbam* to *Pmcb*

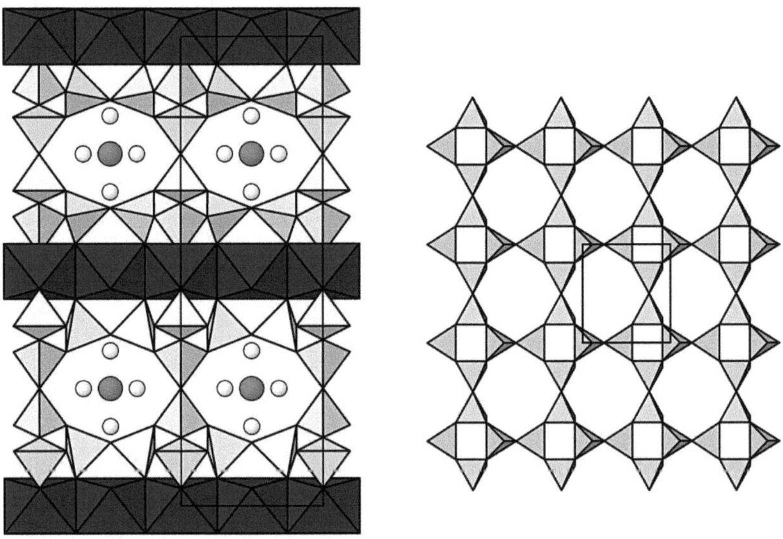

Fig. 1 Crystal structure of rhodesite seen along [010] and (*right side*) the apophyllite-type 4^18^1 silicate sheet occurring in this structure. *Large* and *small circles* represent alkalis and H_2O molecules in the channels, respectively

According to the nomenclature illustrated in the Introduction, the apophyllite-type sheet has topology 4^18^1 being built on four- and eight-rings only. According to the categorization proposed by Makovicky (1997), the rhodesite series is merotypic because the common silicate module alternates with a variable *O* layer; at the same time, the series is plesiotypic because the ratio between the number of up- and down-pointing tetrahedra in the apophyllite-type sheets may vary between members of the series and Al can substitute Si.

The rhodesite-type microporous structure shows at the same time layered and heteropolyhedral-framework features. The variety of chemical compositions that occur in the natural and synthetic members of the rhodesite mero-plesiotype series demonstrates the versatility of this type of heteropolyhedral structure. In fact, the layer anion $(Si_8O_{19})^{6-}$ can be balanced by playing on the following features: (1) type of the *O* layer that connects the double silicate layers into a heteropolyhedral framework; (2) nature (including REE), number and charge of the cations in the *O* layer; (3) presence of either H_2O or F at the corners of the polyhedra in the *O* layer; (4) nature, number and charge of the cations hosted in the channels; (5) presence of either H_2O or Cl in the channels. Further, the negative charge of the anion can be increased via the substitution Al -->Si (Table 1). The alkali cations with a coordination number higher than six which are hosted in the *O* layer establish coordination bonds with oxygen atoms of the silicate double layer. Thus, these cations are in contact with the channels and leaching may be favoured as noted, e.g., in the heterophyllosilicate delindeite (Ferraris et al. 2001).

In particular the modular characteristic of the rhodesite-type structure allows – *via* the synthesis of chemically different compounds that share the silicate layer – the search of compounds suitable for technological applications (*e.g.*, ion exchange and luminescence due to the presence of REE in the framework). Interest in the investigation of the rhodesite-type structures is increasing (cf references given in Table 1) because of their peculiar crystal-chemical, mineralogical and technological aspects. Concerning technology, Rocha et al. (1998) have shown that dehydrated synthetic monteregianite (AV-1) and rhodesite (AV-2) can rehydrate; the same two compounds have been used as catalysers for the isomerisation of D-glucose to D-fructose (Lima et al. 2008); synthetic rhodesite uptakes ammonia exchanging Na and K (Grutzeck and Marks 1999); a rhodesite-type phase likely occurs as a product of alkali-silica reaction in concrete (De Ceukelaire 1991). Recently Cadoni and Ferraris (2009, 2010) have synthesised four new members of the rhodesite series; they are indicated as TR03, TR04. TR09 and TR10 in Table 1. TR03 and TR04 have a mineralogical interest because they can be usefully compared to hydrodelhayelite and to delhayelite (Table 1). TR09 and TR10 show that the presence of the large cation Sr^{2+} in the "octahedral" layer squeezes the eight-membered channels that develop along [001] as mentioned above (Fig. 2).

4 Sazhinite-Related Structures

The ideal chemical formula $A_3REE[Si_6O_{15}]{\cdot}nH_2O$ ($A =$ usually alkalis, $n \sim 2$) is common (Table 1) to the minerals sazhinite-(Ce) (Shumyatskaya et al. 1980) and sazhinite-(La) (Cámara et al. 2006) and to a group of synthetic compounds studied by Haile et al. (1997), Haile and Wuensch (2000), Jeong et al. (2007),

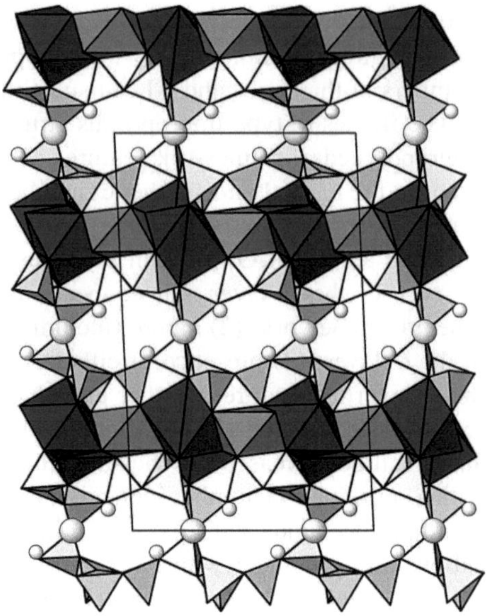

Fig. 2 Crystal structure of TR09 seen along [010] showing the deformation of the channels in comparison with those present in rhodesite (Fig. 1). *Circles* as in Fig. 1

Cadoni et al. (2010) and Cadoni and Ferraris (2011). The structures of all these compounds contain mixed-rings sheets built by the polymerization of a xonotlite-type double chain. The resulting anion $[Si_6O_{15}]^{6-}$ assumes, however, different configurations depending on the mechanism of repetition. Independently of the sheet configuration, the crystal structures of Table 2 are built by stacking one type of (001) sheet along the [001] direction; the sheets are connected by REE coordination polyhedra. The resulting heteropolyhedral frameworks are microporous, being crossed by wide [100] channels; these are delimited by eight-rings formed by six Si tetrahedra and two REE polyhedra. With reference to the unit-cell content, the following structures types occur in the compounds listed in Table 2.

In sazhinite (Fig. 3) the sheet consists of two xonotlite-type double chains related by an *m* (010) mirror plane. Between the two chains a [100] row of alternating four- and six-rings is formed and finally the topology of the sheet is $4^1 6^1 8^1$.

If the unit-cell content of sazhinite is duplicated by an **a**/2 shear the structures of Table 2 with space group *Cmm*2 are obtained (Fig. 4).

Consequently to the shear, a [100] chain of five-rings is formed at the interface between two $4^1 6^1 8^1$ sazhinite-type blocks: the ring symbol of the sheet becomes $4^1 5^1 6^1 8^1$. The resulting structure is in polytipic relation with that of sazhinite; the latter represents the $1O_1$ member of a family of polytypes of which the former structures are the $2O_2$ members (Cadoni and Ferraris 2011).

The structure (Fig. 5) described by Jeong et al. (2007) is obtained if two contiguous xonotlite-type double chains are related by an *a* (010) glide; the resulting topology of the sheet is $5^2 8^1$. This structure represents a polymorph of sazhinite where the different topology of the sheet and the consequent octahedral coordination of Ce are the main modifications.

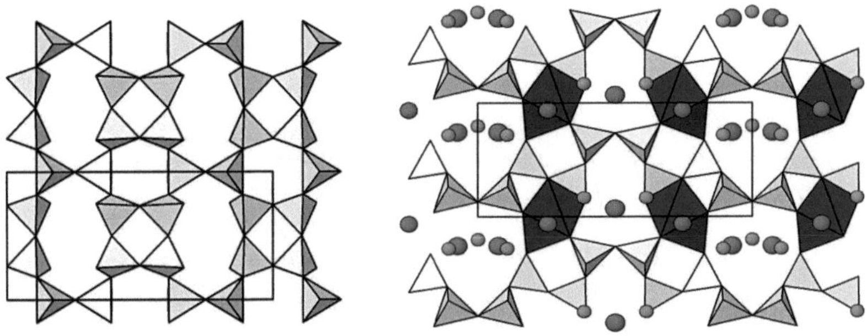

Fig. 3 $4^1 6^1 8^1$ silicate (001) sheet (*left side*) occurring in the structure of sazhinite and (*right side*) view of the structure seen along [100]. *Circles* as in Fig. 1

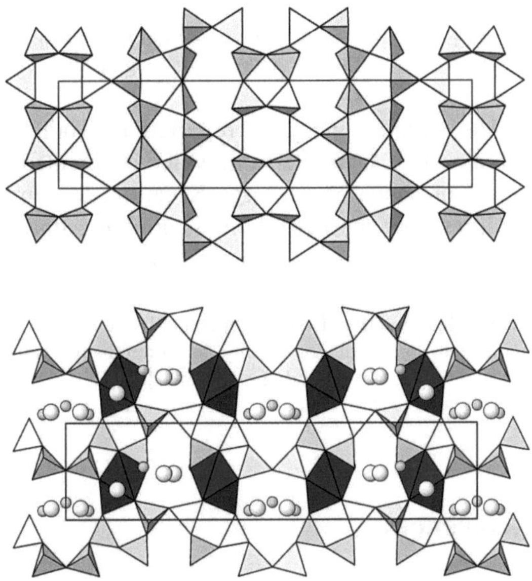

Fig. 4 $4^1 5^1 6^1 8^1$ silicate (001) sheet (*top*) occurring in the structure of $Na_{2.74}K_{0.26}Ce[Si_6O_{15}]{\cdot}2.25H_2O$ and (*bottom*) view of the structure seen along [100]. *Circles* as in Fig. 1

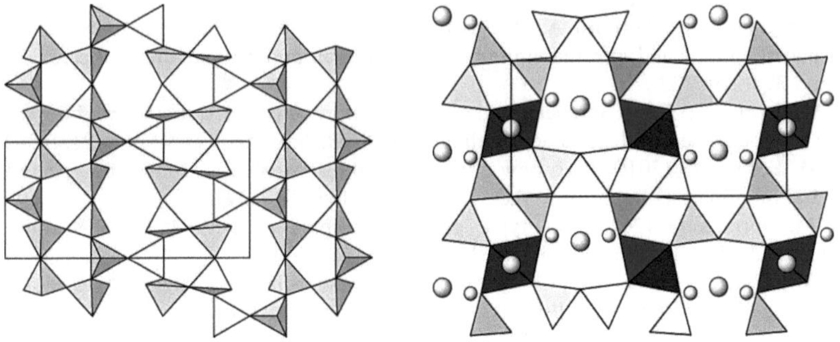

Fig. 5 $5^2 8^1$ silicate (001) sheet (*left side*) occurring in the structure of $Na_{2.4}Ce[Si_6O_{15}]{\cdot}2H_2O$ and (*right side*) view of the structure seen along [100]. *Circles* as in Fig. 1

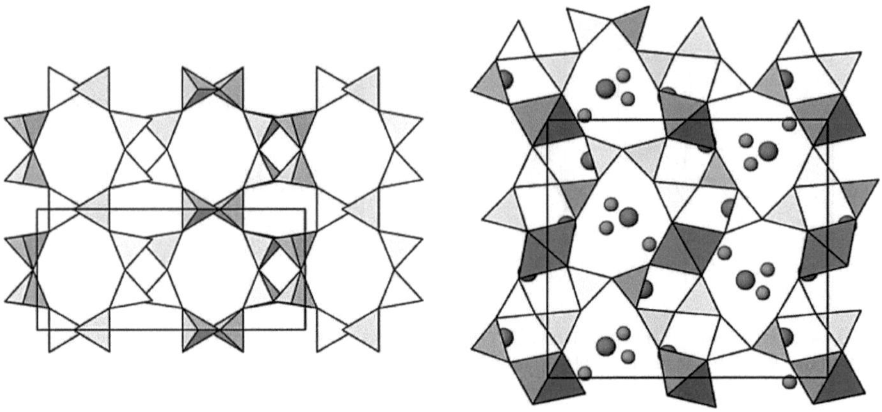

Fig. 6 $4^1 6^1 8^1$ silicate (001) sheet (*left side*) occurring in the structure of $K_3Nd[Si_6O_{15}]{\cdot}2H_2O$ and (*right side*) view of the structure seen along [100]. *Circles* as in Fig. 1

The structure (Fig. 6) reported by Haile and Wuensch (2000) is a second polymorph of sazhinite from which is obtained *via* a substantial distortion required by the presence of the large K cation. In comparison to sazhinite, the distortion modifies both the periodicities, doubling that along [001], and the profile of the silicate sheet, thus favouring octahedral coordination for Nd.

Acknowledgements Research financially supported by MIUR [Roma, PRIN project "Compositional and structural complexity in minerals (crystal chemistry, microstructures, modularity, modulations): analysis and applications"].

References

Ananias D, Ferreira A, Rocha J, Ferreira P, Rainho JP, Morais C, Carlos LD (2001) Novel microporous europium and terbium silicates. J Am Chem Soc 123:5735–5742

Ananias D, Rainho JP, Ferreira A, Rocha J, Carlos LD (2004) The first examples of X-ray phosphors, and C-band infrared emitters based on microporous lanthanide silicates. J Alloys Compd 374:219–222

Cadoni M, Ferraris G (2009) Synthesis and crystal structure of two new members of the rhodesite mero-plesiotype series close to delhayelite and hydrodelhayelite. Eur J Miner 21:485–493

Cadoni M, Ferraris G (2010) Two new silicate structures based on a rhodesite-type heteropolyhedral microporous framework. Acta Crystallogr B66:151–157

Cadoni M, Ferraris G (2011) Polytypic and polymorphic relations between sazhinite and isochemical alkali-REE layer silicates. Eur J Miner 23:85–90

Cadoni M, Cheah YL, Ferraris G (2010) New RE microporous heteropolyhedral silicates containing $4^1 5^1 6^1 8^2$ tetrahedral sheets. Acta Crystallogr B66:158–164

Cámara F, Ottolini L, Devouard B, Garvie LAJ, Hawthorne FC (2006) Sazhinite-(La), $Na_3LaSi_6O_{15}(H_2O)_2$, a new mineral from the Aris phonolite, Namibia: description and crystal structure. Mineral Mag 70:405–418

Cannillo E, Rossi G, Ungaretti L, Carobbi SG (1968) The crystal structure of macdonaldite. Atti Accad Naz Lincei Classe Sci Fis 45:399–414
Cannillo E, Rossi G, Ungaretti L (1970) The crystal structure of delhayelite. Rend Soc Ital Mineral Petrol 26:63–75
De Ceukelaire L (1991) The determination of the most common crystalline alkali-silica reaction product. Mater Struct 24:169–171
Ferraris G, Gula A (2005) Polysomatic aspects of microporous minerals –Heterophyllosilicates, palysepioles and rhodesite-related structures. Rev Mineral Geochem 57:69–104
Ferraris G, Merlino S (eds) (2005) Micro- and mesoporous mineral phases. Mineralogical Society of America, Washington DC
Ferraris G, Ivaldi G, Pushcharovsky DYu, Zubkova N, Pekov IV (2001) The crystal structure of delindeite, $Ba_2\{(Na, K, D)_3(Ti, Fe)[Ti_2(O, OH)_4Si_4O_{14}](H_2O, OH)_2\}$, a member of the mero-plesiotype bafertisite series. Can Mineral 39:1307–1316
Ferraris G, Belluso E, Gula A, Khomyakov AP, Soboleva SV (2003) The crystal structure of seidite-(Ce), $Na_4(Ce, Sr)_2\{Ti(OH)_2(Si_8O_{18})\}(O, OH, F)_4{\cdot}5H_2O$, a modular microporous titanosilicate of the rhodesite group. Can Mineral 41:1183–1192
Ferraris G, Makovicky E, Merlino S (2008) Crystallography of modular materials. IUCr/Oxford University Press, Oxford
Ghose S, Sen Gupta PK, Campana CF (1987) Symmetry and crystal structure of monteregianite, $Na_4F_2Y_2Si_{16}O_{38}{\cdot}10H_2O$, a double-sheet silicate with zeolitic properties. Am Mineral 72:365–374
Grutzeck MW, Marks JA (1999) Synthesis of double-layer silicates from recycled glass cullet: a new type of chemical adsorbent. Environ Sci Technol 33:312–317
Haile SM, Wuensch BJ (1997) Comparison of the crystal chemistry of selected MSi_6O_{15}-based silicates. Am Mineral 82:1141–1149
Haile SM, Wuensch BJ (2000) Structure, phase transitions and ionic conductivity of $K_3NdSi_6O_{15}{\cdot}H_2O$. I. α-$K_3NdSi_6O_{15}{\cdot}2H_2O$ and its polymorphs. Acta Crystallogr B56:335–348
Haile SM, Wuensch BJ, Laudise RA, Maier J (1997) Structure of $Na_3NdSi_6O_{15}{\cdot}H_2O$ - a layered silicate with paths for possible fast-ion conduction. Acta Crystallogr B53:7–17
Hesse KF, Liebau F, Merlino S (1992) Crystal structure of rhodesite, $HK_{1-x}Na_{x+2y}Ca_{2-y}$ $\{lB,3,2_{\infty}^{2}\}[Si_8O_{19}]$ $(6\text{-}z){\cdot}H_2O$, from three localities and its relation to other silicates with dreier double layers. Z Kristallogr 199:25–48
Jeong HK, Chandrasekaran A, Tsapatsis M (2007) Synthesis of a new framework cerium silicate and its structure determination by single crystal X-ray diffraction. Chem Commun 2007:2398–2399
Krivovichev SV (2009) Structural crystallography of inorganic oxysalts. IUCr/Oxford University Press, Oxford
Liebau F (1985) Structural chemistry of silicates: structure, bonding and classification. Springer, Berlin
Lima S, Dias AS, Lin Z, Brandão P, Ferreira P, Pillinger M, Rocha J, Calvino-Casilda V, Valente AA (2008) Isomerization of D-glucose to D-fructose over metallosilicate solid bases. Appl Catal A Gen 339:21–27
Makovicky E (1997) Modularity – different types and approaches. EMU Notes Mineral 1:315–344
McCusker LB, Liebau F, Engelhardt G (2003) Nomenclature of structural and compositional characteristics of ordered microporous and mesoporous materials with inorganic hosts (IUPAC recommendations 2001). Micropor Mesopor Mat 58:3–13
Pekov IV, Zubkova NV, Chukanov NV, Sharygin VV, Pushcharovskii DYu (2009) Crystal chemistry of delhayelite and hydrodelhayelite. Dokl Earth Sci 428:1216–1221
Pekov IV, Zubkova NV, Chukanov NV, Zadov AE, Pushcharovskii DYu (2010) Fivegite, $K_4Ca_2[AlSi_7O_{17}(_{O2-x}OH_x)][(H_2O)_{2-x}OH_x]Cl$, a new mineral from the Khibiny alkaline complex, Kola Peninsula, Russia. Zapiski RMO 2010(4):47–63 (in Russian)
Ragimov KG, Chiragov MI, Mamedov KS, Dorfman MD (1980) Crystal structure of hydrodelhayelite, $KH_2Ca(Si,Al)_8O_{19}.H_2O$. Dokl Akad Nauk Azerbaid SSR 36:49–51 (in Russian)

Rocha J, Lin Z (2005) Microporous mixed octahedral-pentahedral-tetrahedral framework silicates. Rev Mineral Geochem 57:173–201

Rocha J, Ferreira P, Lin Z, Brandão P, Ferreira A, Pedrosa de Jesus JD (1998) Synthesis and structural characterization of microporous yttrium and calcium silicates. J Phys Chem B102:4739–4744

Rocha J, Ferreira P, Carlos LD, Ferreira A (2000) The first microporous framework cerium silicate. Angew Chem Int Ed 39:3276–3279

Rocha J, Carlos LD, Ferreira A, Rainho J, Ananias D, Lin Z (2004) Novel microporous and layered luminescent lanthanide silicates. Mater Sci Forum 455–456:527–531

Shumyatskaya NG, Voronkov AA, YaA P (1980) $Na_2Ce[Si_6O_{14}(OH)]\cdot nH_2O$: a new representative of the dalyite family in crystal chemistry. Sov Phys Crystallogr 25:419–423

Cs-Exchanged Cuprosklodowskite

Andrey A. Zolotarev, Sergey V. Krivovichev, and Margarita S. Avdontseva

In connection with the problem of radioactive waste disposal, investigations of uranium minerals have received a special importance. Significant attention is paid to the studies of minerals capable to incorporation of radionuclides into their structures. Burns (1999) and Burns and Li (2002) studied structures of Cs-boltwoodite and Sr-becquerelite obtained in the course of ion-exchange reactions on single crystals. Ion exchange was also investigated for several synthetic uranium compounds (Shvareva et al. 2005a, b, 2007; Ok et al. 2006).

The purpose of this work is to study Cs-exchanged form of cuprosklodowskite obtained as a result of ion-exchange process between 2 M solution of CsCl and cuprosklodowskite. Cuprosklodowskite $Cu(UO_2)_2(SiO_3OH)_2{\cdot}6(H_2O)$ (Vaes 1933; Novacek 1935; Piret-Meunier and Van Meerssche 1963; Rosenzweig and Ryan 1963) is a layered copper uranyl silicate from Shinkolobve deposit, Democratic Republic of Kongo, where it occurs as bright-green crystals. Cuprosklodowskite is a typical secondary mineral and formed as the result of alteration of primary uranium minerals such as uraninite. Cuprosklodowskite belongs to the group of uranium silicate with ratio uranium/silicon 1 to 1, which also includes uranophane, boltwoodite, sklodowskite, etc. Cuprosklodowskite has a triclinic symmetry (Piret-Meunier and Van Meerssche 1963; Rosenzweig and Ryan 1963), space group *P*-1 with unit cell parameters: a = 6.654(1), b = 7.048(1), c = 9.255(2) Å, $\alpha = 109.9(1)$, $\beta = 109.1(1)$, $\gamma = 89.88(1)°$. The layered structure of cuprosklodowskite based on the $[(UO_2)_2(SiO_4)_2]^{4-}$ layers consisting of edge-sharing UO_7 pentagonal bipyramids and SiO_4 tetrahedra. Cu^{2+} ions are located in the interlayer (Melon and Dejace 1959).

We obtained pale-green crystals of Cs-exchanged cuprosklodowskite as a result of the ion-exchange process. Single crystals of cuprosklodowskite were heated in 2 M solution CsCl at 170°C during 10 days. Structural study was carried out on

A.A. Zolotarev (✉) • S.V. Krivovichev • M.S. Avdontseva
Saint-Petersburg State University, 199034 St. Petersburg, Russia
e-mail: aazolotarev@gmail.com

S.V. Krivovichev (ed.), *Minerals as Advanced Materials II*,
DOI 10.1007/978-3-642-20018-2_15,

diffractometer STOE IPDS II (MoKα), equipped with Image Plate area detector (Table 1). The SHELX program (Sheldrick 1997) was used for the determination and refinement of the crystal structure. The quality of crystal was extremely poor, which explains the high R-value obtained in this study. Thus we were able to determine an approximate structure model only, which, however, provides important information on incorporation of Cs into cuprosklodowskite structure.

Cs-exchanged cuprosklodowskite is triclinic, space group *P*-1, a = 6.606 (1), b = 6.987 (1), c = 22.690 (5) Å, α = 81.4 (2), β = 84.2 (2), γ = 89.86 (1)°. In comparison with cuprosklodowskite, Cs-exchanged cuprosklodowskite has a considerably larger c parameter, which is clearly seen in reconstructed sections of reciprocal diffraction space (Figs. 1 and 2). There are "additional" reflections with diffuse scattering for Cs-exchanged cuprosklodowskite that indicate the increase of parameter c. It can be seen that diffraction streaks are elongated along the c* axis

Table 1 Data of single crystal experiment of the cuprosklodowskite and Cs-exchanged cuprosklodowskite

Paremeter	Cuprosklodowskite	Cs- cuprosklodowskite
a (Å)	6.652(4)	6.606(1)
b (Å)	7.055(5)	6.987(1)
c (Å)	9.261	22.690(5)
α (°)	109.716(1)	81.41(1)
β (°)	109.104(1)	84.20(2)
γ (°)	89.849(1)	89.86(1)
V ($Å^3$)	383.69	1030.10
Space group	P-1	P-1
Z	2	3
Diffractometer	STOE IPDS II	STOE IPDS II
Radiation	MoKα	MoKα
Unique reflections with $\lvert F_o \rvert \geq 4\sigma_F$	1,631	2,807
R_1	0.0506	0.2154
wR_2	0.1459	0.4632

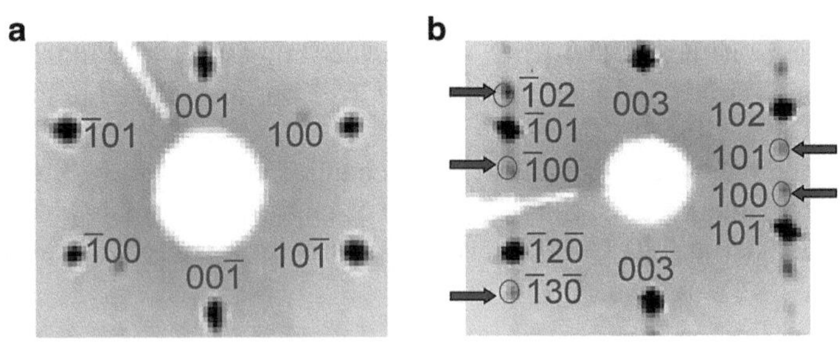

Fig. 1 Reciprocal diffraction space, plane (010), k = 0: (**a**) cuprosklodowskite, (**b**) Cs-exchanged cuprosklodowskite (*arrows* indicate the additional superstructure reflections that appears in the ion-exchanged crystal)

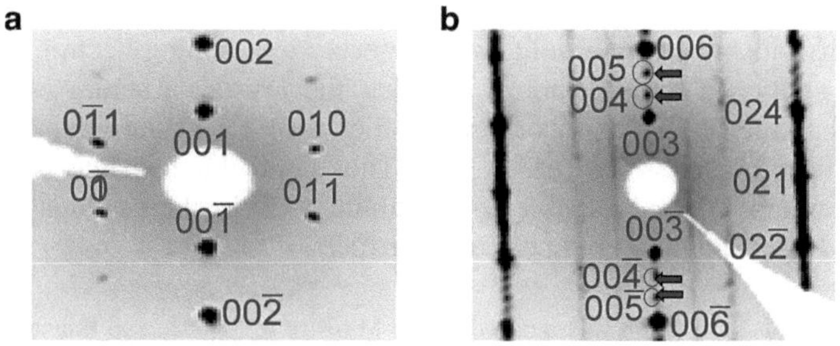

Fig. 2 Reciprocal diffraction space, plane (100), h = 0: (**a**) cuprosklodowskite, (**b**) Cs-exchanged cuprosklodowskite (*arrows* indicate the additional superstructure reflections that appears in the ion-exchanged crystal). Note almost continuous diffuse streaks along the c* axis

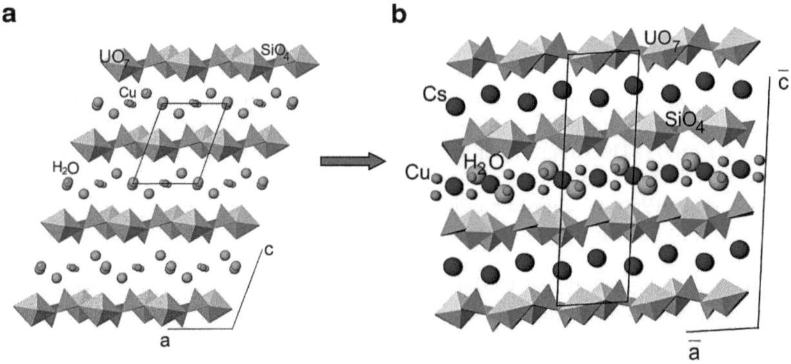

Fig. 3 Crystal structures of: (**a**) cuprosklodowskite, (**b**) Cs-exchanged cuprosklodowskite

that indicates a disorder in the stacking sequence of uranyl silicate layers parallel to (001).

Structural model of Cs-exchanged cuprosklodowskite demonstrates that Cs is incorporated into the interlayer space of the structure by partial replacement of the water molecules adjacent to the Cu^{2+} ions. The process is accompanied by significant distortion of the uranyl silicate layers. The interlayers, which are equivalent in cuprosklodowskite, in its Cs-exchanged form, splits into two interlayers, one solely occupied by Cs^{+} ions and another occupied by both Cs^{+} and Cu^{2+} ions (Fig. 3). This leads to the tripling of the c parameter resulting in the appearance of additional superstructure reflections visible in Figs. 1 and 2.

Powder diffraction study of Cs-exchanged cuprosklodowskite was carried out on diffractometer STOE STADI P (CuKα). In comparison with cuprosklodowskite, powder diffraction pattern of Cs-exchanged form is of very poor quality.

Microprobe chemical analysis was performed using scanning electron microscope CamScan MX2500. The empirical chemical formula of cuprosklodowskite is $Cu_{0,84}[(UO_2)((Si_{0,92}O_{2,52}(OH)_{1,48}))]_2{\cdot}6H_2O$, whereas empirical formula of its Cs-exchanged form can be written as $Cu_{0,32}Cs_{1,48}[(UO_2)((Si_{0,93}O_{2,62}(OH)_{1,38}))]_2$

nH_2O, n = 1–2. It is of interest that, for some cuprosklodowskite crystals, oriented intergrowths with crystals of kasolite were observed by scanning electron microscopy.

In this study we have demonstrated that the crystal structure of cuprosklodowskite has the ability to incorporate radioactive Cs in the course of the ion exchange reaction: $2Cs^+ \rightarrow Cu^{2+} + H_2O$. Thus cuprosklodowskite is yet another layered uranium mineral capable to accumulate Cs in the secondary processes occurring during oxidation of spent nuclear fuel.

Acknowledgments We thank Russian Federation Ministry of Science and Education for financial support through the Federal Program Grant and Grant of President of Russia for young scientists (MK-1783.2010.5). This work was also supported through the internal research programme of St. Petersburg State University.

References

Burns PC (1999) Cs-boltwoodite obtained by ion exchange from single crystals: implications for radionuclide release in a nuclear repository. J Nucl Mater 265:218–223

Burns PC, Li Y (2002) The structures of becquerelite and Sr-exchanged becquerelite. Am Mineral 87:550–557

Melon J, Dejace J (1959) La cuprosclodowskite. Acad Roy Belg Bull Cl Sci 45:507–515

Novacek R (1935) Revise druhotnych uranovych nerostu z Jachimova. Narodni Mus Casopis 109:100–107

Ok KM, Baek J, Halasyamani PS, O'Hare D (2006) New layered uranium phosphate fluorides: syntheses, structures, characterizations, and ion-exchange properties of $A(UO_2)F(HPO_4).xH_2O$ ($A = Cs^+, Rb^+, K^+$; x = 0–1). Inorg Chem 45:10207–10214

Piret-Meunier J, Van Meerssche M (1963) Structure de la jachimovite. Acad Roy Belg Bull Cl Sci 49:181–191

Rosenzweig A, Ryan RR (1963) Refinement of the crystal structure of cuprosklodowskite, Cu $(UO_2)_2$ $(SiO_3OH)_2{\cdot}6(H_2O)$. Am Mineral 60:448–453

Sheldrick GM (1997) SHELXL-97. Program for crystal structure refinement. University of Gottingen, Gottingen

Shvareva TY, Almond PM, Albrecht-Schmitt TE (2005a) Crystal chemistry and ion – exchange of the layered uranyl iodide $K[UO_2(IO_3)_3]$. J Solid State Chem 178:499–504

Shvareva TY, Sullens TA, Shehee TC, Albrecht-Schmitt TE (2005b) Syntheses, structures, and ion-exchange properties of the three-dimensional framework uranyl gallium phosphates, $Cs_4[(UO_2)_2(GaOH)_2(PO_4)_4]H_2O$ and $Cs[UO_2Ga(PO_4)_2]$. Inorg Chem 45:300–305

Shvareva TY, Skanthakumar S, Soderholm L, Clearfield A, Albrecht Schmitt TE (2007) Cs^+ selective ion exchange and magnetic ordering in a three-dimensional framework uranyl vanadium(IV) phosphate. Chem Mater 19:132–134

Vaes JF (1933) Sur un mineral de Kalongwe (Katanga). Ann Soc Geol Belg 56:331–332

Kinetics and Mechanisms of Cation Exchange and Dehydration of Microporous Zirconium and Titanium Silicates

Nikita V. Chukanov, Anatoliy I. Kazakov, Vadim V. Nedelko, Igor V. Pekov, Natalia V. Zubkova, Dmitry A. Ksenofontov, Yuriy K. Kabalov, Arina A. Grigorieva, and Dmitry Yu. Pushcharovsky

1 Introduction

Synthetic and natural aluminosilicate zeolite materials are widely used in industry, agriculture and laboratory practice for water purification, sorption of toxic substances and radionuclides, gas dehydration, and as catalysts and catalyst supports in organic synthesis.

From the beginning of 1990s, synthetic zeolite-like materials of a new type (primarily alkaline titano-, niobo- and zirconosilicates) attach growing interest due to a considerable diversity of their structures and properties. Crystal structures of these microporous materials are based on heteropolyhedral frameworks containing SiO_4 tetrahedra and octahedra centered by "strong" cations like Ti, Nb, Zr, Ta, Sn, W, *etc.* (see Chukanov et al. 2004). Usually, heteroframework microporous materials are synthesized by the hydrothermal method from oxides, or by the dry-gel conversion method. In most cases, fine-powder products or multiphase mixtures are obtained, that complicates the investigation of their crystal structures and topochemical mechanisms of cation-exchange processes proceeding therein. On the other hand, more than 100 alkaline Ti-, Nb- and Zr-silicates with heteropolyhedral crystal structures are known in Nature (mainly in alkaline pegmatites), and the number of these minerals permanently increases (Chukanov and Pekov 2005; Pekov and ChukanovNV 2005). In most cases, crystals of natural microporous heteroframework silicates (HFS) are suitable for detailed crystal-chemical and topochemical investigations. Moreover, many of them have never been synthesized.

N.V. Chukanov (✉) • A.I. Kazakov • V.V. Nedelko
Institute of Problems of Chemical Physics RAS, Chernogolovka, Russia
e-mail: chukanov@icp.ac.ru

I.V. Pekov • N.V. Zubkova • D.A. Ksenofontov • Y.K. Kabalov • A.A. Grigorieva • D.Y. Pushcharovsky
Faculty of Geology, Moscow State University, 119991 Moscow, Russia

S.V. Krivovichev (ed.), *Minerals as Advanced Materials II*,
DOI 10.1007/978-3-642-20018-2_16,

Thus available minerals provide a good possibility to test technologically important properties of various materials before methods of their synthesis are elaborated.

2 Kinetics of Cation-Exchange Reactions of Terskite

Although a large number of data on the synthesis and properties of HFS has been published during the last two decades, kinetics of cation-exchange reactions of these materials remained unstudied until recently. In order to elaborate methodological approach to such investigations, terskite, $Na_4ZrSi_6O_{15}(OH)_2 \cdot H_2O$, was chosen by us as a model object. Unaltered terskite shows low activity towards ion exchange with different cations. Highly-hydrated, amorphized and undergone natural leaching of a part of Na variety of terskite from the Shkatulka pegmatite, Lovozero alkaline complex, Kola peninsula, Russia, is, on the contrary, an effective ionite (Chukanov et al. in press). Its empirical formula is: $H_x(Na_{2.0}K_{0.1}Ca_{0.1}U_{0.04}Th_{0.03})(Zr_{1.03}Ti_{0.10}Nb_{0.07})(Si_{5.94}P_{0.04}Al_{0.02}O_{16}) \cdot nH_2O$. The kinetics of ion exchange of terskite with excess of aqueous solutions of CsF, CsCl, $CuSO_4$, K_2SO_4 and $SrCl_2$, as well as with mixed solutions CsF + $CuSO_4$ and CsCl + $CuSO_4$, has been investigated by the dynamic calorimetry method, using differential calorimeter DAK-1-2 (Chukanov et al. 2010; in press). The following experimental conditions have been realized: temperature range 27.5–55.2°C, concentrations from 0.6 to 2.2 mol/l for CsF; 29.4°C and concentration of 0.6 mol/l for other salts.

The presence of extensive rectilinear segments $dQ/dt = kQ_0\text{-}kQ$ on the dependences of thermal flux on heat release (Figs. 1 and 2) indicates that the reaction follows first-order kinetic low on time: $d\eta/dt = k(1-\eta)$, where

$$\eta = \frac{\int_0^t \frac{dQ}{dt} dt}{\int_0^\infty \frac{dQ}{dt} dt}$$

is a reaction conversion. The refined kinetic order on salt concentration C_{CsF} is equal to 0.92. The total kinetic equation is $k = 3.1\ 10^3\ C_{CsF}^{0.92} \cdot \exp(-21/RT)$, where the value of activation energy of 21 kJ/mol was obtained from the Arrhenius dependence.

Local saturating concentrations of Cs_2O in terskite vary in the limits 25.5–29.1 and do not depend on temperature and concentration of CsF solution; the variations are determined only by spacial nonuniformity of the material. Although heat efficiency of the substitution of Na for Cs is very low (3.4 kJ per mol of cations), under the experimental conditions the equilibrium of ion-exchange is strongly shifted towards Cs-substituted form of terskite. It means that fixation of Cs^+ cations in zeolite channels is controlled mainly by entropy factors.

The above data show that terskite-based sorbents could be of practical interest in technologies of the ^{137}Cs extraction from natural, industrial and drinking water. In spite of the fact that terskite shows high affinity towards cesium, Cs^+ cations do

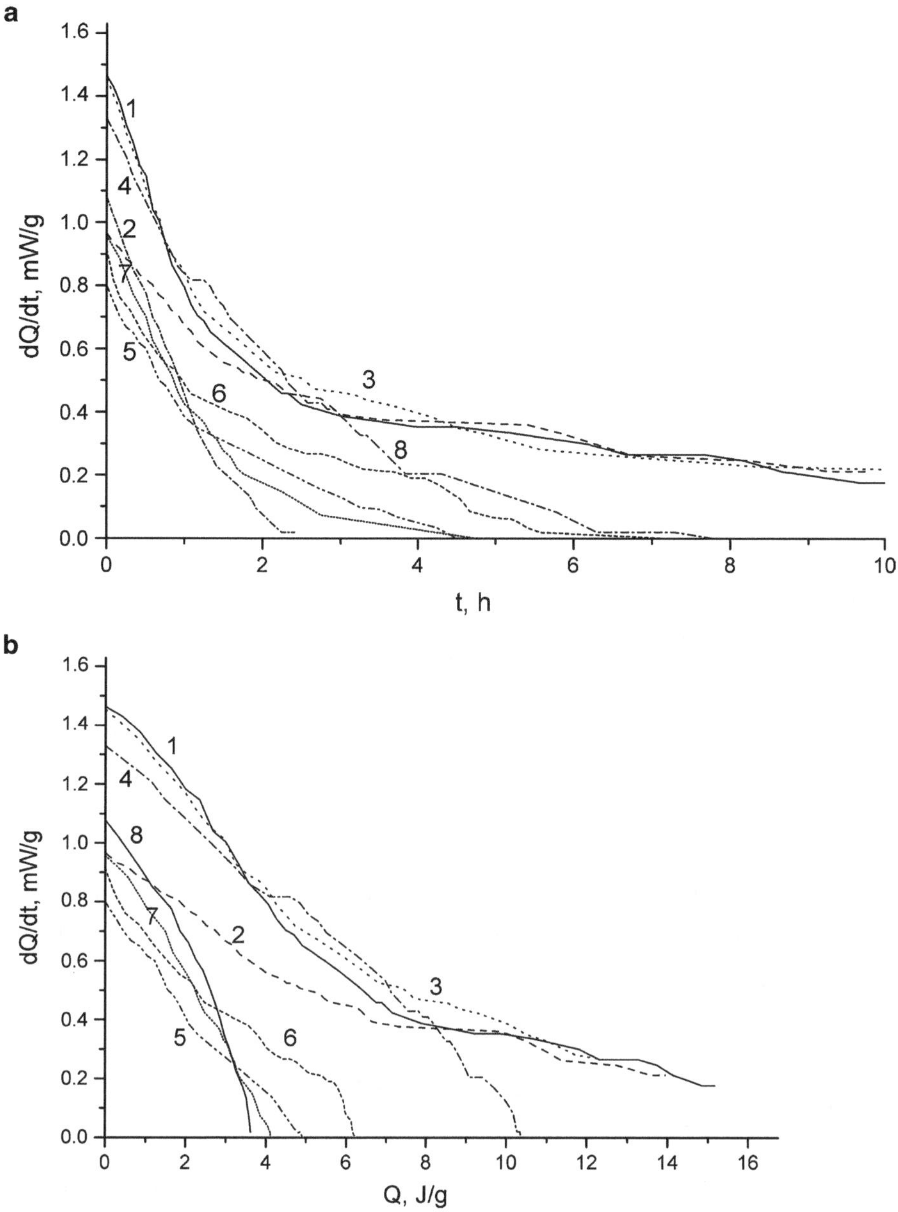

Fig. 1 The dependences of thermal flux on time (**a**) and on heat release (**b**) in the course of ion exchange of Na^+ for Cs^+ in terskite: 1 – 27.5°C, 2 – 29.4°C, 3 – 30.3°C, 4 – 32.4°C, 5 – 36.9°C, 6 – 40.1°C, 7 – 44.9°C, 8 – 55.2°C

not displace U and Th from this material. Moreover, despite intensive natural hydrothermal exposure, U and Th have been preserved in terskite for a long geological time. Taking this into account, terskite can be considered as a possible perspective material for the immobilization of actinides.

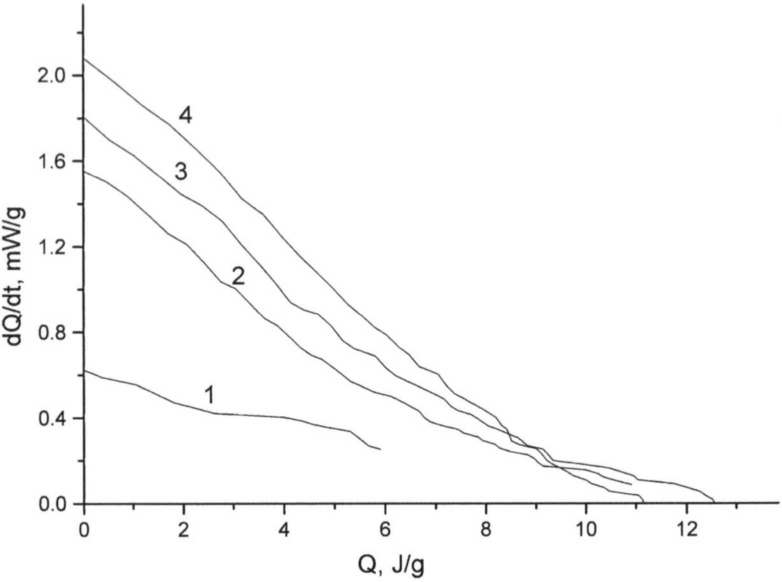

Fig. 2 The dependences of thermal flux on heat release in the course of ion exchange of Na^+ for Cs^+ in terskite at 29.4°C: 1 – 0.6, 2 – 1.0, 3 – 1.5, 4 – 2.2 mol/l CsF

The kinetics of ion-exchange reactions of terskite with aqueous solutions of CsCl, CsF, $CuSO_4$, K_2SO_4 and $SrCl_2$ as well as with mixed solution CsCl + $CuSO_4$ has been investigated under the same conditions (Figs. 3 and 4; Table 1). Reactions with large univalent cations (K^+, Cs^+) are exothermal, whereas substitution of Na^+ for relatively small bivalent cations (Cu^{2+}, Sr^{2+}) proceeds with heat absorption. Endothermic reactions with Cu^{2+} and Sr^{2+} are caused by the increase of entropy upon the entry of these cations in zeolite channels due to the existence of numerous variants of their allocation there. In the presence of different cations in the solution (*e. g.* Cs^+ + Cu^{2+}, Fig. 4), energy factors prevail over entropy ones.

3 Kinetics of Ion-Exchange Reactions of Different Zr- and Ti-silicates with Cs^+

The kinetics of ion-exchange reactions of six minerals with heteropolyhedral framework structures [heterosilicates: sitinakite, $KNa_2Ti_4Si_2O_{13}(OH){\cdot}4H_2O$ (s), penkvilksite, $Na_4Ti_2Si_8O_{22}{\cdot}5H_2O$ (p), zorite, $Na_6Ti(Ti,Nb)_4(Si_6O_{17})_2O(OH)_4{\cdot}11H_2O$ (z), terskite, $Na_4ZrSi_6O_{15}(OH)_2{\cdot}H_2O$ (t), gaidonnayite, $Na_2ZrSi_3O_9{\cdot}2H_2O$ (g), and kuzmenkoite-Mn, $K_2Mn(Ti,Nb)_4(Si_4O_{12})_2(OH,O)_4{\cdot}nH_2O$ (k)], and so-called zirfesite (zf), an amorphous product of natural hydrolysis and cation leaching from eudialyte, with aqueous solutions of salts of CsF was investigated by the dynamic calorimetry method (Chukanov et al. in press). Except the reaction with sitinakite that is accompanied

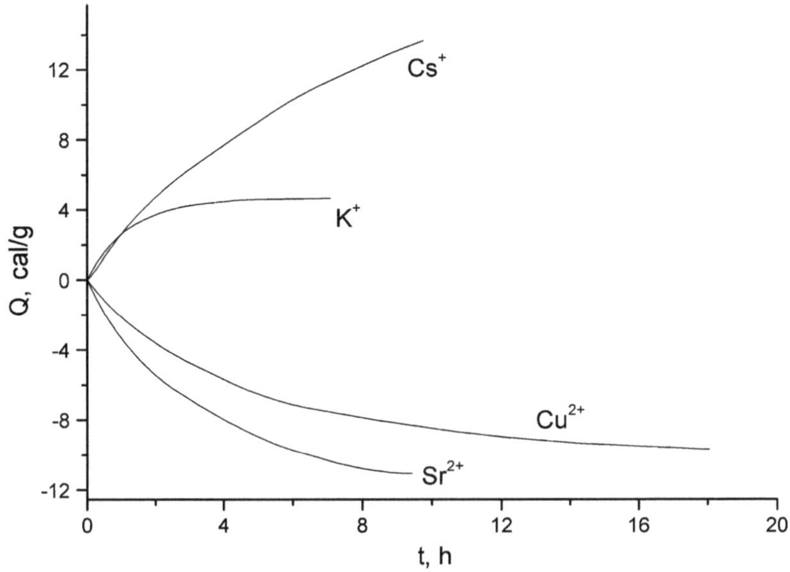

Fig. 3 The time dependences of heat release per gram of terskite in ion-exchange reactions with aqueous solutions of CsF, $CuSO_4$, K_2SO_4 and $SrCl_2$ (29.4°C, salt concentration 0.6 mol/l)

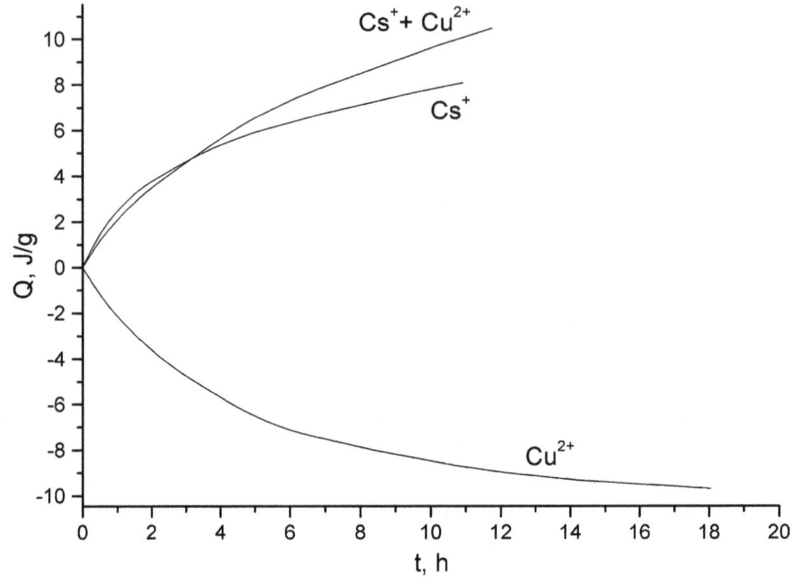

Fig. 4 The time dependences of heat release per gram of terskite in ion-exchange reactions with aqueous solutions of CsCl, $CuSO_4$ and mixed solution CsCl + $CuSO_4$ (29.4°C, salt concentration 0.6 mol/l)

Table 1 The values of first order rate constants and total heat release in ion-exchange reactions of terskite with aqueous solutions of different salts

Salt	$CuSO_4$	CsF	CsCl	$SrCl_2$	K_2SO_4
K, h^{-1}	0.25	0.31	0.64	0.34	0.84
Q, J/g	−9.2	10.9	5.4	−10.9	4.6

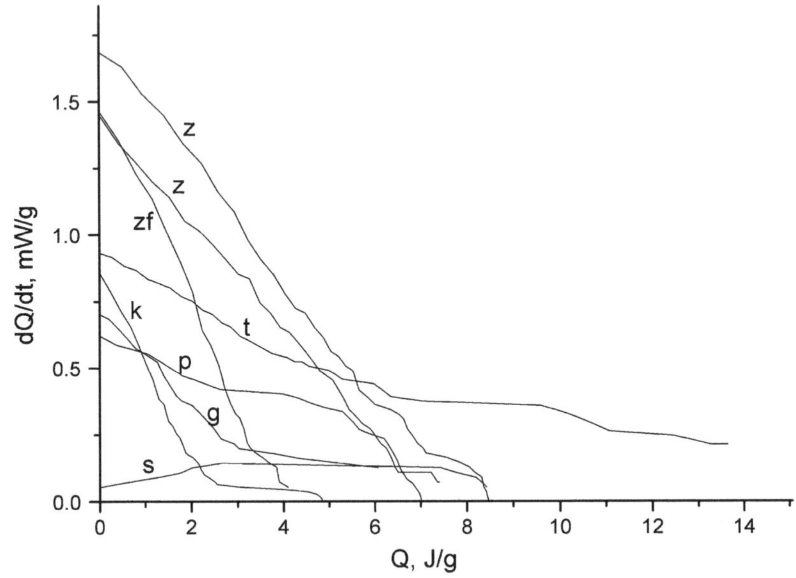

Fig. 5 The dependences of thermal flux on heat release in the course of ion exchange of Na^+ for Cs^+ at 29.4°C in different minerals-sorbents (symbols are introduced in text above)

Table 2 The values of first order rate constants and total heat release in ion-exchange reactions of different sorbents with 0.6 M aqueous solution of CsF

Sorbent	Penkvilksite	Terskite	Gaidonnayite	Zorite	Kuzmenkoite	Zirfesite
k, h^{-1}	0.20	0.31	0.64	0.70	1.30	1.30
Q, J/g	10.7	10.9	4.2	7.5	2.5	4.2

by a very low total heat release, all investigated reactions are exothermal and follow first-order kinetic low on time (Fig. 5, Table 2).

The highest values of rate constants of cation exchange with Cs^+ are observed for zorite (natural analogue of microporous material ETS-4), samples undergone natural leaching of a part of extra-framework cations (terskite, gaidonnayite, kuzmenkoite-Mn) and zirfesite, whereas unaltered sitinakite and penkvilksite are characterized by the lowest k values. Thus partial leaching of extra-framework cations is a possible way of modification of heteroframework microporous materials directed at the enhancement of their activity in ion-exchange processes.

Another possible way of modification of microporous materials could be their thermal dehydration followed by subsequent re-hydration. In particular, elpidite from magmatic rocks is a much more active ionite than elpidite of low-temperature hydrothermal origin. For this reason, kinetics and crystal-chemical mechanism of elpidite thermal dehydration have been investigated.

4 Kinetics and Mechanism of Elpidite Dehydration

The kinetics of dehydration of elpidite, $Na_2ZrSi_6O_{15}\cdot 3H_2O$, has been investigated by TG method, using thermobalance ATV-14 M (Nedelko et al. 2011). The following experimental conditions have been realized: constant temperature (from 110°C to 254°C), argon atmosphere, mean dimensions of elpidite crystals 0.05 × 1.0 mm. Elpidite of hydrothermal genesis from the Elpiditovoye pegmatite, Mt. Alluaiv, Lovozero, was used in all experiments.

Elpidite dehydration proceeds in two stages. The kinetics of the first stage (Fig. 6) follows first-order low, with the rate constant $k_1 = 10^{8.9\pm0.2}\cdot\exp[(-98.74 \pm 3.1)/RT]\ s^{-1}$. The first stage becomes completed after the loss of 50% of H_2O contained in elpidite initially. The second stage proceeds with a low rate whose time dependence can not be described by a simple kinetic low (Figs. 7 and 8). Its activation energy E_a was determined from the temperature dependence of initial rates. The lowering of E_a from 98.74 kJ/mol on the first stage to 76.5 kJ/mol on the second stage can be explained by the weakening of hydrogen bonds, as it is seen from the high-frequency shift of the bands of O-H stretching vibrations in the range 3,400–3,600 cm^{-1} (Figs. 9 and 10).

In accordance with structural data obtained by Rietveld method (Zubkova et al. 2011), partial (50%) dehydration of elpidite ($a = 7.1136(1)$, $b = 14.6764(2)$, $c = 14.5977(2)$ Å; space group *Pbcm*), is accompanied by the transformation of its heteropolyhedral framework [into that with $a = 14.0899(1)$, $b = 14.4983(1)$, 14.3490(1) Å, space group *Cmce*], and subsequent H_2O evolution proceeds from a transformed crystal having contracted channels (Fig. 11; see Chukanov et al. in press) that results in a substantial lowering of the dehydration rate.

5 Cation Exchange in Hilairite: Dependence of Crystal Chemical Mechanism on Type of Cation

Natural zirconosilicate hilairite, $Na_2Zr[Si_3O_9]\cdot 3H_2O$, possesses a low-density Zr, Si-framework: 15.7 Si and Zr atoms per 1,000 Å^3 (Chukanov and Pekov 2005), and has pronounced cation-exchange properties (Zubkova et al. 2009). Our experiments in aqueous solutions of K, Rb, Cs, Ca, Sr, Ba and Pb salts showed strong

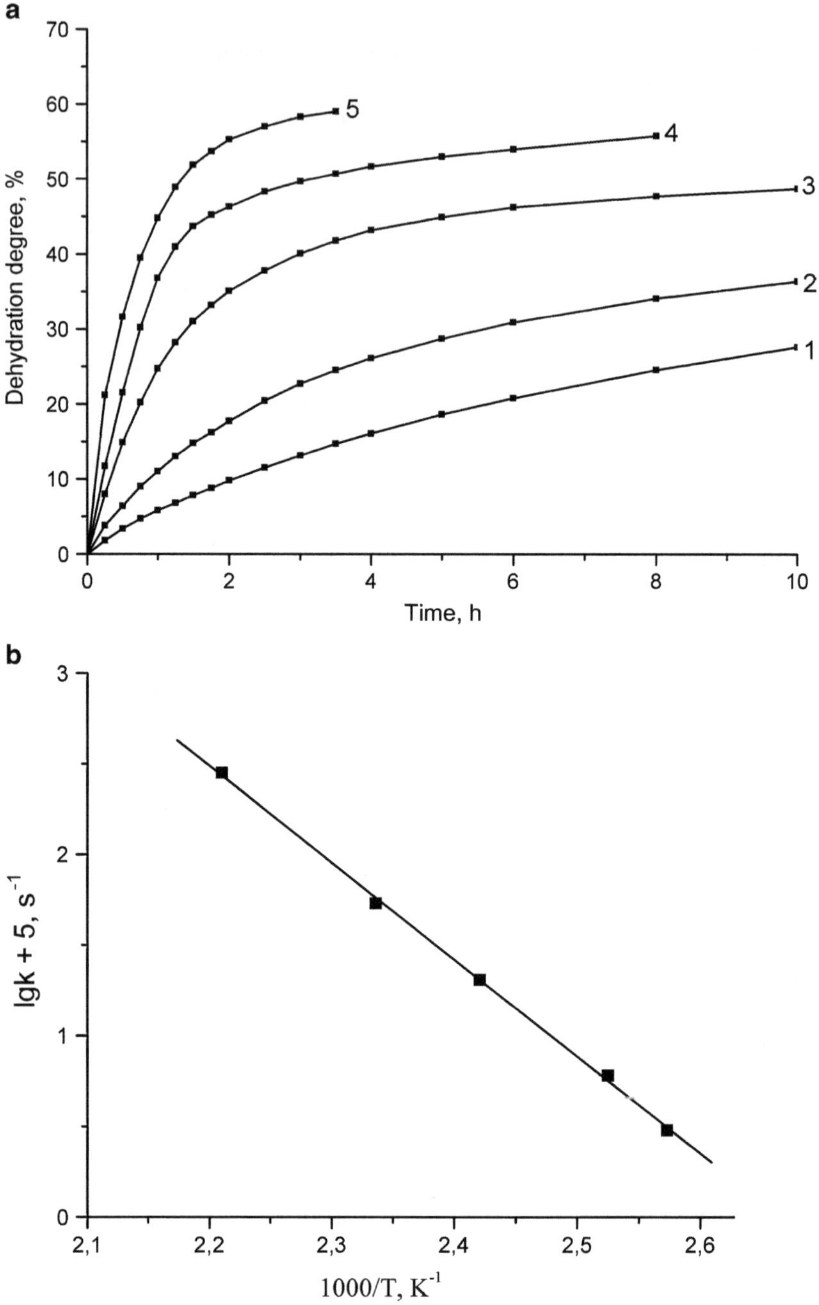

Fig. 6 (**a**) Kinetics of the first stage of elpidite dehydration at 110°C (1), 123°C (2), 140°C (3), 155°C (4) and 180°C (5). Sample weight 80–85 mg. (**b**) Arrhenius dependence for the first stage of elpidite dehydration

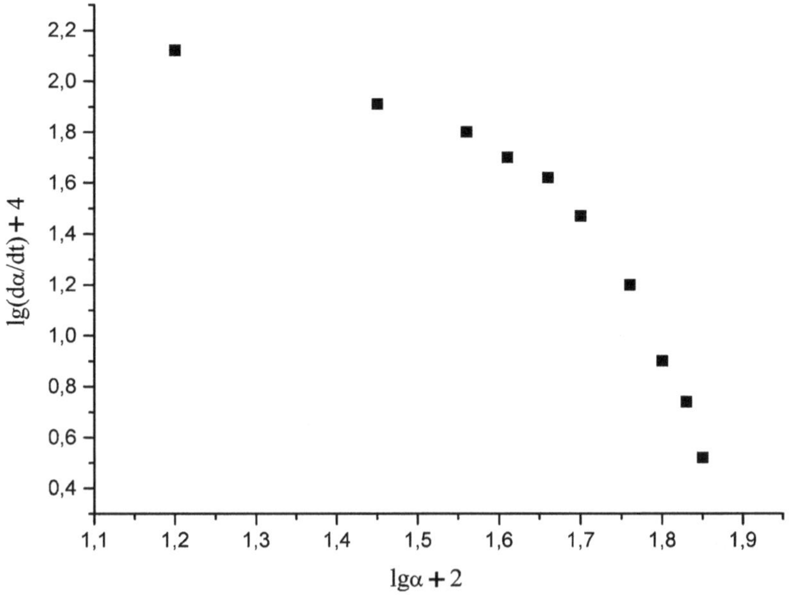

Fig. 7 The dependence of lg(dα/dt) on lgα for the second stage of elpidite dehydration at 198°C

dependence of exchange limit and structural characteristics of cation-substituted forms on type of cation and on temperature.

The initial material for the experiments was hilairite from Kirovskii Mine, Khibiny alkaline complex, Kola peninsula. It is trigonal, $R32$, $a = 10.553$, $c = 7.954$ Å. The base of its structure (Fig. 12a) is a framework formed by twisted tetrahedral chains $[Si_3O_9]^{\infty}$ and isolated octahedra ZrO_6. Na cations occupy sites of two types: between Zr octahedra and in zeolitic channel; H_2O molecules are located only in channel (Grigorieva et al. 2009).

Temperature increase results in the extension of the group of cations that react with hilairite and in the general increase of exchange degree (Grigorieva et al. 2011). Above-listed cations show different behavior tendencies in these reactions. For alkali cations, the exchange weakens with increase of ionic radius (from K to Cs) whereas alkali-earth cations show contrary tendency: at 90°C hilairite absorbs only a little Ba but no exchange with Sr and Ca observed, while at 150°C Ba or Sr replace Na in maximally possible amounts unlike Ca that shows exchange no more than for 80% of its possible limit. Generally, our data show the following empirical sequence of decrease of hilairite affinity to extra-framework cations in cation-exchange reactions: K → Rb → Pb → Ba → Sr → Ca → Cs.

Bivalent cations occupy only the site between Zr octahedra keeping the sites in the zeolitic channel for residual Na and H_2O molecules. On the contrary, univalent (alkali) cations are located in sites of both types. Being larger than Na, they displace H_2O molecules from zeolitic channel and, as a result, H_2O content in cation-substituted forms of hilairite strongly depends on the amount of such cations therein.

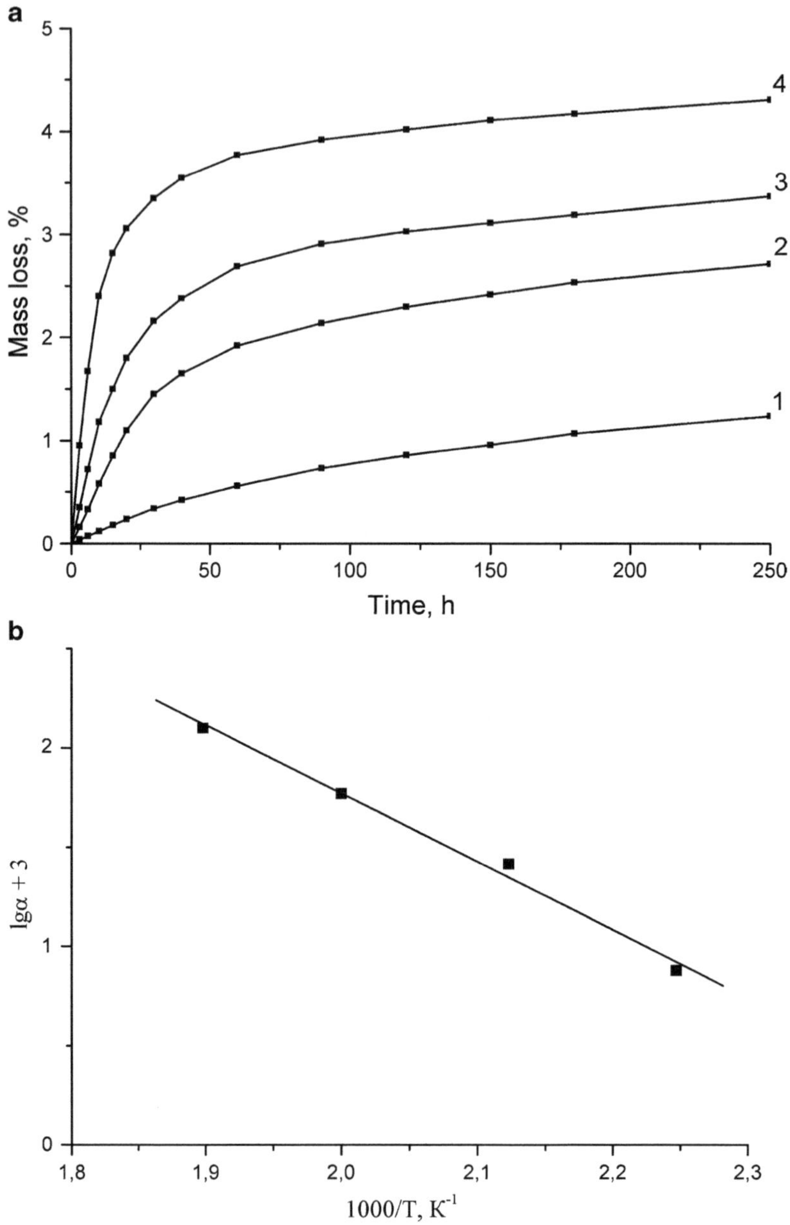

Fig. 8 (**a**) Kinetics of the second stage of elpidite dehydration (for 96–98 mg samples, preliminarily 50% dehydrated at 160°C) at 172°C (1), 198°C (2), 227°C (3) and 254°C (4). (**b**) Arrhenius dependence for initial rates of the second stage of elpidite dehydration

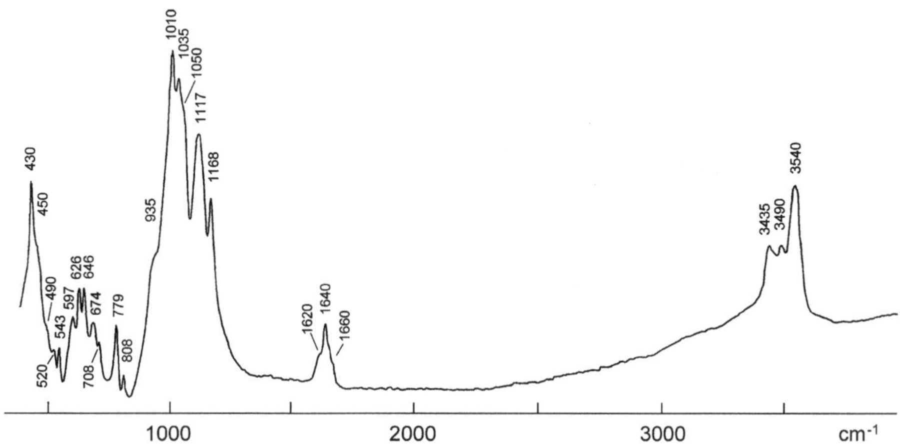

Fig. 9 Infrared (IR) spectrum of elpidite

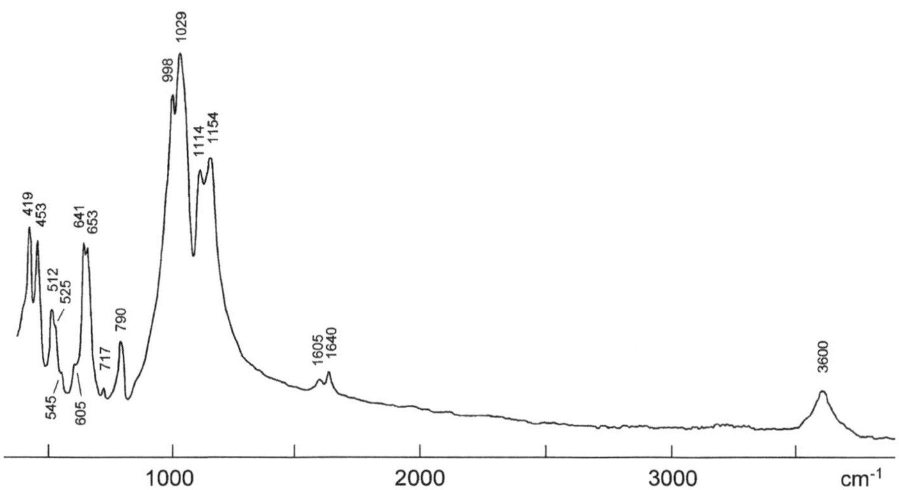

Fig. 10 Infrared (IR) spectrum of the product of partial (50%) dehydration of elpidite

The forms maximally enriched in K and Rb are almost or even completely water-free (Fig. 12b). In the Cs-substituted form, all Cs sites are split into numerous sub-sites with low occupancy (Fig. 12c). Very similar phenomenon was earlier found in Cs-substituted titanosilicate zorite (Zubkova et al. 2005).

Increase of contents of cations larger than Na and change of H_2O content in ion-substituted forms of hilairite provoke the distortion of its heteropolyhedral Zr, Si,O-framework (Fig. 12): stretching or pressing of the twisted tetrahedral Si-chains, tilt of the Zr octahedr and some deformations of polyhedra of both types. These changes cause symmetry lowering from *R*32 to *R*3 (cation-substituted

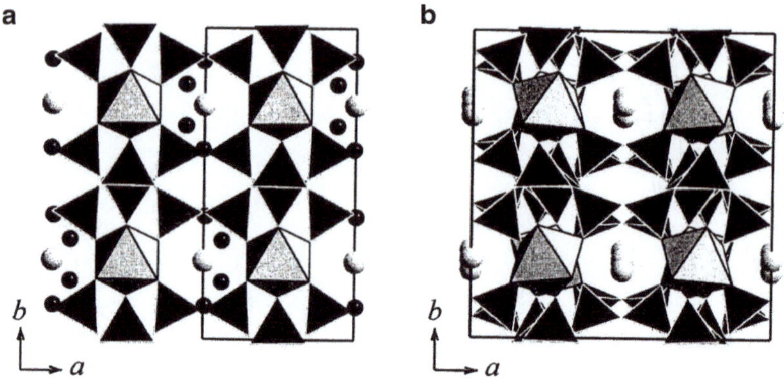

Fig. 11 Crystal structures of initial (**a**) and dehydrated (**b**) elpidite projected along the *c* axis

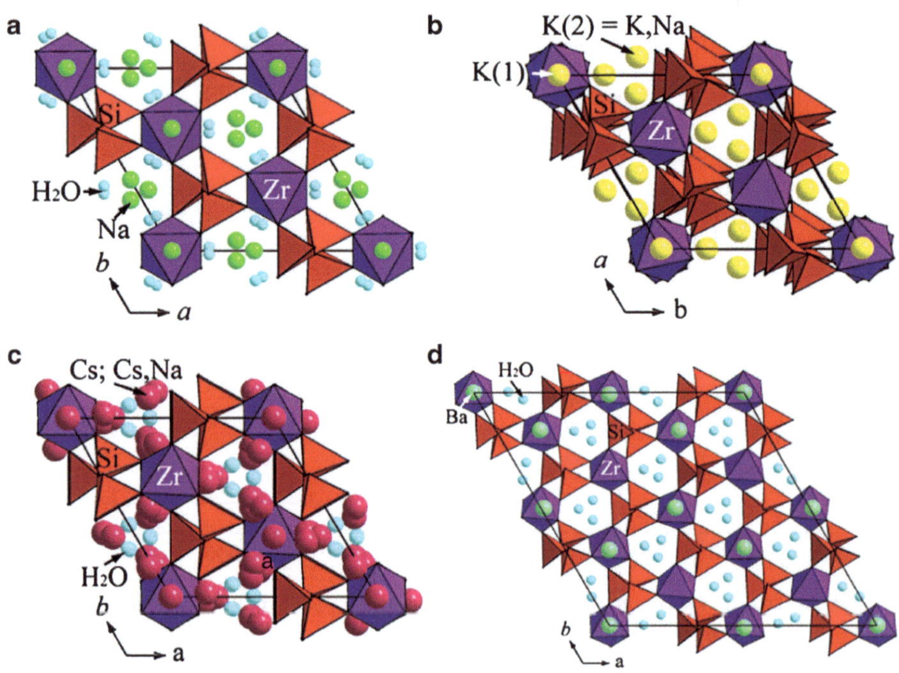

Fig. 12 Crystal structures of hilairite from Khibiny (**a**) and its cation-substituted forms obtained at 150°C: (**b**) K, (**c**) Cs, (**d**) Ba

forms except Cs-exchanged one) and doubling of the *c* (K-, Rb- and Cs-substituted forms) or the *a* (Sr- and Ba-substituted forms: Fig. 12d) parameters of the unit cell.

Hilairite is a very bright example showing significant flexibility of structures of HFS for post-crystallization changes that causes (1) high adaptability to changed conditions and (2) high selectivity to cations forming stable substituted forms.

Acknowledgements The work was supported by grants of the Russian Foundation for Basic Research (09-05-12001-ofi_m and 09-05-00143-a) and grants of President of Russian Federation (MK-320.2010.5 and NSh-4034.2010.5).

References

Chukanov NV, Pekov IV, Rastsvetaeva RK (2004) Crystal chemistry, properties and synthesis of microporous silicates containing transition elements. Russ Chem Rev 73(1):227–246. doi:10.1070/RC2004v073n03ABEH000825

Chukanov NV, Pekov IV (2005) Heterosilicates with tetrahedral-octahedral frameworks: mineralogical and crystal-chemical aspects. Rev Mineral Geochem 57:105–143. doi:10.2138/rmg.2005.57.4 (Micro-and Mesoporous Mineral Phases)

Chukanov NV, Kazakov AI, Pekov IV, Grigorieva AA (2010) Kinetics of cation exchange of amorphized terskite. Russ J Phys Chem A 84(12):2154–2159

Chukanov NV, Kazakov AI, Pekov IV, Grigorieva AA (2011) Kinetics of cation exchange on heteroframework microporous titano- and zirconosilicates. Russ. J Phys Chem B 5:278–283 doi:10.1134/S1990793111020308

Grigorieva AA, Zubkova NV, Pekov IV, Pushcharovsky DYu (2009) Crystal structure of hilairite from Khibiny alkaline massif. Dokl Earth Sci 428(7):1051–1053. doi:10.1134/S1028334X09070010

Grigorieva AA, Pekov IV, Zubkova NV, Turchkova AG, Pushcharovsky DYu (2011) K- and Rb-exchanged forms of hilairite: evolution of crystal-chemical characteristics with the increase of ion exchange temperature. This volume

Nedelko VV, Chukanov NV, Pekov IV (2011) Dehydration kinetics of the microporous zirconosilicate elpidite. Inorganic Materials. 47(5):502–505. doi:10.1134/S0020168511050165

Pekov IV, Chukanov NV (2005) Microporous framework silicate minerals with rare and transition elements: minerogenetic aspects. Rev Mineral Geochem 57:145–171. doi:10.2138/rmg.2005.57.5 (Micro-and Mesoporous Mineral Phases)

Zubkova NV, Pushcharovsky DYu, Giester G, Pekov IV, Turchkova AG, Chukanov NV, Tillmanns E (2005) Crystal structures of K- and Cs-exchanged forms of zorite. Cryst Rep 50:367–373. doi:10.1134/1.1927591

Zubkova NV, Kolitsch U, Pekov IV, Turchkova AG, Vigasina MF, Pushcharovsky DYu, Tillmanns E (2009) Crystal chemistry of Rb-, Sr-, Ba-, Ca- and Pb-exchanged forms of natural hilairite. Eur J Miner 21:495–506. doi:10.1127/0935-1221/2009/0021-1912

Zubkova NV, Ksenofontov DA, Kabalov YuK, Chukanov NV, Nedelko VV, Pekov IV, Pushcharovsky DYu (2011) Dehydration-induced structural transformations of the microporius zirconosilicate elpidite. Inorganic Materials. 47(5):506–512. doi: 10.1134/S0020168511050232

K- and Rb-Exchanged Forms of Hilairite: Evolution of Crystal-Chemical Characteristics with the Increase of Ion Exchange Temperature

Arina A. Grigorieva, Igor V. Pekov, Natalia V. Zubkova, Anna G. Turchkova, and Dmitry Yu Pushcharovsky

Hilairite, $Na_2ZrSi_3O_9{\cdot}3H_2O$, is a trigonal (space group *R*32) zeolite-like zirconosilicate. Its structure is based on the heteropolyhedral framework consisting of isolated $[ZrO_6]$ octahedra and helical chains $[Si_3O_9]_\infty$. Zeolitic cavities are occupied by Na^+ cations and H_2O molecules (Ilyushin et al. 1981; Grigorieva et al. 2009). Hilairite is one of the most wide-porous natural zirconosilicates. The experimental data showed that the mineral has strong ion-exchange properties with different large cations (Turchkova et al. 2006). In this work we study the dependence of crystal-chemical characteristics of its cation-exchanged forms (basing on K and Rb forms) on temperature of ion exchange.

The initial material was hilairite from Kirovskii apatite mine, Mountain Kukisvumchorr, Khibiny alkaline complex, Kola Peninsula, Russia, with averaged chemical composition: $(Na_{1.78}K_{0.06}Mn_{0.03}Ba_{0.01})_{1.88}(Zr_{0.99}Nb_{0.02})_{1.01}Si_3O_{8.99}{\cdot}2.92H_2O$. Its crystal structure was recently studied (Grigorieva et al. 2009). Earlier the crystal structure of hilairite from the Lovozero alkaline complex, Kola Peninsula, was studied by Ilyushin et al. (1981). These samples are identical in the symmetry and structure of framework but different in distribution of extra-framework components - Na cations and H_2O molecules. It explains the doubling of the parameter *c* of the trigonal cell of hilairite from Lovozero in comparison with our Khibiny sample.

Cation-exchange experiments were performed in 1 M aqueous solutions of KCl and $RbNO_3$ at 20°C, 90°C and 150°C. Only K cations substitute Na at 20°C, while at higher temperatures the exchange of K and Rb is observed and their contents increase with temperature increase. Electron microprobe data showed that the content of K_2O after the experiment at 20°C is 7.4–8.4 wt%, at 90°C – 15.4–18.0% and at 150°C – 20.0–21.6%. The content of Rb_2O after the experiment at 90°C is 4.4–15.5% and at 150°C – 51.0–51.4%.

A.A. Grigorieva • I.V. Pekov • N.V. Zubkova • A.G. Turchkova • D.Yu Pushcharovsky
Faculty of Geology, Moscow State University, Moscow, Russia
e-mail: arina1984@bk.ru

S.V. Krivovichev (ed.), *Minerals as Advanced Materials II*,
DOI 10.1007/978-3-642-20018-2_17, © Springer-Verlag Berlin Heidelberg 2012

The crystal structures of cation-exchanged forms of hilairite after the experiments at 90°C and 150°C were studied. A three-dimensional set of diffraction data has been obtained on the single-crystal diffractometer Xcalibur *S* CCD (Mo*K*α-radiation, $\lambda = 0.71073$ Å) at room temperature. Unit-cell parameters of K-exchanged form of hilairite (90°C) are: $a = 10.678(4)$, $c = 7.944(2)$ Å, sp. gr. *R*32, the refined formula is $K_{0.51}ZrSi_3O_9{\cdot}3[(H_2O)_{0.50}K_{0.27}(H_3O)_{0.23}]$ (Zubkova et al. 2007). The similar characteristics were found for the K-exchanged form obtained at 150°C: $a = 10.3804(2)$, $c = 14.9541(6)$ Å, sp. gr. *R*3 and the formula $K_{1.81}Na_{0.09}H_{0.10}ZrSi_3O_9$. For Rb-exchanged forms of hilairite these characteristics are: $a = 10.4743(4)$, $c = 15.5433(7)$, *R*3, $Rb_{1.48}Na_{0.45}H_{0.07}ZrSi_3O_9{\cdot}0.53H_2O$ (90°C); $a = 10.477(1)$, $c = 15.377(2)$, *R*3, $Rb_{1.80}Na_{0.20}Zr[Si_3O_9]{\cdot}0.35H_2O$ (150°C). The crystal structures have been determined on the basis of direct methods and refined using the software SHELX'97.

The heteropolyhedral framework of the initial hilairite consisting of Zr-octahedra and Si,O helical chains remains in the structures of all studied samples However we found the distortion of framework for some ion-exchanged forms.

Space group *R*32 remains for K-exchanged form obtained at 90°C (Fig. 1). Unit cell parameters are comparable with those of the initial sample (Zubkova et al. 2007). The distance Zr-O is 2.114(10) Å. Si-O distance range is from 1.583(17) to 1.666(10) Å.

K-exchanged form obtained at 150°C is characterized by a doubled *c* parameter and sp. gr. *R*3. These changes are caused by considerable deformation of the framework. Zr atoms occupy two nonequivalent sites centering octahedra that are tilted with respect to each other. Si,O chains are also strongly distorted and their period doubles to 6 tetrahedra compare to in initial hilairite (3 tetrahedra) (Fig.2). The distance Zr-O in these polyhedra range from 2.067(8) to 2.089(8) Å for Zr(1) and from 2.053(11) to 2.093(9) Å for Zr(2). Average Si-O distance in these tetrahedra is 1.636 Å for (Si1) and 1.610 Å for (Si2).

In the structure of K-exchanged form of hilairite obtained at 90°C, K cations occupy the Na(1) between $[ZrO_6]$ octahedra with occupancy factor ~0.5. The Na(2) position, located in the zeolitic channel, is occupied by K, H_3O and H_2O with the ratio ~1:1:2 in the K-exchanged form obtained at 90°C (Zubkova et al. 2007). Cations K(1) are located in the ninefold coordination with distances K-O in the

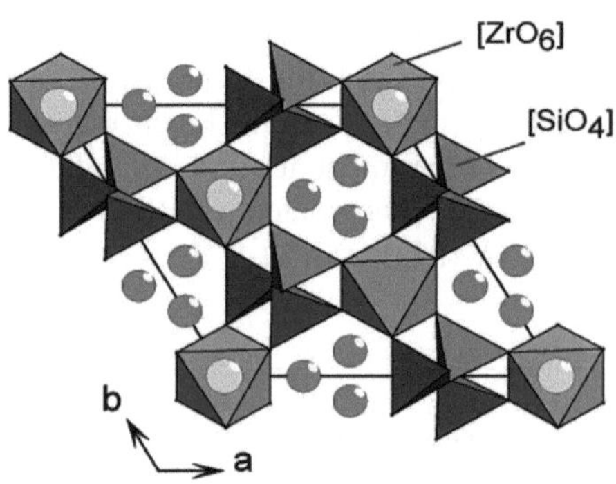

Fig. 1 K-exchanged form of hilairite obtained at 90°C

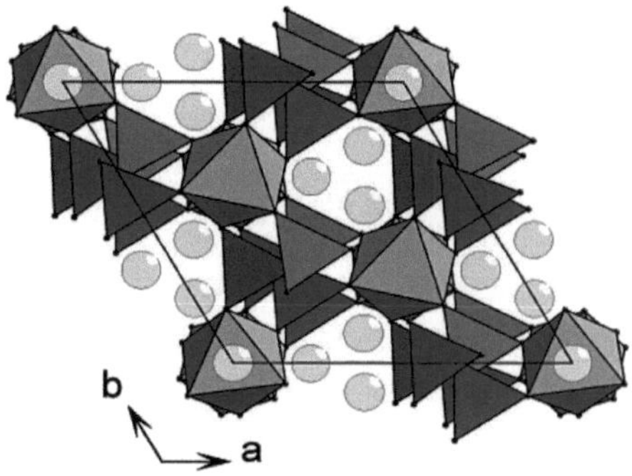

Fig. 2 K-exchanged form of hilairite obtained at 150°C

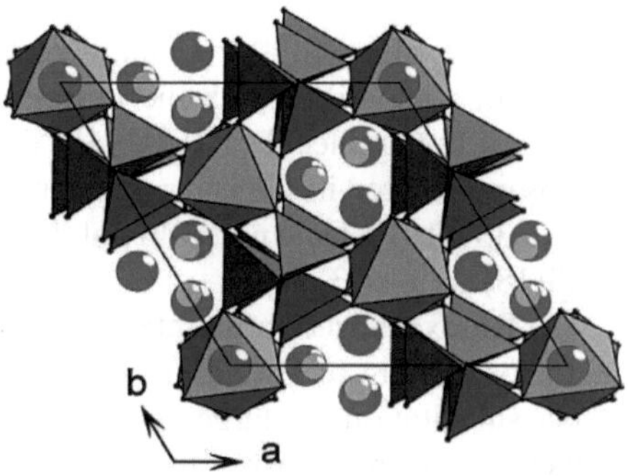

Fig. 3 Rb-exchanged form of hilairite obtained at 90°C

range from 2.57(3) to 3.299(12) Å. Cations K(2) are located in the eightfold coordination with distances K-O 3.12(4)–3.507(11) Å.

In the K-exchanged at 150°C form, K cations occupy both Na(1) and Na(2) positions. The latter site contains small amount of residual Na. This sample is anhydrous. Cations K(1) are located in the ninefold coordination with distances K (1)-O in the range from 2.891(8) to 3.221(12) Å, K(2) occupies the sixfold coordination with distances K(2)-O in the range from 2.789(11) to 3.040(8) Å.

In both Rb-exchanged forms of hilairite obtained at 90°C and 150°C, the heteropolyhedral framework is strongly distorted. Zr octahedra are tilted with respect to each other and Si-O chains are strongly distorted (Figs. 3 and 4). The distortion degree increases with the increase of Rb concentration. In both cases it causes a doubling of the *c* parameter and symmetry lowering in comparison with the initial hilairite. Rb cations occupy two nonequivalent positions in both structures. For Rb-exchanged forms obtained at 90°C the Zr-O distance range is from 2.046(12) to 2.118(14) Å for Zr(1) and from 2.026(13) to 2.146(15) Å for Zr(2). The distance Si-O for both the sites Si(1) and Si(2) is 1.61 Å. For Rb-exchanged forms obtained at 150°C average Zr-O distances are 2.079 and 2.073 Å for Zr(1) and Zr(2), respectively. Average Si-O distances in these tetrahedra are 1.618 Å for the site Si(1) and 1.625 Å-for the Si(2).

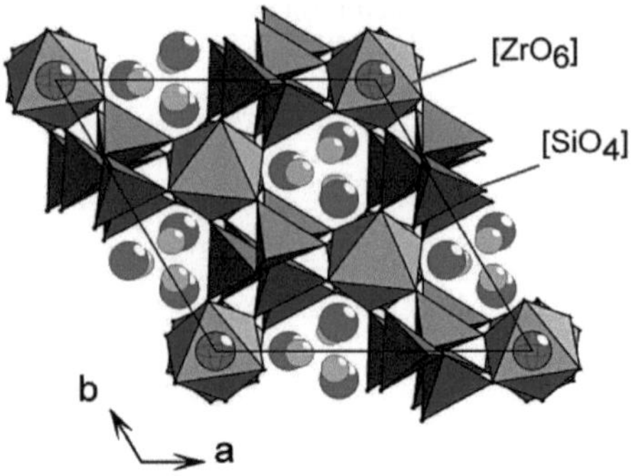

Fig. 4 Rb-exchanged form of hilairite obtained at 150°C

For Rb-exchanged form obtained at 90°C, a considerable decrease of H_2O content in comparison with the initial hilairite is found: the occupancy factor for O atom of H_2O molecule is 0.356(19) that corresponds to $0.53H_2O$ per formula. Rb cations are located in 12-fold polyhedra [Rb(1) with distance Rb-O range from 2.61 (3) to 3.337(19) Å] and ninefold polyhedra [Rb(2) with distance Rb-O varying from 2.943(18) to 3.44(3) Å].

The Rb-exchanged form obtained at 150°C is more H_2O-depleted: the occupancy factor for O atom of H_2O molecule is 0.23(2) = $0.35H_2O$ per formula. The degree of Na substitution by larger alkaline cations increases with the cation-exchange temperature increase. This leads to stronger deformation of the heteropolyhedral framework and the decrease of H_2O content up to anhydrous forms. The latter explains at first glance paradoxical decrease of the unit cell dimensions with the increase of content of cations larger than Na. K or Rb substitutes Na at first in the position between Zr octahedra and further in the position in the wide channel. Rb cations are located in a 12-fold coordination Rb(1) site with an average Rb-O distance of 3.10 Å and a eightfold coordination Rb(2) site with an average distance Rb-O of 3.03 Å.

These examples show that ion exchange in silicates with heteropolyhedral frameworks in chemically identical systems at different temperatures give products with essentially different both chemical and structural characteristics.

Acknowledgements This study was supported by RFBR grants 09-05-00143-a and 09-05-12001-ofi_m and grants of President of Russian Federation nos. MK-320.2010.5, NSh-4034.2010.5 and NSh-3848.2010.5.

References

Grigorieva AA, Zubkova NV, Pekov IV, Pushcharovsky DYu (2009) Crystal structure of hilairite from Khibiny alkaline massif (Kola Peninsula). Dokl Earth Sci 428:1051–1053

Ilyushin GD, Voronkov AA, Nevskii NN, Ilyukhin VV, Belov NV (1981) Crystal structure of hilairite $Na_2ZrSi_3O_9$ $(H_2O)_3$. Sov Phys Dokl 26:916–917

Turchkova AG, Pekov IV, Bryzgalov IA (2006) Cation-exchange properties of natural zeolite-like sodium zirconosilicates: an experimental study in aqueous solutions at 80–90ºC and 1 atm. 19th General Meeting IMA, Kobe, Book of Abstracts, p. 280

Zubkova NV, Pekov IV, Turchkova AG, Pushcharovsky DYu, Merlino S, Pasero M, Chukanov NV (2007) Crystal structures of potassium-exchanged forms of catapleiite and hilairite. Crystallogr Rep 52:65–70

Comparison of Structural Changes upon Heating of Zorite and Na-ETS-4 by In Situ Synchrotron Powder Diffraction

Michele Sacerdoti and Giuseppe Cruciani

1 Introduction

Zorite is a microporous titanosilicate, with ideal chemical formula $Na_6Ti_5Si_{12}O_{36}\cdot 11H_2O$, reported for the first time in the Lovozero massif by (Mer'kov et al. 1973). Sandomirskii and Belov (1979), while recognizing the OD character of zorite, solved its superposition (family) structure in the space group *Cmmm* based on a unit cell with $a = 23.241$ Å, $b = 7.238$ Å and $c = 6.955$ Å. The same authors inferred a likely doubling of b and c parameters, based on precession photographs, and suggested several models for the possible OD maximum degree of order (MDO) polymorphs. According to the average structure model of Sandomirskii and Belov (see projections along [001] and [010] in Fig. 1), zorite may be described as consisting of chains of TiO_6 octahedra parallel to the [010] direction laterally linked by a chain of four SiO_4 tetrahedra along [100] direction. These SiO_4 groups are linked in [010] direction by a TiO_5 semi-octahedron (pyramid) with statistically reverted polarity. Thus, this model postulates the occurrence of a 5-fold coordinated Ti site bridging the titanosilicate chains made by xonotlite-like bands and octahedral Ti chains. The resulting titanosilicate framework is characterised by two orthogonal sets of channels, one delimited by 12-membered rings of four TiO_6 and eight SiO_4 in the [001] direction, and the other by 8-membered rings of SiO_4 tetrahedra in the [010] direction.

A synthetic zeolite-like titanosilicate, called ETS-4 (Engelhard Titanium Silicate-4), was patented by Kuzniki and co-workers (1989, 1990). These authors claimed that the ETS-4 was the synthetic analogue of the mineral zorite where, in particular, both octahedrally and pentahedrally coordinated Ti atoms are linked to SiO_4 tetrahedra to build a microporous structure. Concerning the resemblance between the crystal structures of ETS-4 and zorite, Philippou and Anderson

M. Sacerdoti • G. Cruciani (✉)
Department of Earth Sciences, University of Ferrara, Via Saragat 1, 44100 Ferrara, Italy
e-mail: cru@unife.it

S.V. Krivovichev (ed.), *Minerals as Advanced Materials II*, 187
DOI 10.1007/978-3-642-20018-2_18, © Springer-Verlag Berlin Heidelberg 2012

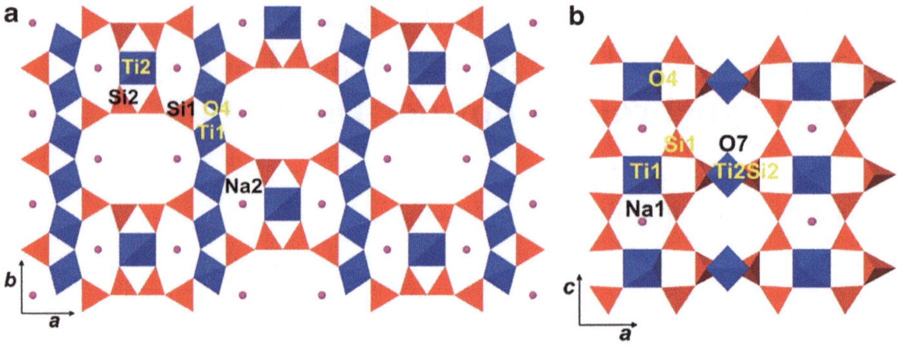

Fig. 1 A view of the zorite/ETS-4 structure projected along [001] (**a**) and along [010] (**b**)

(1996) found remarkable differences in the ^{29}Si MAS NMR spectra of the two materials and concluded that ETS-4 structure is essentially related to that of zorite, but cannot be considered as the synthetic counterpart of this mineral.

The first crystal structure refinement of Na-ETS-4 was performed by Cruciani et al. (1998) using synchrotron powder diffraction. These authors concluded that the *Cmmm* superposition model of zorite was a suitable description for the Na-ETS-4 structure, with the major difference being the 6-fold coordination of Ti at the chain-bridging site. The same authors proposed a real structure model where the partial occupancy of the $[(SiO_4)_4(TiO_6)]$ units was explained by the statistical distribution of empty and blocked 12-membered ring (MR) channels. Such a mixed occupancy by different species (water molecules, sodium ions and hydroxyls coordinated to Ti cations) well explained the refined electron density into the 12-MR channels. The findings of Cruciani et al. (1998) were then questioned by Kuznicki and co-workers who refined the structure of Sr-exchanged ETS-4 (Braunbarth et al. 2000) and Na-ETS-4 (Nair et al. 2001a) from powder diffraction and single crystal data, respectively. These studies, based on the zorite superposition structure model, found evidences supporting the hemi-octahedral coordination of Ti in the chain-bridging site. Similar supporting evidences were also obtained from the single crystal structure refinements of K-, Cs- and Pb-exchanged zorite by Zubkova et al. (2005, 2006) and from the zorite-related structure of chivruaiite (Men'shikov et al. 2006). On the other hand, more recent single crystal refinements of the Tl-, Ag-, Rb-, and Cs-exchanged forms of zorite performed by Spiridonova et al. (2008, 2010) showed in all cases the six-fold coordination of the chain-bridging Ti site, Ti(2). In particular, a superstructure with *Immm* space group and a $23.1 \times 21.6 \times 13.8$ Å unit cell was determined in Tl-exchanged zorite where two split Ti(2) sites, Ti(2a) and Ti(2b), are octahedrally coordinated by four oxygens and two H_2O molecules (Spiridonova et al. 2008).

Concerning the thermal behaviour of zorite and ETS-4, Naderi and Anderson (1996) described for ETS-4 a series of transformations upon heating encompassing a partial structure rearrangement at ca. 200°C, followed by complete amorphization at 500°C, formation of a new layered material at 600°C, and then of highly crystalline narsarsukite at 700°C. According to Rocha and Anderson (2000) the collapse of Na-ETS-4 to an amorphous near 200°C was due to the loss of structural water chains present along the channel systems. A detailed neutron powder diffraction and spectroscopic study was carried out by Kuznicki et al. (2001) and Nair et al. (2001b) on samples of Sr-exchanged ETS-4 heated *ex situ* at selected temperatures between 150°C and 300°C. This study showed that Sr-ETS-4 withstands thermal dehydration up to 350°C. Through the heat treatment the microporous framework of ETS-4 can be systematically contracted to adjust the effective size of the pores and increase its size-selectivity in the gas adsorption applications. The so-obtained material (called CTS-1, Contracted TitanoSilicate-1) has been claimed to be effective for separation of molecules with similar size in gas mixtures (e.g. N_2/CH_4, Ar/O_2 and N_2/O_2), selective adsorption of olefins, and purification of water containing organic and inorganic pollutant (Kuzniki 2000, 2002, 2003).

In spite of the previous studies on the *ex-situ* thermal behaviour of ETS-4 and the reported comparison of the physicochemical properties of ETS-4 and zorite, a parallel study of the structural behaviour of these materials upon heating is still lacking. Our present investigation is aimed to compare the structural modifications, including the transient states, underwent by zorite and Na-ETS-4 when dynamically (continuously) heated. This goal can be achieved by means of temperature-resolved *in situ* powder diffraction using synchrotron radiation and Rietveld refinement. In fact, this kind of time-resolved experiments allow to study the structural modifications upon heating in the so-called ‘far-from-equilibrium’ conditions and provide a complementary view to results obtained in the ‘near-equilibrium’ approximation typically carried out *ex situ* (Cruciani 2006).

2 Experimental

The zorite sample used here is from the Jubilaeum (Yubileinaya) pegmatite, Lovozero Massif, Kola Peninsula (Russia) whereas Na-ETS-4 is the same previously analyzed by Cruciani et al. (1998). Time-resolved X-ray powder diffraction patterns were collected at the GILDA beamline of ESRF (Grenoble, France). The zorite and ETS-4 samples, loaded in capillaries, were heated *in situ* from room temperature up to 500°C with heating rate of about 5°/min, while the diffracted intensity was continuously collected on the slit-delimited portion of a translating imaging plate (IP). The incident beam wavelength was set to $\lambda = 0.68890$ Å. The recorded IP images (Fig. 2) where integrated along stripes to extract two series of 22 diffraction patterns each, in the temperature range 45–410°C. The Rietveld refinements were performed using the GSAS-EXPGUI suite of programs (Larson and Von Dreele 1994; Toby 2001) starting from the superposition orthorhombic structural model in the space group *Cmmm* of Cruciani et al. (1998). Above $T \approx 410$°C, although both zorite and ETS-4 showed some residual crystallinity, the low number of reflections did not allow any Rietveld refinement. The Bragg peak profile was modelled using a pseudo-Voigt function with a 0.01% cut-off of

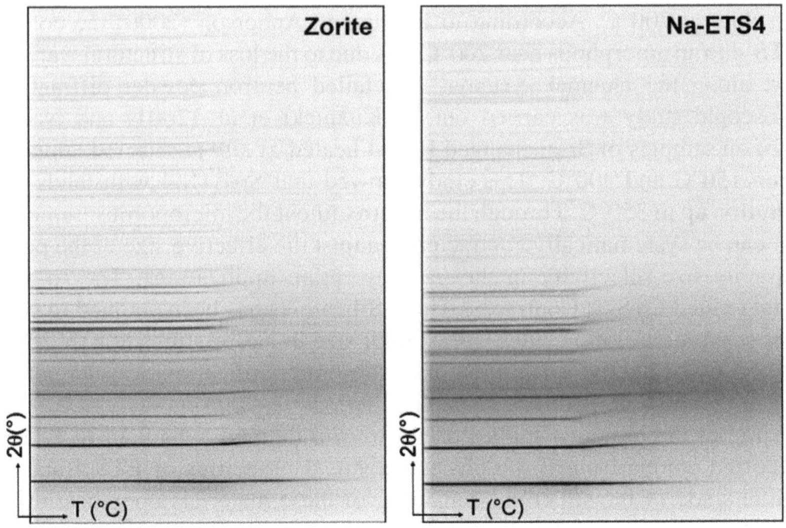

Fig. 2 Continuous diffraction patterns recorded on a translating imaging plate system for zorite (*left*) and Na-ETS-4 (*right*). The temperature range (*bottom axis*) is equal to RT-500°C in both cases

the peak intensity. The background curve was fitted using a Chebyshev polynomial with 24 variable coefficients. Soft constraints were imposed on Ti-O and Si-O distances, and the same weight was used throughout the refinement procedure. Scattering curves of neutral atoms were used.

3 Results

The dissimilar thermal behaviour of zorite and Na-ETS-4 is evident since the preliminary simple inspection of Fig. 2. Not only zorite appears to persist at higher temperature by ca. 50°C more than ETS-4 but the modifications of the crystal lattice of the two structures at the onset of the structural breakdown is significantly different. Further insights on the latter aspect can be gained by a close inspection of the series of diffraction patterns in the temperature range where the major lattice transformations take place in both compounds, as reported in Fig. 3. In particular, Fig. 3 (bottom) shows that a significant anisotropic contraction occurs for the (110) and (310) lattice planes of Na-ETS-4 in the 254–357°C temperature range, while the same effect is not so apparent in zorite.

Comparing the variation of cell parameters from Rietveld refinements of zorite (Fig. 4) and ETS-4 (Fig. 5) the clearly distinct trends are confirmed. In Na-ETS-4 the shortening along the *b*-axis for $T > 280°C$ is much larger than what observed

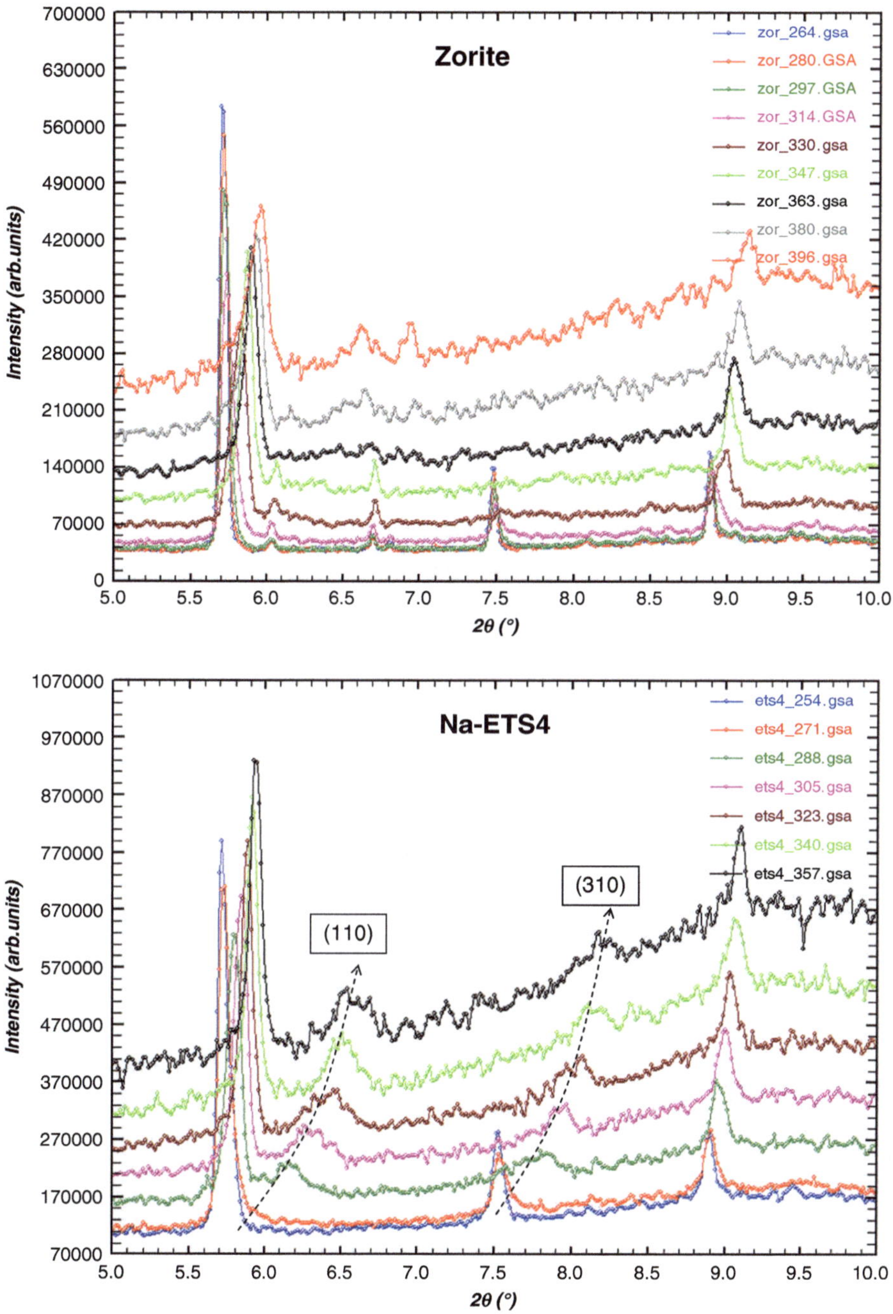

Fig. 3 Diffraction patterns as a function on temperature of zorite in the T range 264–396°C (*top*) and of Na-ETS-4 in the T range 254–357°C

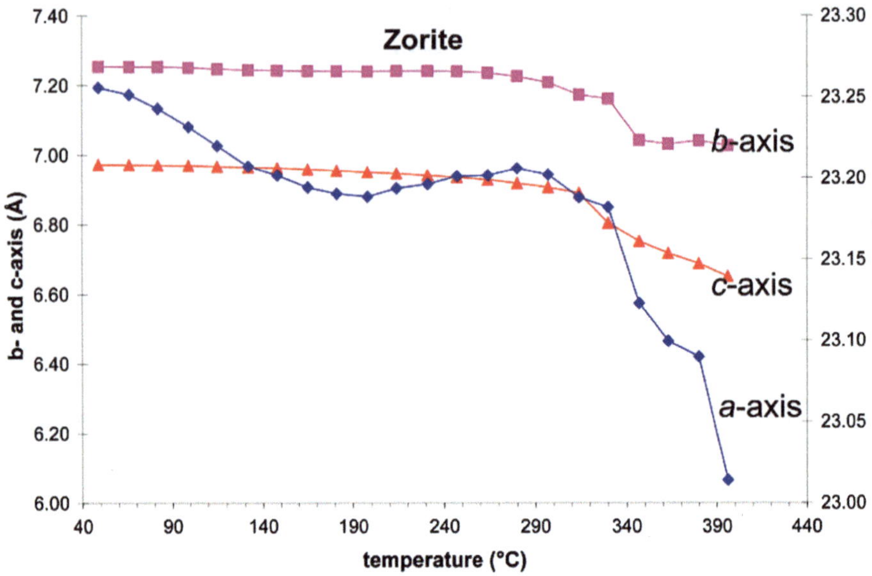

Fig. 4 Variations of the *a*-, *b*-, and *c*-axis lengths as a function of temperature in zorite

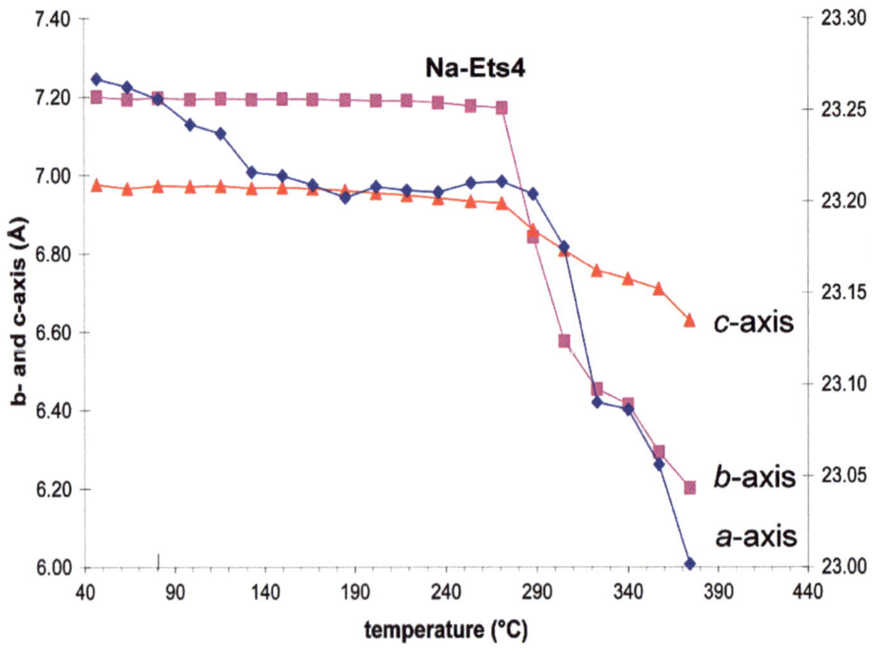

Fig. 5 Variations of the *a*-, *b*-, and *c*-axis lengths as a function of temperature in Na-ETS-4

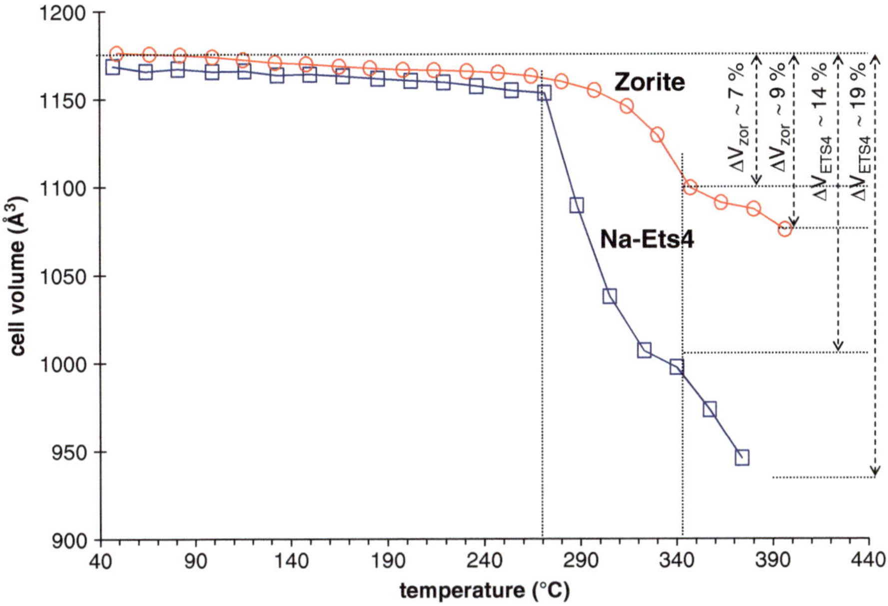

Fig. 6 Comparison of the cell volume variations as a function of temperature in zorite and Na-ETS-4

for the same axis in zorite and is similar to the contraction rate of the *a*-axis in both structures. As a net result, in Na-ETS-4 the *b*-axis length becomes shorter than the *c*-axis for T > 280°C, thus explaining the shifts of 110 and 310 reflections to more high 2-theta values as observed in Fig. 3. A similar behaviour was reported by Nair et al. (2001b) for Sr-exchanged Na-ETS-4, in which the *b*- and *c*-cell parameters were almost equal at 300°C.

The distinct behaviour of cell parameters also makes the cell volume of zorite and Na-ETS-4 to vary in a different way as function of temperature (Fig. 6). Below ca. 270°C the two structures show about the same volume contraction (ca. 2%). Above this temperature, a similar trend is observed for the two compounds while from 270°C to 340°C the cell volume of Na-ETS-4 contracts 14%, remarkably more than zorite whose unit cell contracts only ~4%; from 340°C to 390°C zorite cell volume contracts a further 3% while this contraction is 5% in Na-ETS-4. Altogether, dehydration involves a volume reduction of 9% in zorite while for ETS-4 the unit cell shrinkage is 21%.

In agreement with the previously reported *ex-situ* study (Nair et al. 2001b) the contraction of the structure is related to a reduction of the angle Ti1-O4-Ti1 (see Fig. 7), i.e. to the contraction of the Ti-O chain in the *b*-axis direction. This reduction is very significant in the Na-ETS-4 structure, where the Ti1-O4-Ti1 angle narrows from 128.1(5)° at 270°C to 111.2(5)° at 305°C and to 105.4(5)° at 340°C, while in zorite this change is very little. The different behaviour of Ti1-O4-Ti1

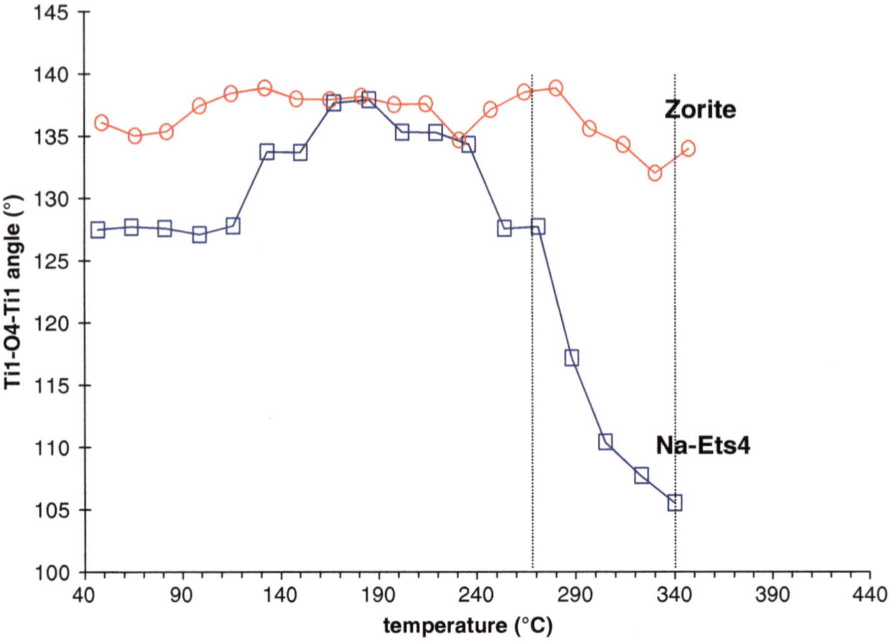

Fig. 7 Change of Ti1-O4-Ti1 angle with temperature in zorite and Na-ETS-4

between the two compound is most evident at about 290°C. The crystal structure refinements of Na-ETS-4 also show (data not plotted here) that the tetrahedral O1–Si1–O2 angle changes from 106.3(11)° at 271°C to 110.1(31)° at 305°C, i.e. the Si1 tetrahedra become more distorted at higher temperature. We suggest that the unlike behaviour of the Ti1 octahedral chains in zorite and Na-ETS-4 upon heating, in particular the extreme crumpling of the chain in the latter, mainly explains the lower thermal stability of ETS-4 compared to zorite.

Whether the larger narrowing of Ti1-O4-Ti1 with temperature in ETS-4 compared to zorite is related to the nature and arrangement of the channel species (including the water molecules) or is due to the local structure (e.g. ordered layer sequences, frequency of stacking faults, and coordination of the Ti2 site) remains an open question here.

As previously discussed by Sandomirskii and Belov (1979) and Cruciani et al. (1998), both in zorite and ETS-4 the distribution of water molecules over four sites (W1, W2, W4, and W5) allows the formation, at room temperature, of an intricate network of hydrogen bonding which is likely to play a fundamental role in defining the channel geometry and controlling the stability to framework collapse of zorite and ETS-4 upon dehydration. Furthermore, Rocha and Anderson (2000) ascribed to breakage of the water chains in the channels of ETS-4 its low temperature of structure collapse. On the other hand, Nair et al. (2001a) concluded that the water chains proposed in zorite do not exist in ETS-4. Unfortunately, our attempts to refine the variation of the occupancy fractions of individual water sites as a function of temperature did not provide any conclusive results on the sequence of water release, due to the mixed and disordered arrangement of the extraframework species. The most reliable information provided by the structure refinements are that the variation in the content of water molecules is very similar in zorite and Na-ETS-4, and that no water is present in ETS-4 above 300°C while in zorite it disappears completely only above 330°C. Such a 30°C temperature gap is consistent with the 50°C gap of the structure breakdown between the two compounds. This can be taken as an indirect evidence that dehydration is indeed the major cause for structural collapse in both zorite and Na-ETS-4.

Another issue concerns the preferential pathway along which the water molecules escape the pores of zorite/ETS-4 structure. In fact it has been long discussed in the literature that stacking faults are responsible for poor adsorption properties of ETS-4. Usseglio et al. (2006) recently reported the N2 adsorption/desorption isotherms at 77 K of ETS-4 showing that this material does not possess a (true) microporous structure despite the nominal occurrence of channels delimited by large 12-membered rings. These authors suggested that the 12-membered rings on the (001) plane is actually blocked in ETS-4 by the $[(SiO_4)_4(TiO_6)]$ groups, destroying the microporous nature of the ideal structure and supporting the ordered stacking model proposed by Cruciani et al. (1998).

The present study provide the possibility to shed some light on this issue because during time-resolved experiments at non-equilibrium conditions, transient states can be recorded. For instance, Martucci et al. (2003) were able to infer the preferential escape path of water molecules during the in situ dehydration study of mordenite. In this zeolite the free diameter of the 12-membered ring (MR) shrank as the temperature increased, while the 8-membered ring was widest at about 200°C and became smaller at 800°C, when mordenite is completely dehydrated. This testified that the 8 MR channels were the preferred exit for water molecules in mordenite. In analcime, Cruciani and Gualtieri (1999) recognized the water escape path in the 8 MR channels by following the changes of the T-O-T angles delimiting these channels: the rings became more widen and regular during the water passage, then turned to the initial irregular geometry after the complete water loss.

Figure 8 shows the variation with temperature of the normalized free areas (A/A_0 where A_0 is the value at room temperature) of the 8- and 12-membered rings in zorite and Na-ETS-4. The 8-membered ring area in both compounds shows an increase in the 100–220°C range before it contracts back to the starting dimensions. This behaviour can be interpreted by the widening of the 8-MR channel openings induced by the passage of water during dehydration, followed by the return to the original size, then by the contraction due to framework collapse. This trends suggest that, although larger openings are ideally present in the zorite and ETS-4 structures, the occurrence of stacking faults in these materials makes the relatively narrow 8-membered rings as the forced path for the exit (access) from (to) the interior of the crystals.

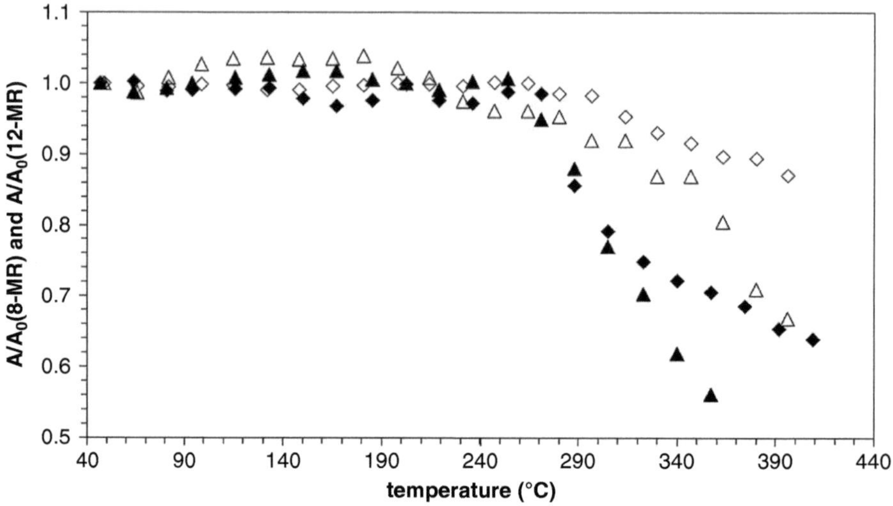

Fig. 8 Variation with temperature of the normalized areas (A/A_0 where A_0 = area at RT) of 8-membered rings (triangles) and 12-membered rings (diamonds) in zorite (open symbols) and Na-ETS-4 (filled symbols)

Acknowledgements We are indebted to Stefano Zanardi (ENI S.p.A.) for the preliminary data analysis he performed at the University of Ferrara. The public beamtime programme of ESRF is acknowledged for granting access to synchrotron radiation. We thank the staff at GILDA for the help during time-resolved experiments.

References

Braunbarth CM, Hillhouse HW, Nair S, Tsapatsis M (2000) Structure of strontium ion-exchanged ETS-4 microporous molecular sieves. Chem Mater 12:1857–1865

Cruciani G, De Luca P, Nastro A, Pattison P (1998) Rietveld refinement of the zorite structure of ETS-4 molecular sieve. Micropor Mesopor Mat 21:143–153

Cruciani G, Gualtieri A (1999) Dehydration dynamics of analcime by in situ synchrotron powder diffraction. Am Mineral 84:112–119

Cruciani G (2006) Zeolites upon heating: factors governing their thermal stability and structural changes. J Phys Chem Solids 67:1973–1994

Kuznicki SM, Bell VA, Nair S, Hillhouse HW, Jacubunas RM, Braunbarth CM, Toby MH, Tsapatsis M (2001) A titanosilicate molecular sieve with adjustable pores for size-selective adsorption of molecules. Nature 412:720–724

Kuzniki SM (1989) US Patent 4:853–202

Kuzniki SM (1990) US Patent 4:938–989

Kuzniki SM (2000) US Patent 6:068–682

Kuzniki SM (2002) US Patent 6:340–433

Kuzniki SM (2003) US Patent 6:517–611

Larson AC, Von Dreele RB (1994) General structure analysis system (GSAS), Los Alamos National Laboratory Report LAUR-86–748

Martucci A, Sacerdoti M, Cruciani G, Dalconi MC (2003) In situ time resolved synchrotron powder diffraction study of mordenite. Eur J Mineral 15:485–493

Men'shikov YuP, Krivovichev SV, Pakhomovsky YaA, Yakovenchuk VN, Ivanyuk GYu, Mikhailova YuA, Armbruster T, Selivanova EA (2006) Chivruaiite, $Ca_4Ti_5[(Si_6O_{17})_2|O_4(OH)]$. $14H_2O$, a new mineral from hydrothermalites of Khibiny and Lovozero alkaline massifs, and its relations with zorite. Am Mineral 91:922–928

Mer'kov AN, Bussen IV, Goiko EA, Kul'chitskaya EA, Men'shikov YuP, Nedorezova AP (1973) Raite and zorite – new minerals from the Lovozero Tundra. Zap VMO 102(1):54–62 (in Russian)

Naderi M, Anderson MW (1996) Phase transformation of microporous titanosilicate ETS-4 into narsarsukite. Zeolites 17:437–443

Nair S, Jeong HK, Chandrasekaran A, Braunbarth CM, Tsapatsis M, Kuznicki SM (2001a) Synthesis and structure determination of ETS-4 single crystals. Chem Mater 13:4247–4254

Nair S, Tsapatsis M, Toby BH, Kuznicki SM (2001b) A study of heat-treatment induced framework contraction in strontium-ETS-4 by powder neutron diffraction and vibrational spectroscopy. J Am Chem Soc 123:12781–12790

Philippou A, Anderson MW (1996) Structural investigation of ETS-4. Zeolites 16:98–107

Rocha J, Anderson MW (2000) Microporous titanosilicates and other novel mixed octahedral-tetrahedral framework oxides. Eur J Inorg Chem 5:801–818

Sandomirskii PA, Belov NV (1979) The OD structure of zorite. Sov Phys Crystallogr 24:686–693, Kristallographiya 24:1198–1210

Spiridonova DV, Britvin SN, Krivovichev SV, Yakovenchuk VN, Armbruster T (2008) Tl-exchange in zorite and ETS-4. In: Krivovichev SV (ed) Minerals as advanced materials I. Springer, Berlin, pp 65–69

Spiridonova DV, Britvin SN, Krivovichev SV, Yakovenchuk VN (2010) Crystal chemistry of ion-exchanged forms of zorite, a natural analogue of the ETS-4 material. In: Minerals as advanced materials II, Kirovsk, Kola Peninsula (Russia) 19–25 July 2010, Book of Abstracts 23–24

Toby BH (2001) EXPGUI, a graphical user interface for GSAS. J Appl Crystallogr 34:210–213

Usseglio S, Calza P, Damin A, Minero C, Bordiga S, Lamberti C, Pelizzetti E, Zecchina A (2006) Tailoring the selectivity of Ti-based photocatalysts (TiO_2 and microporous ETS-10 and ETS-4) by playing with surface morphology and electronic structure. Chem Mater 18:3412–3424

Zubkova NV, Pushcharovsky DYu, Giester G, Pekov IV, Turchkova AG, Chukanov NV, Tillmanns E (2005) Crystal structures of K- and Cs-exchanged forms of zorite. Crystallogr Rep 50:367–373

Zubkova NV, Pushcharovsky DYu, Giester G, Pekov IV, Turchkova AG, Tillmanns E, Chukanov NV (2006) Crystal structure of Pb-exchanged form of zorite. Crystallogr Rep 51:379–382

Crystal Chemistry of Ion-Exchanged Forms of Zorite, a Natural Analogue of the ETS-4 Titanosilicate Material

Dar'ya V. Spiridonova, Sergey V. Krivovichev, Sergey N. Britvin, and Viktor N. Yakovenchuk

1 Introduction

Titanosilicates have important applications in various areas of technology such as catalysis, gas separation, energy storage, optoelectronics, radioactive waste management, etc. Many of these materials are known as minerals, e.g. zorite $Na_6[Ti(Ti,Nb)_4(Si_6O_{17})_2(O,OH)_5].nH_2O$ ($n = 10$–11), a natural analogue of ETS-4 (Engelhard Titanium Silicate-4). The modified form of ETS-4, CTS-1 (contracted titanosilicate-1), is used for separation of mixtures of gases with similar molecular diameters, e.g. N_2/O_2 and CO/H_2. Ba-exchanged ETS-4 is used for separation of mixtures of N_2 and CH_4. Another area of applications of ETS-4 and CTS-1 is purification of water polluted by organic contaminants. Natural Ca-bearing analogues of zorite are chivruaiite $Ca_4[(Ti,Nb)_5(Si_6O_{17})_2|(OH,O)_5].13\text{-}14H_2O$, and haineaultite, $(Na,Ca)_5Ca[(Ti,Nb)_5(Si,S)_{12}O_{34}(OH,F)_8].5H_2O$.

Ion-exchange forms of zorite and ETS-4 have been studied by several authors. The structure of Sr-exchanged ETS-4 was reported by Braunbarth et al. (2000) and Nair et al. (2001). Structures of Cs-, K-, and Pb-exchanged forms of zorite were refined by Zubkova et al. (2005, 2006). Tl-exchange in zorite in acidic solutions was investigated

D.V. Spiridonova (✉) • S.N. Britvin
Department of Crystallography, St. Petersburg State University, University Emb. 7/9, 199034 St. Petersburg, Russia
e-mail: spiridonovadarya@mail.ru

S.V. Krivovichev
Department of Crystallography, St. Petersburg State University, University Emb. 7/9, 199034 St. Petersburg, Russia
and
Nanomaterials Research Center, Kola Science Center, The Russian Academy of Sciences, 14 Fersman Street, Apatity 184209, Russia

V.N. Yakovenchuk
Nanomaterials Research Center, Kola Science Center, The Russian Academy of Sciences, 14 Fersman Street, Apatity 184209, Russia

S.V. Krivovichev (ed.), *Minerals as Advanced Materials II*,
DOI 10.1007/978-3-642-20018-2_19,

and the structure of Tl-zorite $Tl_{3.82}H_{0.45}Ti_{4.22}Nb_{0.57}[(Si_6O_{17})_2(OH)_4]\cdot 3.97H_2O$ was reported by Krivovichev et al. (2007), whereas structure of zorite exchanged with Tl in an alkaline solution was reported by Spiridonova et al. (2008). The structures of Tl-exchanged forms demonstrate that, at least at ambient conditions, degree and completeness of ion exchange in zorite and ETS-4 strongly depend upon the pH of aqueous solution.

In this work, we report on Cs-, Ag-, Rb-exchanged forms of zorite and their structural characteristics and provide their comparison to the previous results.

2 Experimental

Crystals of zorite used in this study originated from "Yubileinaya" vein, the Lovozero massif (Kola peninsula, Russia). Their chemical composition corresponds to the empirical formula $(Na_{6.71}\ K_{0.28}\ Ca_{0.07}\ Sr_{0.05}\ Mn_{0.03}\ Ce_{0.02}\ Mg_{0.01})_{7.17}\ (Ti_{4.05}\ Nb_{0.45}\ Fe^{3+}_{0.11}\ Al_{0.02})_{4.63}\ [Si_{12}O_{34} \mid (OH_{3.79}\ O_{1.21})]\cdot 13\text{-}14H_2O$ as reported by Men'shikov et al. (2006).

Ion-exchange experiments were performed as follows. 50 mg of zorite were immersed into 5 ml 1 M CsBr, 1 M $AgNO_3$, 1 M RbI for getting the corresponding ion-exchanged form. Exchange reaction was done in 5 days, and each day reaction mixture was slightly stirred. The resulting material was washed by deionized water and ethanol followed by drying at room temperature in air.

The exchange experiments were done in basic, neutral and acidic solutions, since our previous results indicated influence of pH upon cation exchange processes in zorite. It was found that the ion-exchange capacity of zorite is higher in neutral and basic solutions compared to the alkaline solutions (Fig. 1).

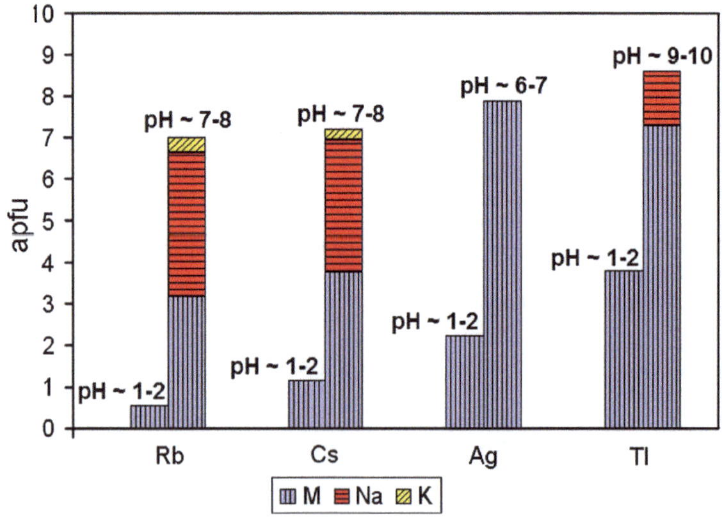

Fig. 1 Dependence of the ion-exchange capacity of zorite in point of cations on pH of the solution used for ion-exchange experiments (chemical analyses data)

Table 1 Crystallographic data and refinement parameters for Cs-, Ag-, Rb-exchanged zorite

	Cs-zorite	Ag-zorite	Rb-zorite
a (Å)	7.229(5)	7.239(1)	7.221(2)
b (Å)	23.277(5)	23.249(5)	23.251(7)
c (Å)	6.969(5)	6.921(1)	6.963(2)
V ($Å^3$)	1172.7(12)	1164.9(4)	1169.1(6)
Space group	*Cmmm*	*Cmmm*	*Cmmm*
ρ_{calc}(g/cm^3)	2.79	2.98	2.57
Radiation	MoKα	MoKα	MoKα
Total Ref.	3,286	3,200	1,802
Unique Ref.	772	616	760
Unique $\|F_o\| \geq 4\sigma_F$	517	448	467
R_1	0.061	0.070	0.058
wR_2	0.139	0.167	0.133
S	1.007	1.082	0.993

Crystal structures of Cs-, Ag-, Rb-exchanged zorite obtained in neutral and basic medium were investigated by single crystal X-ray diffraction. The crystals were mounted on glass fibers. For the Cs- and Rb-exchanged forms, data were collected by means of SMART 1 K CCD diffractometer, for the Ag-exchanged form, data were collected using STOE IPDS II diffractometer. The unit-cell parameters (Table 1) were refined by least-squares techniques. The intensity statistics indicated space group *Cmmm* for each crystals.

Structures of Cs-, Ag-, Rb-zorite were solved by the SHELX97 program. The structural formulas of Cs-, Ag-, Rb-exchanged forms of zorite determined on the basis of the structure refinement can be written as $Cs_{3.61}Na_{3.49}[Ti_{4.41}Nb_{0.54}Si_{12}O_{34}(OH)_{2.56}O_{2.44}].10.43H_2O$, $Ag_{7.89}[Ti_{4.01}Nb_{0.69}Si_{12}O_{34}(OH)_{2.62}O_{2.38}]\cdot H_2O$, and $Rb_{3.69}Na_{3.67}[Ti_{4.43}Nb_{0.57}Si_{12}O_{34}(OH)_{2.07}O_{2.93}].10.08H_2O$, respectively.

3 Discussion

The structure of zorite is based upon mixed octahedral-tetrahedral titanosilicate framework (Sandomirskii and Belov 1979; Cruciani et al. 1998) with two symmetrically independent Ti positions. The Ti1 site has an octahedral coordination (Ti1-O = 1.92-1.97 Å), whereas the Ti2 site has a square pyramidal coordination and is disordered over two sites that are symmetrically equivalent relative to the (001) mirror plane. The $Ti1O_6$ octahedra share trans-corners to form chains running parallel to the *a* axis. The SiO_4 tetrahedra form double chains parallel to the *c* axis. The chains are linked by $Ti2O_5$ square pyramids. The Na^+ cations and H_2O molecules reside in the framework cavities. Na atoms occupy two nonequivalent sites, Na1 and Na2.

Spiridonova et al. (2008) classified extra-framework cation sites into three groups denoted as *A*I, *A*II, *A*III (Table 2, Fig. 2). Figure 2 shows the structure of chivruaiite/zorite with nodal representation of its titanosilicate framework.

Table 2 Classification of extra-framework positions in the structure of zorite-type minerals and materials

	Positions of out of framework cations (according to the classification)		
Sample	*A*I	*A*II	*A*III
Cs-zorite	Cs4	Na1, Na2, Cs1	Cs2, Cs3
Ag-zorite	Ag1, Ag3, Ag6, Ag7, Ag8, Ag10, Ag12	Ag2	Ag4, Ag5, Ag9, Ag11
Rb-zorite	Na2, Rb2	Na1	Rb1
Tl-zorite[a]	Tl2, Tl3, Tl4, Tl7, Tl9, Tl13, Tl15	Tl1, Tl5, Tl6, Tl8, Tl14	Tl10, Tl11, Tl12
Tl-zorite[b]	Tl1, Tl2, Tl2A, Tl5, Tl6	Na	Tl3, Tl3A, Tl4
Cs-zorite[c]	Cs7, Cs8	Cs1, Cs2, Cs6	Cs3, Cs4, Cs5
K-zorite[c]	K	Na	–
Pb-zorite[d]	Pb	Ca	–
Sr-ETS-4[e]	Sr	Na	–
Haineaultite[f]	Na	Ca	–

[a]Krivovichev et al. 2008
[b]Spiridonova et al. 2008
[c]Zubkova et al. 2005
[d]Zubkova et al. 2006
[e]Braunbarth et al. 2000
[f]McDonald and Chao 2004

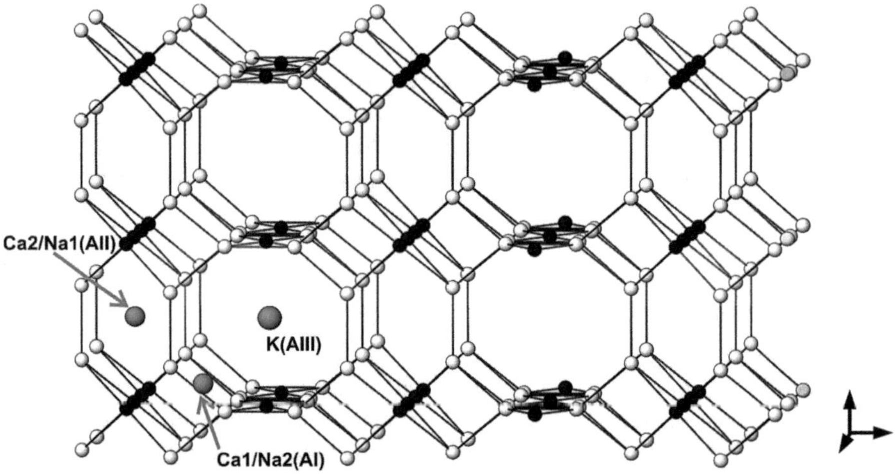

Fig. 2 Nodal representation of titanosilicate framework of chivruaiite (Ca1, Ca2, K)/zorite (Na1, Na2) (see text for details)

The Ti and Si coordination polyhedra are symbolized by black and white vertices, respectively; the presence of an edge indicates linkage of the Ti and Si polyhedra via common oxygen atoms. The *A*I position is located between two chains of white vertices and corresponds to the Ca1 and Na2 sites in chivruaiite and zorite, respectively. The *A*II site is at the centre of a hexagonal ring in the

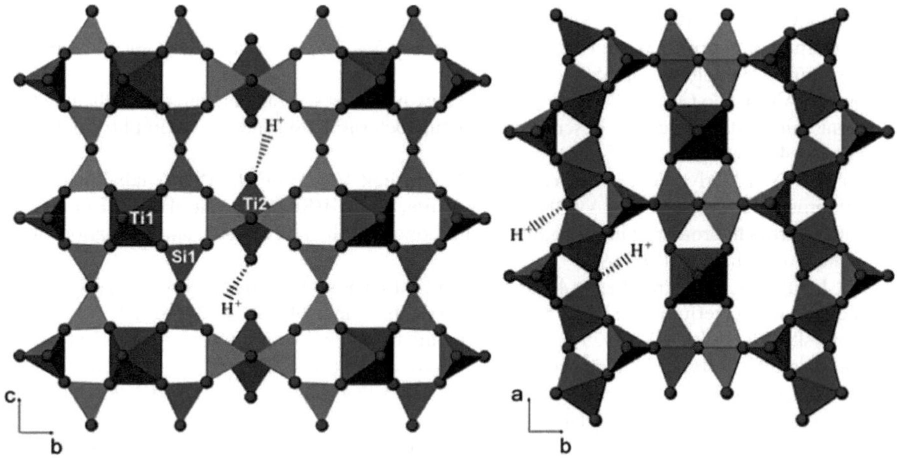

Fig. 3 Protonation of oxygen positions in the titanosilicate framework of zorite

framework and corresponds to the Ca2 and Na1 sites in chivruaiite and zorite, respectively. The *A*III site is located in the center of large framework channel. This site is occupied by water molecules in zorite, whereas, for chivruaiite, its partial occupation by K^+ cations has been suggested. Refinement of the structures of Tl-exchanged form of zorite (Spiridonova et al. 2008) provided direct evidence of the availability of this site in cation exchange reactions. As a rule, ion exchange results in a high degree of disorder of extra-framework cations forming groups of sites around the *A*I, *A*II, *A*III positions.

Crystal chemical analysis of cation-exchanged forms of zorite, which were obtained in basic, neutral, and acidic solutions, demonstrates that the degree of cation exchange depends upon pH of exchange solution: it is higher in neutral and basic solutions (pH $\geq$ 7) and lower in acidic solutions. The most likely explanation of this observation is the reactions of protonation/deprotonation of the titanosilicate framework. There are two different types of sites available for these reactions: oxygen atoms linking adjacent $Ti1O_6$ octahedra and oxygen atoms located at the vertex of the $Ti2O_5$ square pyramid (Fig. 3).

Acknowledgements This work was supported by the Russian Federal Programme 'Scientific and Education Cadres of Innovative Russia' (state contract no. 02.740.11.0326) and President of Russian Federation grant for young doctors of sciences (to SVK, grant no. MD 407.2009.5).

References

Braunbarth C, Hillhouse HW, Nair S, Tsapatis M, Burton A, Lobo RF, Jacubinas RM, Kuznicki SM (2000) Structure of strontium ion-exchanged ETS-4 microporous molecular sieves. Chem Mater 12:1857–1865

Cruciani G, Deluca P, Nastro A, Pattisson P (1998) Rietveld refinement of the zorite structure of ETS-4 molecular sieves. Micropor Mesopor Mater 21:143–153

Krivovichev SV, Britvin SN, Spiridonova DV, Yakovenchuk VN, Armbruster T (2008) Tl-Exchange in zorite and ETS-4. In: Krivovichev SV (ed) Minerals as advanced materials I. Springer, Heidelberg, pp 65–69

McDonald AM, Chao GY (2004) Haineaultite, a new hydrated sodium calcium titanosilicate from Mont Saint-Hilaire, Quebec: description, structure determination and genetic implications. Can Mineral 42:769–780

Men'shikov YuP, Krivovichev SV, YaA P, Yakovenchuk VN, Ivanyuk GYu, Mikhailova YuA, Armbruster T, Selivanova EA (2006) Chivruaiite, $Ca_4Ti_5[(Si_6O_{17})_2| O_4(OH)]·14H_2O$, a new mineral from hydrothermalites of Khibiny and Lovozero alkaline massifs, and its relations with zorite. Am Mineral 91:922–928

Nair S, Jeong HK, Chandrasekaran A, Braunbarth C, Tsapatis M, Kuznicki SM (2001) Synthesis and structure determination of ETS-4 single crystals. Chem Mater 13:4247–4254

Sandomirskii PA, Belov NV (1979) The OD structure of zorite. Sov Phys Crystallogr 24:686–693

Spiridonova DV, Britvin SN, Krivovichev SV, Yakovenchuk VN (2008) Crystal structure of Tl-exchanged alkaline form of zorite. Vestnik Sankt-Peterburgskogo Universiteta. Ser Geol Geogr 2008(3):41–46 (in Russian)

Zubkova NV, Pushcharovsky DYu, Giester G, Pekov IV, Turchkova AG, Chukanov NV, Tillmanns E (2005) Crystal structures of K- and Cs-exchanged forms of zorite. Crystallogr Rep 50:367–373

Zubkova NV, Pushcharovsky DYu, Giester G, Pekov IV, Turchkova AG, Tillmanns E, Chukanov NV (2006) Crystal structure of Pb-exchanged form of zorite. Crystallogr Rep 51:379–382

Ivanyukite-Group Minerals: Crystal Structure and Cation-Exchange Properties

Victor N. Yakovenchuk, Ekaterina A. Selivanova, Sergey V. Krivovichev, Yakov A. Pakhomovsky, Dar'ya V. Spiridonova, Alexander G. Kasikov, and Gregory Yu. Ivanyuk

1 Introduction

Four microporous titanosilicates of the ivanyukite group: ivanyukite-Na-*T*, ivanyukite-Na-*C*, ivanyukite-K, and ivanyukite-Cu – were discovered in a natrolitized microcline-aegirine-sodalite vein within orthoclase-bearing urtite of the Koashva apatite deposit, Khibiny alklaine massif (Yakovenchuk et al. 2009). Ivanyukite-Na-*T* formed as a late-stage, hydrothermal phase resulting from alteration of lamprophyllite, whereas ivanyukite-Na-*C*, ivanyukite-K and ivanyukite-Cu are products of either cation exchange or hydration of ivanyukite-Na-*T*. It is noteworthy that crystals of ivanyukite-Cu were found in close association with corroded grains of djerfisherite and chalcopyrite (Ivanyuk et al. this book).

V.N. Yakovenchuk (✉) • G.Y. Ivanyuk
Nanomaterials Research Center, Kola Science Center, the Russian Academy of Sciences, 14 Fersman Street, Apatity 184209, Russia
e-mail: yakovenchuk@ksc.ru

E.A. Selivanova • Y.A. Pakhomovsky
Geological Institute, Kola Science Center, the Russian Academy of Sciences, 14 Fersman Street, Apatity 184209, Russia

S.V. Krivovichev
Nanomaterials Research Center, Kola Science Center, the Russian Academy of Sciences, 14 Fersman Street, Apatity 184209, Russia
and
Department of Crystallography, Faculty of Geology, St. Petersburg State University, University Emb. 7/9, 199034 St. Petersburg, Russia

D.V. Spiridonova
Department of Crystallography, Faculty of Geology, St. Petersburg State University, University Emb. 7/9, 199034 St. Petersburg, Russia

A.G. Kasikov
Institute of Chemistry and Technology of Rare Elements and Mineral Raw Materials, Kola Science Center, the Russian Academy of Sciences, 14 Fersman Street, Apatity 184209, Russia

S.V. Krivovichev (ed.), *Minerals as Advanced Materials II*,
DOI 10.1007/978-3-642-20018-2_20,

Ivanyukite-Na-*T* occurs as numerous mosaic isometric pseudo-cubic crystals (up to 2 mm in diameter) as well as epitaxial zones around prismatic grains of sitinakite. The latter zones were used in a single-crystal diffraction experiment as well as in ion-exchange studies. Here we report results of chemical and structural investigations of natural and ion-exchanged forms of ivanyukite-group minerals, which provide additional information about their properties as prototypes of materials for various kinds of applications.

2 Experimental

In cation-exchange experiments, grains of ivanyukite-Na-*T* were loaded in 1 M solutions of NH_4Cl, $(N_2H_5)_2SO_4$, CsCl, RbCl, $SrCl_2$, $CoCl_2$, $NiCl_2$, $LaCl_3$ and Clerici liquid at ambient temperature. The experiment time was between 12 and 300 h. The chemical composition of original ivanyukite-Na-*T* and substituted phases has been studied by wave-length dispersion spectrometry using a Cameca MS-46 electron microprobe (Geological Institute, Kola Science Centre of the Russian Academy of Science, Apatity) operating at 20 kV, 20–30 nA and 5–10 μm beam diameter. Content of nitrogen was determined using a Leo-1450 scanning electron microscope with a Röntec energy dispersion spectrometer. Water content was calculated as 100 wt. % minus SUM of analysis.

It was found (Table 1) that three atoms per formula unit (*apfu*) of Na^+ and K^+ in original ivanyukite-Na-*T* can be replaced by 3 *apfu* of Tl^+, 2.8 *apfu* of $N_2H_5{}^+$, 2.5 *apfu* of Rb^+, 2.2 *apfu* of Cs^+, 1.8 *apfu* of $NH_4{}^+$, 0.8 *apfu* of Sr^{2+}, 0.7 *apfu* of Co^{2+}, 0.1 *apfu* of Ni^{2+} and 0.2 *apfu* of La^{3+}. Chemical formulae of the initial phase and the resulting solids are given in Table 2.

Distribution of Tl, Rb, K, Cs, NH_4, N_2H_5, Sr and Co is usually homogeneous in whole volume of crystals of the exchanged forms. Cs substitution for Na and K sometimes produces zoned crystals with cubic Cs-phase in rims and Na-rich trigonal phase in nuclei (Yakovenchuk et al. 2008). Distribution of Ni and La is limited by thin (few tens of micrometers) surface part of the crystals, whereas nuclei of such crystals are usually Na-free.

3 Crystal Structure

Crystal structures of ivanyukite-Na-*T* and its Rb- and Sr-substituted derivatives were investigated using a STOE IPDS II diffractometer operated at 50 kV and 40 mA. The data were collected using monochromatic MoKα X-radiation, with frame widths of 1° in ω, and with 7 min counting times for each frame. The intensity data were reduced and corrected for Lorentz, polarization, and background

Table 1 Chemical composition of ivanyukite-Na-*T* and its cation-exchanged forms (wt. %)

		Exchanged								
Original		NH_4^+	$N_2H_5^+$	Co^{2+}	Ni^{2+}	Rb^+	Sr^{2+}	Cs^+	La^{3+}	Tl^+
Exposition		12 h	12 h	17 h	17 h	48 h	300 h	48 h	168 h	12 h
N_2O_5		13.41	32.15							
Na_2O	7.46	bdl	bdl	bdl	6.52	0.40	bdl	0.39	bdl	bdl
Al_2O_3	0.07	bdl	bdl	bdl	bdl	0.13	0.07	bdl	bdl	bdl
SiO_2	23.75	24.60	22.48	24.63	22.62	23.06	23.94	22.45	24.35	14.36
K_2O	5.89	0.16	0.10	1.93	5.85	0.43	0.19	1.63	2.25	bdl
CaO	0.21	0.29	bdl	bdl	0.14	1.08	0.30	0.21	bdl	0.28
TiO_2	38.89	42.06	36.99	38.12	34.01	37.55	39.81	32.24	41.36	24.08
MnO	0.05	bdl	bdl	bdl	bdl	0.09	bdl	bdl	bdl	0.08
FeO	0.54	0.22	0.19	0.87	0.19	0.51	0.32	0.41	0.17	0.34
CoO				7.41						
NiO					1.17					
Rb_2O						21.79				
SrO	bdl	bdl	bdl	bdl	bdl	0.67	11.18	bdl	bdl	bdl
Nb_2O_5	2.99	4.25	2.09	3.14	3.43	3.25	3.37	3.83	1.51	2.11
Cs_2O								37.71		
BaO	0.14	bdl	bdl	bdl	bdl	0.40	1.43	bdl	bdl	bdl
La_2O_3									4.53	
Tl_2O										56.22
SUM	79.99	84.99	94.00	76.10	73.93	89.36	80.61	98.87	74.17	97.60
H_2O^a	20.01	15.01	6.00	23.90	26.07	10.64	19.39	1.13	25.83	2.40

bdl below detection limit
[a]$H_2O = 100 - SUM$

Table 2 Empirical chemical formulae of ivanyukite-Na-T and its ion-exchange forms

Phase	Empirical formula
ivanyukite-Na-*T*	$(Na_{1.82}K_{0.95}Ca_{0.03}Ba_{0.01})_{\Sigma 2.81}[(Ti_{3.68}Nb_{0.17}Fe_{0.06}{}^{3+}Mn_{0.01})_{\Sigma 3.92}(Si_{2.99}Al_{0.01})_{\Sigma 3.00}O_{14.61}(OH)_{1.39}](H_2O)_{7.2}$
Tl-exchanged phase	$(Tl_{3.32}Ca_{0.06}Na_{0.05})_{\Sigma 3.44}[(Ti_{3.79}Nb_{0.20}Fe_{0.06}{}^{3+}Mn_{0.01})_{\Sigma 4.06}Si_3O_{15.85}(OH)_{0.16}](H_2O)_{1.6}$
N_2H_5-exchanged phase	$(\{N_2H_5\}_{2.83}K_{0.02})_{\Sigma 2.85}[(Ti_{3.71}Nb_{0.13}Fe_{0.02}{}^{3+})_{\Sigma 3.86}Si_3O_{14.48}(OH)_{1.52}]$
Rb-exchanged phase	$(Rb_{1.81}Ca_{0.15}Na_{0.10}K_{0.07}Sr_{0.05}Ba_{0.02})_{\Sigma=2.20}[(Ti_{3.65}Nb_{0.19}Fe_{0.05}Mn_{0.01})_{\Sigma=3.90}O_{2.07}(OH)_{1.93}(Si_{2.98}Al_{0.02})_{\Sigma=3.00}O_{12}](H_2O)_{3.61}$
Cs-exchanged phase	$(Cs_{2.15}K_{0.28}Na_{0.10}Ca_{0.03})_{\Sigma 2.56}[(Ti_{3.24}Nb_{0.23}Fe_{0.05}{}^{3+})_{\Sigma 3.52}Si_3O_{12.84}(OH)_{3.16}]$
NH_4-exchanged phase	$(\{NH_4\}_{1.82}Ca_{0.04}K_{0.03})_{\Sigma 1.89}[(Ti_{3.87}Nb_{0.23}Fe_{0.02}{}^{3+})_{\Sigma 4.12}Si_3O_{14.62}(OH)_{1.38}](H_2O)_{1.7}$
Sr-exchanged phase	$(Sr_{0.81}Ba_{0.07}Ca_{0.04}K_{0.03})_{\Sigma 0.95}[(Ti_{3.74}Nb_{0.19}Fe_{0.03}{}^{3+})_{\Sigma 3.96}(Si_{2.99}Al_{0.01})_{\Sigma 3.00}O_{13.83}(OH)_{2.17}](H_2O)_7$
Co-exchanged phase	$(Co_{0.72}K_{0.30})_{\Sigma 1.02}[(Ti_{3.49}Nb_{0.17}Fe_{0.09}{}^{3+})_{\Sigma 3.75}Si_3O_{12.84}(OH)_{3.16}](H_2O)_{8.1}$
Ni-exchanged phase	$(Na_{1.68}K_{0.99}Ni_{0.13}Ca_{0.02})_{\Sigma 2.82}[(Ti_{3.39}Nb_{0.21}Fe_{0.02}{}^{3+})_{\Sigma 3.62}Si_3O_{13.62}(OH)_{2.38}](H_2O)_{10.34}$
La-exchanged phase	$(K_{0.35}La_{0.21})_{\Sigma 0.56}[(Ti_{3.83}Nb_{0.08}Fe_{0.02}{}^{3+})_{\Sigma 3.93}Si_3O_{12.78}(OH)_{3.22}](H_2O)_9$

effects using the STOE program X-RED. An analytical absorption-correction was made on the basis of measured and optimized crystal shape. The structure was solved and refined using the SHELX system of programs. The final model included all atomic positional parameters, anisotropic displacement parameters for all atoms except two extraframework H_2O molecules, and a weighting scheme applied to the structure factors (Table 3).

Crystal structures of all these phases are based upon 3-dimensional framework of the pharmacosiderite type (Yakovenchuk et al. 2009; Spiridonova et al. 2010). They consist of fundamental building blocks of four edge-sharing Ti-octahedra interlinked by SiO_4 tetrahedra (Fig. 1).

The resulting framework has a 3-dimensional system of channels defined by 8-membered rings (8-MRs) with the crystallographic free diameter of 3.5 Å. The channels are occupied by Na^+, K^+ and Cu^{2+} cations as well as by H_2O molecules.

Table 3 Crystallographic data and refinement parameters for ivanyukite-Na-*T*, and Rb- and Sr-exchanged ivanyukite

Parameter	Ivanyukite-Na-*T*	Rb-phase	Sr- phase
a (Å)	10.921(3)	7.809(5)	7.834(5)
c (Å)	13.885(4)	= *a*	= *a*
V (Å^3)	1434.2(7)	476.2(5)	480.8(5)
Space group	*R*3*m*	*P*-43*m*	*P*-43*m*
Z	3	1	1
D_{calc} (g/cm^3)	2.384	2.805	2.341
Unique reflections $\lvert F_o\rvert \geq 4\sigma_F$	723	188	215
R_1	0.147	0.1281	0.1175
wR_2	0.384	0.2371	0.2240
S	1.330	1.353	1.201

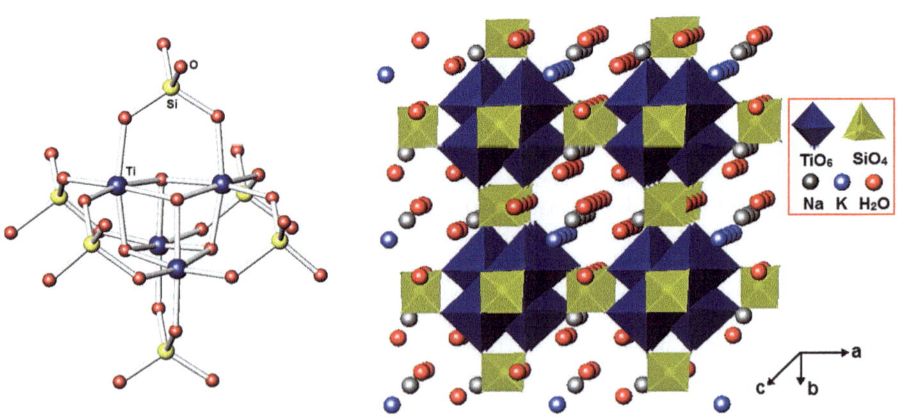

Fig. 1 Basic fundamental unit in titanosilicate framework in the structure of ivanyukite-group minerals (*left*) and structure of ivanyukite-Na-T (*right*)

The crystal chemical formulae of the studied crystals determined on the basis of their structure refinement can be written as:

$$\text{ivanyukite-Na-T} - \mathrm{Na_{1.44}K_{0.52}\left[Ti_4(OH)_{2.04}O_{1.96}(SiO_4)_3\right](H_2O)_6};$$

$$\text{Rb-substituted phase} - \mathrm{Rb_{2.16}\left[Ti_4(OH)_{1.84}O_{2.16}(SiO_4)_3\right](H_2O)_6};$$

$$\text{Sr-substituted phase} - \mathrm{Sr_{0.87}\left[Ti_4(OH)_{2.26}O_{1.74}(SiO_4)_3\right](H_2O)_4}.$$

In ivanyukite-Na-T, potassium cations are located within the framework channels and coordinated by seven anions each with the K-O distances of 2.69–3.15 Å. Na^+ cations are attached to the side of the 8-membered rings and have fivefold coordination (Na-O distance varies from 2.27 to 2.46 Å).

Asymmetric position of K^+ cation within 8-MRs is responsible for rhombohedral distortion of cubanite-like titanosilicate clusters (Fig. 2) and, as a result for the rhombohedral symmetry of the whole framework (Spiridonova et al. 2010). Cation leaching of ivanyukite-Na-*T* results in disappearance of the rhombohedral distortion and relaxation of the ideal cubic symmetry. As a consequence, all exchanged forms of ivanyukite possess cubic space group *P*-43 m, but with different arrangements of extraframework cations. For instance, in Rb-exchanged ivanyukite, 12-coordinated Rb^+ cations occupy centres of the 8-MRs, whereas, in Sr-exchanged ivanyukite, Sr^{2+} cations are either 9- or 7-coordinated (Fig. 3).

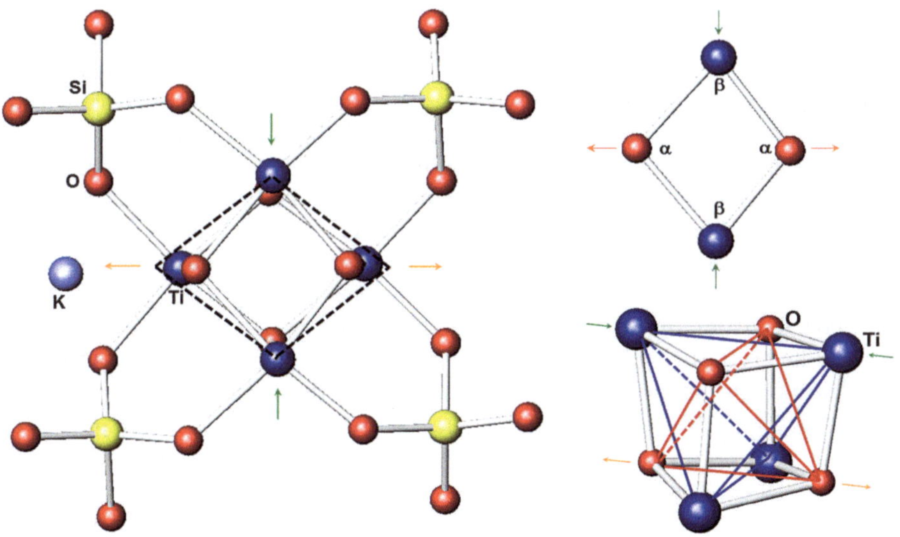

Fig. 2 Distortion of a cubane-like cluster in the rhombohedral structure of ivanyukite-Na-T in comparison to its ideal cubic symmetry can be described as its elongation towards K+ cation (*left*), which results in elongation of the –Ti-O-Ti-O- rhombus (*top right*) along the O-O direction and its shrinking along the Ti-Ti direction (*bottom right*)

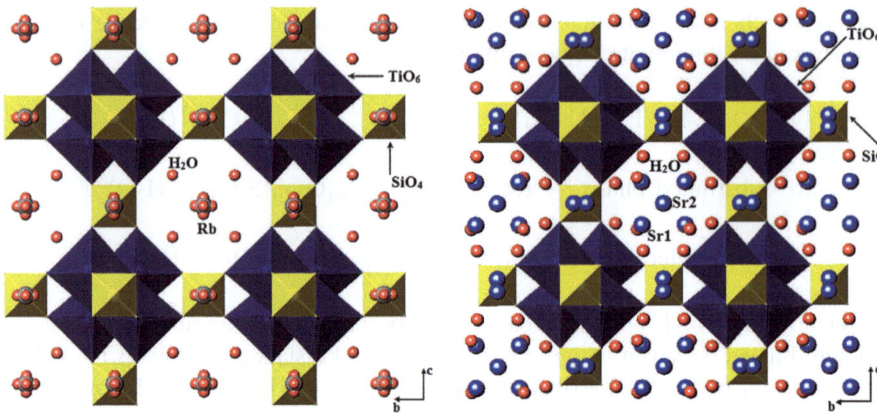

Fig. 3 Crystal structures of Rb- (*left*) and Sr- (*right*) exchanged forms of ivanyukite

4 Perspectives

Rb-, Cs-, Tl-, N_2H_5- and NH_4-substututed phases are H_2O-free and very stable under ambient conditions. We were unable to decationize or exchange Cs-substituted ivanyukite in cold solutions, which means that this mineral can be used not only for the removal of ^{90}Sr and ^{137}Cs from radioactive waste solutions but also for conservation of ^{137}Cs (Yakovenchuk et al. 2009).

The ability of ivanyukite to absorb hydrazine is also very important for the following reasons: (1) hydrazine is a toxic substance used in multipropellants, and ivanyukite can be used for its sorption; (2) hydrazinium-substituted ivanyukite form has a unique combination of reducing and cation-exchange properties that permits it to reduce and absorb many metals and semimetals from water solutions; (3) since hydrazinium-exchanged ivanyukite immediately (in some minutes) reduces gold, platinum, palladium, rhodium and others metals in cold water solution, reduction of noble metals by hydrazinium may lead to manifacturing microporous titanosilicates with noble-metal clusters in the framework channels.

Upon reduction, noble metals and rhodium form porous films on surfaces of ivanyukite crystals (Fig. 4) and partially localize within channels of titanosilicate framework. The empirical chemical formulae of these phases are as follows: $(Pd_{0.21}K_{0.26})_{\Sigma=0.4}7\left[(Ti_{3.51}Nb_{0.33}Fe^{3+}{}_{0.05})_{\Sigma=3.89}Si_3O_{12.94}(OH)_{3.06}\right]\cdot 2.5H_2O$ (Pd-reduced phase) and $(K_{0.14}Rh_{0.06})_{\Sigma=0.20}\left[(Ti_{3.61}Nb_{0.36}Fe^{3+}{}_{0.05})_{\Sigma=4.02}Si_3O_{12.64}(OH)_{3.36}\right]\cdot 8.6H_2O$ (Rh-reduced phase). These phases may possess interesting catalytic properties, especially taking into account their stability over the wide range of *P*, *T* and *p*H.

Fig. 4 Scanning electron microscopy image of ivanyukite crystal exchanged with hydrazine and treated with Pd-containing solution. *White clusters* on the surface of the crystal is metallic palladium

Acknowledgements This work was supported by Russian Federal Programme "Scientific and Education Cadres of Innovative Russia" (to DVS and SVK, state contract no. 02.740.11.0326).

References

Spiridonova DV, Krivovichev SV, Yakovenchuk VN, YaA P (2010) Crystal structures of Rb- and Sr-exchanged forms of ivanyukite-Na-*T*. Zap RMO 139(5):79–88

Yakovenchuk VN, Nikolaev AP, Selivanova EA, YaA P, Korchak JA, Spiridonova DV, Zalkind OA, Krivovichev SV (2009) Ivanyukite-Na-*T*, ivanyukite-Na-*C*, ivanyukite-K, and ivanyukite-Cu: new microporous titanosilicates from the Khibiny massif (Kola Peninsula, Russia) and crystal structure of ivanyukite-Na-*T*. Amer Miner 94:1450–1458

Yakovenchuk VN, Selivanova EA, Ivanyuk GYu, Pakhomovsky YaA, Spiridonova DV, Krivovichev SV (2008) First natural pharmacosiderite-related titanosilicates and their ion-exchange properties. In: Krivovichev S (ed) Minerals as advanced materials I. Springer, Berlin/Heidelberg, pp 27–35

Delhayelite and Mountainite Mineral Families: Crystal Chemical Relationship, Microporous Character and Genetic Features

Igor V. Pekov, Natalia V. Zubkova, Nikita V. Chukanov, Anna G. Turchkova, Yaroslav E. Filinchuk, and Dmitry Yu. Pushcharovsky

Many minerals show the properties commonly named "zeolitic": ion exchange and leaching, sorption of gases and organic molecules, reversible hydration, *etc*. The most well-known examples are aluminosilicate zeolites, clay minerals and pyrochlores. Besides them, several other large groups of minerals are now in focus of research as potential microporous materials or their prototypes. First, there are Ti-, Zr-, Nb-silicates with heteropolyhedral frameworks. This paper is devoted to other minerals of this interest. There are two related families of natural silicates with structures based on Si,O or Al,Si,O tetrahedral layers and columns of edge-shared Ca-centered octahedra.

The delhayelite mineral family can be distinguished within the rhodesite mero-plesiotype series (Ferraris and Gula 2005; Cadoni and Ferraris 2009) as its alumino-silicate branch that includes delhayelite, fivegite and hydrodelhayelite (Table 1). The idealized formula of delhayelite earlier presented as $K_7Na_3Ca_5[AlSi_7O_{19}]_2F_4Cl_2$ ($Z = 1$) (Cannillo et al. 1969) was recently refined: $K_4Na_2Ca_2[AlSi_7O_{19}]F_2Cl$ ($Z = 2$) (Pekov et al. 2009). Its crystal structure is based upon the double-layer tetrahedral blocks $[(Al,Si)_4Si_{12}O_{38}]$ linked by the columns of Ca octahedra (Fig. 1a). Zeolite-like channels inside the tetrahedral blocks host K^+ and Cl^- ions whereas the channels in the interlayer space, between Ca columns, host Na^+ and F^- (Cannillo et al. 1969; Pekov et al. 2009). Fivegite, $K_4Ca_2[AlSi_7O_{17}(O_{2-x}OH_x)][(H_2O)_{2-x}OH_x]Cl$ ($x = 0$–2) (Pekov et al. 2010a), and hydrodelhayelite, $KCa_2[AlSi_7O_{17}(OH)_2]\cdot(6\text{-}x)H_2O$ (Pekov et al. 2009; Ragimov et al. 1980), have the same Ca,Al,Si,O structure motif (rigid 3D part of the structure, $Ca_2[AlSi_7O_{17}\O_2]$, slightly distorted as compared with delhayelite, with protonated "pendent" vertices Ø of Si tetrahedra) but different contents of the channels (Cannillo et al. 1969; Pekov et al. 2009) (Fig. 2). We consider this Ca,Al,Si,O motif as

I.V. Pekov (✉) • N.V. Zubkova • A.G. Turchkova • D.Y. Pushcharovsky
Faculty of Geology, Moscow State University, Moscow, Russia
e-mail: igorpekov@mail.ru

N.V. Chukanov
Institute of Problems of Chemical Physics RAS, Chernogolovka, Russia

Y.E. Filinchuk
Swiss-Norwegian Beam Lines at ESRF, BP-220, 38043 Grenoble, France

S.V. Krivovichev (ed.), *Minerals as Advanced Materials II*,
DOI 10.1007/978-3-642-20018-2_21,

Table 1 Comparative data for minerals of the delhayelite family

Mineral formula	Delhayelite $K_4Na_2Ca_2[AlSi_7O_{19}]F_2Cl$	Fivegite $K_4Ca_2[AlSi_7O_{17}(O_{2-x}OH_x)][(H_2O)_{2-x}OH_x]Cl$	Hydrodelhayelite[a] $KCa_2[AlSi_7O_{17}(OH)_2](H_2O)_{6-x}$
Crystal system space group	Orthorhombic *Pmmn*	Orthorhombic $Pm2_1n$	Orthorhombic $Pm2_1n$
a, Å	24.58	24.34	23.85
b, Å	7.06	7.04	7.07
c, Å	6.58	6.54	6.65
V, Å^3	1,142	1,120	1,121
Z	2	2	2
References	(Cannillo et al. 1969; Pekov et al. 2009)	(Pekov et al. 2010a)	(Pekov et al. 2009; Ragimov et al. 1980)

[a]Hydrodelhayelite unit cell is given in the setting with tetrahedral layers perpendicular to the *a* axis (with corresponding space group), like other minerals in this table

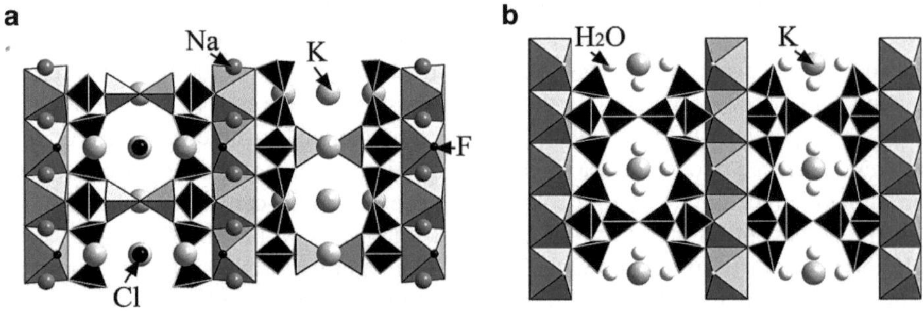

Fig. 1 Crystal structures of delhayelite (**a**): drawn based on the data from Pekov et al. (2009) and rhodesite (**b**): drawn based on the data from Hesse et al. (1992)): *ab* projections. Ca octahedral columns and the tetrahedral motifs are shown in polyhedra; Si tetrahedra are *black*, (Al,Si) tetrahedra in delhayelite are *gray*

a heteropolyhedral quasi-framework. If we do not take into account Ca octahedra then delhayelite-family members could be presented as phyllosilicates. However it seems more correct, from both physical and chemical viewpoints, to include Ca octahedral columns to the main building unit of these minerals.

The Ca,Al,Si,O quasi-framework remains stable in natural transformation series delhayelite → fivegite → hydrodelhayelite (Pekov et al. 2010a) and in products of our experiments with delhayelite for ion leaching, cation exchange and hydration in aqueous systems (Turchkova et al. 2011). Thus the delhayelite family is a new, specific family of natural microporous aluminosilicates with zeolitic properties that is, in fact, intermediate between phyllo – and tecto-aluminosilicates. Its representatives delhayelite and fivegite differ from other microporous minerals by the presence of mobile halogen anions in the same channels as alkali cations. These ions are ordered in accordance with widths of the channels (Figs. 1 and 2). The hydrothermal alteration of delhayelite occurs step-by-step and shows different mobility of ions located in different structural sites: first leaching of F^- and Na^+ from the interlayer space occurs and

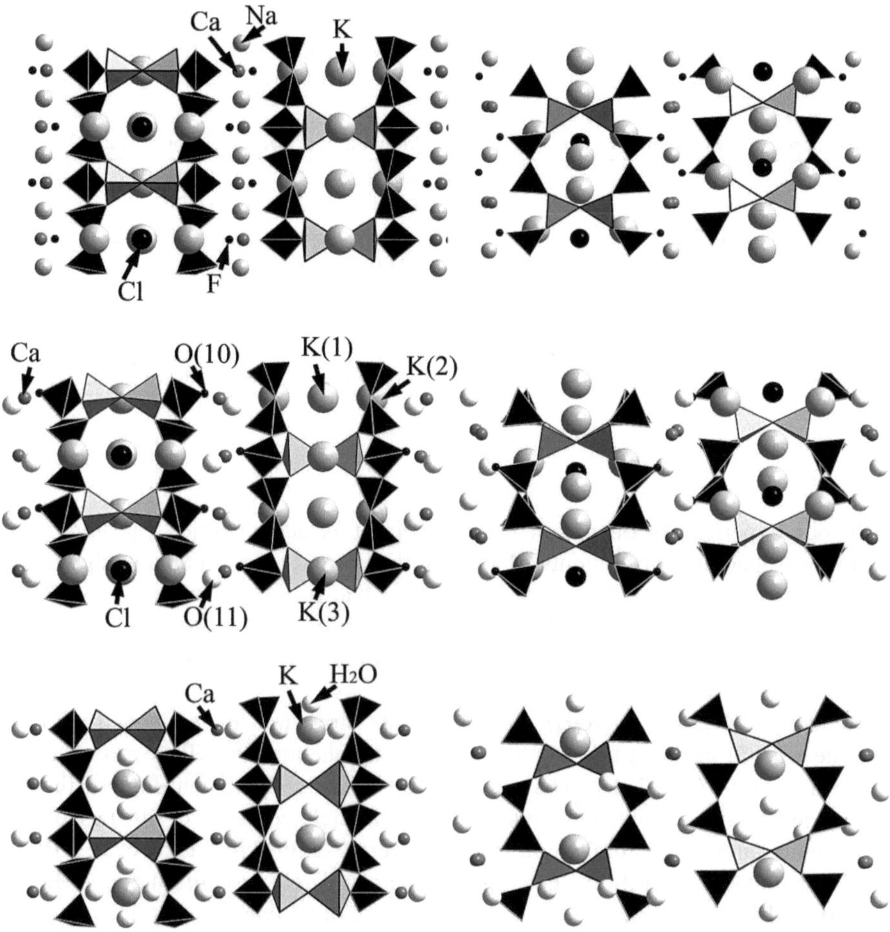

Fig. 2 Crystal structures (from *top* to *bottom*) of delhayelite, fivegite and hydrodelhayelite. Si tetrahedra are *black*, (Al,Si) tetrahedra are *gray*. *Left* column – *ab* projections, *right* column – *ac* projections, in the unified setting (see Table 1)

further Cl^- and K^+ leave their sites within the tetrahedral block. These processes can be presented as following reactions:

(1) Delhayelite $K_4Na_2Ca_2[AlSi_7O_{19}]F_2Cl + 2H_2O \rightarrow$ fivegite $K_4Ca_2[AlSi_7O_{17}(O_{2-x}OH_x)][(H_2O)_{2-x}OH_x]Cl + 2Na^+ + 2F^-$;
(2) Fivegite + $(5\text{-}x)H_2O \rightarrow$ hydrodelhayelite $KCa_2[AlSi_7O_{17}(OH)_2](H_2O)_{6-x}$ + 3 $K^+ + Cl^-$.

Delhayelite, fivegite and hydrodelhayelite are individual, well-stoichiometric minerals and transformations from one to other have discrete character: boarders between the initial and replacing phases in this evolution series are sharp (Pekov et al. 2010a).

Table 2 Comparative data for minerals of the mountainite family

Mineral formula	Mountainite $KNa_2Ca_2[Si_8O_{19}(OH)] \cdot 6H_2O$	Shlykovite $KCa [Si_4O_9(OH)] \cdot 3H_2O$	Cryptophyllite $K_2Ca [Si_4O_{10}] \cdot 5H_2O$
Crystal system space group	Monoclinic *P*2/*c*	Monoclinic $P2_1/c$	Monoclinic $P2_1/n$
a, Å	13.70	6.49	6.49
b, Å	6.58	7.00	6.99
c, Å	13.75	26.71	32.09
β,°	105.75	94.60	94.68
V, $Å^3$	1,193	1,209	1,452
Z	2	4	4
References	(Gard et al. 1957; Zubkova et al. 2009)	(Pekov et al. 2010b)	(Pekov et al. 2010b)

In our experiments on cation exchange (that is accompanied by significant hydration), leaching of contents of the interlayer space and channels shows the same sequence as simple hydration: first F and Na, further Cl and K. These facts indicate high selectivity of "zeolitic" characteristics of delhayelite-family minerals that makes them (and related synthetic compounds) promising as technologically usable microporous materials.

Rhodesite, $KCa_2[Si_8O_{18}(OH)]\cdot(6-x)H_2O$, the structural analogue of hydrodelhayelite with Si instead of Al in corresponding tetrahedral sites (Pekov et al. 2009; Hesse et al. 1992) (Fig. 1b), can be considered as a "bridge" between aluminosilicates of the delhayelite family and crystal chemically related Al-free silicates of the mountainite family of natural phyllosilicates recently defined in (Pekov et al. 2010b).

The mountainite family includes three minerals (Table 2) with unique, different one from others but related structures. The structure of mountainite, a mineral known since 1957 (Gard et al. 1957), was solved in 2008 and its crystallochemically correct formula was determined: $KNa_2Ca_2[Si_8O_{19}(OH)]\cdot 6H_2O$ ($Z = 2$) (Zubkova et al. 2009). Shlykovite, $K_2Ca_2[Si_8O_{18}(OH)_2]\cdot 6H_2O$, and cryptophyllite, $K_4Ca_2[Si_8O_{20}]\cdot 10H_2O$, are new minerals discovered in 2009 in the Khibiny alkaline complex, Kola Peninsula, Russia (Pekov et al. 2010b).

The main structural unit of all three minerals is *TOT* block consisting of two identical tetrahedral Si layers (*T*) $[Si_8O_{18}(O,OH)_2]$ and an octahedral component (*O*), formed by columns of edge-sharing Ca octahedra sandwiched between them (Fig. 3). Single *T*-layer in shlykovite and cryptophyllite is topologically the same as half of the double-layer block of rhodesite, $[Si_{16}O_{36}(OH)_2]$, or delhayelite, $[(Al,Si)_4Si_{12}O_{38}]$ (Fig. 4). *T*-layer in mountainite is very similar and differs from that of shlykovite and cryptophyllite only by the orientation of some Si tetrahedra that causes the difference in configuration of the Ca octahedral columns in these minerals. K^+ cations are located in channels within *TOT* blocks in all three minerals, like delhayelite-family members that also contain only the largest K^+ cations in channels within the tetrahedral block.

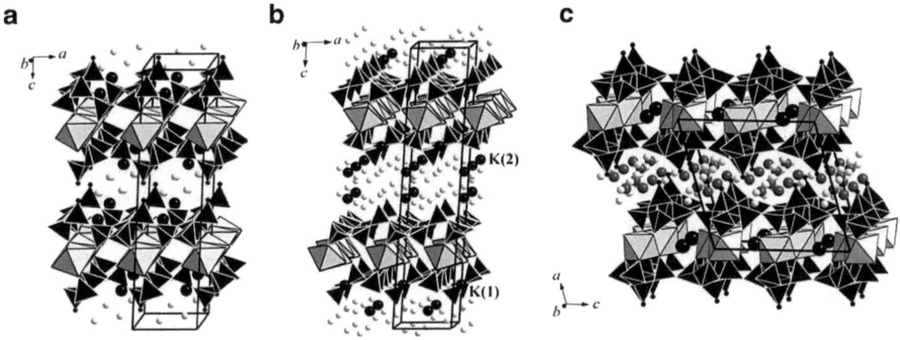

Fig. 3 Crystal structures of shlykovite (**a**), cryptophyllite (**b**), and mountainite (**c**), SiO_4 tetrahedra are *black*, Ca octahedra are *light grey*. K cations are shown as *large black circles*, O atoms of H_2O molecules – as *small grey circles*. Na atoms in mountainite are marked as *large grey spheres*. Positions of OH-groups in shlykovite and (O, OH) in mountainite are shown as *small black circles*

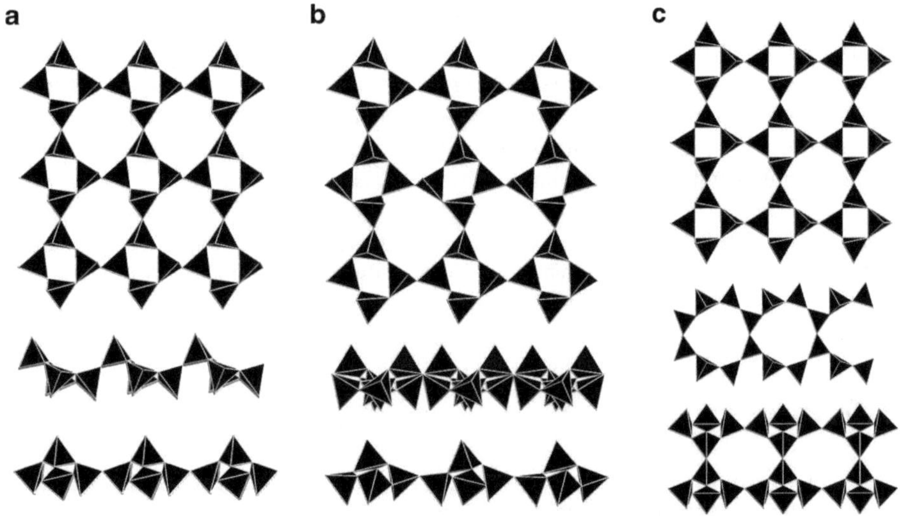

Fig. 4 Silicate layer formed by 4- and 8-membered rings in the structures of shlykovite and cryptophyllite (**a**), mountainite (**b**), and rhodesite (**c**)

Unlike rhodesite mero-plesiotype series minerals with heteropolyhedral quasi-frameworks, members of the mountainite family are true phyllosilicates: *TOT* blocks are isolated from each other and only H_2O molecules (shlykovite) or H_2O molecules and alkali cations (K^+ in cryptophyllite, Na^+ in mountainite) are located between them (Fig. 3). An important feature of the mountainite family is the variability of the distance between *TOT* blocks (that determines the basal parameter of unit cell) in accordance with interlayer contents: $(H_2O)_4$ in shlykovite ($c = 26.7$ Å), $K_2(H_2O)_8$ in cryptophyllite ($c = 32.1$ Å) or $Na_2(H_2O)_4$ in mountainite ($2a = 27.5$ Å).

This variability and significant hydration capacity indicate great potential of mountainite-family minerals as new microporous materials suitable to the exchange of large cations and, probably, their immobilization after heating for H_2O removal and structure contraction.

Minerals of both the delhayelite and the mountainite families were only found in the alkaline formation. Common feature of all of them is species-defining role of K (Tables 1 and 2) and they occur in potassium-enriched rocks or their derivatives. Delhayelite, the only anhydrous mineral among six representatives of these families, is formed under high-temperature conditions in K-rich peralkaline rocks and related pegmatites. All other discussed minerals have hydrothermal origin. The Al-bearing species, fivegite and hydrodelhayelite, are known only as products of transformation of delhayelite with preservation of its stable Ca,Al,Si,O heteropolyhedral quasi-framework whereas Al-free minerals (mountainite family) crystallize directly from low-temperature hydrothermal solutions just as rhodesite. The minerals with single and double tetrahedral layers are formed under the same conditions, sometimes together. Intimate intergrowths of mountainite and rhodesite were found in hydrothermally altered kimberlitic rock at Bultfontein, South Africa (Gard et al. 1957). In one pegmatite body at Mt. Rasvumchorr in Khibiny, we have observed five of six minerals (except mountainite) of both discussed families. In the late-stage, low-temperature hydrothermal assemblage hydrodelhayelite, occurring as partial or complete pseudomorphs after fivegite (that partially or completely replaced delhayelite), is associated with shlykovite and cryptophyllite typically forming intimate intergrowths in cracks and cavities of this pegmatite.

Acknowledgements This work was supported by grants nos. 08-07-00077-a, 09-05-00143-a, and 09-05-12001-ofi_m by RFBR and grants of President of Russian Federation nos. MK-320.2010.5, NSh-4034.2010.5 and NSh-3848.2010.5.

References

Cadoni M, Ferraris G (2009) Two new members of the rhodesite mero-plesiotype series close to delhayelite and hydrodelhayelite: synthesis and crystal structure. Eur J Miner 21:485–493

Cannillo E, Rossi G, Ungaretti L (1969) The crystal structure of delhayelite. Rend Soc Ital Miner Pet 26:63–75

Ferraris G, Gula A (2005) Polysomatic aspects of microporous minerals – heterophyllosilicates, palysepioles and rhodesite-related structures. Rev Miner Geochem 57:69–104

Gard JA, Taylor HFW, Chalmers RA (1957) An investigation of two new minerals: rhodesite and mountainite. Miner Mag 31:611–623

Hesse KF, Liebau F, Merlino S (1992) Crystal structure of rhodesite, $HK_{1-x}Na_{x+2y}Ca_{2-y}$ $\{1B,3,2_{\infty}^{2}\}[Si_8O_{19}]\bullet(6-z)H_2O$, from three localities and its relation to other silicates with dreier double layers. Z Kristallogr 199:25–48

Pekov IV, Zubkova NV, Chukanov NV, Sharygin VV, Pushcharovsky DYu (2009) Crystal chemistry of delhayelite and hydrodelhayelite. Dokl Earth Sci 428(7):1216–1221

Pekov IV, Zubkova NV, Chukanov NV, Zadov AE, Pushcharovsky DYu (2010a) Fivegite, $K_4Ca_2[AlSi_7O_{17}(O_{2-x}OH_x)][(H_2O)_{2-x}OH_x]Cl$, a new mineral from the Khibiny alkaline complex, Kola Peninsula, Russia. Zap RMO 139(4):47–63 (in Russian)

Pekov IV, Zubkova NV, Filinchuk YaE, Chukanov NV, Zadov AE, Pushcharovsky DYu, Gobechiya ER (2010b) Shlykovite, $KCa[Si_4O_9(OH)]\cdot 3H_2O$, and cryptophyllite, $K_2Ca[Si_4O_{10}]\cdot 5H_2O$, two new mineral species from the Khibiny alkaline complex, Kola Peninsula, Russia. Zap RMO 139(1):37–50 (in Russian)

Ragimov KG, Chiragov MI, Mamedov KS, Dorfman MD (1980) Crystal structure of hydrodelhayelite, $KH_2Ca_2(Si,Al)_8O_{19}\cdot 6H_2O$. Doklady AN Azerb SSR 36(12):49–51 (in Russian)

Turchkova AG, Pekov IV, Lykova IS, Chukanov NV, Yapaskurt VO (2011) Delhayelite: ion leaching and ion exchange. In: Krivovichev SV (ed) Minerals as advanced materials II, 221–228

Zubkova NV, Pekov IV, Pushcharovsky DYu, Chukanov NV (2009) The crystal structure and refined formula of mountainite, $KNa_2Ca_2Si_8O_{19}(OH)\cdot 6H_2O$. Z Kristallogr 224:389–396

Delhayelite: Ion Leaching and Ion Exchange

Anna G. Turchkova, Igor V. Pekov, Inna S. Lykova, Nikita V. Chukanov, and Vasiliy O. Yapaskurt

Delhayelite, ideally $K_4Na_2Ca_2[AlSi_7O_{19}]F_2Cl$, is a mineral showing structural features of both phyllosilicates and silicates with heteropolyhedral frameworks. Its crystal structure is based upon the two-layer tetrahedral blocks, or double layers, $[(Al,Si)_4Si_{12}O_{38}]$, linked by the columns of edge-shared Ca-centered octahedra (Cannillo et al. 1969; Pekov et al. 2009). As it is shown in (Pekov et al. 2011), these two main building units form a heteropolyhedral quasi-framework with zeolite-like channels of two types: (1) inside the tetrahedral blocks, hosting K^+ and Cl^-; (2) in the interlayer space, between Ca columns, hosting Na^+ and F^-. Such ordering of three large cations and two halogen anions is a feature of delhayelite making it unique among minerals. Products of the step-by-step hydrothermal alteration of delhayelite in K-enriched peralkaline pegmatites of the Khibiny alkaline complex (Kola Peninsula, Russia) are fivegite, $K_4Ca_2[AlSi_7O_{17}(O_{2-x}OH_x)][(H_2O)_{2-x}OH_x]Cl$ ($x = 0$–2), and hydrodelhayelite, $KCa_2[AlSi_7O_{17}(OH)_2]\cdot(6-x)H_2O$, preserving the stable Ca,Al, Si,O motif of the structure (Dorfman and Chiragov 1979; Pekov et al. 2010). More detailed data on crystal chemistry of delhayelite and its comparison with related minerals are given, including illustrations, in paper (Pekov et al. 2011) published in this volume.

The above-mentioned structural features became a reason for us to assume that delhayelite can have strong zeolitic properties and to check them experimentally. It seems important that, unlike other known microporous silicate minerals showing easy mobility of weakly-bonded cations, delhayelite hosts also halogen anions in the same zeolitic channels. Earlier, the ability of delhayelite to absorb Rb^+ and NH_4^+ cations was reported by Chiragov who detected 0.15 wt.% Rb_2O in delhayelite after the experiments in aqueous solutions with Rb salts and noted the decrease of K and Na contents after the experiments with ammonium salts (Chiragov 1982).

A.G. Turchkova (✉) • I.V. Pekov • I.S. Lykova • V.O. Yapaskurt
Faculty of Geology, Moscow State University, Moscow, Russia
e-mail: annaturchkova@rambler.ru

N.V. Chukanov,
Institute of Problems of Chemical Physics RASChernogolovka, Russia

S.V. Krivovichev (ed.), *Minerals as Advanced Materials II*,
DOI 10.1007/978-3-642-20018-2_22,

Table 1 Chemical composition of initial delhayelite (1) and its Ba-exchanged forms (2–4)

Constituent wt. %	1	2	3	4
Na_2O	6.51	0.58	0.90	0.54
K_2O	19.24	5.10	4.25	3.26
BaO	–	8.50	9.41	11.25
CaO	14.39	14.63	14.17	13.97
Al_2O_3	6.52	7.05	6.81	6.92
SiO_2	47.89	54.34	52.14	53.72
SO_3	0.33	–	–	–
F	4.32	1.34	1.38	–
Cl	3.84	1.03	0.80	0.57
-O = $(Cl,F)_2$	−2.69	−0.79	−0.76	−0.13
Total	100.35	91.78	89.10	90.10
Formula calculated on Si + Al = 8				
Na	1.82	0.14	0.23	0.13
K	3.53	0.83	0.72	0.54
Ba	–	0.42	0.49	0.57
Ca	2.22	2.00	2.02	1.93
Al	1.11	1.06	1.07	1.05
Si	6.89	6.94	6.93	6.95
S	0.04	–	–	–
F	1.97	0.54	0.58	–
Cl	0.94	0.22	0.18	0.12

In our experiments, grains (0.5–3 mm) of fresh, anhydrous delhayelite from Kirovskii mine, Khibiny, were placed in 1 N aqueous solutions of $(NH_4)_2SO_4$, Na_2SO_4, K_2SO_4, $RbNO_3$, $CsNO_3$, $SrCl_2$, $BaCl_2$, $Pb(NO_3)_2$ and $CuSO_4$ at 80–90°C and 1 atm for 3 h and in distilled H_2O under the same conditions for 500 h. Samples after the experiments were examined by infrared spectroscopy (IRS), scanning electron microscopy (SEM) and electron microprobe analysis (EMPA).

In acidic solution of $CuSO_4$, delhayelite partially dissolves and its rest shows no signs of Cu absorption. The mineral remains unaltered in solutions of Na, K, Sr and Pb salts but absorbs NH_4, Rb, Cs and Ba with decrease of K, Na, Cl and F contents. Ratios of Si, Al and Ca remain constant in all cases that confirms stability of the Ca,Al,Si,O motif in delhayelite-like phases considered by us as a heropolyhedral quasi-framework (Pekov et al. 2011). It was confirmed by stable, stoichiometric Ca:Al:Si ratios, the same as in initial delhayelite, in all samples after the experiments (EMPA data: Tables 1–3). SO_4 is typical admixture anion in initial delhayelite (probably substituting Cl: (Pekov et al. 2009)) but in our experiments no signs of absorption of additional SO_4, as well as NO_3 or additional Cl anions, observed.

All processes of ion leaching and ion exchange in delhayelite are accompanied by its hydration that is clearly demonstrated by the IRS data.

Table 2 Chemical composition of initial delhayelite (1) and its Rb-exchanged forms (2–3)

Constituent	1	2	3
wt. %			
Na_2O	6.76	2.48	1.70
K_2O	19.83	7.18	5.53
Rb_2O	–	10.52	11.89
CaO	13.59	15.07	14.70
Al_2O_3	6.72	7.06	6.94
SiO_2	47.17	50.59	51.15
SO_3	0.50	–	0.26
F	4.40	1.86	1.79
Cl	3.86	1.76	1.67
$-O = (Cl,F)_2$	−2.72	−1.22	−1.13
Total	100.11	95.50	94.50
Formula calculated on Si + Al = 8			
Na	1.90	0.65	0.45
K	3.67	1.24	0.95
Rb	–	0.92	1.03
Ca	2.11	2.19	2.12
Al	1.15	1.13	1.10
Si	6.85	6.87	6.90
S	0.05	–	0.03
F	2.02	0.80	0.76
Cl	0.95	0.45	0.38

Table 3 Chemical composition of initial delhayelite (1) and its Cs-exchanged forms (2–8)

Constituent	1	2	3	4	5	6	7	8
wt. %								
Na_2O	6.65	3.43	0.96	0.92	–	–	–	–
K_2O	19.79	15.21	5.48	4.85	2.26	3.36	2.08	0.86
Cs_2O	–	5.21	10.10	11.88	16.77	18.66	19.92	26.95
CaO	14.12	14.37	14.29	14.13	13.97	13.10	13.40	11.86
Al_2O_3	6.55	6.25	6.27	6.78	6.22	5.97	6.34	6.17
SiO_2	47.98	46.93	52.51	54.57	52.33	51.76	51.48	47.76
SO_3	0.24	0.30	–	–	–	–	–	–
F	4.36	1.60	–	–	–	–	–	–
Cl	3.92	2.88	0.75	0.92	0.37	0.73	0.60	–
$-O = (Cl,F)_2$	−2.72	−1.31	−0.17	−0.21	−0.08	−0.16	−0.14	–
Total	100.89	94.87	90.19	93.84	91.84	93.42	93.08	93.60
Formula calculated on Si + Al = 8								
Na	1.85	0.98	0.25	0.23	–	–	–	–
K	3.63	2.86	0.93	0.79	0.39	0.58	0.36	0.16
Cs	–	0.33	0.57	0.65	0.96	1.08	1.15	1.67
Ca	2.17	2.27	2.04	1.94	2.01	1.91	1.95	1.85
Al	1.11	1.09	0.99	1.02	0.98	0.96	1.01	1.06
Si	6.89	6.91	7.01	6.98	7.02	7.04	6.99	6.94
S	0.03	0.03	–	–	–	–	–	–
F	1.98	0.75	–	–	–	–	–	–
Cl	0.95	0.72	0.17	0.20	0.08	0.16	0.14	–

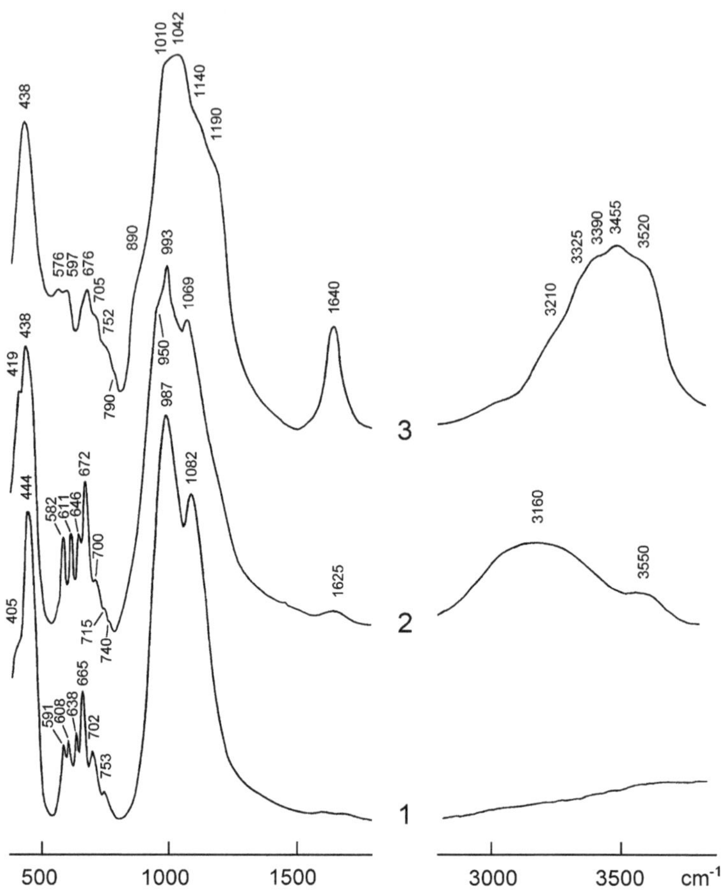

Fig. 1 IR spectra of delhayelite (*1*), fivegite (*2*) and hydrodelhayelite (*3*) from the Khibiny alkaline complex, Kola Peninsula, Russia

The IR spectrum of initial delhayelite (Fig. 1, curve1) is characterized by the absence of bands corresponding to O-H vibrations. The spectrum of fivegite (Fig. 1, curve 2) contains bands due to O-H stretching (3,000–3,600 cm^{-1}) and H-O-H bending (1,625 cm^{-1}) vibrations. Bands at 3,550 and 3,160 cm^{-1} correspond to (OH) groups forming weak hydrogen bonds and weakly acidic OH groups forming strong hydrogen bonds, respectively. An analogous band due to weakly acidic Si-OH (silanol) groups in the IR spectrum of hydrodelhayelite is observed as a shoulder at 3,210 cm^{-1} (Fig. 1, curve 3). Unlike the case with hydrodelhayelite, which displays a strong band due to bending vibrations of H_2O at 1,640 cm^{-1}, the spectrum of fivegite shows relatively a weak H-O-H bending vibration band at 1,625 cm^{-1}. The bands of stretching vibrations of H_2O molecules are presumably overlapped by that of weakly acidic OH groups. Generally, the process of the

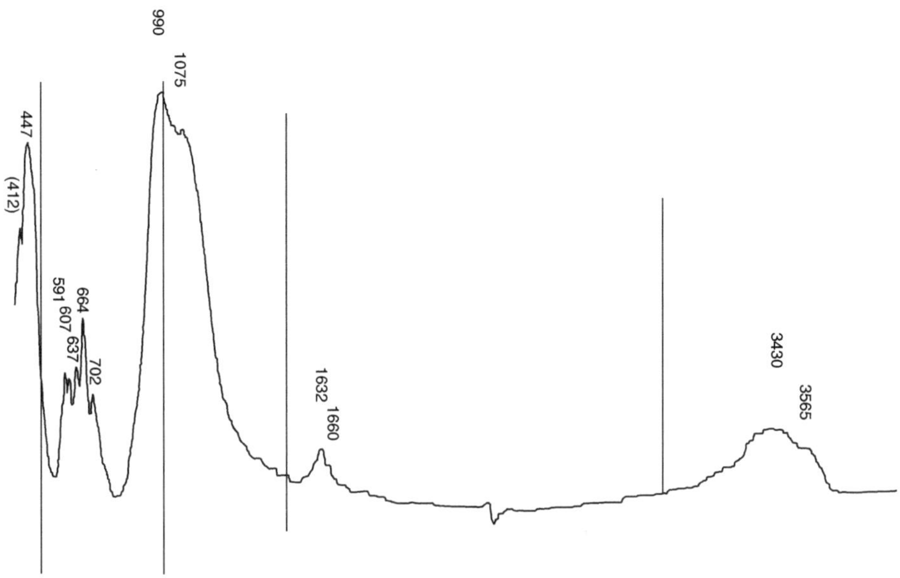

Fig. 2 IR spectrum of the fivegite-like phase obtained from delhayelite after the experiment in distilled H_2O: 500 h, 80–90°C, 1 atm

transition of delhayelite to fivegite could be presented as leaching of Na and F from the interlayer space of delhayelite conjugated with hydration, protonation of "pendent" vertices of Si tetrahedra and entry of H_2O molecules and OH groups into former F sites. It was definitely confirmed by bond valence calculations in the fivegite structure (Pekov et al. 2010).

Our experiment in distilled H_2O also demonstrates partial leaching of Na and F resulting in the formation of a fivegite-like phase. Its IR spectrum (Fig. 2) shows the appearance of H-bearing groups of different kinds: H_2O molecules (band of H-O-H bending vibrations at 1,632 cm^{-1} with shoulder at 1,660 cm^{-1} and band of O-H stretching vibrations at 3,430 cm^{-1} corresponding to H-bonds of medium strength); band at 3,565 cm^{-1} corresponds to (OH) groups forming weak hydrogen bonds.

The experiment in solution of $(NH_4)_2SO_4$ shows absorption of NH_4^+ cation with partial leaching of univalent components and hydration. The hydration process here is more developed than in distilled water (probably because of higher acidity of the solution) and the sample shows the IR spectrum (Fig. 3) close to one of hydrodelhayelite (Fig. 1, curve 3) in the regions of O-H stretching (broad, intensive band with maximum at 3,424 cm^{-1} and shoulder at 3,585 cm^{-1}) and H-O-H bending vibrations (band at 1,641 cm^{-1}). Strong band with maximum at 1,437 cm^{-1} undoubtedly corresponds to bending vibrations of the NH_4 group as well as shoulder at 3,081 cm^{-1} – to its stretching vibrations.

Thus, the first stage of delhayelite alteration in aqueous solutions is partial or complete leaching of Na^+ and F^- from the interlayer space accompanying by hydration.

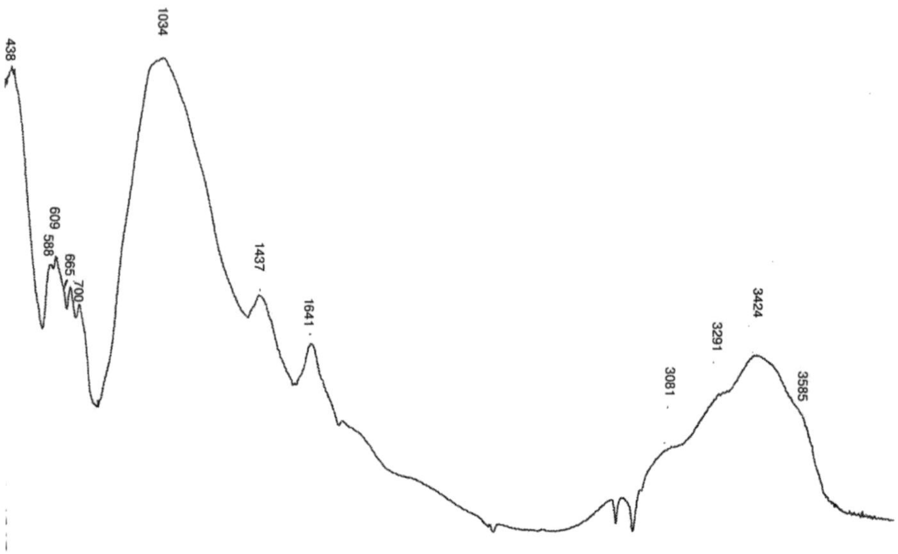

Fig. 3 IR spectrum of the NH_4-exchanged and hydrated phase obtained from delhayelite after the experiment in 1 N aqueous solution of $(NH_4)_2SO_4$: 3 h, 80–90°C, 1 atm

The second stage is leaching of univalent ions from channels inside the tetrahedral block: K^+ (partially or almost completely) and Cl^- (partially or completely) with additional hydration and/or absorption of large cations (Rb^+, Cs^+, Ba^{2+}, NH_4^+) from a solution. The second stage can start when the first stage is still not finished in other parts of a crystal. The data in Tables 1–3 show gradual decrease of univalent components of delhayelite in the sequence F → Na → Cl → K with the increase of contents of each of exchanged large cations: Ba, Rb or Cs those probably substitute K. Lowering of analytical total values for all cation-exchanged forms compare to initial mineral indicates their hydration (note that in Tables 1–3, column 1, chemical composition of initial delhayelite for grains used in corresponding experiments – with Ba, Rb or Cs – is given). The distribution of cation-exchanged areas in delhayelite crystals is uneven. As it is observed on SEM images (Fig. 4), exchanged cations are typically concentrated in local areas near the cracks or in porous zones.

Among all tested cations, delhayelite demonstrates the strongest affinity to Cs^+. The most Cs-enriched exchanged form of this mineral obtained in our experiments is Na-, F- and Cl-free, contains about 27 wt.% Cs_2O and corresponds to the following empirical formula: $Cs_{1.67}K_{0.16}Ca_{1.85}Al_{1.06}Si_{6.94}O_{17.47}(OH)_{1.53}(H_2O)_n$ (Table 3, #8).

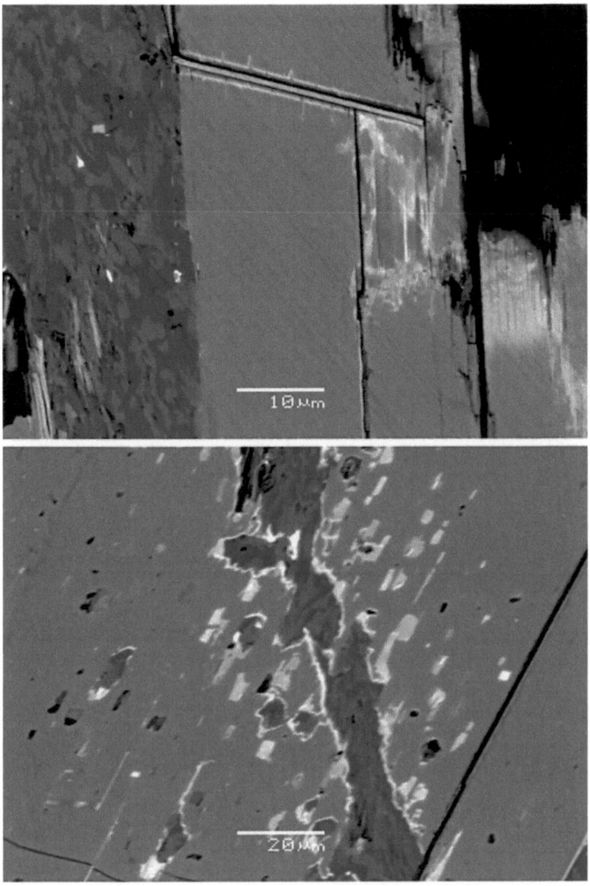

Fig. 4 Distribution of Cs-exchanged areas (*light*) in delhayelite after the experiment in $CsNO_3$ solution: 80–90°C, 1 atm, 3 h. Polished sections, SEM (BSE) image

Acknowledgements This work was supported by grants nos. 08-07-00077-a, 09-05-00143-a, and 09-05-12001-ofi_m by RFBR and grants of President of Russian Federation nos. NSh-4034.2010.5 and NSh-3848.2010.5.

References

Cannillo E, Rossi G, Ungaretti L (1969) The crystal structure of delhayelite. Rend Soc Ital Miner Pet 26:63–75

Pekov IV, Zubkova NV, Chukanov NV, Sharygin VV, Pushcharovsky DYu (2009) Crystal chemistry of delhayelite and hydrodelhayelite. Dokl Earth Sci 428(7):1216–1221

Pekov IV, Zubkova NV, Chukanov NV, Turchkova AG, Filinchuk YaE, Pushcharovsky DYu (2011) Delhayelite and mountainite mineral families: crystal chemical relationship,

microporous character and genetic features. In Krivovichev SV (ed) Minerals as advanced materials II, 213–219
Dorfman MD, Chiragov MI (1979) Hydrodelhayelite, a product of hypergene alteration of delhayelite. Novye Dannye o Mineralakh 28:172–175 (in Russian)
Pekov IV, Zubkova NV, Chukanov NV, Zadov AE, Pushcharovsky DYu (2010) Fivegite, $K_4Ca_2[AlSi_7O_{17}(O_{2-x}OH_x)][(H_2O)_{2-x}OH_x]Cl$, a new mineral from the Khibiny alkaline complex, Kola Peninsula, Russia. Zap RMO 139(4):47–63 (in Russian)
Chiragov MI (1982) Study of cation-exchanged forms of delhayelite and a new synthetic zeolite. Problems of mineralogy and geochemistry of ore and non-ore deposits of Azerbaijan. Azerbaijan State University, Baku, pp 24–33 (in Russian)

Microporous Titanosilicates of the Lintisite-Kukisvumite Group and Their Transformation in Acidic Solutions

Viktor N. Yakovenchuk, Sergey V. Krivovichev, Yakov A. Pakhomovsky, Ekaterina A. Selivanova, and Gregory Yu. Ivanyuk

1 Introduction

Alkaline massifs of the Kola Peninsula are the well-known sources of new minerals that became the prototypes of new advanced materials owing to variety of their unique technological properties. Such minerals as zorite, sitinakite, minerals of ivanyukite group, etc. demonstrate extremely high ability to cation exchange in combination with stability in a wide range of conditions. Other microporous titanosilicates, e.g. minerals of the labuntsovite group do not have such properties, however, in some cases the restrictions can be lifted by subtle modifications of their structures and/or chemical composition. As a rule, these minerals are metastable compounds: relatively small chemical and/or structural fluctuations induce significant changes of their physical properties as well as their cation-exchange capabilities. Such sensitivity to fine compositional and structural changes is characteristic for minerals of the lintisite-kukisvumite group structurally related to vinogradovite.

Sodium titanosilicate vinogradovite, $Na_5Ti_4(Si_7Al)O_{26}{\bullet}3H_2O$, is quite stable in most acids and alkalis and does not exchange Na for any cations in aqueous solutions. Paravinogradovite, $NaTi_4(Si_7Al)(OH)_4O_{22}\ {\bullet}H_2O$, is a sodium deficient

V.N. Yakovenchuk (✉) • G.Y. Ivanyuk
Nanomaterials Research Center, Kola Science Center, the Russian Academy of Sciences, 14 Fersman Street, Apatity 184209, Russia
e-mail: yakovenchuk@geoksc.apatity.ru

S.V. Krivovichev
Nanomaterials Research Center, Kola Science Center, the Russian Academy of Sciences, 14 Fersman Street, Apatity 184209, Russia
and
Department of Crystallography, Faculty of Geology, St. Petersburg State University, University Emb. 7/9, 199034 St. Petersburg, Russia

Y.A. Pakhomovsky • E.A. Selivanova
Geological Institute, Kola Science Center, the Russian Academy of Sciences, 14 Fersman Street, Apatity 184209, Russia

S.V. Krivovichev (ed.), *Minerals as Advanced Materials II*,
DOI 10.1007/978-3-642-20018-2_23,

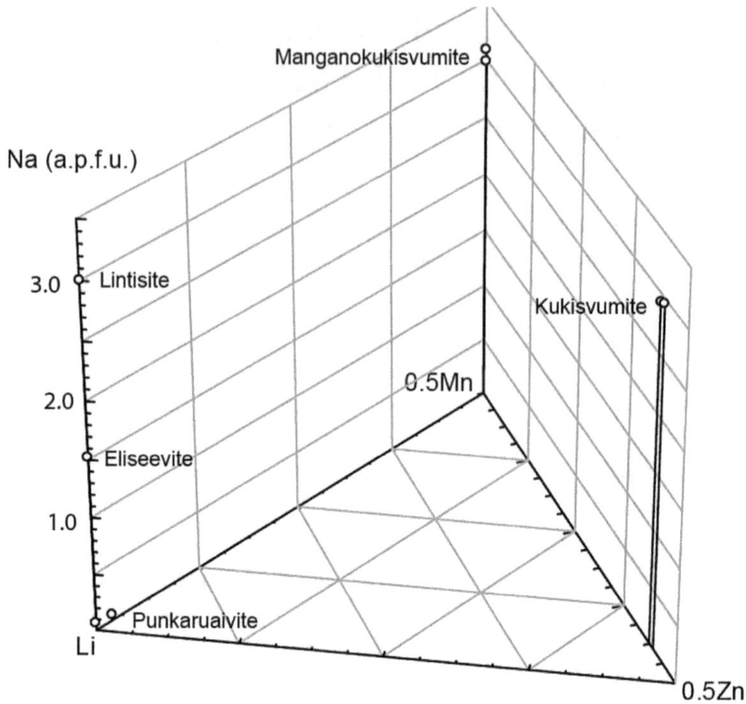

Fig. 1 Chemical compositions of the lintisite-kukisvumite group minerals

analogue of vinogradovite (Khomyakov et al. 2003). A.P. Khomyakov et al. (2003) consider paravinogradovite as a secondary mineral formed by sodium leaching from hypothetic "proto-vinogradovite" that is not as stable as vinogradovite due to slightly higher Fe^{2+} content.

Another sodium titanosilicate, lintisite, $Na_3LiTi_2Si_4O_{14} \cdot 2H_2O$, is closely related to vinogradovite and has the constant sodium content of 3 *apfu* in all known localities (Fig. 1). Recently we have found two new microporous titanosilicates, eliseevite, $Na_{1.5}LiTi_2[Si_4O_{12.5}(OH)_{1.5}]{\bullet}3H_2O$, and punkaruaivite, $LiTi_2[Si_4O_{11}(OH)_3]{\bullet}H_2O$, that can be considered as products of natural decationization of lintisite (Yakovenchuk et al. 2010). In contrast to paravinogradovite, both these minerals do not contain sufficient amount of impurities in their compositions. Mechanisms of cation leaching needs further understanding, since it can be considered as a pathway to new titanosilicate materials.

The lintisite-kukisvumite group includes five structurally related minerals: lintisite, $Na_3LiTi_2[Si_4O_{14}]{\bullet}2H_2O$ (Khomyakov et al. 1990), kukisvumite, $Na_3Zn_{0.5}Ti_2[Si_4O_{14}]$ $\bullet 2H_2O$ (Yakovenchuk et al. 1991), manganokukisvumite, $Na_3Mn_{0.5}Ti_2[Si_4O_{14}]{\bullet}2H_2O$ (Gault et al. 2004), punkaruaivite, $LiTi\text{-}_2[Si_4O_{11}(OH)_3]$ $\bullet H_2O$ (Yakovenchuk et al. 2010), and eliseevite, $Na_{1.5}LiTi_2[Si_4O_{12.5}(OH)_{1.5}]{\bullet}3H_2O$ (Yakovenchuk et al. 2010). All these species are late-stage hydrothermal minerals formed as a result of alteration of earlier titanosilicates: lorenzenite, lamprophyllite, lomonosovite, murmanite, etc. (Ivanyuk et al., this book).

Comparison of punkaruaivite, lintisite, kukisvumite and manganokukisvumite composition (see Fig. 1) permits us to consider these minerals as members of a separate group (Yakovenchuk et al. 2010). Merlino et al. (2000) mentioned possibility to transform lintisite into kukisvumite by the coupled substitution in the tetrahedrally coordinated lithium position: $2Li^+ \rightarrow Zn^{2+}$. Gault et al. (2004) considered manganokukisvumite as a result of homovalent substitution of Zn^{2+} into Mn^{2+} in kukisvumite. Our data showed that lintisite transforms into eliseevite and then into punkaruaivite due to the partial or total removal of Na^+ cations balanced according to the substitution mechanism: $Na^+ + O^{2-} \leftrightarrow (OH)^-$. Sodium content in this case changes in a stepwise fashion: from 3 apfu to 1.5 apfu and then to 0 apfu.

Intermediate Li-Zn, Li-Mn, Mn-Zn as well as vacant and semivacant phases have not yet been found in nature. Absence of the intermediate phases may be predetermined by the structural features of these minerals.

2 Crystal Structure

The crystal structures of punkaruaivite and lintisite were solved using a STOE IPDS II Image-Plate based X-ray diffractometer with monochromatic MoKα radiation operated at 50 kV and 40 mA, with frame widths of 2° in ω, and with 180 s counting times for each frame. The unit-cell parameters (Table 1) were refined using least-squares techniques.

The crystal structures of punkaruaivite and eliseevite as well as structures of lintisite and kukisvumite are based upon a three-dimensional (3-D) framework consisting of pyroxene-like Si_2O_6 chains linked in the framework by brookite-like chains of edge-sharing TiO_6 octahedra and corner-sharing LiO_4, ZnO_4 or MnO_4 tetrahedra (Fig. 2). Titanosilicate blocks of the framework are similar to those found in vinogradovite, except, in vinogradovite, they are condensed into 3-D titanosilicate framework. The blocks are separated by LiO_4, ZnO_4 or MnO_4 tetrahedra that results in formation of channels with free crystallographic diameter of about 4 Å.

There are two Na positions in the structures of lintisite, kukisvumite, manganokukisvumite and punkaruaivite. Octahedrally coordinated Na(1) adjoins on both sides to the brookite-like chain in the b direction, whereas Na(2) atom occupies large channels of the framework. Na(1) site has 100% occupancy in lintisite, kukisvumite and manganokukisvumite, and is empty in punkuruaivite. Na(2) site is occupied by Na^+ cations in lintisite, kukisvumite and manganokukisvumite and is vacant in punkuruaivite.

Table 1 Crystallographic data for punkaruaivite, eliseevite and K3

	Punkaruaivite	Eliseevite	K3
Space group	*C*2/*c*	*C*2/*c*	$P2_1/c$
a (Å)	26.688(2)	27.483(12)	8.765(2)
b (Å)	8.7568(7)	8.669(4)	5.2199(14)
c (Å)	5.2188(5)	5.246(2)	23.493(6)
β(°)	91.189(2)	90.782(8)	90
V (Å^3)	1219.38(18)	1249.7(9)	1074.9(5)

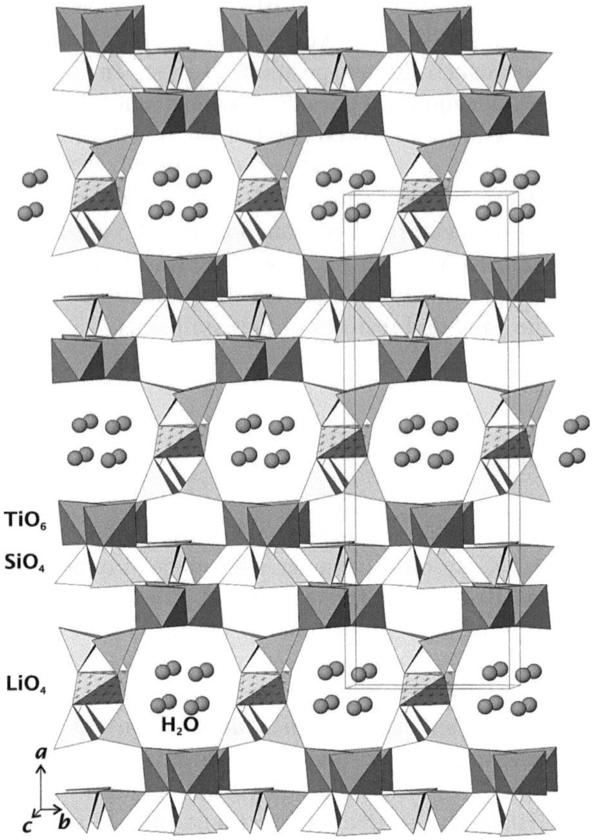

Fig. 2 The crystal structure of punkaruaivite

Eliseevite is a compositional intermediate phase between lintisite and punkaruaivite. Topology of titanosolicate block in its crystal structure is identical with those in the other minerals of the group. All Na cations are localized in the Na(1) site that has the 75% occupancy. The most unexpected feature of the eliseevite crystal structure is location of Li in octahedral coordination instead of tetrahedral one (as in lintisite and punkaruavite). Thus, Li occupies the site equivalent to the Na(2) site in lintisite and punkaruaivite. It means that consequent transformation of lintisite into eliseevite and then punkaruaivite by means of Na leaching is impossible without respective migration of Li in the structure.

3 Cation-Exchange Experiments

To check the cation exchange properties of the lintisite-kukisvumite-group minerals we have performed experiments with different minerals and solutions:

1. punkaruaivite, lintisite and kukisvumite was treated in 1 M solutions of KCl, RbCl and CsCl at 30°C during 48 h;

2. lintisite and kukisvumite was treated in 0.5 M solution of HCl at 30°C during 3, 6, 15, 24 and 48 h;
3. punkaruaivite, lintisite and kukisvumite were treated in 0.2 M solution of $N_2H_6SO_4$ at 30°C during 3 and 24 h.

After treatment, all the samples were washed by distilled water and dried at ambient conditions. Chemical composition of the crystals has been studied by A. Bazai using a Cameca MS-46 electron microprobe operating in WDS mode at 20 kV and 20–30 nA (Geological Institute, Kola Science Center, Russian Academy of Sciences). The Li content was determined by S. Drogobuzhskaya using a Perkin Elmer ELAN 9000 DRC-e mass-spectrometer equipped with an UP-266 MACRO laser probe (Institute of Chemistry and Technology of Rare Elements and Mineral Resources, Kola Science Center, Russian Academy of Sciences). Water content was calculated to balance the formula. X-ray powder-diffraction patterns were obtained by means of the URS-1 instrument operated at 40 kV and 16 mA with X-ray 114.7 mm Debye-Scherrer camera and Fe$K\alpha$-radiation (Geological Institute, Kola Science Center, Russian Academy of Sciences).

It was found that punkaruaivite is stable under selected conditions and does not absorb any cations from aqueous solutions. Both chemical composition and X-ray powder pattern of this mineral are identical before and after the treatment. Both lintisite and kukisvumite do not exchange Na^+ into another cations in cold water solutions, but display surprisingly intense changes in chemical composition and crystal structure after treating with hydrochloric acid.

Microprobe analyses of treated crystals of lintisite and kukisvumite (Table 2) showed the full loss of sodium, lithium and zinc even after 3-h exposure to weak HCl. Resulting products of lintisite and kukisvumite alteration, called L3- and K3-phases, respectively, have similar chemical compositions:

$$(Na_{0.04}K_{0.01})_{\Sigma 0.05}(Ti_{1.82}Nb_{0.05}Fe^{3+}{}_{0.03})_{\Sigma 1.90}Si_4O_{9.67}(OH)_{4.33} - \text{L3}$$

$$(Na_{0.01}K_{0.01})_{\Sigma 0.02}(Ti_{1.94}Nb_{0.01}Fe^{3+}{}_{0.02})_{\Sigma 1.97}Si_4O_{9.88}(OH)_{4.12} - \text{K3}$$

Both phases gave quite acceptable X-ray powder diffraction patterns, drastically different from their parent products. The unit cell parameters of phases produced from kukisvumite treated with 0.5 M HCl during 6, 15, 24 and 48 h respectively are very close to parameters of K3.

The infrared-absorption spectra of lintisite, kukisvumite, L3 and K3 (Fig. 3) were obtained by O. Zalkind using a Nicolet 6700 FT-IR spectrometer at the Institute of Chemistry and Technology of Rare Elements and Mineral Resources (Kola Science Center). The IR spectra of L3 and K3 are similar to the spectra of original minerals.

The IR spectra of kukisvumite and K3 have approximately equal intensities of absorption bands, while the IR spectrum of L3 has significantly deeper band of O-H vibrations at 3,400 cm^{-1} than the spectrum of original lintisite has.

Table 2 Chemical composition (wt. %) and empiric formula (*apfu*, Si = 4) of lintisite and kukisvumite treated with 0.5 M HCl and 0.2 M $N_2H_6SO_4$

	Lintisite			Kukisvumite		
	Original	Treated with		Original	Treated with	
		HCl	$N_2H_6SO_4$		HCl	$N_2H_6SO_4$
Li_2O	2.70	bdl	bdl			
N_2O_5			bdl			5.00
Na_2O	14.71	0.30	1.84	16.10	0.09	13.36
SiO_2	45.58	56.09	53.41	42.75	54.47	43.93
K_2O	bdl	0.06	bdl	bdl	0.09	0.03
TiO_2	28.83	33.99	31.96	26.86	35.08	27.74
MnO	bdl	bdl	bdl	0.74	bdl	0.89
FeO	0.30	0.49	0.38	0.25	0.26	0.36
ZnO	bdl	bdl	bdl	5.88	bdl	5.29
Nb_2O_5	0.52	1.49	0.79	0.86	0.37	0.83
H_2O	7.36*	9.10*	8.80*	6.50	8.40*	2.57*
SUM	100.00	101.52	97.18	99.94	98.76	100.00
Na	2.50	0.04	0.27	2.92	0.01	2.36
K		0.01			0.01	
N^{2-}						0.50
Li	0.95					
Zn				0.41		0.36
Mn^{2+}				0.06		0.07
Ti	1.90	1.82	1.80	1.89	1.94	1.90
Nb	0.02	0.05	0.01	0.04	0.01	0.03
Fe^{3+}	0.02	0.03	0.02	0.02	0.02	0.03
Si	4.00	4.00	4.00	4.00	4.00	4.00
H	4.32	4.33	4.40	4.07	4.12	1.94
O	15.78	14.00	14.00	15.86	14.00	14.00

bdl below detection limit
* Calculated to balance the formula

Treatment of lintisite with 0.2 M solution of hydrazinium sulfate during 3 and 24 h gave again L3 phase (Table 2):

$$Na_{0.27}(Ti_{1.80}Nb_{0.01}Fe^{3+}{}_{0.02})_{\Sigma 1.83}Si_4O_{9.60}(OH)_{4.40}.$$

Kukisvumite after 24–48 h of exposition in $N_2H_6SO_4$ exchanged up to 0.5 *apfu* of Na^+ into the same quantity of $N_2H_5{}^+$ (Table 2), but Zn^{2+} content remains quite stable and the framework has not collapsed.

The crystal structure of K3 phase was solved using a Bruker APEX II diffractometer operated at 50 kV and 40 mA and equipped with CCD area detector. The unit-cell parameters of this phase (Table 1) are different from those of initial kukisvumite by 6 Å contraction of unit cell parameter perpendicular to the

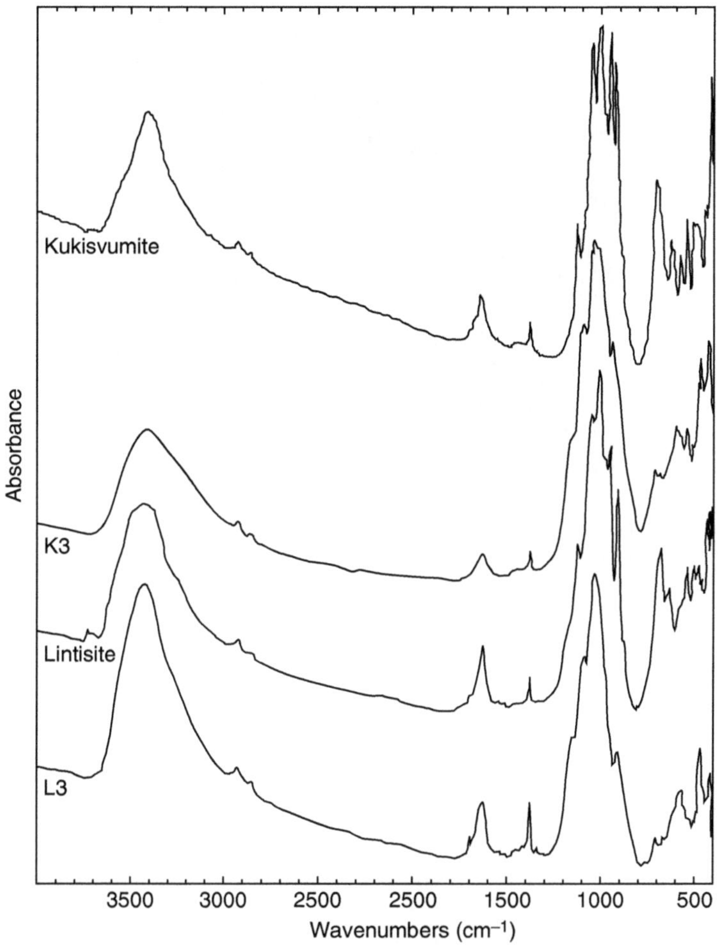

Fig. 3 Infrared-absorption spectra of kukisvumite, lintisite, K3 and L3

titanosilicate block. The data were of poor quality (Fig. 4), but still allowed to decipher basic features of the structure. It is based upon electroneutral $\{Ti_2(OH)_2[Si_4O_{10}(OH)_2]\}$ titanosilicate blocks that pack together closely to form a dense layered structure (Fig. 5). Thus, transformation 'kukisvumite-K3' involves removal of Na and Zn from the structure and sliding of titanosilicate layers by ~4.3 Å. It is quite remarkable that, after this transformation, crystals still possess their three-dimensional integrity. This is a very rare case of single-crystal-to-single-crystal transformation in inorganic crystals (in recent years, many examples of such behaviour was reported for polymers and metal-organic compounds with rather soft structures (Halasz 2010)).

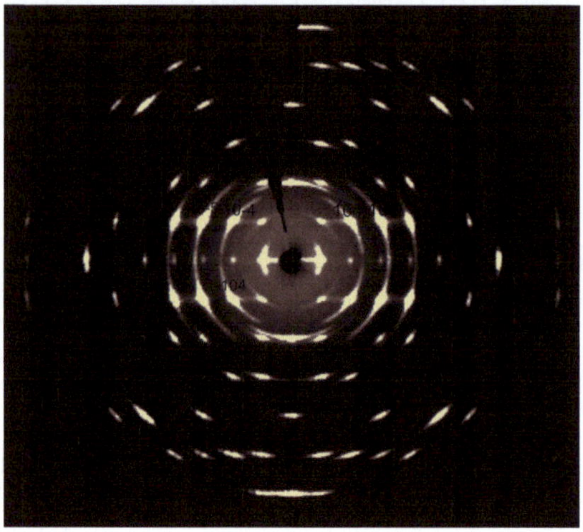

Fig. 4 The (010) section (k = 0) of reciprocal diffraction space for K3, showing poor diffraction spots with strong diffuse scattering features

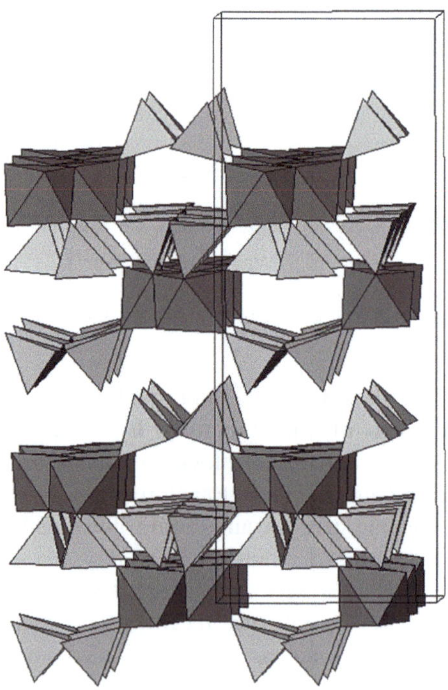

Fig. 5 The crystal structure of K3 (compare with Fig. 2) consisting of electroneutral $\{Ti_2(OH)_2[Si_4O_{10}(OH)_2]\}$ titanosilicate blocks

4 Discussion

Our data allow formulating the following conclusions.

1. Separated two-dimensional titanosilicate blocks of the lintisite-kukisvumite group minerals are not only the basic elements of these mineral crystal structures but also quite stable and quite rigid elements. Our experiments gave an evidence of these blocks stability in acid environment.
2. Li^+, Mn^{2+} or Zn^{2+} cations interconnecting titanosilicate blocks represent "weak point" of lintisite-related structures and can be easily removed from the interblock space. Moreover, Li^+ cation can shift from the tetrahedral site and occupy the octahedral Na(2) site, as it has been in the eliseevite structure.
3. Presence of such "weak points" in considered mineral structures as well as high ability of titanosilicate block surface to form hydrogen bonds causes formation of new condensed structure by means of a single-crystal-to-single-crystal transformation. It is important that new crystal substance was obtained by means of self-organization from already existing titanosilicate blocks instead of crystallization from separated ions. This process happens at ambient conditions.

The structures of lintisite-kukisvumite-group minerals consist of two different structural elements: two-dimensional titanosilicate blocks and interconnecting cations of Li^+, Mn^{2+}, or Zn^{2+}. We believe that titanosilicate blocks were formed in dry hyperalkaline environment and then they played a role of tectons for formation of various structures at hydrothermal conditions.

Since titanosilicates with kukisvumite-like structures can be prepared under laboratory conditions (Dadachov et al. 1997), their acid treatment may lead to formation of specific layered protonated titanosilicates that may serve as a precursor for fabrication of novel titanosilicate nanomaterials.

Acknowledgements We thank O. Zalkind who obtained the IR spectrums of investigated phases, A. Bazai who made their microprobe analysis, and S. Drogobuzhskaya who determined Li content. This work was supported by Russian Federal Programme "Scientific and Education Cadres of Innovative Russia" (to SVK and MSA, state contract no. 02.740.11.0326) and internal grant of St. Petersburg State University.

References

Dadachov MS, Rocha O, Ferreira A, Lin Z, Anderson MW (1997) Ab initio structure determination of layered sodium titanium silicate containing edge-sharing titanate chains (AM-4) $Na_3(Na, H)Ti_2O_2[Si_2O_6]\cdot 2.2H_2O$. Chem Commun 2371–2372

Gault RA, Ercit TS, Grice JD, Velthuizen JV (2004) Manganokukisvumite, a new mineral species from Mont Saint-Hilaire, Quebec. Can Miner 42:781–785

Halasz I (2010) Single-crystal-to-single-crystal reactivity: gray, rather than black or white. Cryst. Growth Des 10:2817–2823

Khomyakov AP, Polezhaeva LP, Merlino S, Pasero M (1990) Lintisite $Na_3LiTi_2Si_4O_{14} \cdot 2H_2O$ – a new mineral. ZVMO 3:76–79 (in Russian)

Khomyakov AP, Kulikova IE, Sokolova E, Hawthorn FC, Kartashov PM (2003) Paravinogradovite, $(Na,\square)_2((Ti, Fe)_4(Si_2O_6)_2(Si_3AlO_{10})(OH)_4)\cdot H_2O$, a new mineral species from the Khibina alkaline massif, Kola Peninsula, Russia: description and crystal structure. Can Miner 41:989–1002

Merlino S, Pasero M, Ferro O (2000) The crystal structure of kukisvumite, $Na_6ZnTi_4(Si_2O_6)_4O_{4.4}(H_2O)$. Z Kristallogr 215:352–356

Yakovenchuk VN, YaA P, Bogdanova AN (1991) Kukisvumite – a new mineral from alkaline pegmatites of the Khibiny massif (Kola Peninsula). Miner Zh 13:63–67 (in Russian)

Yakovenchuk VN, Ivanyuk GYu, Pakhomovsky YA, Selivanova EA, Men'shikovYuP KJA, Krivovichev SV, Spiridonova DV, Zalkind OA (2010) Punkaruaivite, $LiTi_2[Si_4O_{11}(OH)](OH)_2\cdot H_2O$, a new mineral species from hydrothermal assemblages, Khibiny and Lovozero alkaline massifs, Kola peninsula, Russia. Can Miner 48:41–50

Microporous Vanadylphosphates – Perspective Materials for Technological Applications

Olga V. Yakubovich

1 Introduction

The presently known world resources of vanadium are estimated as 63 million tons. There is no single mineral ore from which vanadium is recovered because it does not occur in highly concentrated forms. However, it is known as a trace element in a number of different rock materials and is a by-product of other mining operations.

Vanadium is found in about 90 different minerals among which important sources of this metal can be mentioned: patronite VS_4, vanadinite $Pb_5(VO_4)Cl$, camotite $K_2(UO_2)_2(VO_4)_2 \cdot 3H_2O$. It also occurs in phosphate rock and certain iron ores, and it is present in some crude oils in the form of organic complexes. Among more than 65 vanadium ores, all except five or six are of secondary origin formed by oxidation or weathering. The primary approximately 20 vanadium-bearing minerals are generally associated with basic igneous rock, basic alkaline igneous rock, carbonatite complexes, titaniferous magnetite complexes and chromite, uranium, iron and manganese deposits.

The rich chemistry of vanadium includes a range of coordination geometries with oxygen in crystal structures as tetrahedral, pentahedral, trigonal pyramidal, tetragonal pyramidal and octahedral. Combined with its variable oxidation states of 2, 3, 4 and 5, this leads to a large diversity of vanadium-containing structures. The amphoteric character of vanadium oxocomplexes explains the peculiarities of its crystal chemistry: these oxocomplexes can have both cation and anion-forming functions in mineral and biological processes (Baran 2003; Yakubovich 2009). As a cation (V^{3+}, VO^{2+}, VO_2^{+}, VO^{3+}), vanadium acts like a typical transition metal, while its anionic form $(VO_4)^{3-}$ resembles phosphorous in phosphates (the latter are typical for low-temperature exogenic conditions of mineral formation in the zone

O.V. Yakubovich (✉)
Moscow State Lomonosov University, Vorob'evy Gory, Moscow 119992, Russia
e-mail: yakubol@geol.msu.ru

S.V. Krivovichev (ed.), *Minerals as Advanced Materials II*,
DOI 10.1007/978-3-642-20018-2_24, © Springer-Verlag Berlin Heidelberg 2012

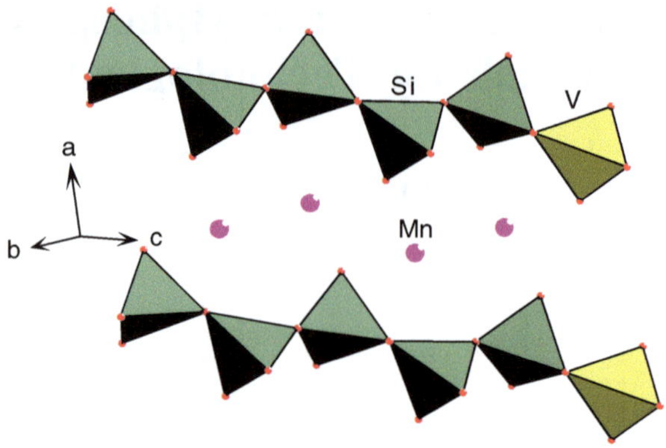

Fig. 1 Chain fragments built from Si- and V-tetrahedra in the crystal structure of medaite

of oxidation of various deposits). In a tetrahedral surrounding of oxygen atoms, V^{5+} ion forms minerals – vanadates. Some of them are isotypic with silicates, like this schäferite $NaCa_2Mg_2(VO_4)_3$ crystallizes in a garnet structure type, and mineral dreyerite $Bi(VO_4)$ crystallizes in the structure type of zircon.

In more complicated cases vanadium tetrahedra built complex anionic constructions in association with silicate complexes. Thus, the anionic layers in the crystal structure of scheuchzerite, $Na(Mn,Mg)_9(VSi_9O_{28}(OH))(OH)_3$ are formed from $(Si_9O_{25}(OH))$ loop-branched chains of silicate tetrahedra. The loops contain six tetrahedra and are separated by three tetrahedra in a broken four-loop arrangement. A vanadate $(VO_4)^{3-}$ tetrahedron branches off the six-tetrahedra loop, and hence the overall formula of the tetrahedral chains is $(VSi_9O_{28}(OH))$ (Brugger et al. 2006).

Another example of this kind presents the structure of medaite, $Mn_6(VSi_5O_{18}(OH))$, containing a vanadatopentasilicate anion (with some substitution of As for V) $(VSi_5O_{18}(OH))^{12-}$, comprising six tetrahedra linked together to form a chain fragment (Fig. 1). This ion is a representative of a series of heteropolysilicate ions, the conformation of which resemble polyphosphates (Gramaccioli et al. 1981). Solid solutions between VO_4 and AsO_4 oxocomplexes are known; besides medaite, the volcanic mineral coparsite $Cu_4ClO_2((As_{0.5}V_{0.5})O_4)$ (Starova et al. 1998) can be mentioned in this context.

However, in spite of the fact that phosphate and vanadate minerals often have isotypic crystal structures, isomorphous substitution between VO_4 and PO_4 tetrahedra is very rare (mineral bushmakinite (Yakubovich et al. 2002) may be referred as an example) most certainly due to the large difference in the sizes of V^{5+} and P^{5+} ions.

Five- or six-coordinated vanadium is often involved in the formation of anionic structures of a mixed type, along with other acid anionic (most often phosphate) complexes. Vanadium polyhedra are characterized by one or two shortened V=O vanadyl bonds among five (trigonal bipyramid) or six (distorted octahedron) bonds in the first coordination sphere. Some silicate minerals containing VO^{2+} cations are known. As an example the dimorph pair cavansite – pentagonite $Ca((VO)Si_4O_{10})\cdot 4H_2O$ can be refer to.

The dimorphism of cavansite and pentagonite is based on the difference in linkages in the silicate layer. Zig-zag $(SiO_3)^{2-}$ chains form layers in both crystal structures, but in cavansite these chains are joined laterally into sheets made up of fourfold and eightfold rings, while in pentagonite they are differently joined in sixfold ring layers. In both structures the unlinked tetrahedral apices of the chains are coordinated in pairs by VO^{2+}- groups, which tie the layers in an anionic framework. Thus, the vanadium atoms achieve their characteristic fivefold, square-pyramidal coordination. Large cavities in the structure are filled zeolitically by Ca^{2+}- ions and H_2O molecules (Evans 1973). According to resent data pentagonite is formed when the hydrothermal fluid is in supercritical condition (>300°C), while cavansite is formed when the hydrothermal fluid is not in supercritical condition. Thus, cavansite is identified as a low-temperature form and pentagonite as a high-temperature one (Ishida et al. 2009).

Mineral and synthetic phases with complex anions and open framework structures have been intensively studied over the last two decades. Among them, vanadyl phosphates seem to be promising due to their potential applications as catalytic materials, sorbents, molecular sieves or ion-exchange materials similar to zeolites (Centi et al. 1988), or as positive electrode materials for Li-ion and Na-ion energy storage devices (Sauvage et al. 2006; Chen et al. 2009). Here, we present recently studied crystal structures of microporous vanadyl-phosphates $Na_3(V_2O_2F(PO_4))_2$ (Massa et al. 2002), $Cs_2(Ti(VO_2)_3(PO_4)_3)$ (Yakubovich et al. 2006), $Rb_2(Cu(VO_2)_2(PO_4)_2)$ (Yakubovich et al. 2008) and $Rb_2(Ti(VO_2)_3(PO_4)_3)$ (Yakubovich et al. 2010) synthesized under hydrothermal conditions, in context of their physical properties and structure relations to minerals and other synthetic phases.

2 The Sodium Vanadyl(IV) Fluorophosphate, $Na_3(V_2O_2F(PO_4))_2$ – Positive Electrode Material for Li-ion and Na-ion Energy Storage Devices

Materials present extremely important component in development of advanced electricity storage and attendant technologies. Global interest in batteries, particularly those for electric vehicles, has intensively grown in recent years, because batteries allow vehicle operation from renewable resources of energy. Batteries contain three essential parts: an anode, a cathode, and an electrolyte. Most of the current Li-ion batteries use layered transition metal oxides as cathode material. These oxides provide excellent cycle ability, but it comes along with limited electrochemical capacity, poor performance in high-power applications and safety problems caused by oxygen release at the top of charge (Whittingham 2004). It has been shown that transition metal phosphates are attractive as possible cathode materials due to their enhanced thermodynamic and kinetic stability compared to oxides, higher operating potentials due to the inductive effect of the phosphate group and environmental friendliness evidenced by the great number of naturally occurring phosphate minerals (Whittingham et al. 2005). The discovery of the ability of olivine-type phosphate – triphylite $LiFePO_4$ to reversibly intercalate and deintercalate lithium ions at potential around 3.4 V has raised a new wave of interest in transition metal phosphates as cathode materials for lithium-ion batteries (Padhi et al. 1997). Modern $LiFePO_4$ technology for lithium batteries is in our days in commercial growth owing to the excellent electrochemical performance, high-rate capability and structural stability of this cathode material.

Conventional lithium-ion cells employ a lithium-containing cathode material used in conjunction with a suitable carbon anode. The properties of these commercial cells are thus limited by the availability and performance characteristics of the lithium-based cathode phases. However, in the paper of J. Barker with co-authors (Barker et al. 2006) was demonstrated the feasibility of a novel lithium-ion cell based on the application of a sodium-based cathode phase, a metallic lithium or graphite as the negative electrode, along with a lithium salt electrolyte. The properties of this system – graphite/Li^+ electrolyte/$Na_3V_2(PO_4)_2F_3$ compare favorably with commercial lithium ion cells. It has been shown (Barker et al. 2006) that sodium is deintercalated from the positive electrode upon the initial charge, resulting in a mixed alkali electrolyte composition. The presence of sodium did not appear to degrade the electrolyte or the negative electrode. According to (Barker et al. 2006), "the basic cell fabrication concept is not limited to this particular active material." As has been underline in (Sauvage et al. 2006) a key development of J. Barker et al. vastly expanded the range of compounds potentially possible for usage in a lithium ion cell, especially with concerns over the global supply of economically recoverable lithium.

One more vanadium-containing phase – a sodium vanadylfluorophosphate $Na_3(V_2O_2F(PO_4))_2$ was shown to electrochemically react with 0.56 Na ions per formula unit, through a step-wise process that enlists two plateaus located at 3.6 and 4.0 V, with the lower one having the best sustained reversibility upon cycling (Sauvage et al. 2006). This phase has been received by solid state reaction method and studied by X-ray powder diffraction. Previously, we have synthesized in hydrothermal conditions and structurally investigated the same compound $Na_3(V_2O_2F(PO_4))_2$ (Massa et al. 2002) using its monocrystals .

In the crystal structure of $Na_3\{V_2O_2F(PO_4)_2\}$, V^{4+} ions occupy distorted octahedra with D_{4v} symmetry formed by O and F atoms. The shortest V – O distance in the octahedron of 1.625(2) Å indicates a vanadyl group, four equal 'equatorial' V – O distances to the bridging (PO_4) tetrahedra are of 2.004(1) Å and the F atom in *trans*-position to the vanadyl group shows a very long V – F bond of 2.1081(5) Å consistent with an expected *trans*-effect. The P – O distances in the regular orthophosphate tetrahedra are 1.531(1) Å. The Na^+ ions are disordered over two independent positions with close distances (Na1 – Na2 = 0.42 Å and Na2 – Na2 = 0.74 Å). The refinement process showed an equivalent amount of Na atoms in each of Na1 and Na2 positions in the structure: three Na1 atoms instead of eight occupy an 8*h* position and likewise there are three Na2 atoms in a 16*l* position. As the refined occupations (using equal anisotropic displacement parameters) did not deviate significantly from 3/8 to 3/16 (Na1 0.37(2), Na2 0.19(2)), the idealized values were kept fixed in the final refinement

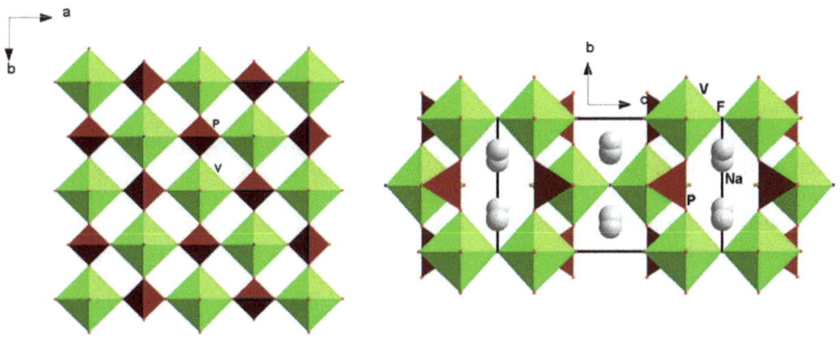

Fig. 2 Crystal structure of $Na_3\{V_2O_2F(PO_4)_2\}$in *ab* and *bc* projections

cycles. In the observed seven-vertex Na1 polyhedron the distances vary from 2.30(2) Å (Na1 – F) to 2.71(2) Å (Na1 – O).

The same spread of cation – anion distances is observed for the seven-vertex Na2 polyhedron, but the Na2 – F bond in this case is not shortest in the polyhedron, its value is 2.53(1) Å (Massa et al. 2002). The crystal structure obtained in (Sauvage et al. 2006) differs from ours regarding the sodium sites that were found located in 8h and 8j Wyckoff positions.

The crystal structure of $Na_3\{V_2O_2F(PO_4)_2\}$ is formed by layers of alternating (VO_5F) octahedra and (PO_4) tetrahedra sharing O vertices (Fig. 2a). These layers are loosely interconnected along the *c* axis over the F ligands while the vanadyl O2 atoms remain terminal. Thus, the structure is described by a mixed paraframework of octahedra and tetrahedra $\{V_2O_2F(PO_4)_2\}_{\infty\ \infty\ \infty}$ with disordered Na atoms in the interstices (Fig. 2b).

As we showed in (Massa et al. 2002) the new vanadyl fluoride phosphate is isotypic to three fluoride phosphates of the common formula $Na_3\{M_2F_3(PO_4)_2\}$, M = Fe, Al, Cr (Yakubovich and Mel'nikov 1996). Susceptibility measurements reveal an antiferromagnetic behavior for each of the paramagnetic compounds in this family with $M = V^{3+}$, Cr^{3+}, Fe^{3+} (Le Meins et al. 1999). All these compounds can be considered as derivatives of the mineral natisite $Na_2\{TiO(SiO_4)\}$. The three Fe-, Al- or Cr-phosphate structures differ from that of $Na_3\{V_2O_2F(PO_4)_2\}$ by the degree of the octahedron distortion. Strongly distorted VO_5F octahedra with one short distance V – O = 1.625 Å, characterizing the vanadyl group, in our structure (Fig. 3) correspond to rather regular MO_4F_2 (Fe, Al and Cr) octahedra in $Na_3\{M_2F_3(PO_4)_2\}$. For instance, in the Al containing phase these octahedra are described by two Al – F bonds of 1.824(1) (terminal) and 1.869(1) Å (bridging), and by four equidistant Al – O bonds of 1.889(1) Å (Yakubovich and Mel'nikov 1996). One more difference between these structures occurs in a distribution of Na atoms in the framework interstices. In $Na_3M_2F_3(PO_4)_2$ compounds (M = Fe, Al, Cr) three sites 8h, 8j and 16l with close coordinates are differently populated for Na atoms.

We have also shown that crystal structures of two Ge analogues of natisite: $Na_2\{TiO(GeO_4)\}$ and $Li_2\{TiO(GeO_4)\}$ on one hand, and the structures of Fe and Al phosphate representatives in the natisite raw of compounds on the other hand, differ

Fig. 3 A distribution of bond lengths in Al, Fe, Cr or V-octahedra in the phosphate crystal structures in comparison with distances' distribution in the structure of natisite Na_2TiSIO_5. Number of bonds are placed along x axis, bond lengths – along y axis

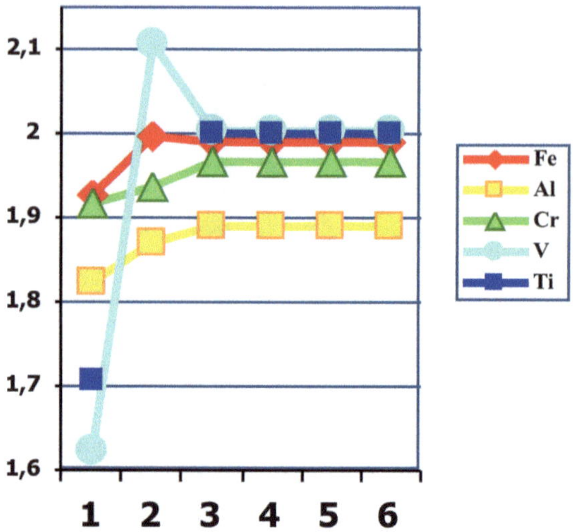

mainly by the degree of association of the mixed anionic substructure (layer or paraframework) and show symmetry relations of the "group-subgroup" type.

For germanium compounds related to natisite $A_2\{TiO(GeO_4)\}$ ionic conductivity was observed with high anisotropy (Yakubovich et al. 2000), though the alkali positions show now disorder there. Thus, we have expected a possibility of ionic transport in Na vanadylfluorophosphate and our supposition has been confirmed in (Sauvage et al. 2006). However, a sodium ion RT conductivity was found equal to 1.8×10^{-7} S/cm. As a conclusion the authors (Sauvage et al. 2006) wrote: "While these new fluorophosphate phases represent new interesting host materials in terms of voltage and reactivity, there remains much work to be done to control and optimize their synthesis as well as their electrochemical properties"

Special cases present crystal structures where additional cations along with $(VO)^{2+}$, $(VO_2)^{+}$, or $(VO)^{3+}$ form three-dimensional mixed-anionic frameworks in combination with PO_4 tetrahedra. Further, new vanadylphosphates of this type with additional Ti and Cu oxo-complexes will be discussed.

3 The Titanium Vanadyl(V) Phosphates, $Cs_2(Ti(VO_2)_3(PO_4)_3)$ and $Rb_2(Ti(VO_2)_3(PO_4)_3)$ – Microporous Phases with Intercrossing Channels

In spite of identical electron configurations of Ti^{4+} and V^{5+}, these cations never meet together in mineral phases. The reason is in their completely different geochemical ways: while Ti^{4+} is well known as an accessory in rock forming minerals, V^{5+} is absolutely untypical for magmatic conditions. Our attempt to synthesize compounds containing these two elements was successful and we found

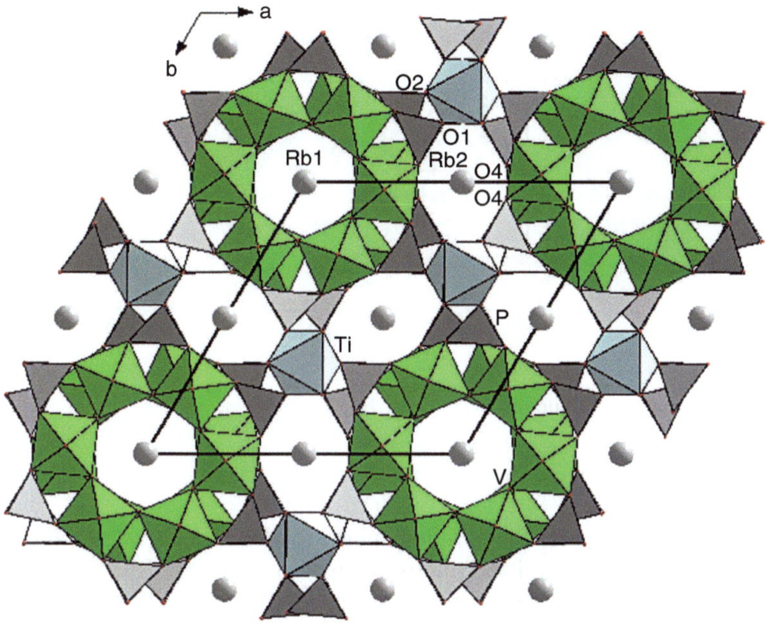

Fig. 4 The crystal structure of $Rb_2(Ti(VO_2)_3(PO_4)_3)$ in *xy* projection

new microporous structures with open mixed framework made by Ti, V and P oxo-complexes (Yakubovich et al. 2006, 2010).

The (VO_5) polyhedra approach square pyramidal coordination and form six-membered rings around six-fold axes by sharing vertices (Fig. 4). Each two neighboring V polyhedra in the ring are additionally bridged by (PO_4) tetrahedra. These large ring constructions are packed in layers parallel to the *ab* plane at z ≈ ¼ and with slight rotations against each other. All vanadyl groups (V=O 1.586(5) and 1.574(4) in Cs and Rb compounds accordingly) remain terminal and are pointing toward the *c*-axis direction (Fig. 5) giving rise to a strongly polar structure.

The Ti octahedra are between these V,P rings (in heights z = 0 and ½) and link them by sharing all six vertices with P tetrahedra. Thus, a three-dimensional anionic framework with the formula $(Ti(VO_2)_3(PO_4)_3)_{\infty\ \infty\ \infty}{}^{2-}$ forms; this framework contains channels of six and of eight polyhedra in the circumference, as seen from the (001) direction. Intercrossing channels intersect the structure as well in the (100), (010), and (110) directions, that suggests to consider the new compounds as possible 3D ion-conductors.

Rb^+ cations in the $Rb_2(Ti(VO_2)_3(PO_4)_3)$ crystal structure occupy two types of sites in these channels. Rb1 is located on axis 6; 12 oxygen atoms in its nearest environment form a hexagonal antiprism, where the Rb1–O distances are 6 × 3.115(4) and 6 × 3.571(3) Å. Although the Rb1 polyhedron (average Rb1–O distance is 3.343 Å) is somewhat smaller than the similar Cs1 polyhedron (average Cs1–O distance is 3.372 Å) in the $Cs_2(Ti(VO_2)_3(PO_4)_3)$ structure, it

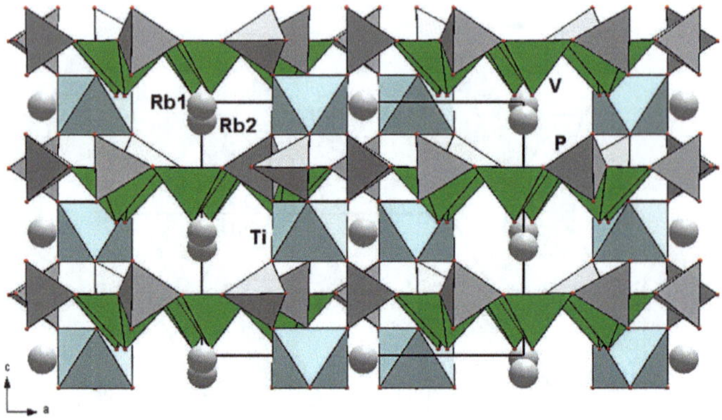

Fig. 5 V/P-layers linked along the *c* axis by Ti-octahedra in the crystal structure of $Rb_2(Ti(VO_2)_3(PO_4)_3)$

is nevertheless too large for the rubidium atom. The latter is fairly "mobile" in wide channels, which manifests itself in its thermal parameters. Rb2 atoms on axis 2 are coordinated by ten oxygen atoms at distances of 2.877(3)–3.401(3) Å (the average value is 3.084 Å). As one would expect, the Rb2 polyhedron is much smaller than the 14-vertex Cs2 polyhedron (the average Cs2–O distance is 3.328 Å). The isotypic $Cs_2(Ti(VO_2)_3(PO_4)_3)$ and $Rb_2(Ti(VO_2)_3(PO_4)_3)$ compounds with structures where large Cs + and Rb + cations, having fairly different sizes, occupy spacial positions 2*a* and 6*c* of the sp. gr. *P*6*cc* (allowing only one degree of freedom along the *z* axis) can be formed due to the flexibility of the mixed anionic framework, in which three types of polyhedra (octahedra, five_vertex polyhedra, and tetrahedra) are linked through only vertex bridge contacts. The framework flexibility manifests itself in variations in the V–O–V, V–O–P, and Ti–O–P angles, whose increase or decrease leads to the formation of channels of a necessary diameter for incorporating Cs or Rb atoms.

Topologically identical units of Ti octahedra sharing vertices with six tetrahedra $\{TiT_6O_{24}\}$, which form the title structures ($T = P$), are the essential fragments of the framework in mineral benitoit $BaTiSi_3O_9$ with $T = Si$, crystallizing in the hexagonal space group $P\bar{6}c2$ (Fig. 6). Two oxygen vertices of each Si tetrahedron that are not shared with Ti octahedra are connected with two tetrahedra from the neighboring Ti,Si oxo-complexes under formation of three-membered rings of Si tetrahedra in this trisilicate. In our structures of $Cs(Ti(VO_2)_3(PO_4)_3)$ and $Rb_2(Ti(VO_2)_3(PO_4)_3)$, hexagonal rings of V polyhedra are inserted between the $\{TiP_6O_{24}\}$ complexes, hence these compounds are orthophosphates. In accordance with these structural peculiarities, the *a* and *b* unit cell parameters of the titanium vanadyl phosphates (13.765 and 13.604 Å) are more than twice larger than the corresponding parameters in benitoite (6.641 Å). The similar *c* parameters of the structures (9.475, 9.386 and 9.760 Å) reflect the double sum of the edges of a Ti octahedron and a *T* tetrahedron (Figs. 5 and 7), and differ only according to the sizes of P and Si tetrahedra.

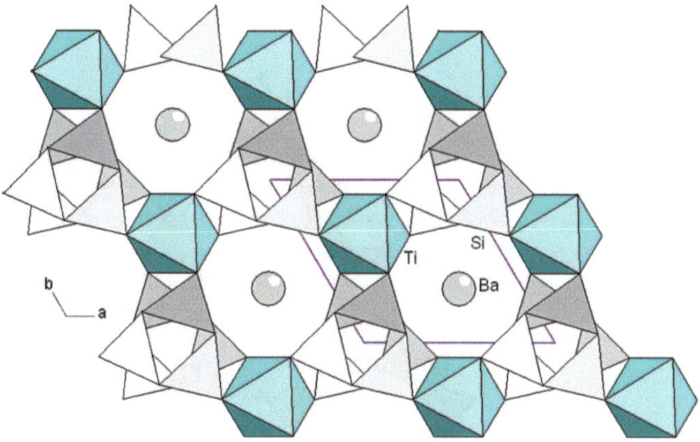

Fig. 6 The basic projection of the benitoite crystal structure (Massa et al. 2002)

Fig. 7 The benitoite crystal structure in *ac* projection (Massa et al. 2002)

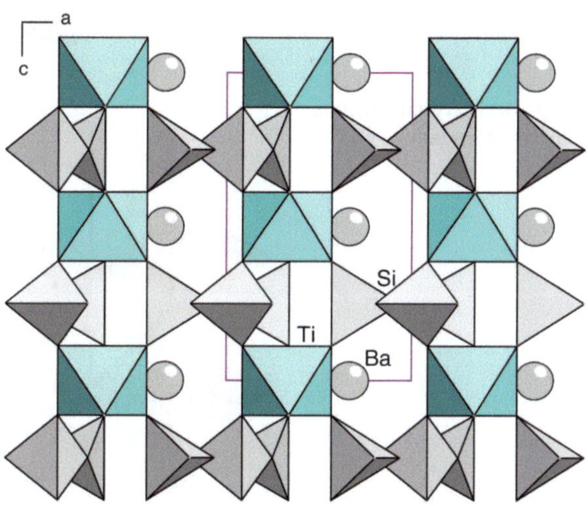

The topologically identical (MT_6O_{24}) complexes (M = Al, T = Si) from octahedron and six tetrahedral also form the crystal structure of beryl $Be_3Al_2Si_6O_{18}$ (Fig. 8), characterized by the sp. gr. *P6/mcc*, which differs from the space group of titanium vanadyl phosphates by the presence of an inversion center. As in the benitoite structure, twovertices in each tetrahedron of the ($AlSi_6O_{24}$) complex are shared with the tetrahedron of the neighboring complex, which leads to the formation of an anionic mixed framework. However, the groups of an octahedron and six tetrahedra in benitoite are twisted around the threefold axis (sp. gr.) with the formation of three-membered rings Si_3O_9, whereas similar groups in beryl are multiplied by the sixfold axis in a unit cell vertex with the formation of six-membered Si_6O_{18} rings, limiting the wide channels that are oriented parallel to the *z* axes. At a formal transition from the beryl structure to the $A_2(Ti(VO_2)_3(PO_4)_3)$ structure (Fig. 4), the

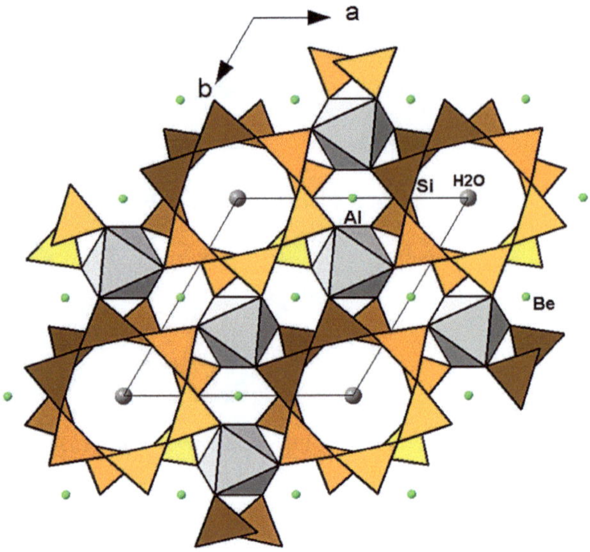

Fig. 8 Mixed anionic framework built from Si-tetrahedra and Al octahedra in the crystal structure of beryl

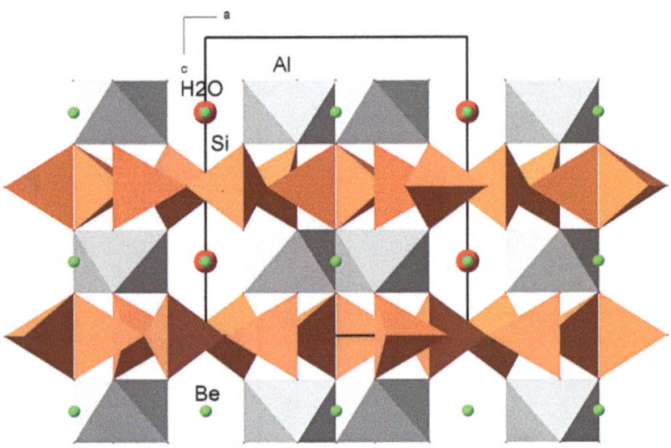

Fig. 9 Beryl crystal structure in the *xz* projection

neighboring tetrahedra in a six-membered ring are moved apart as a result of the insertion of an "internal" ring of six tetragonal pyramids (VO_5) which share oxygen vertices of neighboring polyhedra. Such an insertion violates the horizontal reflection plane, which is inherent in the beryl structure (Fig. 9), and the centrosymmetric sp. gr. *P*6/*mcc* is transformed into the polar sp.gr. *P*6*cc*: the horizontal *m* plane in the A_2(Ti $(VO)_3(PO_4)_3$) structure is violated by the position of the oxygen atom in the vertex of VO_5 tetragonal pyramid (Fig. 5). The incorporation of rings of six vanadium polyhedra also leads to a significant increase in the diameter of channels at the center and at the midpoints of the unit cell edges. These "minichannels" in the beryl structure contain beryllium atoms with an admixture of lithium in vorobyevite (Yakubovich et al. 2009); the similar positions in the moved apart channels in vanadylphosphate structures are occupied by cesium or rubidium atoms, which are large. These atoms also occupy the wide channels centered at the vertices of

$Cs_2(Ti(VO_2)_3(PO_4)_3)$ and $Rb_2(Ti(VO_2)_3(PO_4)_3)$ unit cells; the same channels in beryl derivatives can contain Cs and Na atoms and water molecules.

4 The Microporous Copper Vanadyl(V) Phosphate, $Rb_2(Cu(VO_2)_2(PO_4)_2)$ with Rb Atoms in Intersecting Channels

The V^{5+} ions in the structure occupy strongly distorted five-vertex (VO_5) polyhedra. The two shortest V=O bonds of 1.620(2) and 1.621(2)Å mark vanadyl groups, while the three longer V–O distances of 1.944(2), 1.986(2) and 2.090(2) Å correspond to the oxygen atoms shared with (PO_4) tetrahedra. The (VO_5) polyhedra approach a trigonal bipyramidal configuration and, together with the (PO_4) tetrahedral, form mixed anionic layers (Fig. 10). Alternating vertex-sharing (VO_5) bipyramids and (PO_4) tetrahedra form both eight- and four-membered rings within these layers. The vanadyl groups (V=O) remain terminal and are pointing into the interlayer space, or toward the eight-membered rings. Along the (100) direction V,P-layers alternate with layers of Rb and Cu atoms (Fig. 11), and (CuO_4) quadrilaterals link the V/P layers by sharing two vertices with P tetrahedra and two others with P and V polyhedra. Thus, a three-dimensional anionic framework with the formula

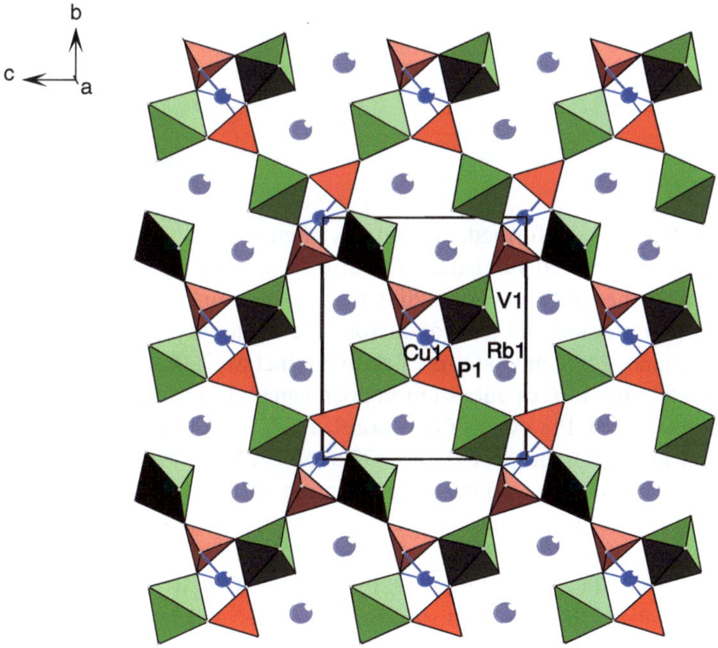

Fig. 10 The $Rb_2(Cu(VO_2)_2(PO_4)_2)$ crystal structure in the (100) projection

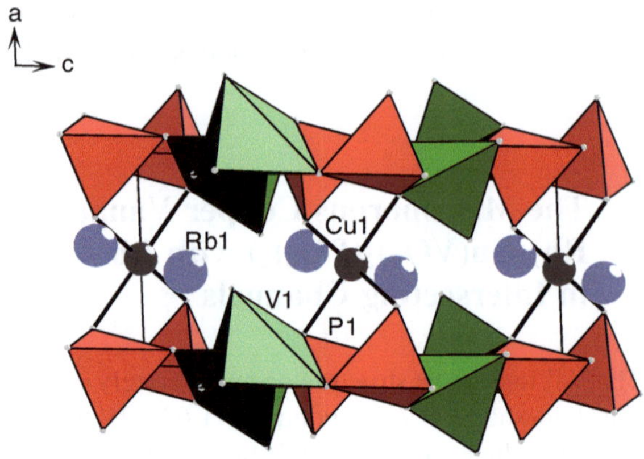

Fig. 11 Layers of VO_5 bipyramids and PO_4 tetrahedra linked by Cu atoms in a three-dimentional framework in the structure of $Rb_2(Cu(VO_2)_2(PO_4)_2)$

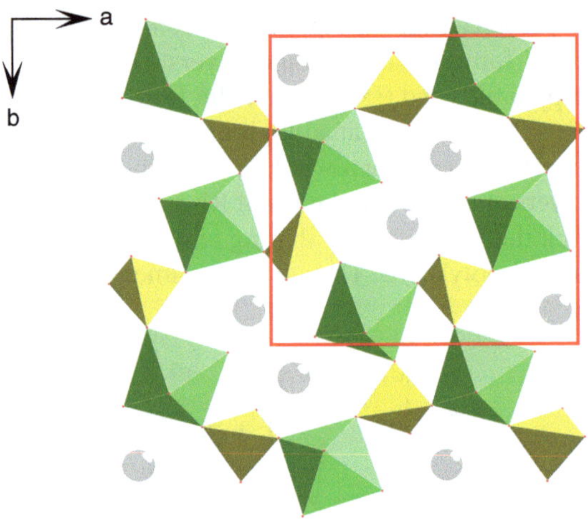

Fig. 12 Microporous layers built from V and P polyhedra in the $Pb(VO_2)(PO_4)$ crystal structure

$(Cu(VO_2)_2(PO_4)_2)^{2-}$ is formed. It contains channels with eight polyhedra in the circumference, as viewed along the (100) direction. Crossing channels also intersect the structure in the (001) and (011) directions, which suggests considering the new compound as a possible 3D ion-conductor. The Rb atoms reside in these channels and are surrounded by ten O atoms (Yakubovich et al. 2008).

The crystal structure of rubidium copper vanadyl phosphate is closely related to the structures of $Pb(VO_2)(PO_4)$ (Borel et al. 2000) and $Pb(VO_2)(PO_4){\cdot}H_2O$ (Leclaire et al. 2001), and $BaCrF_2LiF_4$ (Babel 1974). In Figs. 12–14 one can see eight-member windows formed by alternating octahedra (VO_6 and CrF_6) and tetrahedra (PO_4 and LiF_4) sharing vertices. These windows are topologically very similar to the windows walled-in by VO_5 bipyramids and PO_4 tetrahedra in the $RbCu_{0.5}(VO_2)(PO_4)$ structure described above (Fig. 10). In all four structures, the eight-member windows enclose large channels which contain Rb, Ba or Pb atoms. The main topological difference between these structures occurs along the axis perpendicular to the layers with eight-membered windows. In the $RbCu_{0.5}(VO_2)(PO_4)$

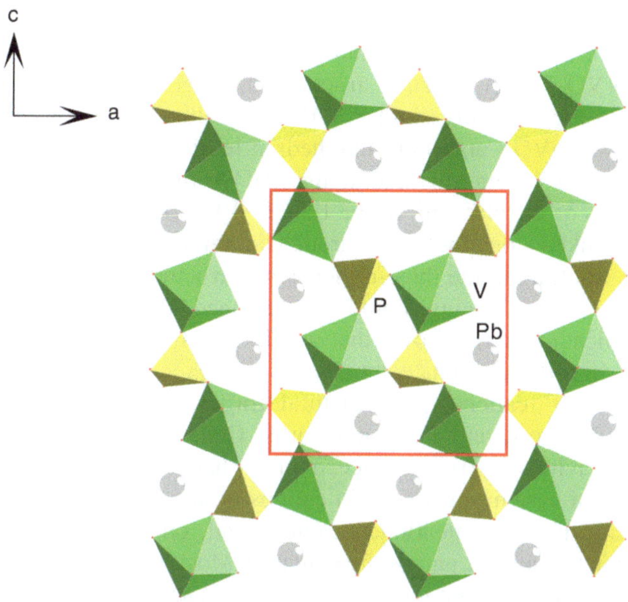

Fig. 13 Quasi layers built from VO_6 octahedra and PO_4 tetrahedra in the $Pb(VO_2)(PO_4)\cdot H_2O$ crystal structure

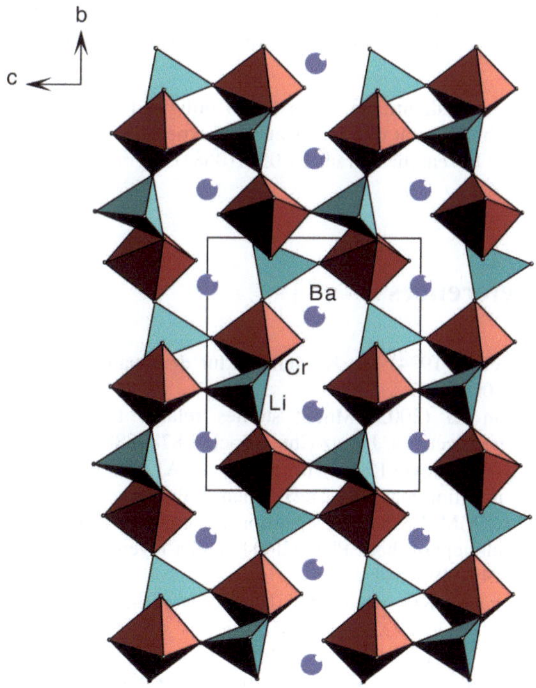

Fig. 14 Quasi layers built from CrF_6 octahedra and LiF_4 tetrahedra in the $Ba(CrF_2LiF_4)$ crystal structure

structure, Cu atoms with square-planer coordination link the V/P layers along the *a* axis to form the 3D mixed anionic framework (Fig. 11).

In the $Pb(VO_2)(PO_4)$ crystal structure, the windows walled-in by VO_5 bipyramids and topologically similar V/P layers alternate along the *b* axis with layers of isolated VO_4 and PO_4 tetrahedra. In the hydrated phase, $Pb(VO_2)(PO_4)\cdot H_2O$, the V/P layers alternate along (001) with chains of V/P composition parallel to *b* axis of the unit cell. In the $Ba(CrF_2)(LiF_4)$ structure, eight-membered ring layers are linked along the *a* axis through common vertices of Cr octahedra and Li tetrahedra, resulting in the three-dimensional structure.

In conclusion, the following should be mentioned. Our experimental data and a review of publications have shown that usually the anionic structures of a mixed type, formed by orthophosphate tetrahedra and oxocomplexes around $(VO)^{2+}$, $(VO)^{3+}$ and $(VO_2)^+$ cations, are structural fragments of different dimensions: 1D ribbons, 2D layers, and 3D frameworks. Generally, these are microporous structures with cavities, interlayer space, or open channels. To form these structures, the crystallization system must contain cations of an appropriate size, on the one hand, to occupy the aforementioned cavities and, on the other, to compensate for the charge in the anionic part of the structure. Most often, atoms of alkali metals K, Rb, and Cs and ammonium ions (more rarely, Na atoms of smaller size), as well as Ba, Sr, Pb, Tl, Ag, and Cd atoms, serve to this end (Yakubovich 2009). Currently the synthesis of hybrid crystal structures, where organic cations (for example, diprotonated piperazine ions $C_4H_{12}N_2$ (Zhou et al. 2010)) serve as compensators for the inorganic anionic structure of a mixed type, is being actively developed. The above mentioned features of the vanadyl phosphate crystal structures permit to believe in their potential applications in different technologies.

Acknowledgements I am much obliged to W. Massa, I. Steele and O. Dimitrova for their collaboration at different stages of experimentations. Financial support by the Russian Fund for Basic Researches (grant 10-05-01068a) is gratefully acknowledged.

References

Babel D (1974) Die Kristallstruktur des quaternaeren Fluorids $BaLiCrF_6$. Z Anorg Allg Chem 406:23–37

Baran EJ (2003) Model studies related to vanadium biochemistry: recent advances and perspectives. J Braz Chem Soc 14:878–888

Barker J, Gover RKB, Burns P, Bryan AJ (2006) Hybrid-ion a lithium-ion cell based on a sodium insertion material. Electrochem Solid State Lett 9:A190–A192

Borel MM, Leclaire A, Chardon J, Daturi M, Raveau B (2000) Dimorphism of the vanadium(V) monophosphate PbVO2PO4: alpha-layered and beta-tunnel structures. J Solid State Chem 149:149–154

Brugger J, Krivovichev SV, Meisser N, Ansermet S, Armbruster T (2006) Scheuchzerite, Na(Mn, Mg)9(VSi9O28(OH))(OH)3, a new single-chain silicate. Amer Miner 91:937–943

Centi G, Trifiro F, Ebner JR, Franchetti VM (1988) Mechanistic aspects of maleic-anhydride synthesis from C4-hydrocarbons over phosphorus vanadium-oxide. Chem Rev 88:55–80

Chen Y, Zhao Y, An X, Liu J, Dong Y, Chen L (2009) Preparation and electrochemical performance studies on Cr-doped $Li_3V_2(PO_4)_3$ as cathode materials for lithium-ion batteries. Electrochim Acta 54:5844–5850

Evans HT Jr (1973) The crystal structures of cavansite and pentagonite. Amer Miner 58:412–424

Gramaccioli CM, Liborio G, Pilati T (1981) Structure of medaite, $Mn_6(VSi_5O_{18}(OH))$: the presence of a new kind of heteropolysilicate anion. Acta Crystallogr B37:1972–1978

Ishida N, Kimata M, Nishida N, Hatta T, Shimizu M, Akasaka T (2009) Polymorphic relation between cavansite and pentagonite: genetic implications of oxonium ion in cavansite. J Mineral Petrol Sci 104:241–252

Le Meins JM, Crosnier-Lopez MP, Hemon-Ribaud A, Courbion G (1999) Phase transitions in the $Na_3M_2(PO_4)_2F_3$ family ($M = Al^{3+}$, V^{3+}, Cr^{3+}, Fe^{3+}, Ga^{3+}): synthesis, thermal, structural, and magnetic studies. J Solid State Chem 148:260–277

Leclaire A, Chardon J, Raveau B (2001) Hydrothermal synthesis and crystal structure of a lead vanadium monophosphate hydrate with an open structure: $Pb_2V_2O_4(PO_4)_2.2(H_2O)$. J Mater Chem 11:1482–1485

Massa W, Yakubovich OV, Dimitrova OV (2002) Crystal structure of a new sodium vanadyl(IV) fluoride phosphate $Na_3\{V_2O_2F(PO_4)_2\}$. Solid State Sci 4:495–501

Padhi AK, Nanjundaswamy KS, Goodenough JB (1997) Phospho-olivines as positive-electrode materials for rechargeable lithium batteries. J Electrochem Soc 144:1188–1194

Sauvage F, Quarez E, Tarascon JM, Baudrin E (2006) Crystal structure and electrochemical properties vs. Na^+ of the sodium fluorophosphate $Na_{1.5}VOPO_4F_{0.5}$. Solid State Sci 8:1215–1221

Starova GL, Krivovichev SV, Filatov SK (1998) Crystal chemistry of inorganic compounds based on chains of oxocentered tetrahedra. II. Crystal structure of Cu4O2((As, V)O4). Z Kristallogr 213:650–653

Whittingham MS (2004) Lithium batteries and cathode materials. Chem Rev 104:4271–4302

Whittingham MS, Song Y, Lutta S, Zavalij PY, Chernova NA (2005) Some transition metal (oxy) phosphates and vanadium oxides for lithium batteries. J Mater Chem 15:3362–3379

Yakubovich OV (2009) Structural typomorphism of amphoteric oxocomplexes in a lithophilous system of crystal genesis: genetic crystal chemistry of vanadium. In: Urusov VS (ed) The problems of crystallology. GEOS, Moscow, in Russian

Yakubovich OV, Mel'nikov OK (1996) Anionic framework of a mixed type in $Na_3\{Al_2F_3(PO_4)_2\}$ crystalline structure. Crystallogr Rep 41:663–668 (in Russian)

Yakubovich OV, Kireev VV, Mel'nikov OK (2000) Refinement of crystal structure of a Ge-analogue of natisite Na2{TiGeO4} and prediction of new phases with anionic MTO5 radicals. Crystallogr Rep 45:578–584

Yakubovich OV, Massa W, Pekov IV (2002) Crystal structure of the new mineral bushmakinite, Pb2{(Al, Cu)(PO4)((V, Cr, P)O4)(OH)}. Dokl Earth Sci 382:100–105

Yakubovich OV, Massa W, Dimitrova OV (2006) A new type of mixed anionic framework in microporous caesium titanium vanadyl(V) phosphate $Cs_2(Ti(VO_2)_3(PO_4)_3)$. Solid State Sci 8:71–76

Yakubovich OV, Steele IM, Dimitrova OV (2008) A new type of mixed anionic framework in microporous rubidium copper vanadyl(V) phosphate, $Rb_2Cu(VO_2)_2(PO_4)_2$. Acta Crystallogr C64:62–65

Yakubovich OV, Pekov IV, Steele IM, Massa W, Chukanov NV (2009) Alkali metals in beryl and their role in the formation of derivative structural motifs: comparative crystal chemistry of vorobyevite and pezzottaite. Crystallogr Rep 54:399–412

Yakubovich OV, Yakovleva EV, Dimitrova OV (2010) Crystal structure of $Rb_2(Ti(VO_2)_3(PO_4)_3)$. Crystallogr Rep 55:210–215

Zhou Y, Ming P, Liu J (2010) Hydrothermal synthesis and structure of an open-framework, large-channel vanadium–cobalt phosphate $(C_4H_{12}N_2)_2(Co^{II}(H_2O)_2(V^VO_2)_2(V^{IV}O)_2(PO_4)_4)\cdot 2H_2O$. Inorg Chem Commun 13:1–4

Thermal Expansion of Aluminoborates

Martin Fisch and Thomas Armbruster

1 Introduction

Crystal structures containing triangular BO_3 and tetrahedral BO_4 units are known for low or even negative thermal expansion behavior. Filatov and Bubnova (2008) conclude that (1) BO_3 or BO_4 units behave as rigid bodies, comparable to SiO_4 tetrahedra (Hazen and Prewitt 1977; Tucker et al. 2000); (2) BO_3 and BO_4 units act as hinges, which is a reason for anisotropic thermal expansion; (3) the more cations with low bond strength a borate structure contains, the higher is its thermal expansion.

Thermal expansion of borate structures is therefore controlled by translational and vibrational motions of BO_3 and BO_4 groups combined with expansion and/or tilting of other polyhedra in the structure.

Temperature dependent behavior of two synthetic aluminoborates Al_5BO_9 (Sokolova et al. 1978; Fisch et al. 2011 and refs. therein) and $Al_4B_2O_9$ (Fischer et al. 2008 and refs. therein) and the minerals grandidierite $(Mg,Fe)Al_3BSiO_9$ (Stephenson and Moore 1968) and jeremejevite $Al_6B_5O_{15}(F,OH)_3$ (Golovastikov et al. 1955; Foord et al. 1981) were investigated by high-temperature X-ray diffraction (XRD) methods. Al_5BO_9, $Al_4B_2O_9$ and grandidierite belong to the group of boron-mullites. They have mullite-type crystal structures characterized by unfolded, infinite edge-connected chains of AlO_6 octahedra and crystallize in subgroups of space group *P*4/*mbm* (Fischer and Schneider 2008). Jeremejevite crystallizes in space group *P*6$_3$/*m* and has no structural relation to the mullite-types. The compound was included in this study because it only contains Al and B cations, as in synthetic Al_5BO_9 and $Al_4B_2O_9$. The four investigated structures contain chains of rather undistorted edge-connected AlO_6 octahedra and planar coordinated BO_3 triangles together with AlO_5 and AlO_4

M. Fisch (✉) • T. Armbruster
Mineralogical Crystallography, Institute of Geological Sciences,
University of Bern, Bern, Switzerland
e-mail: fisch@krist.unibe.ch.

S.V. Krivovichev (ed.), *Minerals as Advanced Materials II*,
DOI 10.1007/978-3-642-20018-2_25, © Springer-Verlag Berlin Heidelberg 2012

(Al_5BO_9), AlO_5, AlO_4 and BO_4 ($Al_4B_2O_9$), and AlO_5, SiO_4 and MgO_5 (grandidierite). In jeremejevite, Al octahedra have five O and one (F,OH) ligand.

2 Sample Description

The Al_5BO_9 single-crystal ($0.13 \times 0.07 \times 0.4\ mm^3$) was obtained from a mixture of $9Al_2O_3$:$2B_2O_3$ in a K_2CO_3:$3MoO_3$ melt (slowly cooled from 1373 K) whereas the powder sample resulted from a solid-state reaction of $9Al_2O_3$ with $2B_2O_3$ heated at 1473 K. $Al_4B_2O_9$ powder was prepared according to Fischer et al. (2008).

The grandidierite fragment ($0.17 \times 0.34 \times 0.27\ mm^3$) was cut from a raw, bluish gemstone from Madagascar. The jeremejevite crystal ($0.14 \times 0.14 \times 0.07\ mm^3$) was separated from a pale light-blue sample from Erongo region, Namibia. The remaining jeremejevite material was powdered and mixed with corundum as internal standard for high-temperature powder XRD measurements. Grandidierite could not be measured with powder XRD due to the small sample available. All single-crystals were squeezed into quartz glass capillaries and subsequently mounted on a goniometer head for high-temperature data collection.

3 X-ray Diffraction Methods

Single-crystal XRD data were collected with a Bruker Smart APEXII CCD diffractometer (graphite monochromatized Mo$K\alpha$) at 298 and 873 K. During high-temperature measurements, crystals were heated using a hot N_2-gas blower, calibrated to known phase transitions. Data were collected with ω-φ-scans and subsequently treated using Apex2 v. 2009–11.0 software package (Bruker 2009). The structural model by Fisch et al. (2011) was used for refinement of Al_5BO_9 data whereas grandidierite and jeremejevite structures were solved by direct methods. Details of single-crystal XRD data treatment are given in Table 1. Atomic coordinates and anisotropic displacement parameters are listed in Fisch (2011).

According to Hazen and Prewitt (1977), average bond-expansion in polyhedra obeys the following rule: $<\alpha> = 32.9 \cdot (0.75 - z/\rho) \cdot 10^{-6}\ K^{-1}$, where z is the cation charge and ρ the coordination number. If $z/\rho \geq 0.75$ (e.g. AlO_4, SiO_4, BO_3), $<\alpha>$ is predicted as $0\ K^{-1}$. Average α values were calculated from mean polyhedral bond distances $<d>$ according to: $<\alpha> = (<d_{873K}> - <d_{298K}>)/<d_{298K}> \cdot (873\text{–}298\ K)$.

Bond-lengths data were not corrected for rigid body thermal motion effects (Johnson 1970) because the differences between corrected and uncorrected data were only slightly larger than uncorrected data within standard deviations. Average bond expansion in polyhedra calculated from uncorrected data sometimes resulted in negative thermal expansion for BO_3 units, with errors almost as large as the value itself.

Negative expansion with increasing temperature for MO_x polyhedra is physically not feasible, thus, negative values are artifacts of not applied corrections for thermal

Table 1 Details of crystal structure data collection and structure refinements

Sample	Al_5BO_9		Jeremejevite		Grandidierite	
Temperature	298 K[a]	873 K	298 K	873 K	298 K	873 K
Space group	*Cmc*2$_1$	*Cmc*2$_1$	*P*6$_3$/*m*	*P*6$_3$/*m*	*Pnma*	*Pnma*
a-axis	5.6689(6) Å	5.6825(6) Å	8.5526(3) Å	8.5840(16) Å	10.9967(1) Å	11.0193(1) Å
b-axis	15.0045(13) Å	15.0754(16) Å	8.5526(3) Å	8.5840(16) Å	5.7634(1) Å	5.7884(1) Å
c-axis	7.6897(11) Å	7.7121(7) Å	8.1793(4) Å	8.2187(15) Å	10.3321(1) Å	10.3434(1) Å
Volume	654.07(13) Å^3	660.67(12) Å^3	518.13(4) Å^3	524.46(17) Å^3	653.05(3) Å^3	659.74(3) Å^3
Z	4	4	2	2	4	4
Time p. frame	10 s	30 s	40 s	40 s	10 s	20 s
Refl. coll.	4919	1277	5491	2522	10569	14415
Max. 2θ	69.91°	65.64°	69.21°	60.86°	63.20°	89.18°
Index range *h*	−8...8	−7...7	−12...11	−9...12	−15...14	−16...21
Index range *k*	−23...23	−21...20	−11...13	−11...11	−8...8	−10...10
Index range *l*	−12...12	−10...10	−12...8	−11...11	−14...15	−15...18
Unique refl.	1110	595	781	555	1105	2468
Refl. > 4σ(*I*)	1056	581	673	465	1118	2283
R(int)	0.0423	0.0414	0.0324	0.027	0.0167	0.0162
R(σ)	0.0349	0.0529	0.0222	0.0251	0.0095	0.011
L. S. param.	83	83	50	50	87	87
GOF	1.001	1.076	1.064	1.236	1.278	1.262
*R*1 *I* > 4σ(*I*)	0.0195	0.0242	0.0258	0.0308	0.0182	0.0207
*R*1, all data	0.0211	0.0247	0.0326	0.0434	0.0186	0.023
*wR*2 on F^2	0.0351	0.0557	0.0627	0.1145	0.0563	0.0636
$\Delta\rho_{max}$, near	0.23 e^-•Å^{-3}, B1	0.26 e^-•Å^{-3}, Al1	0.49 e^-•Å^{-3}, O1	0.69 e^-•Å^{-3}, O3	0.46 e^-•Å^{-3}, O6	0.53 e^-•Å^{-3}, O3
$\Delta\rho_{min}$, near	−0.18 e^-•Å^{-3}, O4	−0.26 e^-•Å^{-3}, Al1	−0.40 e^-•Å^{-3}, Al1	−0.73 e^-•Å^{-3}, O1	−0.39 e^-•Å^{-3}, Si1	−0.37 e^-•Å^{-3}, Si1

[a]Lattice parameters measured with Enraf-Nonius CAD4 diffractometer

rigid body librational effects and may be interpreted as zero expansion between 298 and 873 K.

Powder XRD data were measured with a PANalytical X'Pert PRO MPD diffractometer (Cu radiation) equipped with an Anton-Paar HTK 200 high-temperature goniometer attachment. Samples were measured at 298, 323 and then in 50 K steps up to 1273 K (Al_5BO_9), 1323 K ($Al_4B_2O_9$) and 1073 K (jeremejevite). Diffraction patterns were measured from 10° to 70° 2θ at 0.008°/step with 30–90 s/step. To eliminate outliers, lattice parameters were parameterized by a second order polynomial: $\mathrm{length}_i(T) = x_{0i} + x_{1i} \cdot T + x_{2i} \cdot T^2$. Polynomial parameters x_{0i}, x_{1i} and x_{2i} (i are lattice parameters a, b and c) were refined simultaneously for all diffraction patterns using Topas-Academic v. 4.1 (Coelho 2007). Sample displacement was refined as a function of temperature due to its correlation with lattice parameters. From polynomials expressing temperature dependent lattice parameters, thermal expansion parameters $\alpha(T)$ were derived according to the formula $\alpha(T) = a(T)^{-1} \cdot \partial a(T)/\partial T$. From $\alpha(T)$, the anisotropy factor A can be obtained as follows: $A = (|\alpha_a - \alpha_b| + |\alpha_a - \alpha_c| + |\alpha_b - \alpha_c|) \cdot 10^6$ (Dreyer 1974; Schneider and Eberhard 1990). For hexagonal jeremejevite ($a = b$), $A = 2 \cdot |\alpha_a - \alpha_c| \cdot 10^6$.

4 Results and Discussion

4.1 Thermal Evolution of Al_5BO_9

Main feature of the Al_5BO_9 structure are mullite-type chains of edge-connected AlO_6 octahedra extending parallel to the a-axis. Adjacent AlO_6 chains are interconnected by two alternating segments parallel (010). One contains $Al2O_5$ and AlO_4 polyhedra whereas the other consists of $Al3O_5$ and BO_3 polyhedra, which build corner-linked chains ($-BO_3-Al3O_5-BO_3-$) parallel b linking adjacent octahedra. Both AlO_5 polyhedra are more accurately described as AlO_{4+1} polyhedra as one Al-O bond is significantly longer than the other four. The latter segment containing the BO_3 unit is less dense than the former one, resulting in a cavity parallel to the AlO_6 chains (Fig. 1).

Temperature dependent lattice parameters and $\alpha(T)$ are shown in Tables 2 and 3, respectively. Between 298 and 1273 K, the b-axis expands strongly, followed by the c- and the a-axis (the latter parallel to chains of AlO_6 octahedra). The volumetric expansion is +1.56%. In the same temperature interval, α_a increases most, followed by α_c and α_b, indicating that the shortest crystal axis (a-axis) has the steepest gradient, followed by the c-axis, whereas the longest crystal axis (b-axis) displays the lowest slope (Fig. 2). As a result, thermal expansion of Al_5BO_9 trends to more isotropic expansion behavior with increasing temperature, which is confirmed by a decrease of A from $A_{298K} = 6.3$ to $A_{1273K} = 2.4$.

Inter-polyhedral angles remain the same within standard deviations in the investigated temperature interval, except the O5-Al3-O4 angle. This angle changes

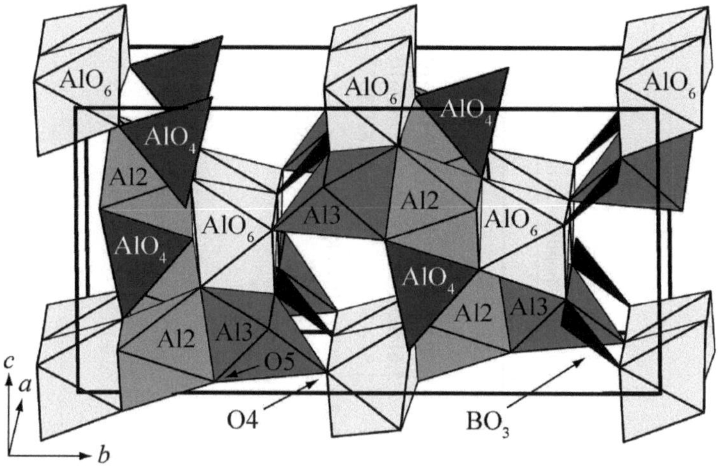

Fig. 1 The structure of Al_5BO_9. The distance O4–O5 (marked) shows a large increase in the investigated temperature regime indicating that the corner linkages between $Al3O_5$ and BO_3 and the adjacent chain of octahedra are the weakest part of the structure

Table 2 Coefficients for the second order polynomial $\text{length}_i(T) = x_{0i} + x_{1i}\cdot T + x_{2i}\cdot T^2$ expressing lattice parameters i as function of temperature

Parameter	x_0	x_1	x_2
Al_5BO_9 for 298 K ≤ T ≤ 1273 K[a]			
a(T) [Å]	5.66407	$1.44340\cdot10^{-5}$	$6.02392\cdot10^{-9}$
b(T) [Å]	14.98650	$9.45645\cdot10^{-5}$	$9.24350\cdot10^{-10}$
c(T) [Å]	7.68450	$3.61387\cdot10^{-5}$	$3.71557\cdot10^{-9}$
V(T) [Å^3]	652.299	$8.829\cdot10^{-3}$	$1.112\cdot10^{-6}$
$Al_4B_2O_9$ for 298 K ≤ T ≤ 1323 K[a]			
a(T) [Å]	14.76128	$1.49152\cdot10^{-4}$	$-1.33666\cdot10^{-8}$
b(T) [Å]	5.54025	$1.49592\cdot10^{-5}$	$4.33164\cdot10^{-9}$
c(T) [Å]	15.05004	$8.43267\cdot10^{-5}$	$1.70444\cdot10^{-8}$
V(T) [Å^3]	1230.685	$2.263\cdot10^{-2}$	$1.417\cdot10^{-6}$
Grandidierite for 298 K ≤ T ≤ 973 K[b]			
a(T) [Å]	10.9548	$6.2274\cdot10^{-5}$	$2.4334\cdot10^{-8}$
b(T) [Å]	5.7557	$2.6772\cdot10^{-5}$	$1.5226\cdot10^{-8}$
c(T) [Å]	10.3292	$5.7471\cdot10^{-6}$	$1.6289\cdot10^{-8}$
V(T) [Å^3]	651.29	$7.03\cdot10^{-3}$	$4.30\cdot10^{-6}$
Jeremejevite for 298 K ≤ T ≤ 1073 K[a]			
a(T) [Å]	8.53719	$6.51545\cdot10^{-5}$	$1.05757\cdot10^{-9}$
c(T) [Å]	8.16183	$6.20481\cdot10^{-5}$	$-6.17363\cdot10^{-10}$
V(T) [Å^3]	515.169	$1.178\cdot10^{-2}$	$1.834\cdot10^{-7}$

[a]From powder XRD, [b]from single-crystal XRD

Table 3 Coefficients for polynomials $\alpha_i(T) = a_{0i} + a_{1i} \cdot T$ expressing thermal expansion as a function of temperature (*i* corresponds to lattice parameters)

Parameter	a_0	a_1	Parameter	a_0	a_1
Al_5BO_9 for 298 K $\leq$ T $\leq$ 1273 K			$Al_4B_2O_9$ for 298 K $\leq$ T $\leq$ 1323 K		
$\alpha_a(T)$ [K^{-1}]	$2.55439 \cdot 10^{-6}$	$2.10359 \cdot 10^{-9}$	$\alpha_a(T)$ [K^{-1}]	$1.00906 \cdot 10^{-5}$	$-1.87160 \cdot 10^{-9}$
$\alpha_b(T)$ [K^{-1}]	$6.31045 \cdot 10^{-6}$	$8.21191 \cdot 10^{-11}$	$\alpha_b(T)$ [K^{-1}]	$2.70463 \cdot 10^{-6}$	$1.54378 \cdot 10^{-9}$
$\alpha_c(T)$ [K^{-1}]	$4.70661 \cdot 10^{-6}$	$9.33647 \cdot 10^{-10}$	$\alpha_c(T)$ [K^{-1}]	$5.61555 \cdot 10^{-6}$	$2.19838 \cdot 10^{-9}$
$\alpha_V(T)$ [K^{-1}]	$1.357 \cdot 10^{-5}$	$3.114 \cdot 10^{-9}$	$\alpha_V(T)$ [K^{-1}]	$1.842 \cdot 10^{-5}$	$1.870 \cdot 10^{-9}$
Grandidierite for 298 K $\leq$ T $\leq$ 973 K			Jeremejevite for 298 K $\leq$ T $\leq$ 1073 K		
$\alpha_a(T)$ [K^{-1}]	$5.3387 \cdot 10^{-7}$	$2.9700 \cdot 10^{-9}$	$\alpha_a(T)$ [K^{-1}]	$7.63280 \cdot 10^{-6}$	$1.86261 \cdot 10^{-10}$
$\alpha_b(T)$ [K^{-1}]	$1.0853 \cdot 10^{-5}$	$8.2929 \cdot 10^{-9}$	$\alpha_c(T)$ [K^{-1}]	$7.60138 \cdot 10^{-6}$	$-2.06187 \cdot 10^{-10}$
$\alpha_c(T)$ [K^{-1}]	$2.5960 \cdot 10^{-6}$	$2.9337 \cdot 10^{-9}$	$\alpha_V(T)$ [K^{-1}]	$2.286 \cdot 10^{-5}$	$1.730 \cdot 10^{-10}$
$\alpha_V(T)$ [K^{-1}]	$1.09 \cdot 10^{-5}$	$1.27 \cdot 10^{-5}$			

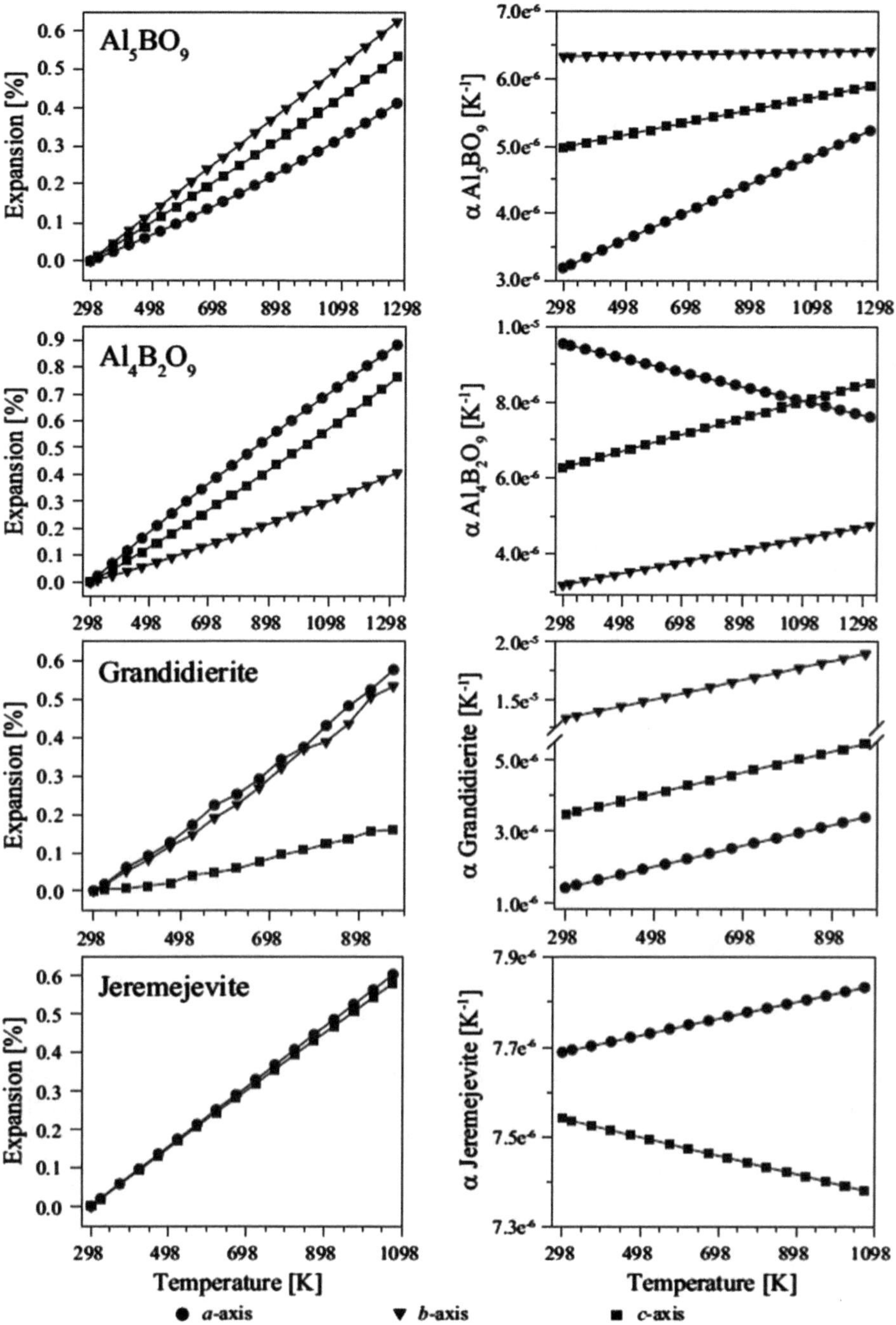

Fig. 2 Normalized lattice parameters vs. temperature (*left*) and thermal expansion parameters α as a function of temperature (*right*)

Table 4 Average polyhedral bond lengths at 298 and 873 K, their average thermal expansion coefficient <α> and expected <α> values according to Hazen and Prewitt (1977)

Polyhedron	298 K [Å]	873 K [Å]	<α> [K^{-1}]
Al_5BO_9			
AlO_6	1.9004(9)	1.909(2)	$7.7(1.8)\cdot 10^{-6}$
$Al2O_5$	1.8642(11)	1.869(2)	$4.5(2.3)\cdot 10^{-6}$
$Al3O_5$	1.8822(12)	1.892(3)	$8.9(2.5)\cdot 10^{-6}$
AlO_4	1.7472(11)	1.752(2)	$4.6(2.4)\cdot 10^{-6}$
BO_3	1.3769(14)	1.377(3)	$-0.7(3.8)\cdot 10^{-6}$
Grandidierite			
$Al1O_6$	1.8981(7)	1.9076(4)	$8.2(0.8)\cdot 10^{-6}$
$Al2O_6$	1.9094(8)	1.9171(4)	$7.1(0.8)\cdot 10^{-6}$
AlO_5	1.8548(11)	1.8604(6)	$5.3(1.2)\cdot 10^{-6}$
MgO_5	2.0311(19)	2.0407(6)	$8.3(1.0)\cdot 10^{-6}$
SiO_4	1.6290(10)	1.6289(6)	$-0.2(1.2)\cdot 10^{-6}$
BO_3	1.3656(14)	1.3631(8)	$-3.2(2.0)\cdot 10^{-6}$
Jeremejevite			
$AlO_5(F,OH)$[a]	1.8861(8)	1.8961(16)	$9.2(2.0)\cdot 10^{-6}$
$B1O_3$	1.3855(13)	1.3851(40)	$-0.5(4.4)\cdot 10^{-6}$
$B2O_3$	1.3646(7)	1.3623(19)	$-2.9(2.6)\cdot 10^{-6}$
Expected values for <α>			
<α> AlO_6:	$8.23\cdot 10^{-6}\ K^{-1}$	<α> MgO_5:	$11.5\cdot 10^{-6}\ K^{-1}$
<α> AlO_5:	$4.94\cdot 10^{-6}\ K^{-1}$	<α> BO_3:	$0\ K^{-1}$
<α> AlO_4:	$0\ K^{-1}$	<α> SiO_4:	$0\ K^{-1}$

[a]Treated as AlO_6

from 112.51(06)° to 113.56(13)°, resulting in an increase of the O4–O5 distance from 2.9236(17) to 2.9470(36) Å (Fig. 1).

Expansion of the AlO_6, BO_3 and $Al2O_5$ polyhedra is in perfect agreement with the predicted values of Hazen and Prewitt (1977). However, for the $Al3O_5$ polyhedron, $<\alpha> = 8.9(2.5)\cdot 10^{-6}\ K^{-1}$ is about 180% of the expected value. In addition, the AlO_4 tetrahedron has $<\alpha> = 4.6(2.4)\cdot 10^{-6}\ K^{-1}$ instead of the expected value of 0 K^{-1} (Table 4). Thermal expansion of Al_5BO_9 is therefore controlled by expansion of the $Al3O_5$ polyhedron and the AlO_4 tetrahedron.

One side of the -BO_3-$Al3O_5$-BO_3- chain linking two adjacent octahedral chains running parallel *a* is only corner-linked to AlO_6 polyhedra (Fig. 1). As this is the only part of the structure where only corner-linkages exist, it is obvious that this joint is the weakest part of the structure, which agrees with $Al3O_5$ having the largest expansion of all polyhedra in the structure (Table 4).

4.2 Thermal Evolution of $Al_4B_2O_9$

Several ambiguous structure solutions for mullite-type $Al_4B_2O_9$ exist. The most promising one is the monoclinic model (space group *C2/m*) of Fischer et al. (2008) based on powder XRD, ^{11}B and ^{27}Al MAS-NMR data. This structure is related to the one of boralsilite (Peacor et al. 1999) but could not be completely solved from available data.

Three types of edge-connected chains of AlO_6 octahedra extend parallel to the *b*-axis. The chains are linked among each other by three different polyhedral configurations. One of them is a complex arrangement of distorted AlO_5 (AlO_{4+1}), AlO_4 polyhedra and BO_3 triangles. The second type comprises AlO_{4+1} and BO_4 tetrahedra which are disordered with AlO_4 tetrahedra and BO_3 triangles due to a partially occupied oxygen site. The third type contains BO_3 and/or BO_4 units. However, its exact configuration has not been resolved yet.

Lattice parameters as function of temperature and thermal expansion coefficients $\alpha(T)$ are listed in Tables 2 and 3. The *a*-axis expands most, followed by the *c*- and the *b*-axis (the latter parallel to AlO_6 chains). The volumetric expansion is +2.06% and the monoclinic angle decreases from 90.8290(10)° at 298 K to 90.7808(13)° at 1323 K. α_b of the shortest crystal axis increases most (in percent), followed by α_c whereas α_a decreases, meaning that the length-change of the *a*-axis diminishes with increasing temperature (Fig. 2). The overall anisotropy of expansion decreases with temperature (A decreases from $A_{298K} = 12.6$ to $A_{1323K} = 7.6$). However, from room-temperature to about 1098 K ($\alpha_a = \alpha_c$), anisotropy of thermal expansion is most pronounced along the *a*-axis whereas at temperatures higher than 1098 K, expansion is dominated along the *c*-axis.

4.3 Thermal Evolution of Grandidierite (Mg, Fe^{2+})Al_3BSiO_9

Grandidierite (Mg, Fe^{2+})Al_3BSiO_9, is the Mg-rich member of the grandidierite-ominelite series (Dzikowski et al. 2007; Hiroi et al. 2001) and is of particular interest because it contains both Al and (Mg, Fe) in fivefold coordination (MacKenzie and Meinhold 1997; Dzikowski et al. 2007; Farges 2001). Structure refinement of single-crystal XRD room-temperature data indicated an almost pure Mg endmember with 97.3(3)% Mg and 2.7(3)% Fe for the selected sample.

The grandidierite structure (space group *Pnma*) contains two different chains of edge-connected AlO_6 octahedra parallel to the *b*-axis. The $Al1O_6$ chain is rather straight whereas chains of $Al2O_6$ octahedra are strongly folded (Fig. 3). Parallel (100), adjacent Al octahedra are directly connected by corner-linked chains of -BO_3-AlO_5-BO_3- extending along *b*. Parallel (001), the chains of octahedra are joined by corner-linked chains of -SiO_4-MgO_5-SiO_4- extending parallel *b*. Along *b*, channels between adjacent chains of octahedra are either occupied by arrangements of SiO_4, AlO_5 and MgO_5 polyhedra, or are locally empty if BO_3 polyhedra link octahedra parallel to (100). AlO_5 and MgO_5 polyhedra are regular bipyramids.

Polynomials expressing temperature dependence of lattice parameters and $\alpha(T)$ are shown in Tables 2 and 3. Between 298 and 973 K, expansion of the *a*-axis is dominant, followed by the *b*-axis (parallel to the AlO_6 chains) and the *c*-axis. The volumetric expansion is +1.28%. Temperature dependence of expansion parameters is $\alpha_a > \alpha_c > \alpha_b$ (Fig. 2). Expansion of grandidierite is highly

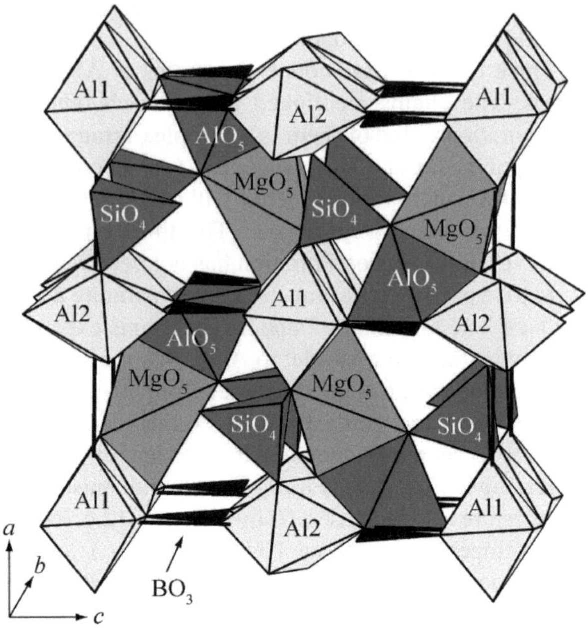

Fig. 3 The crystal structure of grandidierite in space group *Pnma*

anisotropic. A increases from $A_{298K} = 23.8$ to $A_{873K} = 31.0$. This high anisotropy results from α_b being a factor of ten larger than α_a and α_c. The increase in anisotropy is therefore due to the strong expansion along the b-axis.

Polyhedral bond lengths and average thermal expansion values obtained from the single-crystal structural refinements are listed in Table 4. Average thermal expansion of the $Al1O_6$, AlO_5 and SiO_4 polyhedra are in perfect agreement with values of Hazen and Prewitt (1977), Table 4. Values for $Al2O_6$ and BO_3 fit to predicted values within two standard deviations. The expansion of the MgO_5 polyhedron is too low ($<\alpha> = 8.3(1.0) \cdot 10^{-6}$ K^{-1} instead of $11.5 \cdot 10^{-6}$ K^{-1}).

The most significant changes in inter-polyhedral angles occur at Al2-O5-Al2 (stretching from 100.48(5)° to 101.10(3)°) and at Al2-O4-Al2 (bending from 94.01 (5)° to 93.40(3)°), which results from stretching of the folded chain of $Al2O_6$ octahedra. Further angular changes occur at Mg-O2-Al3 (124.32(6)° to 124.69(3)°), Mg-O5-Al2 (129.00(3)° to 128.73(1)°), Si1-O4-Al2 (132.91(2)° to 133.28(1)°) and at B1-O7-Al3 (130.82(8)° to 131.22(5)°). Except unfolding of the $Al2O_6$ chain, the largest structural changes are observed in the configuration of SiO_4, AlO_5 and MgO_5 polyhedra within the channels along b. Expansion of this channel is limited to expansion of the AlO_6 octahedra at its corners because the channel walls are held together by non-expanding BO_3 units parallel to the (100) plane and by a face of the non-expanding SiO_4 tetrahedra parallel to the (001) plane. MgO_5 has the lowest bond strength in the structure ($z/\rho = 0.4$) making it the most preferable buffer to compensate expansion of other polyhedra in the structure.

4.4 Thermal Evolution of Jeremejevite $Al_6B_5O_{15}(F,OH)_3$

Jeremejevite $Al_6B_5O_{15}(F,OH)_3$ was the first known aluminoborate mineral (Mallard 1887). The unit-cell volume of 518.13(4) $Å^3$ (Table 1) indicates that our sample is close to a F-endmember of jeremejevite, which is a common composition for natural jeremejevite (Foord et al. 1981).

Viewed along the *c*-axis, the jeremejevite structure (Fig. 4) is characterized by two types of channels. An empty hexagonal channel is located at 0, 0, *c*, whereas the other is triangular and located at $^2/_3$, $^1/_3$, *c* and $^1/_3$, $^2/_3$, *c*. The triangular channels are enclosed by chains of edge sharing AlO_5F octahedra. These chains are not straight as in mullite-type structures but consist of staggered arrangements of edge connected dimers of AlO_5F octahedra expanding along *c*. $B1O_3$ triangles (face parallel to the *c*-direction) support the chains along *c*. $B2O_3$ units (face parallel to the (001) plane) within the triangular channel connect adjacent octahedra apexes at the same *c*-height. Along *c*, every third octahedral apex pointing towards the center of a triangular channel consists of F instead of O and is thus not connected to a BO_3 unit.

Polynomials expressing temperature dependent lattice parameters are shown in Table 2 and polynomials expressing $\alpha(T)$ in Table 3. Between 298 and 1073 K, all axes expand similarly. The volumetric expansion is +1.8%. At 298 K α_a is slightly larger than α_c and with increasing temperature α_a expands and α_c decreases slightly (Fig. 2) resulting in an increase of the anisotropy factor A from $A_{298K} = 0.3$ to $A_{1073K} = 0.9$. Jeremejevite has the lowest A of all investigated structures. Expansion of BO_3 triangles and AlO_5F (Table 4) is in good agreement with the expected values of Hazen and Prewitt (1977).

In the investigated temperature regime, structural changes in jeremejevite are controlled by inter-polyhedral folding at Al1-O3-Al1 from 140.93(2)° to 141.11(6)° and at Al1-O2-Al1 from 124.19(2)° to 123.86(6)° (Al1-O2-Al1 is the angle between two connected dimers of octahedra). Angular changes are a result of

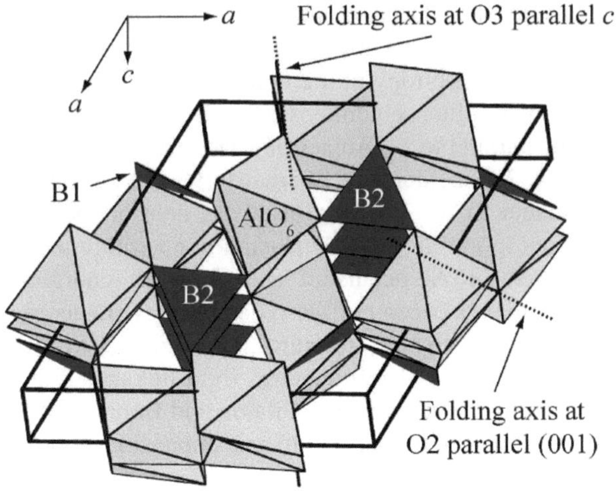

Fig. 4 The crystal structure of jeremejevite. Dashed lines indicate axes at which M-O-M angular changes are observed. B1 and B2 indicate $B1O_3$ and $B2O_3$ triangles, respectively

AlO_6 expansion, which is attenuated by non-expanding BO_3 units connected to them. Larger expansion of AlO_5F octahedra (compared to zero expansion of BO_3 units) can only be compensated by rotation. Dimers of octahedra are rotated at the edge joining two dimers (Fig. 4), resulting in decrease of the Al1-O2-Al1 angle. In addition, at the linking corner of two adjacent octahedral chains the Al1-O3-Al1-angle expands.

5 Discussion

Thermal behavior of the investigated boron-mullite compounds can be compared to those of the Al_2SiO_5 polymorphs andalusite and sillimanite and mullite ($Al_2(Al_{2+2x}Si_{2-2x})O_{10-x}$, $0.18 \leq x \leq 0.88$). All three have mullite-type structures with AlO_6 chains, interlinked by single chains of alternating Si and Al tetrahedra (sillimanite), double-chains of AlO_5 and SiO_4 polyhedra (andalusite) and by disordered arrangements of Al and Si in mullite (Fischer and Schneider 2005).

Winter and Ghose (1979) investigated thermal expansion of sillimanite and andalusite. In sillimanite, from 298 to 1273 K, lattice parameter expansion with temperature is $b > c > a$ (c is parallel to the chains of octahedra) and A decreases from 11.3 to 10.8. The most distinctive expansion along the b-axis is due to anisotropic expansion of octahedra. Even though the mullite-type chain is held together by chains of rigid non-expanding SiO_4/AlO_4 tetrahedra, octahedral rotation into the channels is responsible for expansion $c > a$. The expansion of andalusite is most expressed along the a-axis, followed by the b- and c-axis (mullite-type chains parallel to c). Average expansion of AlO_6 is slightly higher than expected ($<\alpha> = 12.6(3.2)\cdot10^{-6}\ K^{-1}$ instead of $8.2\cdot10^{-6}\ K^{-1}$). Anisotropic expansion of AlO_6 units is has its strongest contribution to the a-axis whereas expansion along the c-axis is constrained by the double-chain of rigid AlO_5/SiO_4 polyhedra. Due to the double-chains, no open channels exist and AlO_6 rotation is limited leading to the lowest expansion along the c-axis. Andalusite shows high and increasing anisotropy of thermal expansion ($A_{298K} = 20.6$, $A_{1273K} = 21.75$).

In 3:2 mullite, cell dimensions increase in the sequence $b > c > a$ (mullite-type chains parallel to c). Anisotropy factors are rather low, $A = 6.2$ between 298 and 1173 K (Schneider and Eberhard 1990) and $A = 3.8$ between 298 and 1273 K (Brunauer et al. 2001a). Based on neutron powder diffraction data, structural changes in mullite with increasing temperature are due to rotation and increasing distortion of Al octahedra and shifts in coordinates and varying Si and Al populations between mullite-type chains (Brunauer et al. 2001b).

In all investigated structures, BO_3 units (and SiO_4 in grandidierite) behave as rigid bodies and therefore act as rigid linkages (hinges) between other polyhedra, confirming the statement of Filatov and Bubnova (2008). Between 298 and 973 K, Al_5BO_9 has the lowest volumetric expansion, followed by grandidierite, $Al_4B_2O_9$ and jeremejevite.

The smallest volumetric expansion of Al_5BO_9 can be explained by its structural relation to sillimanite: Expansion of polyhedra is balanced by the empty channels parallel to the mullite-type chains. As one wall of the channel is formed by the O4–O5 edge of $Al3O_5$, the channel is rather flexible. However, in contrast to sillimanite, mullite-type chains are only close to two channels instead of four, and therefore, rotation of octahedra is limited. To explain the smallest expansion behavior along the *a*-axis, the relation to the andalusite structure becomes relevant: Expansion along the *a*-axis is limited by the chain of -BO_3-$Al3O_5$-BO_3- consisting of rigid BO_3 units (rigid AlO_4 and SiO_4 in andalusite) parallel to the mullite-type chains (Wada et al. 1993). Combined with limited rotation of octahedra, expansion along the *a*-axis (parallel to the mullite-type chains) has the strongest restrictions. For structures containing AlO_4 and AlO_6 polyhedra Hazen and Prewitt (1977) state that structural changes with increasing pressure correspond to those with decreasing temperature. Elastic behavior of Al_5BO_9 was investigated by Gatta et al. (2010) and the behavior of lattice parameters with increasing pressure (compression is highest along the *b*-axis, followed by the *c*- and *a*-axis) is in agreement with our results for expansion of lattice parameters with temperature.

In grandidierite, all polyhedra except MgO_5 expand as expected according to Hazen and Prewitt (1977). The MgO_5 polyhedron buffers thermal expansion due to its low bond strength ($z/\rho = 0.4$). Furthermore, slightly anisotropic expansion of the Al2 octahedra combined with unfolding of the corresponding chain is responsible for $b > c$ expansion (mullite-type chains parallel *b*).

All polyhedra in jeremejevite behave as rigid bodies. No flexible parts in the structure can compensate expansion and the structure reacts to rising temperature with folding and tilting of polyhedral arrangements.

The observed sequence of volumetric expansion may seem to contradict the conclusion of Filatov and Bubnova (2008) stating that thermal expansion of borates decreases with the number of cations with low bond strength (according to which grandidierite should undergo the largest expansion). It has to be considered that all of the structures investigated in this study are of similar chemical composition. In other words, AlO_4 behaves not much different to SiO_4, as do MgO_5 and AlO_5. Expansion properties of the investigated compounds are therefore mainly controlled by structural flexibility characteristic of each individual structure, i.e. their ability to buffer expansion by soft links between polyhedra rather than by differences in bond-strengths.

References

Bruker AXS Inc (2009) Apex2 v. 2009–11.0 software package. Madison, Wisconsin USA

Brunauer G, Frey F, Boysen H, Schneider H (2001a) High temperature thermal expansion of mullite: an in situ neutron diffraction study up to 1600°C. J Eur Ceram Soc 21:2563–2567

Brunauer G, Boysen H, Frey F, Hansen T, Kriven W (2001b) High temperature crystal structure of 3:2 mullite from neutron diffraction data. Z Kristallogr 216:284–290

Coelho Software (2007) Topas-Academic v. 4.1 software. Brisbane, Australia

Dreyer W (1974) Materialverhalten Anisotroper Festkörper. Springer, Wien/New York

Dzikowski TJ, Groat LA, Grew ES (2007) The geometric effects of $Fe^{V(2+)}$ for Mg^{V} substitution on the crystal structures of the grandidierite-ominelite series. Am Miner 92:863–872

Farges F (2001) Crystal chemistry of iron in natural grandidierites: an X-ray absorption fine-structure spectroscopy study. Phys Chem Miner 28:619–629

Filatov SK, Bubnova RS (2008) Structural mineralogy of borates as perspective materials for technological applications. In: Krivovichev SV (ed) Minerals as advanced materials I. Springer, Berlin

Fisch M (2011) Crystal chemistry of boron-bearing mullite-type compounds. Dissertation, University of Bern

Fisch M, Armbruster T, Rentsch D, Libowitzky E, Pettke T (2011) Crystal-chemistry of mullite-type aluminoborates $Al_{18}B_4O_{33}$ and Al_5BO_9: a stoichiometry puzzle. J Solid State Chem 184:70–80

Fischer RX, Schneider H (2005) Crystal chemistry of mullite and related phases. In: Schneider H, Komarneni S (eds) Mullite. Wiley-VCH, Weinheim

Fischer RX, Schneider H (2008) Crystal chemistry of borates and borosilicates with mullite-type structures: a review. Eur J Miner 20:917–933

Fischer RX, Kahlenberg V, Voll D, MacKenzie JDK, Smith ME, Schnetger B, Brumsack HJ, Schneider H (2008) Crystal structure of synthetic $Al_4B_2O_9$: a member of the mullite family closely related to boralsilite. Am Miner 93:918–927

Foord EE, Erd RC, Hunt GR (1981) New data for jeremejevite. Can Miner 19:303–310

Gatta GD, Rotiroti N, Fisch M, Armbruster T (2010) Stability at high pressure, elastic behavior and pressure-induced structural evolution of 'Al_5BO_9', a mullite-type ceramic material. Phys Chem Miner 37:227–236

Golovastikov NI, Belova EN, Belov NV (1955) The crystal structure of jeremejevite. Dokl Akad Nauk SSSR 104:78–81

Hazen RM, Prewitt CT (1977) Effects of temperature and pressure on interatomic distances in oxygen-based minerals. Am Miner 62:309–315

Hiroi Y, Grew ES, Motoyoshi Y, Peacor DR, Rouse EC, Matsubara S, Yokohama K, Miyawaki R, McGee JJ, Su SC, Hokada T, Furuwaka N, Shibasaki H (2001) Ominelite, (Fe, Mg)Al3BSiO9 (Fe2+ analogue of grandidierite), a new mineral from porphyritic granite in Japan. Am Miner 87:160–170

Johnson CK (1970) An introduction to thermal motion analysis. In: Ahmed FR (ed) Crystallographic computing. Munksgaard, Copenhagen

MacKenzie JDK, Meinhold RH (1997) MAS NMR study of pentacoordinated magnesium in grandidierite. Am Miner 82:479–482

Mallard E (1887) Sur diverses substances cristallisées qu' Ebelmen avait préparées et non decrites. Compt Rend Franç 105:1260–1265

Peacor DR, Rouse RC, Grew ES (1999) Crystal structure of boralsilite and its relation to a family of boroaluminosilicates, sillimanite, and andalusite. Am Miner 84:1152–1161

Schneider H, Eberhard E (1990) Thermal expansion of mullite. J Am Ceram Soc 73:2073–2076

Sokolova EV, Azizov AV, Simonov MA, Leoniuk NI, Belov ND (1978) Crystal-structure of synthetic ortho-3-borate, $Al_5(BO_3)O_6$. Dokl Akad Nauk SSSR 243:655–658

Stephenson DA, Moore PB (1968) The crystal structure of grandidierite, (Mg, Fe)Al3SiBO9. Acta Cryst B24:1518–1522

Tucker MG, Dove MT, Keen DA (2000) Direct measurement of the thermal expansion if the Si-O bond by neutron total scattering. J Phys Condens Matter 12(425–430):71–675

Wada H, Sakane K, Kitamura T (1993) Thermal expansion of aluminum borate. J Mater Sci Lett 12:1735–1737

Winter JK, Ghose S (1979) Thermal expansion and high-temperature crystal chemistry of the Al_2SiO_5 polymorphs. Am Miner 64:573–586

High-Temperature Crystal Chemistry of Cs- and Sr-Borosilicates

Maria Krzhizhanovskaya, Rimma Bubnova, and Stanislav Filatov

1 Introduction

Although the crystal chemical studies of titled borosilicate system have about 40 year's history, recent interest has been activated by their potential application for producing of low thermal expansion materials ($CsBSi_2O_6$), photo luminescent materials ($SrB_2Si_2O_8$). Since the immobilization of cesium and strontium radioactive isotopes (Cs-137 and Sr-90) are the main problem during the high level radioactive wastes decay the study of phase relations and phase diagrams of Cs and Sr systems are of special interest. Moreover the alkaline borosilicates are widely used now for producing low-temperature cofired ceramic (LTCC) materials for microwave, wireless, sensor and other devices [http://www.ltcc.de/en/publications.php].

In this paper we summarize our data on phase formation, crystal structures, thermal and compositional structural changes together with that of collected to the date by other authors.

2 $Cs_2O-B_2O_3-SiO_2$ System

Up to now the phase diagram of $Cs_2O-B_2O_3-SiO_2$ system was neither reported. The crystal chemistry and thermal phase transformations of Cs-borosilicates are shortly reviewed in Bubnova and Filatov (2008a). Two synthetic compounds $CsBSi_2O_6$ (Richerson and Hummel 1972; Voldan 1979; Mazza and Lucco-Borlera 1997; Huebner et al. 2002; Bubnova et al. 2004) and $CsBSi_5O_{12}$ (Bubnova et al. 2007;

M. Krzhizhanovskaya (✉) • S. Filatov
Department of Crystallography, St. Petersburg State University, St. Petersburg, Russia
e-mail: Krzhizhanovskaya@mail.ru

R. Bubnova
Grebenshchikov Institute of Silicate Chemistry RAS, St. Petersburg, Russia

S.V. Krivovichev (ed.), *Minerals as Advanced Materials II*,
DOI 10.1007/978-3-642-20018-2_26, © Springer-Verlag Berlin Heidelberg 2012

Krzhizhanovskaya et al. 2008) are noted in the system; their crystal structures are described in Mazza and Lucco-Borlera (1997), Huebner et al. (2002), Bubnova et al. (2004) and Krzhizhanovskaya et al. (2008) correspondingly. The data on liquidus line along $CsBO_2$-Cs_2SiO_3 pseudobinary system are presented in Stolyarova et al. (2006).

$CsBSi_2O_6$ stoichiometry. First $CsBSi_2O_6$ compound with a pollucite-like structure has been prepared by solid state reaction as well as by glass crystallization at above 800°C by Richerson and Hummel (1972). It was shown that short heat-treatments at about 1,100°C or longer heat-treatments below 1,000°C resulted in a disappearance of the pollucite-like phase and the formation of new phases (Richerson and Hummel 1972).

In comments of Voldan (1979) it was noted that $CsBSi_2O_6$ had three modifications and the modification similar to the pollucite structure prepared by solid-state reaction was stable above 900°C. Powder pattern of a pollucite-like phase has been presented in Voldan (1979). The pollucite-like structure of $CsBSi_2O_6$ has been refined by the Rietveld method in the *Ia*-3*d* space group (Mazza and Lucco-Borlera 1997; Huebner et al. 2002). Mazza and Lucco-Borlera (1997) have examined also the boron coordination by oxygen atoms using IR spectra: the IR pattern of the $CsBSi_2O_6$ did not reveal any detectable peak in the BO_3 stretching region whereas a wide absorption band was found in BO_4 stretching region (Mazza and Lucco-Borlera 1997). Finally crystal structure of $CsBSi_2O_6$ has been determined by us from single crystal X-ray diffraction in the space group *Ia*-3*d* (a = 13.009(1) Å, Z = 16) (Bubnova et al. 2004). The crystal structure is identical to that of pollucite $CsAlSi_2O_6$ (Inorganic crystal structure database 2010). It is based on a three-dimensional framework of TO_4 (T = Si, B) tetrahedra linked to four-, six- and eightfold rings (Fig. 1a). The Cs atoms are placed within the large cavities in the channels along [111] between sixfold rings. Boron and silicon atoms are statistically distributed over tetrahedral sites. Cs-deficient character of

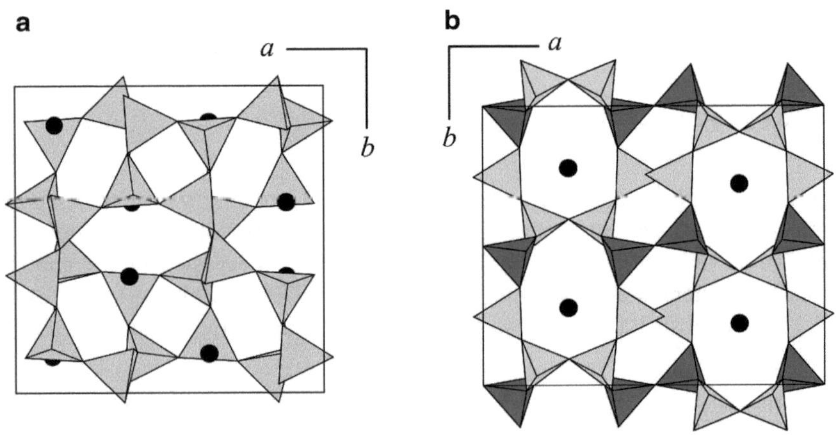

Fig. 1 The projections of crystal structures of cubic $CsBSi_2O_6$ (**a**) and orthorhombic $CsBSi_5O_{12}$ (**b**) along [001] direction. For the $CsBSi_2O_6$ a half of unit cell x/y/0.5z is shown only. The spheres represent cesium atoms. Dark tetrahedra in $CsBSi_5O_{12}$ have only 5% of boron, whereas light tetrahedra have 25% and 12% of B

$CsBSi_2O_6$ crystal structure is reported for boropollucite in Huebner et al. (2002) and Bubnova et al. (2004) with up to 15 and 19 mol.% of vacancies in Cs sites, respectively. In the relative aluminosilcate system Cs-deficient solid solutions $Cs_{1-x}Al_{1-x}Si_{2+x}O_6$ ($x \leq 0.25$) are reported in Kobayashi et al. (2006).

In Huebner et al. (2002) the metastable $CsBSi_2O_6$ modification prepared by glass crystallization for the short time at about 800°C was indexed in orthorhombic symmetry with possible groups *Pmc*2$_1$, *Pma*2 or *Pmma* (a = 6.592(3) Å, b = 11.825 (5) Å and c = 12.620(6) Å). The DSC investigation (Huebner et al. 2002) has shown that the orthorhombic phase is transformed irreversibly into the cubic boroleucite phase at 850°C.

Recently another polymorph of $CsBSi_2O_6$, mineral kirchhoffite, (tetragonal, *I*4$_1$*/acd*; a = 13.019(2), c = 12.900(2) Å) was found in Darai-Pioz alkaline massif, Tadzhikistan (Agakhanov et al. 2010). Interesting to note that cubic $CsBSi_2O_6$ was synthesized several times in a wide range of temperature and compositions using both solid state reaction and glass crystallization by many authors and we were neither able to find the notification on its tetragonal modification.

Although in nature Cs-K substitution is rare observed, we have studied cubic ↔ cubic structural transformation of *Ia-3d* ($CsBSi_2O_6$) ↔ *I-43d* ($KBSi_2O_6$) under the Cs-K substitutions (Bubnova et al. 2010; Derkacheva et al. 2011) as well as *Ia-3d* ($CsBSi_2O_6$) ↔ *I-43d* ($RbBSi_2O_6$) transformation under the Cs-Rb (Krzhizhanovskaya et al. 2006, 2007) (Fig. 2). A narrow range of immiscibility (Fig. 2) in both solution series has been revealed using the full profile analysis.

Thermal phase transformations of synthetic $CsBSi_2O_6$ compound were investigated using DTA, DSC and TG, annealing at different temperatures and high-temperature X-ray powder diffraction methods in Bubnova et al. (2004, 2007). In Bubnova et al. (2004) using DTA, TG and wet chemical analysis it was shown that cubic $CsBSi_2O_6$ loses mass before melting. In the range of 1,030–1,080°C it decomposes to form a new phase enriched in silica with approximate composition $CsBSi_3O_8$. Later in Bubnova et al. (2007) it was shown that the composition of this resulting phase is corresponded to the $CsBSi_5O_{12}$ and the decomposition scheme could be written as follows: $2CsBSi_2O_6 \rightarrow CsBSi_5O_{12} + CsBO_2\uparrow$.

$CsBSi_5O_{12}$ stoichiometry. In Bubnova et al. (2007) and Krzhizhanovskaya et al. (2008) the formation and decomposition of $CsBSi_5O_{12}$ zeolite like phase

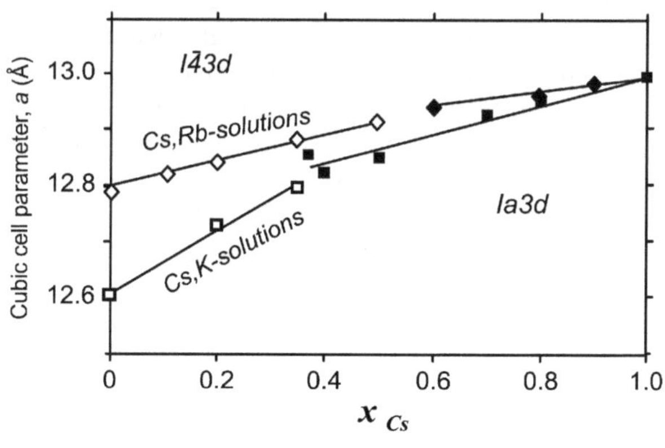

Fig. 2 Cubic cell parameter vs chemical composition for $(Cs,K)BSi_2O_6$ and $(Cs,Rb)BSi_2O_6$ solid solutions

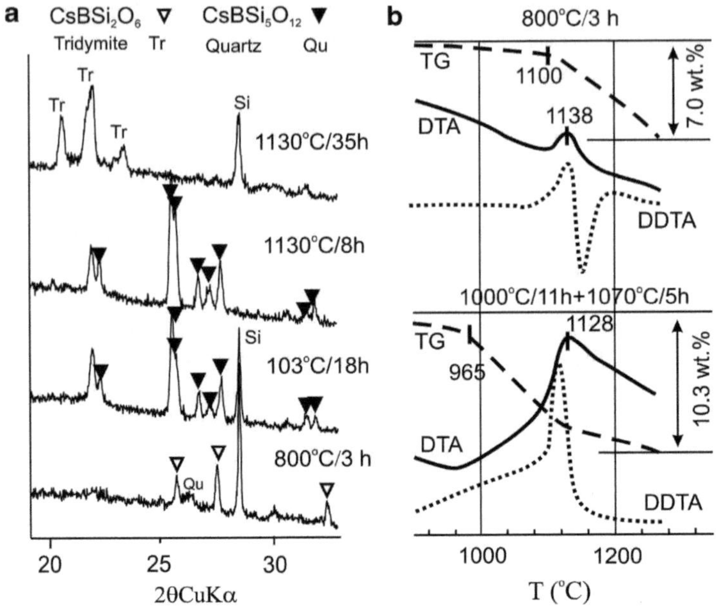

Fig. 3 XRD (**a**) and DTA study (**b**) of $CsBSi_5O_{12}$ high-temperature behavior

isostructural to $CsAlSi_5O_{12}$ (Inorganic crystal structure database 2010) was studied by performing annealing series in the temperature range from 800°C up to 1,100°C (Fig. 3a). According to these data the following consequence of transformations takes place during the solid-state synthesis and subsequent heat treatments of stoichiometric mixture. First boropollucite $CsBSi_2O_6$ crystallises as an intermediate product of the solid state reaction. $CsBSi_5O_{12}$ forms above 900°C and decomposes continuously above 1,070°C via liberating a gas phase. After heat treatment at 1,100°C for 35 h the XRD pattern contained the diffraction maxima of SiO_2 phase only, with no traces of $CsBSi_5O_{12}$ being left (Fig. 3a).

The DTA study as well as heat treatment investigations showed that the $CsBSi_5O_{12}$ decomposes in the solid state at about 1,100°C according to the reaction $CsBSi_5O_{12} \rightarrow 5SiO_2 + CsBO_2\uparrow$ (Fig. 3b). The mechanism of the $CsBSi_5O_{12}$ decomposition with formation of a gas phase and a solid phase is similar to that of $CsBSi_2O_6$, however, the latter compound decomposes at lower temperature.

The crystal structure of $CsBSi_5O_{12}$ (Krzhizhanovskaya et al. 2008) was refined by Rietveld method from X-ray powder diffraction data using the model of isostructural aluminosilicate $CsAlSi_5O_{12}$ (CAS zeolite type). $CsBSi_5O_{12}$ is orthorhombic, *Ama*2, Z = 4, a = 16.2312(5), b = 13.3655(4), c = 4.8733(1) Å. In the three dimensional framework tetrahedra share corners to form five- and eightfold rings, which define infinite channels along the [001] direction and measuring approximately 7 × 5 Å in cross-section. The cesium cations are located in these channels (Fig. 1b). The boron and silicon atoms in the tetrahedral framework are distributed not ideally random over the tetrahedral positions (Fig. 1b). Due to a cesium deficiency the approximate chemical formula of $CsBSi_5O_{12}$ could be assigned as $Cs_{0.9}B_{0.9}Si_{5.1}O_{12}$.

Thermal expansion of Cs borosilicates. Using the high-temperature X-ray powder diffraction (HTXRD) data the average linear coefficients of thermal expansion (α) of $CsBSi_2O_6$ are determined within ranges 20–500°C and from 500°C to 730°C, α are equal to 2×10^{-6} and $5 \div 8 \times 10^{-6}$°C^{-1}, respectively (Bubnova et al. 2004). Low coefficient of thermal expansion is characteristic for boropollucite at low temperatures. The substitution of cesium with potassium or rubidium leads to increase of thermal expansion of boropollucite structure (Bubnova et al. 2010; Krzhizhanovskaya et al. 2007).

The anisotropic character of the thermal expansion of $CsBSi_5O_{12}$ with maximum expansion along the infinite channels revealed by HTXRD is typical for tunnel structures. The thermal expansion coefficients are $\alpha_a = 4 \times 10^{-6}$°C^{-1}, $\alpha_b = -1.43 \times 10^{-6} + 44.485 \times 10^{-9} \times T$ ($\alpha_b = -1, 12$ and 25×10^{-6}°C^{-1} at 20°C, 350°C and 700°C correspondingly), $\alpha_c = 20 \times 10^{-6}$°C^{-1}, $\alpha_V = 23.4 \times 10^{-6} + 44.066 \times 10^{-9} \times T$ ($\alpha_V = 20$, 39 and 54×10^{-6}°C^{-1} at 20°C, 350°C and 700°C, correspondingly). The maximal direction of thermal expansion in the structure coincides with the direction [001] of the infinite channels, where the cesium atoms form Cs-columns with Cs-Cs distances of 4.9 Å. On heating, the magnitude of thermal vibration rises resulting in expansion of the structure along the [001] direction. The volume expansion of $CsBSi_5O_{12}$ is lower than that of isotypical $CsAlSi_5O_{12}$: α_v for aluminosilicate is about 65 (Fisch et al. 2008) and α_v for borosilicate is about 50×10^{-6}°C^{-1}.

3 $SrO–B_2O_3–SiO_2$ System

In the SrO-B_2O_3-SiO_2 system, glass forming and liquid immiscibility regions have been studied (Baylor and Brown 1976a; Turnina et al. 2010). First the subsolidus relations in $SrO–B_2O_3–SiO_2$ system were investigated in 1976 (Baylor and Brown 1976b): a stable $Sr_3B_2SiO_8$ compound and a metastable $SrB_2Si_2O_8$ were reported and unindexed XRD data were presented. The structure of $SrB_2Si_2O_8$ was solved 20 years later in Berger and Range (1996) and that of $Sr_3B_2SiO_8$ was done recently by us (Krzhizhanovskaya et al. 2010).

$SrB_2Si_2O_8$ stoichiometry. The structure of synthetic $SrB_2Si_2O_8$ isotypical to danburite was solved in Berger and Range (1996). It is orthorhombic, *Pnma*, $Z = 4$, $a = 8.1891(5)$, $b = 7.9311(6)$, $c = 8.9284(5)$ Å. The three dimensional framework is based on SiO_4 and BO_4 tetrahedra forming four-, six- and eightfold rings. Sr atoms are located in columns along [010] direction in large infinite channels built up by eightfold rings (Fig. 4). Not long ago naturally occurring Sr-dominant analogue of danburite (pekovite, $SrB_2Si_2O_8$) was found in alkaline massif and structurally characterized in Pautov et al. (2004).

$Sr_3B_2SiO_8$ stoichiometry. Single crystals of $Sr_3B_2SiO_8$ were obtained by solid state reaction of stoichiometric mixture at 1,200°C (Krzhizhanovskaya et al. 2010). The crystal structure is orthorhombic, *Pnma*, $a = 12.361(4)$, $b = 3.927(1)$,

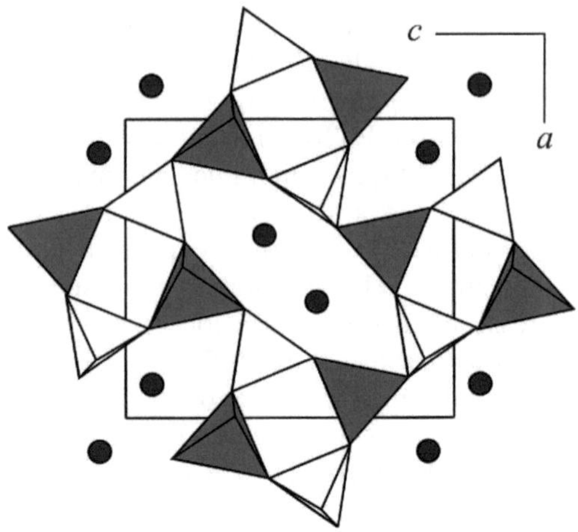

Fig. 4 Crystal structure of $SrB_2Si_2O_8$ (Berger and Range 1996). Dark polyhedra are corresponded to SiO_4 tetrahedra, white – to BO_4

$c = 5.419(1)$ Å, $V = 263.05(11)$ Å^3. The structure contains zigzag pseudo-chains running along the *b* axis and built up from corner sharing (Si,B)–O polyhedra (Fig. 5). Boron and silicon are statistically distributed over one site with their coordination strongly disordered. Sr atoms are located between the chains providing three-dimensional linkage of the structure.

The XRD patterns after the annealing stages are presented in Fig. 6. During the solid state reaction, borate phases crystallize first as intermediate products. In particular, $Sr_3B_2O_6$ crystallizes with admixture of $Sr_2B_2O_5$ and traces of Sr_2SiO_4 by keeping the starting mixture 900°C for 3 h (Fig. 6, bottom XRD pattern). After the following short treatment at 1,000°C for 3 h the intensity of $Sr_3B_2O_6$ peaks decreases; weak traces of $Sr_3B_2SiO_8$ appeared. Long-term treatment of the initial mixture performed at 900°C for 3 days (72 h) showed that the title compound starts to form at this temperature. The broadened maxima of this phase could be identified, although they are remarkably shifted in 2θ from the theoretical positions calculated from structural data (Fig. 6), which possibly indicates a non-stoichiometric nature of the phase. The approximate unit cell dimensions calculated for this phase treated at lower temperatures are less than that of after the treatment at 1,100°C. It is very likely the evidence that the phase forms by the step-by-step process. The decrease of the cell parameters indicates that SiO_2-poor phase crystallizes first and then gradually transforms to $Sr_3B_2SiO_8$. Over 1,100°C, the probe contains pure $Sr_3B_2SiO_8$; the unit cell parameters do not change after the treatment at 1,200°C. Below 900°C the compound is unstable.

The conclusions mentioned above are confirmed by occurrence of solid solutions in the pseudobinary $Sr_2B_2O_5$–$Sr_3B_2SiO_8$ system. Thus a new series of solid solutions based on $Sr_3B_2SiO_8$ have been crystallized from melt in the pseudobinary $Sr_2B_2O_5$–$Sr_3B_2SiO_8$ system (Krzhizhanovskaya et al. 2010).

The thermal behaviour of $Sr_3B_2SiO_8$ was investigated using powder high-temperature X-ray diffraction (HTXRD) in the temperature range 20–900°C (Krzhizhanovskaya et al. 2010). The anisotropic character of thermal expansion has

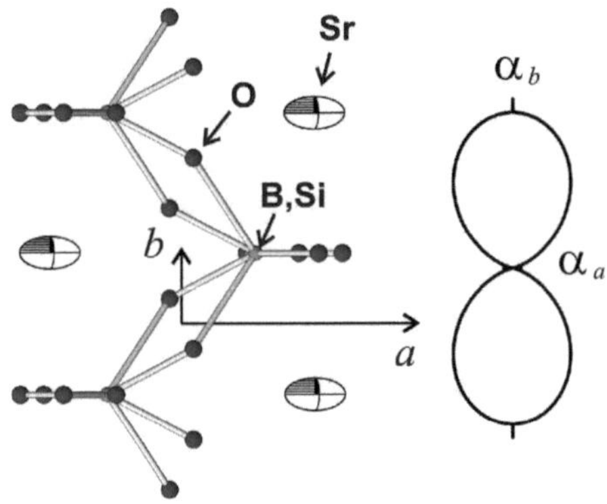

Fig. 5 A fragment of $Sr_3B_2SiO_8$ structure showing isolated B,Si-O pseudo-chain and Sr atom ellipsoids comparing with pole figure of thermal expansion coefficients (Krzhizhanovskaya et al. 2010)

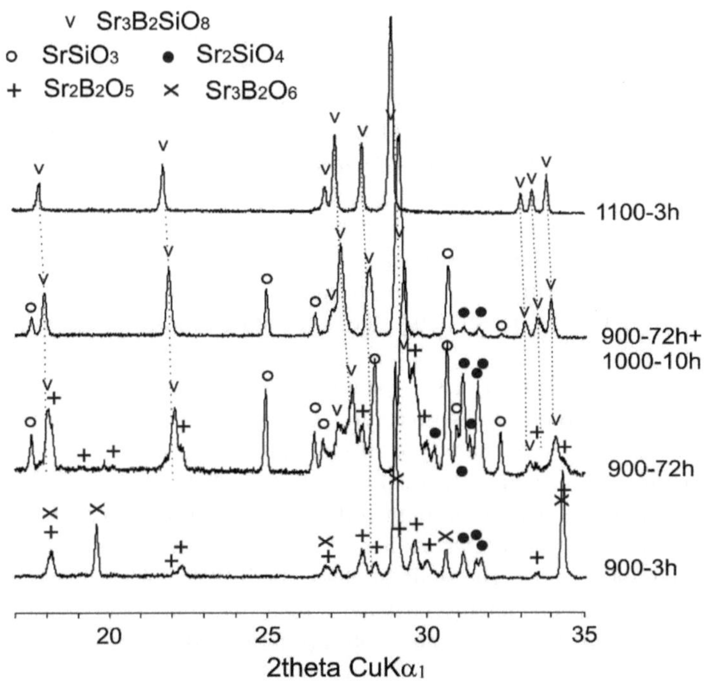

Fig. 6 Formation of $Sr_3B_2SiO_8$ during solid state reaction

been observed: $\alpha_a = -1.3$, $\alpha_b = 23.5$, $\alpha_c = 13.9$, and $\alpha_V = 36.1 \times 10^{-6}$ °C^{-1} (25 °C); $\alpha_a = -1.3$, $\alpha_b = 23.2$, $\alpha_c = 5.2$ and $\alpha_V = 27.1 \times 10^{-6}$ °C^{-1} (650°C). Maximal thermal expansion of the structure along of the chain direction [010] (Fig. 5) is caused by the partial straightening of chain zigzag. The chain expansion is governed by the so- called hinge mechanism (Bubnova and Filatov 2008a, b) – if the structure expands along chain axis [010] ($\alpha_b = 23.5 \times 10^{-6}$°C^{-1}), it has to contract in the perpendicular plane *ac* ($\alpha_a = -1.3 \times 10^{-6}$°C^{-1}). The maximal expansion along the chain direction could be explained by relatively weak bonding caused by incomplete occupancies of bridging oxygen sites. Another reason for straightening of the chain zigzag could be strong vibrations of Sr atoms in the *a* direction (Fig. 5). The volumetric expansion coefficient is practically the same as the average one for borosilicates (36×10^{-6}°C^{-1}) and slightly higher than that of silicates (23×10^{-6}°C^{-1}) (Bubnova and Filatov 2008a).

4 Conclusions

Up to now two compounds $CsBSi_2O_6$ and $CsBSi_5O_{12}$ are presented in the Cs_2O-B_2O_3-SiO_2 system. $CsBSi_2O_6$ has three polymorphs. Cubic *Ia*-3*d* modification isostructural to cubic pollucite has the lowest thermal expansion among borosilicates. Its crystal chemistry, structural deformations on temperature and cationic substitutions are studied in details. The crystal structure of another low temperature orthorhombic polymorph is not studied jet. The third polymorph, naturally occurring mineral kirchhoffite, presumably isotypical to tetragonal pollucite, is reported in tetragonal $I4_1/acd$ space group. The structure of zeolite-like phase $CsBSi_5O_{12}$ is isostructural with $CsAlSi_5O_{12}$.

Comparison of the coefficients of thermal expansion for $CsBSi_5O_{12}$ and $CsAlSi_5O_{12}$ evidenced that the exchange of aluminum for boron in the structure leads to lowering of thermal expansion. Both $CsBSi_2O_6$ and $CsBSi_5O_{12}$ compounds are unstable over 1,000°C presumably due to the ability of the alkali cation to leave the structure.

In the moment in SrO-B_2O_3-SiO_2 system two compounds are known: tetrahedral framework phase $SrB_2Si_2O_8$, isostructural to danburite and a high-temperature compound $Sr_3B_2SiO_8$ with chain type of structure. There are the data on mineral, pekovite, having the structure identical to that of synthetic $SrB_2Si_2O_8$.

Comparing the data mentioned for the titled systems one could conclude that there are still lacunas in the area. Our studies of Cs_2O-B_2O_3-SiO_2 and SrO-B_2O_3-SiO_2 system are in progress now. The study of phase stability and phase diagrams of these systems remains of special interest because of their potential usage for the immobilization of cesium and strontium radioisotopes (Cs-137 and Sr-90).

Acknowledgments The studies are supported by Russian Foundation for Basic Research (project # 10-03-00732).

References

Agakhanov AA, Pautov LA, Sokolova E, Hawthorne FC, Karpenko VYu (2010) Kirchhoffite, IMA 2009–094. CNMNC Newsletter. Miner Mag 74:375–377

Baylor R, Brown JJ (1976a) Phase separation of glasses in the system SrO-B_2O_3-SiO_2. J Am Ceram Soc 59:131–136

Baylor R, Brown JJ (1976b) Subsolidus phase equilibria in the system $SrO\text{-}B_2O_3\text{-}SiO_2$. J Am Ceram Soc 59:21–23

Berger T, Range KJ (1996) Hochdrucksynthese und strukturverfeinerung von strontium-Danburit, $SrB_2Si_2O_8$. Z Naturforsch B 51:172–174

Bubnova RS, Derkacheva ES, Filatov SK, Ugolkov VL (2010) Thermal expansion and phase transitions in $K_{1-x}Cs_xBSi_2O_6$ borosilicate solid solutions. Glass Phys Chem 36:61–69

Bubnova RS, Filatov SK (2008a) High-temperature crystal chemistry of borates and borosilicates. Nauka, St. Petersburg (in Russian)

Bubnova RS, Filatov SK (2008b) Strong anisotropic thermal expansion in borates. Phys Status Solidi B 245:2469–2476

Bubnova RS, Krzhizhanovskaya MG, Filatov SK, Ugolkov VL, Paufler P (2007) XRD and DSC study of the formation and the melting of a new zeolite like borosilicate $CsBSi_5O_{12}$ and $(Cs,Rb)BSi_5O_{12}$ solid solutions. Z Kristallogr 222:83–88

Bubnova RS, Levin AA, Stepanov NK, Meyer DC, Filatov SK, Paufler P (2004) Crystal structure and thermal behaviour of boropollucite $CsBSi_2O_6$. Solid State Sci 6:629–637

Derkacheva ES, Krzhizhanovskaya MG, Bubnova RS, Filatov SK (2011) Structural transformations in $K_{1-x}Cs_xBSi_2O_6$ borosilicate solid solutions. Glass Phys Chem 37, Issue 5

Fisch M, Armbruster Th, Kolesov B (2008) Temperature-dependent structural study of microporous $CsAlSi_5O_{12}$. J Solid State Chem 181:423–431

Huebner R, Belger IA, Meyer DC, Paufler P, Polyakova IG (2002) Crystallisation of caesium borosilicate glasses with approximate boroleucite composition. Z Kristallogr 217:223–232

Inorganic crystal structure database (2010) Fachinformationszentrum. Karlsruhe, Germany

Kobayashi H, Sumino S, Tamai S, Yanase I (2006) Phase transition and lattice thermal expansion of Cs-deficient pollucite, $Cs_{1-x}Al_{1-x}Si_{2+x}O_6$ ($x \leq 0.25$), compounds. J Am Ceram Soc 89:3157–3161

Krzhizhanovskaya MG, Bubnova RS, Depmeier W, Filatov SK, Ugolkov VL (2008) Crystal structure and thermal behavior of a new borosilicate with the CAS framework type. Micropor Mesopor Mat 116:569–574

Krzhizhanovskaya MG, Bubnova RS, Filatov SK, Meyer DC, Paufler P (2006) Crystal structure and thermal behaviour of (Rb, Cs)BSi2O6 solid solutions. Cryst Res Technol 41:285–292

Krzhizhanovskaya MG, Bubnova RS, Krivovichev SV, Belousova OL, Filatov SK (2010) Synthesis, crystal structure and thermal behavior of $Sr_3B_2SiO_8$ borosilicate. J Solid State Chem 183:2352–2357

Krzhizhanovskaya MG, Bubnova RS, Ugolkov VL, Filatov SK (2007) Thermal expansion and polymorphism in a series of rubidium cesium boroleucites. Glass Phys Chem 33:242–249

Mazza D, Lucco-Borlera M (1997) On the substitution of Fe and B for Al in the pollucite $(CsAlSi_2O_6)$ structure. J Eur Ceram Soc 17:1767–1772

Pautov LA, Agakhanov AA, Sokolova EV, Hawthorne FC (2004) Maleevite, $BaB_2Si_2O_8$, and pekovite $SrB_2Si_2O_8$, new mineral species from the Dara-i-Pioz alkaline massif, northern Tajikistan: description and crystal structure. Can Miner 42:107–119

Richerson DW, Hummel FA (1972) Synthesis and thermal expansion of polycrystalline cesium minerals. J Am Ceram Soc 55:269–273

Stolyarova VL, Lopatin SI, Belousova OL, Grishchenko LV (2006) Phase equilibria and thermodynamic properties of components in the $Cs_2O\text{-}B_2O_3\text{-}SiO_2$ system at high temperatures. Glass Phys Chem 32:55–62

Turnina NG, Belousova OL, Domanskii AI, Doronina LA, Ugolkov VL (2010) Glass formation region and order of formation of crystalline phases in the $Sr_2O\text{-}B_2O3\text{-}SiO_2$ system. Glass Phys Chem 36:294–303

Voldan J (1979) Crystallisation of a three-component compound in the system K2O–B2O3—SiO2. Silikaty 23:133–141

Iron-Manganese Phosphates with the Olivine – and Alluaudite-Type Structures: Crystal Chemistry and Applications

Frédéric Hatert

1 Introduction

Iron-manganese phosphates are common accessory minerals occurring in granitic pegmatites, in metamorphic rocks, and in meteorites. In rare-elements pegmatites, primary phosphates of the triphylite-lithiophilite series [$Li(Fe^{2+},Mn^{2+})(PO_4)$-$Li(Mn^{2+},Fe^{2+})(PO_4)$] form masses that can reach several meters in diameter, enclosed in silicates. During the oxidation processes affecting the pegmatites, these olivine-type phosphates progressively transform to ferrisicklerite-sicklerite [$Li_{1-x}(Fe^{3+},Mn^{2+})(PO_4)$-$Li_{1-x}(Mn^{2+},Fe^{3+})(PO_4)$] and to heterosite-purpurite [$(Fe^{3+},Mn^{3+})(PO_4)$-$(Mn^{3+},Fe^{3+})(PO_4)$], according to the substitution mechanism $Li^+ + Fe^{2+} \leftrightarrow \square + Fe^{3+}$. This oxidation sequence was first observed by Quensel (1937), and then confirmed by Mason (1941).

Minerals of the alluaudite group, with ideal chemical compositions ranging from $Na_2Mn(Fe^{2+}Fe^{3+})(PO_4)_3$ to $\square NaMnFe_2^{3+}(PO_4)_3$, are generally produced from primary triphylite-lithiophilite, by oxidation coupled with a $Li \leftrightarrow Na$ metasomatic exchange (Moore 1971). More recently, the existence of primary alluaudites was confirmed by Fransolet et al. (1997, 1998, 2004), thus indicating that the metasomatic replacement process proposed by Moore (1971) cannot be generalized. As observed for phosphates of the triphylite-lithiophilite series, primary alluaudites, which are weakly oxidized, progressively transform into oxidized secondary alluaudites, according to the substitution mechanism $Na^+ + Fe^{2+} \leftrightarrow \square + Fe^{3+}$ (Mason 1941; Fransolet et al. 1985, 1986, 2004). This mechanism explains the transformation of hagendorfite [$Na_2MnFe^{2+}Fe^{3+}(PO_4)_3$] into alluaudite [$\square NaMnFe_2^{3+}(PO_4)_3$], and of ferrohagendorfite [$Na_2Fe_2^{2+}Fe^{3+}(PO_4)_3$] into ferroalluaudite [$\square NaFe^{2+}Fe_2^{3+}(PO_4)_3$].

The past 20 years have seen an increasing number of publications devoted to olivine – and alluaudite-type phosphates. On the one hand, solid-state synthesis

---

F. Hatert (✉)
Laboratory of Mineralogy B18, University of Liège, B-4000 Liège, Belgium
e-mail: fhatert@ulg.ac.be

S.V. Krivovichev (ed.), *Minerals as Advanced Materials II*,
DOI 10.1007/978-3-642-20018-2_27, © Springer-Verlag Berlin Heidelberg 2012

experiments, coupled with structural and spectroscopic investigations, have shown an unexpected flexibility for the alluaudite structure (Hatert et al. 2000; Hatert 2004, 2008), which can accommodate divalent and trivalent cations in its octahedral sites, as well as large monovalent cations in its channels parallel to the *c* axis. These properties will, no doubt, permit the use of alluaudite-type phosphates for practical applications, such as corrosion inhibition, passivation of metal surfaces, and catalysis (Korzenski et al. 1998; Kacimi et al. 2005). On the other hand, the exceptional performance of triphylite as cathode material for Li-ion batteries has initiated an increasing interest for these phosphates, which are now produced industrially and used in batteries for many applications such as electric cars, electric bicycles, or for the storage of green energy. This chapter will summarize the crystal-chemical features of alluaudite- and olivine-type phosphates, and will present their unusual properties and applications.

2 Olivine-Type Phosphates

The most common olivine-type phosphates, observed in granitic pegmatites, are triphylite and lithiophilite (triphylite: $a = 4.690$, $b = 10.286$, $c = 5.987$ Å, *Pbnm*). Their crystal structure has been investigated from synthetic samples (Geller and Durand 1960; Yakubovich et al. 1977) and natural minerals (Finger and Rapp 1969; Losey et al. 2004; Fehr et al. 2007), and is characterized by two chains of edge-sharing octahedra parallel to the *c* axis. The first chain is constituted by the *M*(1) octahedra occupied by Li, while the second chain is formed by the *M*(2) sites occupied by Fe and Mn (Fig. 1). The chains are connected in the *b* direction by sharing edges of their octahedral sites, and the resulting planes are connected in the *a* direction by the PO_4 tetrahedra. Natrophilite, $NaMnPO_4$, is another pegmatite

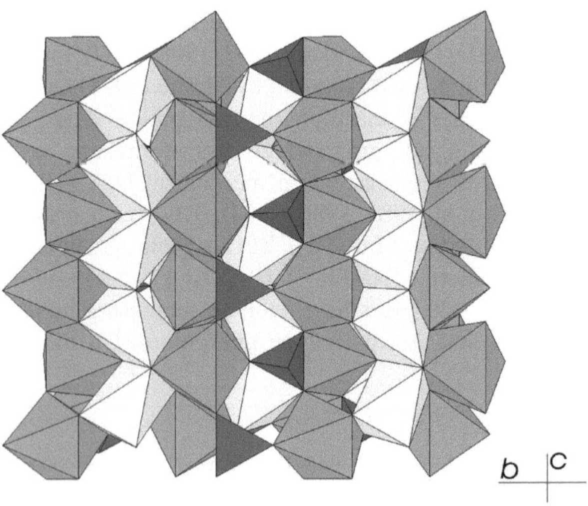

Fig. 1 The crystal structure of triphylite projected along the *a* axis. The *M*(1) octahedra are *white*, the *M*(2) octahedra are *light grey*, and the PO_4 tetrahedra are *dark grey*

phosphate with the olivine structure, in which the *M*(1) site is occupied by Na while the *M*(2) site contains the smaller divalent cations (Moore 1972).

The crystal structure of natural triphylite was solved by Finger and Rapp (1969), who confirmed that the phosphate is isostructural with minerals of the olivine group. Fransolet et al. (1984) investigated twenty natural and five synthetic olivine-type phosphates of the triphylite-lithiophilite series by X-ray powder diffraction, and established correlations between the unit-cell parameters and the chemical composition. More recently, Losey et al. (2004) determined the structural variations along the triphylite-lithiophilite series, and showed that the bond lengths involving *M*(1) and *M*(2) decrease as the concentration of Fe increases, with the exception of the *M*(1)-O(3) bond. An examination of numerous olivine-type structures also demonstrates that if the size difference in the radius of the cations at the *M*(1) and *M*(2) sites is larger than 0.17 Å, then the distortion is greater in the octahedron that is occupied by the larger cation (Losey et al. 2004).

A detailed mineralogical study of ferrisicklerite from Sidi Bou Othmane, Morocco, was performed by Fontan et al. (1976), who indicated a partial replacement of (PO_4) groups by (H_4O_4) groups. The crystal structure of the same sample showed the presence of Li and Na on the *M*(1) site, and of Mn^{2+} and Fe^{3+} on the *M*(2) site (Alberti 1976). In the crystal structure of heterosite, Eventoff and Martin (1972) noticed that the *M*(1) site was empty, and that the complete oxidation of iron and manganese induced a Jahn-Teller distortion of the $M(2)O_6$ octahedron, characterized by two short bonds (1.912–1.914 Å) and four long bonds (2.030–2.163 Å). Mössbauer spectral investigations of natural olivine-type phosphates show that triphylite does not contain any trace of Fe^{3+}, whereas ferrisicklerite and purpurite contain 17 and 6% Fe^{2+}, respectively (Li and Shinno 1997; Fehr et al. 2007).

In order to shed some light on the structural modifications induced by the oxidation processes affecting lithiophilite, Hatert (2010) investigated a natural sample from the Altai Mountains, China, in which a progressive transition from lithiophilite to sicklerite is observed. Under the polarizing microscope, lithiophilite is colorless, whereas sicklerite shows a deep orange color. Several grains also show intermediate colors, suggesting a progressive transition from lithiophilite to sicklerite. This progressive transition if confirmed by SIMS (Secondary Ion Mass Spectrometry) analyses, which indicate Li values from 0.96 to 0.69 Li atoms per formula unit (*p.f.u.*). Five single-crystals, corresponding to zones with different colors, were extracted from the sample and investigated by single-crystal X-ray diffraction techniques. The samples are orthorhombic, space group *Pbnm*, with unit-cell parameters ranging from $a = 4.736(1)$, $b = 10.432(2)$, $c = 6.088(1)$ Å (lithiophilite) to $a = 4.765(1)$, $b = 10.338(2)$, $c = 6.060(1)$ Å (sicklerite). The structures were refined to R_1 values ranging from 2.19% to 2.94%, and show that Li occurs on the *M*(1) site, with occupancy factors from 0.99 Li *p.f.u.* (lithiophilite) to 0.75 Li *p.f.u.* (sicklerite). These values are in good agreement with the values measured by SIMS.

Hatert et al. (2009) reported SIMS analyses of six natural triphylite samples, which show lithium contents from 9.51 to 9.88 wt.% Li_2O, while the analyses of four lithiophilites show higher Li_2O contents ranging from 10.23 to 11.15 wt.% (Fig. 2).

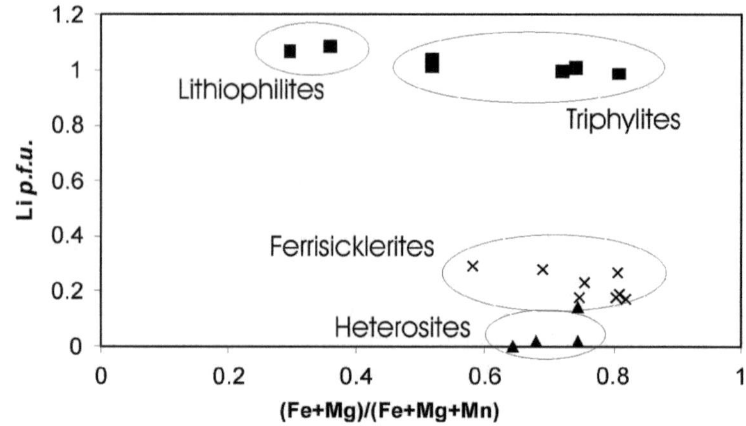

Fig. 2 Diagram showing the chemical composition of natural olivine-type phosphates

These compositions correspond to 0.99–1.04 Li *a.p.f.u.* in triphylites, and 1.07–1.15 Li *a.p.f.u.* in lithiophilites. The significant Li enrichment of lithiophilites indicates that Li can also occur in the *M*(2) site of the olivine structure. Eleven ferrisicklerite samples show Li_2O-contents from 1.65 to 2.84 wt.% (= 0.17–0.29 *a.p.f.u.*), and three heterosite samples contain 0.16–0.21 wt.% Li_2O (= 0.02 *a.p.f.u.*) (Hatert et al. 2009). The presence of significant amounts of Li in heterosites was unexpected, and the low Li content of ferrisicklerites indicates that trivalent manganese also occurs in this mineral. The formula of Li-poor ferrisicklerite corresponds to $Li_{0.17}(Fe^{3+}{}_{0.75}\ Mn^{3+}{}_{0.10}\ Mn^{2+}{}_{0.08}\ Mg_{0.06})(PO_4)$.

In the literature are present many synthetic phosphates with the olivine structure, among which $LiCo(PO_4)$ (Kubel 1994; Bramnik et al. 2004; Penazzi et al. 2004), $LiNi(PO_4)$ (Warda and Lee 1998; Abrahams and Easson 1993), $LiMg(PO_4)$ (Hanic et al. 1982), and $NaCd(PO_4)$ (Ivanov et al. 1974). More recently, Hatert et al. (2010) investigated the alluaudite + triphylite assemblage by hydrothermal technique, and observed a significant amount of Na in triphylite. This amount evolves from 0.04 to 0.08 wt.% Na_2O at 400°C, to 1.25–1.58 wt.% Na_2O at 800°C, according to the substitution mechanism Na → Li. This first experimental evidence of Na incorporation into triphylite is not really surprising, since the *M*(1) site of the olivine structure can be completely filled by Na in natrophilite (Moore 1972).

As shown on Fig. 3, we attempted to correlate the unit-cell parameters of Li-bearing, vacancy-free olivine-type phosphates, with the mean ionic radius of the cations occurring on the *M*(2) site. The correlations are excellent and generally follow Vegard's law, except for the *a* parameter which has been fitted with a second order polynomial curve. These correlations can be used to predict the unit-cell parameters of olivine-type phosphates, but also to estimate the Fe/(Fe + Mn) ratio within the triphylite-lithiophilite solid solution. Starting from the unit-cell parameters published by Fransolet et al. (1984) and by Losey et al. (2004) on natural minerals of the triphylite group, we estimated the Fe/(Fe + Mn) ratio according to

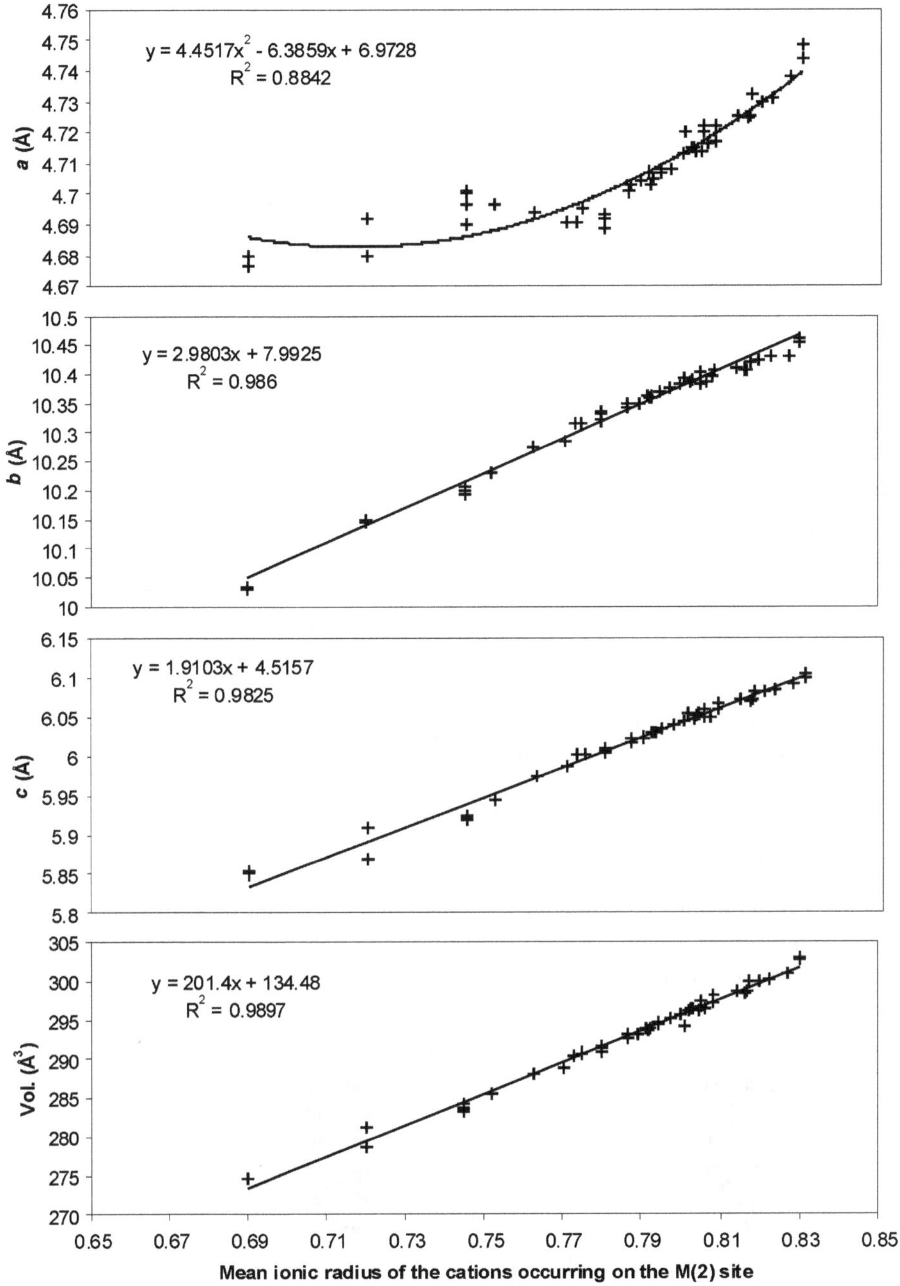

Fig. 3 Variations of the unit-cell parameters in natural and synthetic olivine-type phosphates

the equations of Fig. 3 for *b*, *c*, and *V*. The results are satisfactory when the Mg content of the phosphates is lower than 0.016 *a.p.f.u.*, as shown by a Fe/(Fe + Mn) ratio estimated with an accuracy of +/− 7%. However, an increase of the Mg content induces a significant overestimation of the Fe/(Fe + Mn) ratio, which is characterized by an error reaching 35% for a Mg content of 0.23 *a.p.f.u.*

Since Padhi et al. (1997) reported the reversible electrochemical extraction of lithium from $LiFePO_4$, the olivine-type phosphates $LiMPO_4$ (M = Fe, Mn, Co, Ni) have received a strong attention as candidates for lithium batteries (*e.g.* Andersson et al. 2000; Okada et al. 2001; Ravet et al. 2003; Prosini et al. 2002; Song et al. 2002; Takahashi et al. 2002; Yang et al. 2003; Bramnik et al. 2004; Deniard et al. 2004; Fehr et al. 2007). The number of publications devoted to these compounds increases linearly since 2001, and has reached 300 publications per year in 2010. Recently, Kang and Ceder (2009) showed that $LiFePO_4$-based batteries can achieve ultrafast charging and discharging in 10–20 s, thus reaching the performance of supercapacitors.

The principle of a $LiFePO_4$-based battery is shown on Fig. 4. During the charging process (Fig. 4a), electrons are removed from the triphylite electrode, and Li^+ is consequently extracted from this electrode to maintain charge balance. Li^+ then migrates through an electrolyte to reach a metallic Li electrode. It is noteworthy that this Li^+ extraction process provokes a progressive oxidation of triphylite into heterosite during the charge. During discharge (Fig. 4b), the process is inverted, and electrons are extracted from the metallic Li electrode. Li^+ then migrates to the heterosite electrode, which progressively transforms again into triphylite.

The performance of these $LiFePO_4$-based batteries is exceptional, with a potential of 3.4 V, a very high capacity around 160–170 mAh/g, and a good resistance to cycling. Moreover, this electrode material can be manufactured at low cost from abundant chemicals, and is characterized by a very low toxicity. These features make of $LiFePO_4$ the best candidate for producing batteries for many applications, such as electric bicycles, electric boats, electric cars, or for the storage of green energy. These recent technological advances, which will significantly contribute to improve our future, are one of the best examples of how mineralogical sciences can stimulate the development of new advanced materials.

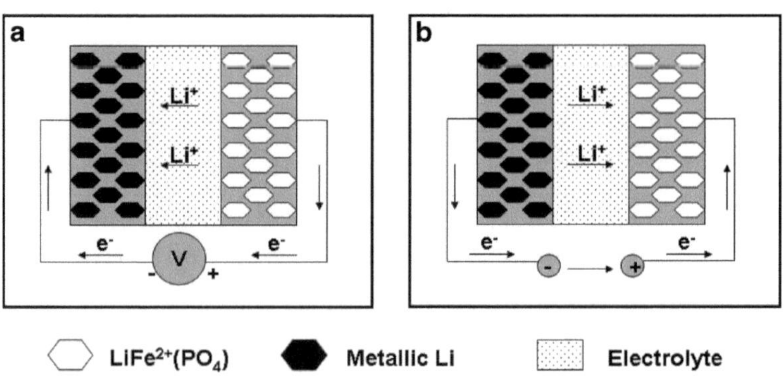

Fig. 4 Principle of a $LiFePO_4$-based battery during charge (**a**) and discharge (**b**)

3 Alluaudite-Type Phosphates

Phosphates of the alluaudite group occur as accessory minerals in granitic pegmatites, particularly in the beryl-columbite-phosphate subtype of the rare-element pegmatites. By using a single-crystal from the Buranga pegmatite, Rwanda, Moore (1971) determined the crystal structure of alluaudite in the monoclinic $C2/c$ space group ($a = 12.004(2)$, $b = 12.533(4)$, $c = 6.404(1)$ Å, $\beta = 114.4(1)°$), and proposed the general structural formula $X(2)X(1)M(1)M(2)_2(PO_4)_3$, with $Z = 4$. The structure consists of kinked chains of edge-sharing octahedra stacked parallel to {101}. These chains are formed by a succession of $M(2)$ octahedral pairs linked by highly distorted $M(1)$ octahedra. Equivalent chains are connected in the b direction by the $P(1)$ and $P(2)$ phosphate tetrahedra to form sheets oriented perpendicular to [010]. These interconnected sheets produce channels parallel to the c axis, channels which contain the distorted cubic $X(1)$ site and the four-coordinated $X(2)$ site (Fig. 5).

According to Moore (1971), the cations are distributed among the different crystallographic sites as a function of their ionic radii. Accordingly, the large $X(2)$ site contains Na, K and vacancies; $X(1)$ contains Na, Mn and Ca; $M(1)$ contains Mn and Fe^{2+}; and the small $M(2)$ site contains Fe^{3+}, Fe^{2+}, Mn, Mg and Li. Because Mn dominates on the $M(1)$ site, and Fe^{2+} and Fe^{3+} dominate on the $M(2)$ site, Moore (1971) proposed the ideal formula $Na_2MnFe^{2+}Fe^{3+}(PO_4)_3$, from which the majority of natural alluaudites can be derived.

Moore and Ito (1979) investigated the crystal chemistry of several natural alluaudite samples, and also proposed a systematic nomenclature for the alluaudite group, which is based on the cation distribution among the different crystallographic sites.

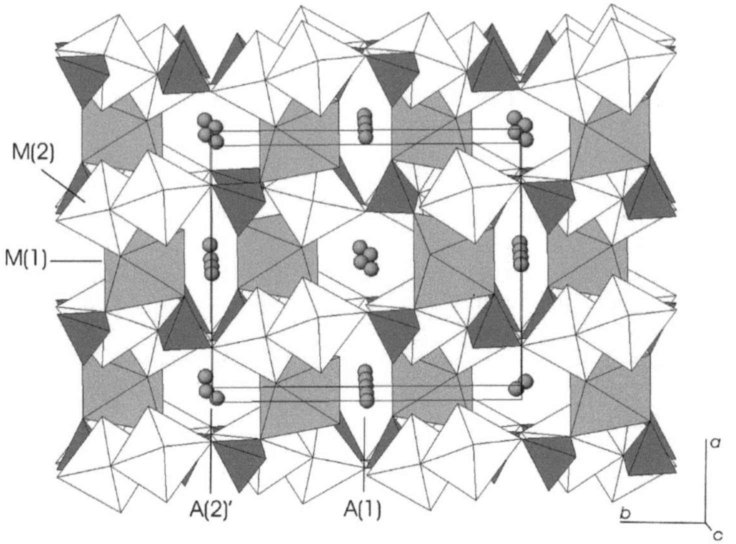

Fig. 5 The alluaudite structure, projected along the c axis. The $M(2)$ octahedra are *white*, the $M(1)$ octahedra are *light grey*, and the PO_4 tetrahedra are *dark grey*

The mineral is given a generic name, that depends on the predominant *M*(2) content (Mn = varulite, Fe^{2+} = hagendorfite, Fe^{3+} = alluaudite), and characterized by a prefix reflecting the *M*(1) content (Fe^{2+} = ferro-, and Mg = mag-). For example, the compositions $Na_2Fe_2^{2+}Fe^{3+}(PO_4)_3$, $Na_2MnFe^{2+}Fe^{3+}(PO_4)_3$, and $Na_2Mn_2Fe^{3+}(PO_4)_3$ correspond to the minerals ferrohagendorfite, hagendorfite, and varulite, respectively, whereas the more oxidized compositions $\square NaFe^{2+}Fe_2^{3+}(PO_4)_3$ and $\square NaMnFe_2^{3+}(PO_4)_3$ correspond to ferroalluaudite and alluaudite, respectively.

Over the past 20 years, many phosphates, arsenates, molybdates, vanadates, and tungstates with the alluaudite structure have been synthesized (*e.g*. Auernhammer et al. 1993; Khorari 1997; Solodovnikov et al. 1998; Hatert et al. 2000; Tsyrenova et al. 2000; Hatert 2004, 2008 , Redhammer et al. 2005). The structural investigation of these synthetic compounds showed the existence of new crystallographic sites localized in the channels of the structure, on positions which are different from those of *X*(1) and *X*(2). In order to take these new crystallographic sites into account, Hatert et al. (2000) proposed the new general formula $[A(2)A(2)'][A(1)A(1)'A(1)''_2]M(1)M(2)_2[PO_4]_3$ for alluaudite-type compounds. In this formula, *A*(1) and *A*(2)' correspond to *X*(1) and *X*(2), respectively (Fig. 5).

In general, the cation distribution in synthetic alluaudite-type phosphates is controlled by the ionic radii of the cations, as suggested by Moore (1971) and Moore and Ito (1979) (Table 1). However, Li is surprisingly not localized on the small *M*(2) crystallographic site, but rather on the large *A*(1) site (Hatert et al. 2000, 2002; Hatert 2004), and a partially disordered distribution of cations with similar

Table 1 Cation distribution in alluaudite-type phosphates

	Ionic radii (Å)		Crystallographic site			
Cation	[VI]	[VIII]	*A*(2)'	*A*(1)	*M*(1)	*M*(2)
Ag^+	1.15	1.28	X	X		
Na^+	1.02	1.18	X	X	X	
Cu^+	0.77	–	P	P		
Li^+	0.76	0.92	P	P		
Ca^{2+}	1.00	1.12	P	P	P	
Cd^{2+}	0.95	1.10		P	X	P
Mn^{2+}	0.830	0.96	P	P	X	X
Fc^{2+}	0.780	0.92			X	X
Co^{2+}	0.745	0.90			X	X
Zn^{2+}	0.740	0.90			X	P
Cu^{2+}	0.73	–		P		
Mg^{2+}	0.720	0.89			X	X
Ni^{2+}	0.690	–			X	X
In^{3+}	0.800	0.92			P	X
Fe^{3+}	0.645	0.78		P		X
Ga^{3+}	0.620	–				P
Cr^{3+}	0.615	–				P
Al^{3+}	0.535	–				P

X: Complete occupancy of the site, *P*: Partial occupancy of the site ionic radii has been observed in the compounds $Na_2Cd_2M^{3+}(PO_4)_3$ (M^{3+} = Ga, Fe^{3+}, Cr) (Antenucci 1992), $NaMn(Fe^{3+}{}_{1-x}In_x)_2(PO_4)_3$ (Hatert et al. 2003), $Na_2(Mn_{1-x}Fe^{2+}{}_x)_2Fe^{3+}(PO_4)_3$ (Hatert et al. 2005), $Na_{1.5}(Mn_{1-x}M^{2+}{}_x)_{1.5}Fe_{1.5}(PO_4)_3$ (M^{2+} = Cd, Zn) (Hatert 2008), and $Na_2(Mn_{1-x}M^{2+}{}_x)Fe^{2+}Fe^{3+}(PO_4)_3$ (M^{2+} = Cd, Mg) (Rondeux and Hatert 2010). Extensive Mössbauer spectral investigations have also been realized (Hermann et al. 2002; Hatert et al. 2003, 2004, 2005; Redhammer et al. 2005), showing the existence of next-nearest neighbour interactions affecting the iron atoms localized in the *M* sites.

The alluaudite structure is characterized by the existence of two channels parallel to the *c* axis, in which occur the *A* crystallographic sites containing monovalent cations, such as Na, Li, Cu, and Ag (Fig. 5, Table 1). Moreover, these channels can host significant amounts of hydrogen, as for example in the synthetic alluaudite-type phosphates $MCo_3(PO_4)[PO_3(OH)]_2$ and $MMn_3(PO_4)[PO_3(OH)]_2$ (*M* = Na, Ag) (Lii and Shih 1994; Leroux et al. 1995a, b; Guesmi and Driss 2002), as well as in mineral groatite, $NaCaMn_2{}^{2+}(PO_4)[PO_3(OH)]_2$ (Cooper et al. 2009). The channels of the alluaudite structure show a diameter from 2.5 to 4.0 Å (Fig. 5), corresponding to the diameter of pores observed in microporous phases. As a consequence, alluaudite-type phosphates are potential materials for practical applications, such as corrosion inhibition, passivation of metal surfaces, catalysis, and energy storage (Korzenski et al. 1998).

Many papers were published on the magnetic (Warner et al. 1993; Chouaibi et al. 2001; Durio et al. 2002; Hatert et al. 2004; Hidouri et al. 2004) and electrochemical properties (Warner et al. 1994; Daidouh et al. 2002; Durio et al. 2002; Richardson 2003; Trad et al. 2010) of alluaudite-type phosphates. Indeed, these compounds are of interest as candidates for lithium battery cathodes because of their expected high electronic conductivity, their high intercalation mobility, and their ease of preparation. Richardson (2003) consequently investigated the $Li_{0.75}Na_{0.25}MnFe_2(PO_4)_3$, $Na_2Mn_2Fe(PO_4)_3$, and $LiNaMn_2Fe(PO_4)_3$ compounds, which showed poor performance as intercalation electrodes in non-aqueous lithium cells. More recently, Trad et al. (2010) tested the intercalation/deintercalation properties of $NaMnFe_2(PO_4)_3$ as positive electrode in lithium and sodium batteries. The cycling curves exhibit a significant polarization, with 1.5 Li^+ ions and electrons intercalated in the structure at low rate. However, the very difficult electronic transfer, as well as the 1D diffusion character in alluaudites, tends to make difficult the alkaline metal diffusion. To improve the performance of alluaudite-type compounds, nanomaterials are thus certainly required (Trad et al. 2010).

The catalytic activity of $AgCaCdMg_2(PO_4)_3$ and $AgCd_2Mg_2(PO_4)_3$ was measured by Kacimi et al. (2005), by using as probe reaction the conversion of butan-2-ol in methyl ethyl ketone. These promising catalytic results are similar to those of well-known nasicon-type phosphates, thus confirming that the large family of alluaudites certainly constitutes a rich source for the engineering of new challenging materials.

Acknowledgments Many thanks are due to the FRS-F.N.R.S. (Belgium) for a position of "Chercheur Qualifié".

References

Abrahams I, Easson KS (1993) Structure of lithium nickel phosphate. Acta Cryst C49:925–926

Alberti A (1976) The crystal structure of ferrisicklerite, Li < 1(Fe3+, Mn2+)PO4. Acta Cryst B32:2761–2764

Andersson AS, Kalska B, Häggström L, Thomas JO (2000) Lithium extraction/insertion in $LiFePO_4$: an X-ray diffraction and Mössbauer spectroscopy study. Solid State Ionics 130:41–52

Antenucci D (1992) Synthèse et cristallochimie de composés à structure alluaudite. Incidences dans les processus d'altération des phosphates Fe-Mn des pegmatites granitiques. Unpublished Ph. D. thesis, University of Liège

Auernhammer M, Effenberger H, Hentschel G, Reinecke T, Tillmanns E (1993) Nickenichite, a new Arsenate from the Eifel, Germany. Mineral Petrol 48:153–166

Bramnik NN, Bramnik KG, Buhrmester T, Baehtz C, Ehrenberg E, Fuess H (2004) Electrochemical and structural study of $LiCoPO_4$-based electrodes. J Solid State Electrochem 8:558–564

Chouaibi N, Daidouh A, Pico C, Santrich A, Veiga ML (2001) Neutron diffraction, Mössbauer spectrum, and magnetic behavior of $Ag_2FeMn_2(PO_4)_3$ with alluaudite-like structure. J Solid State Chem 159:46–50

Cooper MA, Hawthorne F, Ball NA, Ramik RA, Roberts AC (2009) Groatite, $NaCaMn_2^{2+}(PO_4)$ $[PO_3(OH)]_2$, a new mineral species of the alluaudite group from the Tanco pegmatite, Bernic lake, Manitoba, Canada: description and crystal structure. Can Miner 47:1225–1235

Daidouh A, Durio C, Pico C, Veiga ML, Chouaibi N, Ouassini A (2002) Structural and electrical study of the alluaudites $(Ag_{1-x}Na_x)_2FeMn_2(PO_4)_3$ ($x = 0, 0.5$ and 1). Solid State Sci 4:541–548

Deniard P, Dulac AM, Rocquefelte X, Grigorova V, Lebacq O, Pasturel A, Jobic S (2004) High potential positive materials for lithium-ion batteries: transition metal phosphates. J Phys Chem Solids 65:229–233

Durio C, Daidouh A, Chouaibi N, Pico C, Veiga ML (2002) Electrical behavior of new orthophosphates $Na_2M_3(PO_4)_3$ ($M_3 = GaMn_2$, $GaCd_2$, $InMn_2$ and FeMnCd) with alluaudite-like structure. J Solid State Chem 168:208–216

Eventoff W, Martin R (1972) The crystal structure of heterosite. Am Miner 57:45–51

Fehr KT, Hochleitner R, Schmidbauer E, Schneider J (2007) Mineralogy, mössbauer spectra and electrical conductivity of triphylite, Li(Fe2+, Mn2+)PO4. Phys Chem Miner 34:485–494

Finger LW, Rapp GR (1969) Refinement of the structure of triphylite. Carnegie Inst Year B 68:290–293

Fontan F, Huvelin P, Orliac M, Permingeat F (1976) La ferrisicklérite des pegmatites de Sidi-Bou-Othmane (Jebilet, Maroc) et le groupe des minéraux à structure de triphylite. Bull Soc française Minéral Cristallo 99:274–286

Fransolet A-M, Antenucci D, Speetjens J-M, Tarte P (1984) An X-ray determinative method for the divalent cation ratio in the triphylite-lithiophilite series. Miner Mag 48:373–381

Fransolet A-M, Abraham K, Speetjens J-M (1985) Evolution génétique et signification des associations de phosphates de la pegmatite d'Angarf-Sud, plaine de Tazenakht, Anti-Atlas, Maroc. Bull Minéral 108:551–574

Fransolet A-M, Keller P, Fontan F (1986) The phosphate mineral associations of the Tsaobismund pegmatite, Namibia. Contrib Miner Petrol 92:502–517

Fransolet A-M, Keller P, Fontan F (1997) The alluaudite group minerals: their crystallochemical flexibility and their modes of formation in the granite pegmatites. Abstracts of the meeting "Phosphates: biogenic to exotic", London

Fransolet A-M, Fontan F, Keller P, Antenucci D (1998) La série johnsomervilleite-fillowite dans les associations de phosphates de pegmatites granitiques de l'Afrique centrale. Can Miner 36:355–366

Fransolet A-M, Hatert F, Fontan F (2004) Petrographic evidence for primary hagendorfite in an unusual assemblage of phosphate minerals, Kibingo granitic pegmatite, Rwanda. Can Miner 42:697–704

Geller S, Durand JL (1960) Refinement of the structure of $LiMnPO_4$. Acta Cryst 13:325–331

Guesmi A, Driss A (2002) $AgCo_3PO_4(HPO_4)_2$. Acta Cryst C58:i16–i17

Hanic F, Handlović M, Burdová K, Majling J (1982) Crystal structure of lithium magnesium phosphate, $LiMgPO_4$: crystal chemistry of the olivine-type compounds. J Crystallo Spectro Res 12(2):99–127

Hatert F (2004) Etude cristallochimique et synthèse hydrothermale des alluaudites: contribution nouvelle au problème génétique des phosphates de fer et de manganèse dans les pegmatites granitiques et, partant, à celui de l'évolution de ces gisements. Mém Acad royale Sci Belgique, Cl Sci, Coll in-8°, 3ème série XXI

Hatert F (2008) The crystal chemistry of the divalent cation in alluaudite-type phosphates: a structural and infrared spectral study of the $Na_{1.5}(Mn_{1-x}M_x^{2+})_{1.5}Fe_{1.5}(PO_4)_3$ solid solutions ($x = 0$ to 1, $M^{2+} = Cd^{2+}, Zn^{2+}$). J Solid State Chem 181:1258–1272

Hatert F (2010) A structural study of the lithiophilite-sicklerite series. 26th European Crystallographic Meeting, Abstract book: s184

Hatert F, Keller P, Lissner F, Antenucci D, Fransolet A-M (2000) First experimental evidence of alluaudite-like phosphates with high Li-content: the $(Na_{1-x}Li_x)MnFe_2(PO_4)_3$ series ($x = 0$ to 1). Eur J Miner 12:847–857

Hatert F, Antenucci D, Fransolet A-M, Liégeois-Duyckaerts M (2002) The crystal chemistry of lithium in the alluaudite structure: a study of the $(Na_{1-x}Li_x)CdIn_2(PO_4)_3$ solid solution ($x = 0$ to 1). J Solid State Chem 163:194–201

Hatert F, Hermann RP, Long GJ, Fransolet A-M, Grandjean F (2003) An X-ray Rietveld, infrared, and Mössbauer spectral study of the $NaMn(Fe_{1-x}In_x)_2(PO_4)_3$ alluaudite-like solid solution. Am Mineral 88:211–222

Hatert F, Long GJ, Hautot D, Fransolet A-M, Delwiche J, Hubin-Franskin MJ, Grandjean F (2004) A structural, magnetic, and Mössbauer spectral study of several Na-Mn-Fe-bearing alluaudites. Phys Chem Miner 31:487–506

Hatert F, Rebbouh L, Hermann RP, Fransolet A-M, Long GJ, Grandjean F (2005) Crystal chemistry of the hydrothermally synthesized $Na_2(Mn_{1-x}Fe_x^{2+})_2Fe^{3+}(PO_4)_3$ alluaudite-type solid solution. Am Miner 90:653–662

Hatert F, Ottolini L, Keller P, Fransolet A-M (2009) Crystal chemistry of lithium in pegmatite phosphates: A SIMS investigation of natural and synthetic samples. Estudos Geológicos 19 (2):131–134

Hatert F, Ottolini L, Schmid-Beurmann P (2010) Experimental investigation of the alluaudite + triphylite assemblage, and development of the Na-in-triphylite geothermometer: applications to natural pegmatite phosphates. Contrib Mineral Petrol, in press

Hermann RP, Hatert F, Fransolet A-M, Long GJ, Grandjean F (2002) Mössbauer spectral evidence for next-nearest neighbor interactions within the alluaudite structure of $Na_{1-x}Li_xMnFe_2(PO_4)_3$. Solid State Sci 4:507–513

Hidouri M, Lajmi B, Wattiaux A, Fournés L, Darriet J, Amara MB (2004) Characterization by X-ray diffraction, magnetic susceptibility and Mössbauer spectroscopy of a new alluaudite-like phosphate: $Na_4CaFe_4(PO_4)_6$. J Solid State Chem 177:55–60

Ivanov YA, Simonov MA, Belov NV (1974) Crystal structure of the Na, Cd orthophosphate NaCd [PO_4]. Soviet Phys Cryst 19:96–97

Kacimi M, Ziyad M, Hatert F (2005) Structural features of $AgCaCdMg_2(PO_4)_3$ and $AgCd_2Mg_2(PO_4)_3$, two new compounds with the alluaudite-type structure, and their catalytic activity in butan-2-ol conversion. Mater Res Bull 40:682–693

Kang B, Ceder G (2009) Battery materials for ultrafast charging and discharging. Nature 458:190–193

Khorari S (1997) Cristallochimie des arséniates de structure alluaudite. Unpublished Ph. D. thesis, University of Liège

Korzenski MB, Schimek GL, Kolis JW, Long GJ (1998) Hydrothermal synthesis, structure, and characterization of a mixed-valent iron (II/III) phosphate, $NaFe_{3.67}(PO_4)_3$: a new variation of the alluaudite structure type. J Solid State Chem 139:152–160

Kubel F (1994) Crystal structure of lithium cobalt double orthophosphate, $LiCoPO_4$. Z Krist 209 (9):755

Leroux F, Mar A, Payen C, Guyomard D, Verbaere A, Piffard Y (1995a) Synthesis and structure of $NaMn_3(PO_4)(HPO_4)_2$, an unoxidized variant of the alluaudite structure type. J Solid State Chem 115:240–246

Leroux F, Mar A, Guyomard D, Piffard Y (1995b) Cation substitution in the alluaudite structure type: synthesis and structure of $AgMn_3(PO_4)(HPO_4)_2$. J Solid State Chem 117:206–212

Li Z, Shinno I (1997) Next nearest neighbor effects in triphylite and related phosphate minerals. Miner J 19(3):99–107

Lii K-H, Shih P-F (1994) Hydrothermal synthesis and crystal structures of $NaCo_3(PO_4)(HPO_4)_2$ and $NaCo_3(AsO_4)(HAsO_4)_2$: synthetic modifications of the mineral alluaudite. Inorg Chem 33:3028–3031

Losey A, Rakovan J, Hughes JM, Francis CA, Dyar MD (2004) Structural variation in the lithiophilite-triphylite series and other olivine-group structures. Can Miner 42:1105–1115

Mason B (1941) Minerals of the Varuträsk pegmatite. XXIII. Some iron-manganese phosphate minerals and their alteration products, with special reference to material from Varuträsk. Geol För Stockholm För 63:117–175

Moore PB (1971) Crystal chemistry of the alluaudite structure type: Contribution to the paragenesis of pegmatite phosphate giant crystals. Am Miner 56:1955–1975

Moore PB (1972) Natrophilite, $NaMnPO_4$, has ordered cations. Am Mineralog 57:1333–1344

Moore PB, Ito J (1979) Alluaudites, wyllieites, arrojadites: crystal chemistry and nomenclature. Miner Mag 43:227–235

Okada S, Sawa S, Egashira M, Yamaki J, Tabuchi M, Kageyama H, Konishi T, Yoshino A (2001) Cathode properties of phosphoolivine $LiMPO_4$ for lithium secondary batteries. J Power Sources 97–98:430–432

Padhi AK, Nanjundaswamy KS, Goodenough JB (1997) Phosphoolivines as positive materials for rechargeable lithium batteries. J Electrochem Soc 144:1188–1194

Penazzi N, Arrabito M, Piana M, Bodoardo S, Panero S, Amadei I (2004) Mixed lithium phosphates as cathode materials for Li-Ion cells. J Eur Ceram Soc 24:1381–1384

Prosini PP, Lisi M, Zane D, Pasquali M (2002) Determination of the chemical diffusion coefficient of lithium in $LiFePO_4$. Solid State Ionics 148:45–51

Quensel P (1937) Minerals of the Varuträsk Pegmatite. I. The lithium-manganese phosphates. Geol För Stockholm För 59(1):77–96

Ravet N, Abouimrane A, Armand M (2003) On the electronic conductivity of phospho-olivines as lithium storage electrodes. Nat Mater 2:702

Redhammer GJ, Tippelt G, Bernroider M, Lottermoser W, Amthauer G, Roth G (2005) Hagendorfite (Na, Ca)MnFe2(PO4)3 from type locality Hagendorf (Bavaria, Germany): crystal structure determination and 57Fe Mössbauer spectroscopy. Eur J Miner 17:915–932

Richardson TJ (2003) Phosphate-stabilized lithium intercalation compounds. J Power Sources 119–121:262–265

Rondeux M, Hatert F (2010) An X-ray Rietveld and infrared spectral study of the $Na_2(Mn_{1-x}M_x^{2+})Fe^{2+}Fe^{3+}(PO_4)_3$ ($x = 0$ to 1, M^{2+} = Mg, Cd) alluaudite-type solid solutions. Am Miner 95:844–852

Solodovnikov SF, Klevtsov PV, Solodovnikova ZA, Glinskaya LA, Klevtsova RF (1998) Binary molybdates $K_4M^{2+}(MoO_4)_3$ (M^{2+} = Mg, Mn, Co) and crystal structure of $K_4Mn(MoO_4)_3$. J Struct Chem 39:230–237

Song Y, Yang S, Zavalij PY, Whittingham MS (2002) Temperature dependent properties of $FePO_4$ cathode materials. Mater Res Bull 37:1249–1257

Takahashi M, Tobishima S, Takei K, Sakurai Y (2002) Reaction behavior of $LiFePO_4$ as a cathode material for rechargeable lithium batteries. Solid State Ionics 148:283–289

Trad K, Carlier D, Croguennec L, Wattiaux A, Ben Amara M, Delmas C (2010) $NaMnFe_2(PO_4)_3$ alluaudite phase: synthesis, structure, and electrochemical properties as positive electrode in lithium and sodium batteries. Chem Mater 22(19):5554–5562

Tsyrenova GD, Solodovnikov SF, Zolotova ES, Tsybikova BA, Bazarova ZG (2000) Phase formation in the $K_2O(K_2CO_3)$-CdO-MoO_3 system. Russ J Inorg Chem 45:103–108
Warda SA, Lee S-L (1998) Refinement of the crystal structure of lithium nickel phosphate, $LiNiPO_4$. Z Krist NCS 212(3):319
Warner TE, Milius W, Maier J (1993) Synthesis and structure of $Cu_{1.35}Fe_3(PO_4)_3$ and $Cu_2Mg_3(PO_4)_3$: new mixed valence compounds of the alluaudite structure type. J Solid State Chem 106:301–309
Warner TE, Milius W, Maier J (1994) New copper phosphates with the NASICON or alluaudite-type structures as ionic or mixed conductors. Solid State Ion 74:119–123
Yakubovich OV, Simonov MA, Belov NV (1977) The crystal structure of a synthetic triphylite, $LiFe[PO_4]$. Sov Phys Dokl 22:347–350
Yang S, Song Y, Ngala K, Zavalij PY, Whittingham MS (2003) Performance of $LiFePO_4$ as lithium battery cathode and comparison with manganese and vanadium oxides. J Power Sources 119–121:239–246

Crystal Structure of Murataite *Mu*-5, a Member of the Murataite-Pyrochlore Polysomatic Series

Sergey V. Krivovichev, Vadim S. Urusov, Sergey V. Yudintsev, Sergey V. Stefanovsky, Oksana V. Karimova, and Natalia N. Organova

1 Introduction

One of the most important problems related to the use of nuclear energy and of advanced nuclear fuel cycle is the immobilization of minor actinides as well as of rare earth fractions and corrosion products present in high level radioactive waste (HLW) (Lutze and Ewing 1988; Ewing 1999, 2005; Stefanovsky et al. 2004; Yudintsev et al. 2007a; Weber et al. 2009; Livshits and Yudintsev 2008). The most enduring proposal was made by Ringwood et al. (1979), who invented Synroc, a polyphase waste form consisting of titanates. In Synroc, actinides are incorporated mainly in pyrochlore and zirconolite, two closely related complex oxides based upon 3D frameworks of TiO_6 octahedra (Subramanian et al. 1983; Zhang et al. 2009). Within recent years, pyrochlore – and zirconolite-type oxides have been extensively investigated in terms of their chemical diversity, physical and chemical durability and radiation resistance (Sickafus et al. 2007; Zhang et al. 2008; Jiang

S.V. Krivovichev (✉)
Department of Crystallography, Faculty of Geology,
St. Petersburg State University, 199034 St. Petersburg, Russia
and
Nanomaterials Centre, Kola Science Centre, Russian Academy of Science, Apatity, Russia
e-mail: skrivovi@mail.ru

V.S. Urusov
Institute of Geology of Ore Deposits, Petrography, Mineralogy and Geochemistry, Russian Academy of Sciences, 119017 Moscow, Russia
and
Department of Crystallography, Faculty of Geology, Moscow State University, 119992 Moscow, Russia

S.V. Yudintsev • O.V. Karimova • N.N. Organova
Institute of Geology of Ore Deposits, Petrography, Mineralogy and Geochemistry, Russian Academy of Sciences, 119017 Moscow, Russia

S.V. Stefanovsky
MosNPO Radon, 119121 Moscow, Russia

S.V. Krivovichev (ed.), *Minerals as Advanced Materials II*,
DOI 10.1007/978-3-642-20018-2_28, © Springer-Verlag Berlin Heidelberg 2012

et al. 2009; Belin et al. 2009; Lumpkin et al. 2007, 2009; Strachan et al. 2008). In 1982, Morgan and Ryerson (1982) identified another titanate phase, a synthetic analogue of murataite (Adams et al. 1974), in a titanate ceramic with HLW imitators produced at the Savannah River nuclear plant in the U.S. In 1997, Laverov et al. (1997, 1998a; Sobolev et al. 1997) identified murataite in the Synroc matrix with imitators of HLRW wastes from PO Mayak, a radiochemical facility for production and reprocessing of nuclear fuel in Russian Federation. Five volume percent of this phase accumulated about 40% of the total uranium present in the sample, which led to detailed investigations of chemistry and properties of murataite, in particular, of its chemical durability and radiation resistance (Laverov et al. 1998b, 2006; Urusov et al. 2002, 2005, 2007; Karimova et al. 2002; Lian et al. 2005a, b; Perevalov et al. 2006; Stefanovsky et al. 2007a, 2007b; 2009; Yudintsev et al. 2007b). However, structural mechanisms of incroporation of actinides into murataite remained unknown up to date.

In this paper, we develop upon our recent results of structural investigations of murataite *Mu*-5 (Krivovichev et al. 2010), one of the murataite varieties, that provide direct evidence for existence of a family of complex Ti-based oxides consisting of nanoscale pyrochlore clusters connected through murataite-related structural units.

Murataite itself is a natural mineral with the ideal formula $(Y,Na)_6(Zn,Fe)_5Ti_{12}O_{29}(O,F)_{10}F_4$ (Adams et al. 1974; Ercit and Hawthorne 1995). Its structure is based upon a nanoporous 3D framework consisting of polymerized α-Keggin $[Zn^{[4]}Ti_{12}{}^{[6]}O_{40}]^{30-}$ clusters with T_d symmetry (Fig. 1a). Polymerization of Keggin units results in creation of two types of voids that can be characterized as a truncated tetrahedron $3^4 6^4$ (Fig. 1b) and cubooctahedron $4^6 6^8$ (Fig. 1c). In terms of nodal representation, the Keggin cluster can be characterized as a $3^8 4^6$ cubo-octahedron (Fig. 1a). The topology of linkage of polyhedral units in murataite structure is depicted in Fig. 1d, e.

The framework accommodates complex fluorite-like substructure of Y, Fe, and Na cations linked by extra-framework O^{2-} and F^- anions. It has been recognized (Sobolev et al. 1997) that murataite (as well as pyrochlore) can be considered as anion-deficient fluorite structure with ordered arrangement of cations. Murataite has a cubic unit cell with *a* parameter tripled in comparison to fluorite, thus representing a 3 × 3 × 3 fluorite superstructure. By analogy, pyrochlore has a 2 × 2 × 2 fluorite superstructure, $A_2B_2O_6O'$, based upon interpenetrating $[B_2O_6]$ framework of corner-sharing (BO_6) octahedra and $[A_2O']$ framework of corner-sharing oxocentered $O'A_4$ tetrahedra (Fig. 2).

Transmission electron microscopy studies (Laverov et al. 1998b) allowed to identify synthetic murataite varieties with 3 × 3 × 3, 5 × 5 × 5, 7 × 7 × 7 and 8 × 8 × 8 cubic supercells referred to as *Mu*-3, *Mu*-5, *Mu*-7 and *Mu*-8 phases with unknown structures. Experimental studies demonstrated that the crystallization sequence of phases in the U(Pu)-Zr-Mn-Fe-Ti-Al-O complex system can be expressed as: Pyrochlore – *Mu*-7 – *Mu*-5 – *Mu*-8 – *Mu*-3. Along this line, the amount of U(Pu) and other heavy cations decreases, whereas the Ti, Mn, Fe and Al content increases (Laverov et al. 2006). In murataite ceramic, typical grains contain

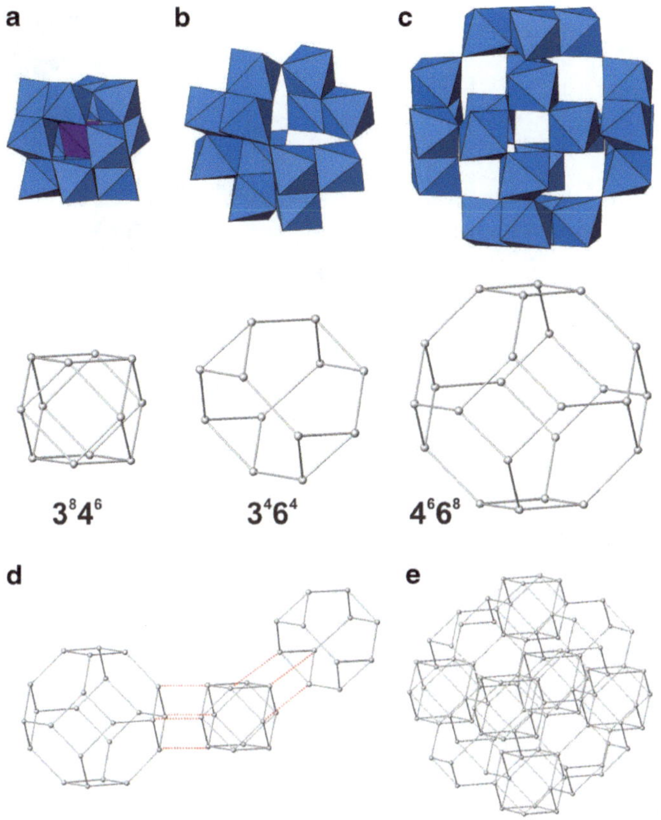

Fig. 1 Basic polyhedral units comprising octahedral framework in the structure of natural murataite (murataite *Mu*-3): cubooctahedron 3^84^6 (**a**), truncated tetrahedron 3^46^4 (**b**), cubo-octahedron 4^66^8 (**c**), the mode of their linkage (**d**) and the resulting 3D network (**e**). In (**a–c**), *light-blue* and *dark-blue* polyhedra are TiO_6 octahedra and ZnO_4 tetrahedra, respectively

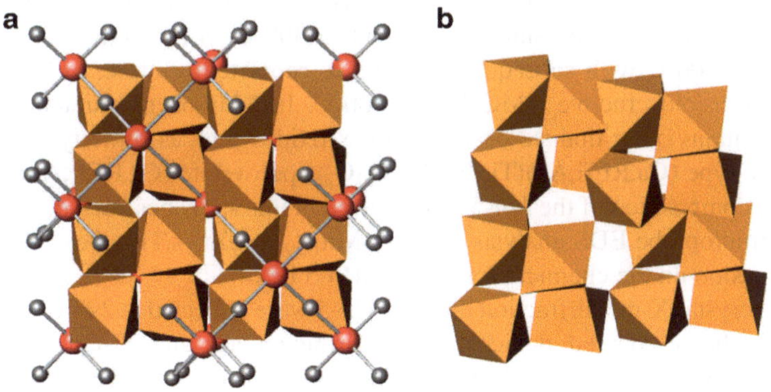

Fig. 2 The structure of pyrochlore $A_2B_2O_6O'$ as consisting of octahedral $[B_2O_6]$ framework (*light-brown*) and $[A_2O']$ framework of corner-sharing oxocentered $O'A_4$ tetrahedra (O' and A atoms are shown as red and grey circles, respectively) (**a**), and the unit-cell portion of the octahedral framework (**b**)

Fig. 3 Scanning electron microscopy (SEM) images of octahedral murataite crystals and their cross section. 1–3 – murataite *Mu*-5, *Mu*-8, and *Mu*-3 varieties with ThO_2 contents 12.6, 10.9, and 5.1 wt.% correspondingly

pyrochlore and *Mu*-5 at the core surrounded by *Mu*-8 and *Mu*-3 phases (Fig. 3). Since the most actinide-bearing phases are encapsulated by low-actinide varieties, this creates an additional barrier for actinide leaching and increases chemical durability of murataite ceramics. In some cases the central parts of such complex grains are composed of zirconolite. However, this also makes structural investigations of murataite crystals a challenging task, taking into account presence of several closely related phases within a single grain.

2 Experiment

In the current work, we have studied sample of murataite ceramic consisting of zoned crystals with *Mu*-5 as a major grain component. In a typical synthesis of murataite ceramics, the initial mixture contains (in wt.%): 55 TiO_2, 10 MnO, 10 CaO, 5 Al_2O_3, 5 Fe_2O_3, 5 ZrO_2, and 10 UO_2 or 10 PuO_2. The ceramic is obtained either by cold compaction combined with sintering at 1,100–1,300°C, or by melting in an electric furnace at 1,500°C with subsequent cooling to room temperature, or by induction melting in a cold crucible (IMCC) at 1,600°C. In addition to murataite, ceramic usually contains other titanate-based complex oxides such as rutile-type $(Ti,Zr)O_2$, crichtonite-type $(Ca,REE,An)(Ti,Zr,Al,Fe)_{21}O_{38}$, perovskite $(Ca,REE,An)TiO_3$, etc. Chemical composition of the *Mu*-5 sample used in this study has been determined by electron microprobe EDS analysis as $Ca_{3.17}Mn_{1.39}U_{0.92}Zr_{0.79}Ti_{8.57}Fe_{0.30}Al_{0.34}O_{26.09}$ (see discussion on the chemical formula of *Mu*-5 below).

Single crystal X-ray diffraction data for *Mu*-5 were collected at 293 K on a Bruker APEX II charge-coupled device (CCD) diffractometer and STOE Image Plate Diffraction System II (STOE IPDS II). The image plate data were used for reconstruction of reciprocal diffraction space sections. The data are as follows (Bruker APEX II): cubic, F-43 m, $a = 24.564(7)$ Å, $V = 14{,}822(7)$ Å^3, $d_{calc} = 4.98$ g/cm^3, $\mu = 14.797$ mm^{-1}, $2\theta_{max} = 56°$, $\lambda = 0.71073$ Å; total reflections – 41,848;

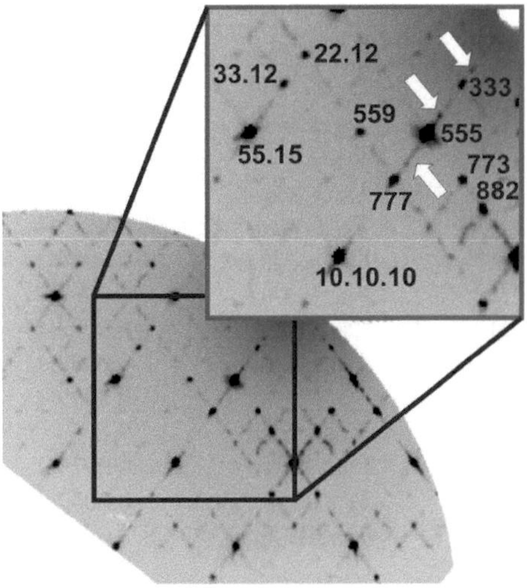

Fig. 4 Reciprocal diffraction space section (110) reconstructed from the image-plate detector data collected on zones crystal of *Mu*-5. *White arrows* in the inbox indicate diffuse streaks in between major reflections that may be attributed either to a coherent intergrowth with *Mu*-3 or *Mu*-8 phase or to incommensurate modulation induced by complex pattern of cation substitution

unique reflections – 1,762 [$R_{int} = 0.0470$]; R_1 [$I > 2\sigma I$)] = 0.074, $wR_2 = 0.231$; GooF = 0.936. The model was accounted for racemic twinning (Flack parameter is 0.45(3)).

Inspection of reciprocal diffraction space (Fig. 4) demonstrated that fluorite-subcell *hkl* reflections with $h,k,l = 5n$ are the strongest. The weaker Bragg reflections divide the spacing between the 'fluorite' reflections in the ratio 2:3, i.e. in the proportion of lattice spacings in the pyrochlore and murataite *Mu*-3 structures.

This observation, made earlier by electron microdiffraction (Laverov et al. 1998b), has served as a basis for Urusov et al. (2005, 2007) to propose that the structures of *Mu*-5 and other varieties could be described as alternating 2D layers excised from the structures of pyrochlore and murataite *Mu*-3. In addition, diffraction pattern contains diffuse streaks in between adjacent *nnn* reflections extended along the [111] direction. These streaks are either the evidence of the zonal structure of crystals studied or are due to modulations induced by complex pattern of cation substitutions.

The crystal chemical formula derived from structure refinement and determined on the basis of site-scattering power of cation sites (see Supporting Information) is $Ca_{24.47}Mn_{21.97}U_{7.71}Al_{2.46}Ti_{59.84}Zr_{8.56}O_{172}$, which is in reasonable agreement with the formula derived from electron microprobe analysis, taking into account possibility of substitution of Fe^{3+} and Ti^{4+} at Mn sites. Note that empirical and structural formula provide 125.19 and 125.00 cations per 172 O atoms, respectively. It should be noted that *Mu*-5 crystals have been obtained that have less than 125 cations per 172 O atoms, which may indicate formation of vacancies at the cation sites, similar to that observed for pyrochlore-type compounds. Structure refinement on the basis of data collected using STOE IPDS II resulted in the crystallographic agreement index R_1 [$I > 2\sigma I$)] = 0.097. Final atom coordinates and displacement parameters are given in Table 1, selected bond lengths are listed in Table 2.

Table 1 Atom coordinates, occupancies and displacement parameters (Å^2) for murataite *Mu*-5

Atom	Occupancy	x	y	z	U_{eq}
*Ti*1	$Ti_{0.90}Zr_{0.10}$	0.0993(1)	0.2017(1)	0.9007(1)	0.011(1)
*Ti*2	$Ti_{0.84}Zr_{0.16}$	0.1984(1)	0.6984(1)	0.8949(1)	0.009(1)
*Ti*3	$Ti_{1.00}$	0	0.1975(2)	0	0.012(1)
*Ti*4	$Ti_{0.84}Zr_{0.16}$	0.2001(1)	0.2001(1)	0.7999(1)	0.012(1)
*Ti*5	$Ti_{0.80}Zr_{0.20}$	0.0975(1)	0.7001(1)	0.9955(1)	0.014(1)
*Mn*1	$Mn_{0.60}Al_{0.40}$	0.0984(1)	0.5984(1)	0.9016(1)	0.016(1)
*Mn*2	$Mn_{0.79}Al_{0.21}$	0.2998(1)	0.7002(1)	0.7998(1)	0.022(1)
*M*1	$Ca_{0.86}U_{0.14}$	−0.0022(1)	0.1998(1)	0.8002(1)	0.035(1)
*M*2	$Ca_{0.84}U_{0.16}$	0.0993(1)	0.4007(1)	0.9007(1)	0.111(3)
*M*3	$Ca_{0.93}U_{0.07}$	0.3999(1)	0.7026(1)	0.8999(1)	0.016(1)
*M*4	$Mn_{0.89}U_{0.14}$	0	1/2	0.8958(1)	0.019(1)
*M*5	$Ca_{0.82}U_{0.18}$	0.0979(1)	0.3002(1)	0.8002(1)	0.028(1)
*M*6	$Ti_{0.87}U_{0.13}$	−0.0037(1)	0.1026(1)	0.1026(1)	0.010(1)
*M*7	$Ca_{0.98}U_{0.03}$	0	0	0	0.031(3)
O1	$O_{1.00}$	0.1447(5)	0.3553(5)	0.8553(5)	0.043(5)
O2	$O_{1.00}$	0.1457(3)	0.6457(3)	0.0198(4)	0.022(2)
O3	$O_{1.00}$	0.0447(8)	1/4	3/4	0.052(5)
O4	$O_{1.00}$	0.0560(5)	0.4440(5)	0.9440(5)	0.047(6)
O5	$O_{1.00}$	0.3493(5)	0.6507(5)	0.8493(5)	0.045(5)
O6	$O_{1.00}$	0.0427(2)	0.6696(3)	0.0427(2)	0.010[a]
O7	$O_{1.00}$	0.0465(3)	0.1655(4)	0.9535(3)	0.020(2)
O8	$O_{1.00}$	0.1247(2)	0.7497(2)	0.0441(2)	0.013(1)
O9	$O_{1.00}$	0.0666(3)	0.1475(2)	0.8525(2)	0.011(2)
O10	$O_{1.00}$	1/4	3/4	0.9123(5)	0.017(2)
O11	$O_{1.00}$	0.2290(2)	0.6461(2)	0.9389(2)	0.013(1)
O12	$O_{1.00}$	0.1477(2)	0.2295(3)	0.8523(2)	0.010[a]
O13	$O_{1.00}$	0.3346(3)	0.7545(4)	0.8346(3)	0.019(2)
O14	$O_{1.00}$	0.1673(5)	1/4	3/4	0.010[a]
O15	$O_{1.00}$	0.0622(2)	0.5622(2)	0.8503(3)	0.012(2)
O16	$O_{1.00}$	1/4	3/4	3/4	0.017(6)
O17	$O_{1.00}$	−0.0404(2)	0.2479(3)	0.9596(2)	0.010[a]
O18	$O_{1.00}$	0.1425(3)	0.6425(3)	0.8575(3)	0.010[a]
O19	$O_{1.00}$	−0.0597(3)	0.0597(3)	0.0597(3)	0.014(3)
O20	$O_{1.00}$	−0.0429(7)	0.1515(4)	0.8485(4)	0.066(5)
O21	$O_{0.68}$	0.0478(6)	0.1359(8)	0.0478(6)	0.053(5)
O21A	$O_{0.32}$	0.0448(7)	0.082(1)	0.0448(7)	0.020[a]
O22	$O_{1.00}$	−0.0486(6)	0.4514(6)	0.9514(6)	0.068(8)

[a]fixed during refinement

3 Results and Discussion

Solution of the crystal structure of *Mu*-5 demonstrated that it contains five octahedral (*Ti*), two trigonal bipyramidal (*Mn*), and seven 7- and 8-coordinated *M* sites (Fig. 5). Octahedral sites display different degrees of distortion typical for Ti^{4+} coordination (Kunz and Brown 1995) with Ti-O bond lengths in the range from

Table 2 Selected bond lengths (Å) in the structure of murataite Mu-5

*Ti*1-O12	1.815(8)	*Mn*1-O15	1.780(9) 3×	*M*4-O22	2.172(8) 2×
*Ti*1-O8	1.914(6) 2×	*Mn*1-O18	1.873(15)	*M*4-O6	2.186(8) 2×
*Ti*1-O9	1.953(4) 2×	*Mn*1-O22	2.12(3)	*M*4-O4	2.278(9) 2×
*Ti*1-O7	2.039(9)			*M*4-O15	2.434(8) 2×
		*Mn*2-O13	1.800(9) 3×	*M*5-O3	2.179(12)
*Ti*2-O11	1.841(6) 2×	*Mn*2-O5	2.11(2)	*M*5-O1	2.233(9)
*Ti*2-O10	1.843(3)	*Mn*2-O16	2.118(5)	*M*5-O14	2.439(8)
*Ti*2-O13	2.046(6) 2×			*M*5-O12	2.480(6) 2×
*Ti*2-O18	2.149(7)	*M*1-O20	1.954(16)	*M*5-O8	2.579(6) 2×
		*M*1-O3	2.090(11)	*M*5-O2	2.686(9)
*Ti*3-O7	1.797(9) 2×	*M*1-O9	2.484(9)	*M*6-O21A	1.921(7)
*Ti*3-O17	1.874(9) 2×	*M*1-O8	2.497(6) 2×	*M*6-O21	2.021(8)
*Ti*3-O21	2.25(2) 2×	*M*1-O11	2.624(6) 2×	*M*6-O19	2.028(3)
		*M*1-O10	2.729(9)	*M*6-O20	2.048(15)
*Ti*4-O14	1.911(5) 3×	*M*2-O4	1.85(2)	*M*6-O9	2.194(8)
*Ti*4-O12	1.957(8) 3×	*M*2-O1	1.93(2)	*M*6-O7	2.321(6) 2×
		*M*2-O2	2.533(9) 3×	*M*7-O19	2.539(15) 4×
*Ti*5-O8	1.832(6)	*M*2-O6	2.616(8) 3×	*M*7-O21A	2.55(3) 4×
*Ti*5-O2	1.882(3)	*M*3-O5	2.171(7)		
*Ti*5-O6	1.929(4)	*M*3-O11	2.240(6) 2×		
*Ti*5-O11	1.964(6)	*M*3-O20	2.268(10) 2×		
*Ti*5-O15	2.072(6)	*M*3-O17	2.353(8)		
*Ti*5-O17	2.091(5)	*M*3-O21	2.45(2)		

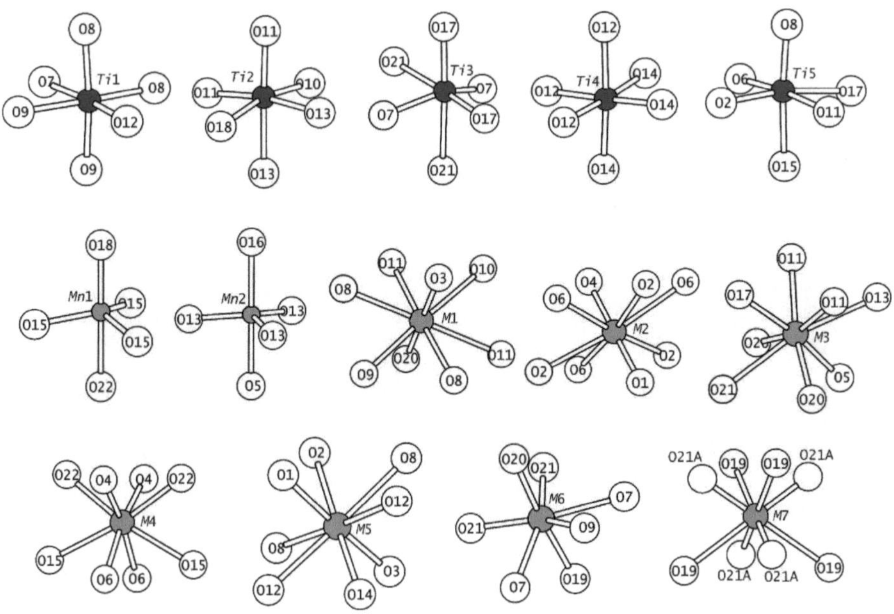

Fig. 5 Coordination of cations in the structure of murataite Mu-5

1.797 to 2.246 Å (individual average bond lengths for TiO_6 octahedra vary from 1.931 to 1.972 Å). Refinement of site-occupation factors indicated that the *Ti* sites are over-occupied, most probably due to the incorporation of Zr. The *Mn*1 and *Mn*2 sites are 5-coordinated by O atoms to form MnO_5 trigonal bipyramids; these sites also accommodate Al (and to some extent excess Ti atoms). There are seven sites with 7- and 8-eightfold coordination. In the refinement procedure, it was assumed that these sites accommodate Ca, Mn and U, except for the 7-coordinated *M*6 site, that is rather compact and probably contains excess Ti.

By analogy with pyrochlore and natural murataite, the structure of *Mu*-5 can be described as a framework of octahedra with cavities occupied by non-octahedral cations and 'additional' O atoms. The octahedral framework is of unusual complexity. Its most outstanding feature is that it consists of cubic pyrochlore clusters with diameter of 1.5 nm. These clusters represent pyrochlore cubic unit cells (Fig. 2b) and are arranged within the *Mu*-5 cubic unit cell in a 3D chessboard-like fashion

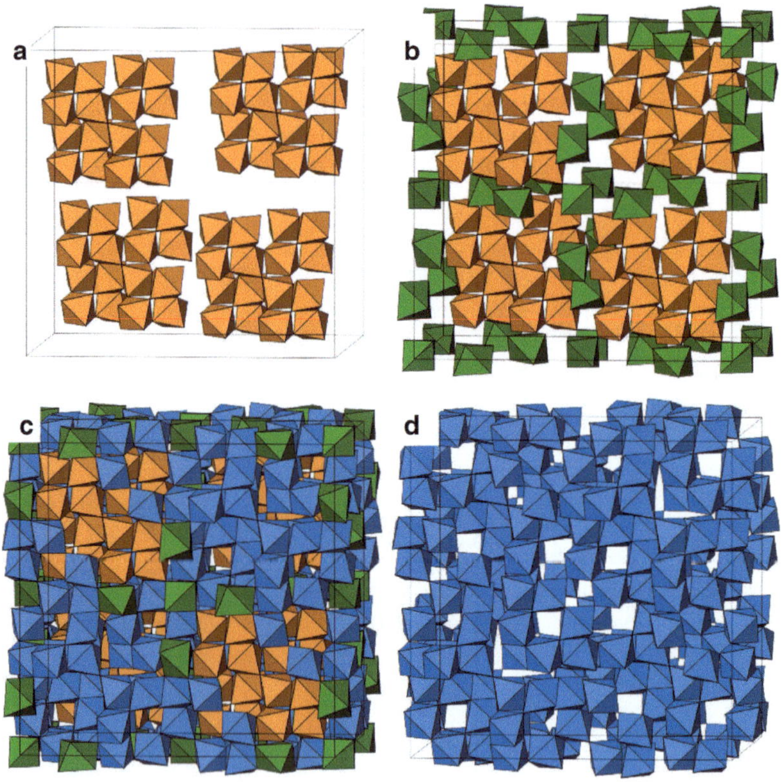

Fig. 6 Octahedral framework in the structure of Mu-5: arrangement of pyrochlore clusters formed by corner sharing of $Ti1O_6$ and $Ti4O_6$ octahedra (**a**); linkage of pyrochlore clusters by $Ti3O_6$ octahedra (**b**); whole framework as combination of linked pyrochlore clusters and murataite-like framework formed by $Ti2O_6$ and $Ti5O_6$ octahedra (**c**); murataite-like framework (**d**)

(Fig. 6a). The pyrochlore clusters are formed by $Ti1O_6$ and $Ti4O_6$ octahedra and are linked together by $Ti3O_6$ octahedra (Fig. 6b).

The $Ti2O_6$ and $Ti5O_6$ octahedra are located in the 'empty' space within the unit cell (Fig. 6c) and form a framework on their own (Fig. 6d). In order to analyse this framework, one has to use a nodal representation. The framework is based upon two polyhedral units derived from the structure of natural murataite *Mu*-3 (see above): a truncated tetrahedron $3^4 6^4$ (Fig. 1b) formed by *Ti*2 and cubooctahedron $4^6 6^8$ (Fig. 1c) formed by *Ti*5 atoms (Fig. 7). In contrast to the structure of murataite *Mu*-3, $3^4 6^4$ and $4^6 6^8$ units do not share hexagonal faces, but unite through linkage of triangular face of $3^4 6^4$ and hexagonal face of $4^6 6^8$ to form a truncated pyramid $3^4 4^3 6^1$ (Fig. 7b). The basic polyhedral units $3^4 6^4$ and $4^6 6^8$ in *Mu*-5 are geometrically identical to those observed in *Mu*-3. However, it is noteworthy that, in contrast to *Mu*-3, the structure of *Mu*-5 contains no Keggin polyanionic clusters.

In the structure of *Mu*-5, the subframework of linked pyrochlore clusters and the subframework of murataite-like units communicate by sharing those corners of TiO_6 octahedra, which are not involved in the subframework constructions.

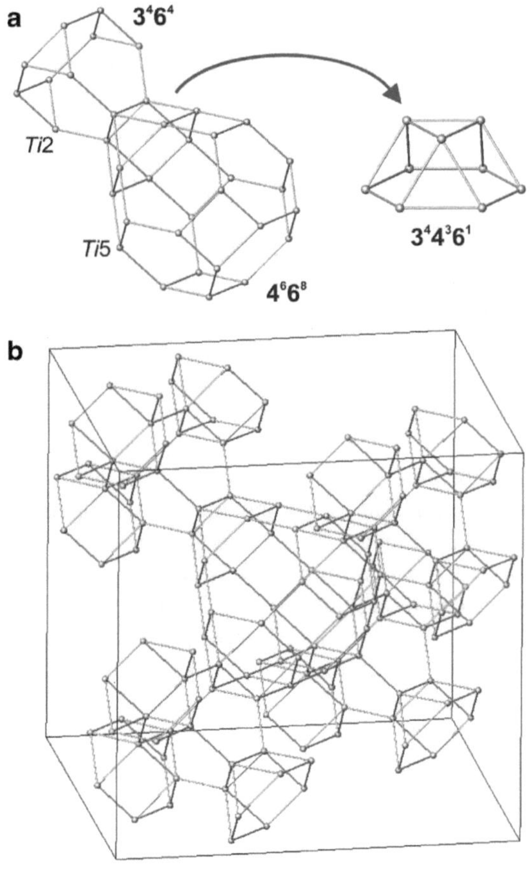

Fig. 7 To the description of murataite-like subframework in the structure of Mu-5: linkage of the $3^4 6^4$ and $4^6 6^8$ units through trigonal face of the former and hexagonal face of the latter results in formation of $3^4 4^3 6^1$ unit (**a**); nodal representation of murataite-like subframework (**b**)

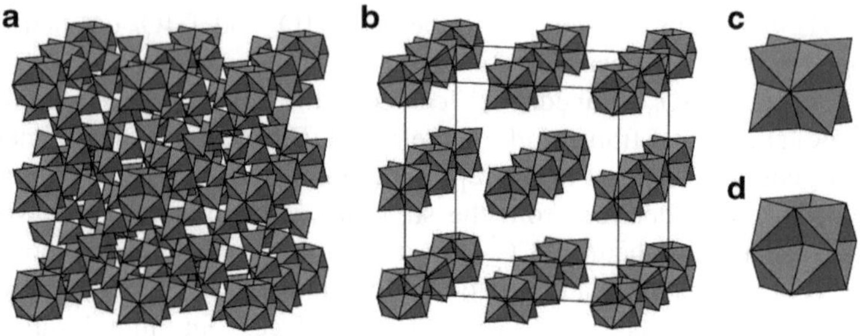

Fig. 8 Structure of oxocentered framework in murataite Mu-5 in terms of oxocentered O'A_4 tetrahedra (**a**) can be described as containing two types of building units (**b**) consisting of edge-sharing tetrahedra (**c**, **d**)

The stoichiometry of octahedral framework is rather complex and has no analogues among known octahedral frameworks. Its formula can be written as $[(Ti,Zr)_{29}O_{86}]^{56-}$. The *Mn* and *M* cations as well as O′ atoms not linked to the *Ti* sites form a fluorite-type substructure with each O′ atom being at the centre of $O'A_4$ oxocentered tetrahedron (Krivovichev and Filatov 1999), similar to the situation observed in pyrochlore and other fluorite-related framework materials (Krivovichev 1999a, b). Taking into consideration this substructure, a general structural formula of *Mu*-5 can be written as $Mn_8M_{59}O_{39}[(Ti,Zr)_{29}O_{86}]_2$. The structure of oxocentered tetrahedral framework is shown in Fig. 8. It can be described as a highly deficient fluorite-type framework based upon linkage of two basic clusters shown in Fig. 8c,d. These clusters consist of eight edge-sharing oxocentered tetrahedra.

As seen from Fig. 6, the whole framework of the *Mu*-5 structure contains modules of essentially pyrochlore (Fig. 6b) and murataite *Mu*-3 (Fig. 6d) structures as proposed by Urusov et al. (2005, 2007). Though this idea was not confirmed in this study in the exact sense, it appears to be correct in general. Namely, since pyrochlore is a $2 \times 2 \times 2$ fluorite superstructure and murataite is a $3 \times 3 \times 3$ fluorite superstructure, their isometric combination should result in the $(2n + 3\,m) \times (2n + 3\,m) \times (2n + 3\,m)$ superstructures, from which $5 \times 5 \times 5$ ($n = 1$; $m = 1$), $7 \times 7 \times 7$ ($n = 2$; $m = 1$) and $8 \times 8 \times 8$ ($n = 1$; $m = 2$) superstructures are the first simplest members of the polysomatic series. Indeed, the structure of *Mu*-5 described in this study represents a combination of single pyrochlore unit cells with recombined murataite framework and thus is a rare case of a 3D polysomatic structure constructed from two parent frameworks (note that structures with 1D and 2D modules are much more common (Ferraris et al. 2004)).

As it was mentioned above, pyrochlore is one of the most promising hosts for actinide immobilization both in terms of actinide load, radiation resistance and chemical durability. The current study reveals that the murataite-pyrochlore polysomatic series is based upon incorporation of high-actinide pyrochlore nanoclusters into modified murataite-like frameworks with cavities that may accommodate not only actinides but also Mn, Fe, Al, Ca, Sr, Na and other cations present in HLWs as admixtures and corrosion products. In this context, complex oxides from the murataite-pyrochlore series may be viewed as a complex waste form and probably the most complicated nanoscale-tuned complex oxide family known to date.

Acknowledgements This work was supported through the internal grant of St. Petersburg State University ('Mineralogy and crystal chemistry of radioactive minerals and materials').

References

Adams JW, Botinelly T, Sharp WN, Robinson K (1974) Murataite, a new complex oxide from El Paso County, Colorado. Amer Mineral 59:172–176

Belin RC, Martin PM, Valenza PJ, Scheinost AC (2009) Experimental insight into the radiation resistance of zirconia-based americium ceramics. Inorg Chem 48:5376–5381

Ercit TS, Hawthorne FC (1995) Murataite, a UB_{12} derivative structure with condensed Keggin molecules. Can Mineral 33:1223–1229

Ewing RC (1999) Nuclear waste forms for actinides. Proc Natl Acad Sci USA 96:3432–3439

Ewing RC (2005) Plutonium and "minor" actinides: safe sequestration. Earth Planet Sci Lett 229:165–181

Ferraris G, Makovicky E, Merlino S (2004) Crystallography of modular materials. Oxford University Press, Oxford

Jiang C, Stanek CR, Sickafus KE, Uberuaga BP (2009) First-principles prediction of disordering tendencies in pyrochlore oxides. Phys Rev B19:104203

Karimova OV, Organova NI, Balakirev VG (2002) Modulation in the murataite structure. Crystallogr Rep 47:957–960

Krivovichev SV (1999a) Systematics of fluorite-related structures. I. General principles. Solid State Sci 1:211–219

Krivovichev SV (1999b) Systematics of fluorite-related structures. II. Structural diversity. Solid State Sci 1:221–231

Krivovichev SV, Filatov SK (1999) Structural principles for minerals and inorganic compounds containing anion-centered tetrahedra. Amer Mineral 84:1099–1106

Krivovichev SV, Yudintsev SV, Stefanovsky SV, Organova NI, Karimova OV, Urusov VS (2010) Murataite–pyrochlore series: a family of complex oxides with nanoscale pyrochlore clusters. Angew Chem Int Ed. doi:http://dx.doi.org/10.1002/anie.201005674

Kunz M, Brown ID (1995) Out-of-center distortions around octahedrally coordinated d^0 transition metals. J Solid State Chem 112:395–406

Laverov NP, Omel'yanenko BI, Yudintsev SV, Nikonov BI, Sobolev IA, Stefanovsky SV (1997) Mineralogy and geochemistry of immobilization matrices for high level wastes. Geol Ore Deposits 39:211–228 (in Russian)

Laverov NP, Sobolev AI, Stefanovsky SV, Yudintsev SV, Omel'yanenko BI, Nikonov BI (1998a) Synthetic murataite, a new mineral for actinide immobilization. Dokl Russ Acad Sci 362:670–672 (in Russian)

Laverov NP, Gorshkov AI, Yudintsev SV, Sivtsov AV, Lapina MI (1998b) New structural species of the synthetic murataite. Dokl Earth Sci 363:1272–1274

Laverov NP, Yudintsev SV, Stefanovsky SV, Omel'yanenko BI, Nikonov BS (2006) Murataite as a universal matrix for immobilization of actinides. Geol Ore Deposits 48:335–356

Lian J, Wang LM, Ewing RC, Yudintsev SV, Stefanovsky SV (2005a) Thermally induced phase decomposition and nanocrystal formation in murataite ceramics. J Mater Chem 15:709–714

Lian J, Wang LM, Ewing RC, Yudintsev SV, Stefanovsky SV (2005b) Ion-beam-induced amorphization and order-disorder transition in the murataite structure. J Appl Phys 97:113536

Livshits TS, Yudintsev SV (2008) Natural and synthetic minerals – matrices (forms) for actinide waste immobilization. In: Krivovichev S (ed) Minerals as advanced materials I. Springer, Heidelberg, pp 193–207

Lumpkin GR, Pruneda M, Rios S, Smith KL, Trachenko K, Whittle KR, Zaluzec NJ (2007) Nature of the chemical bond and prediction of radiation tolerance in pyrochlore and defect fluorite compounds. J Solid State Chem 180:1512–1518

Lumpkin GR, Smith KL, Blackford MG, Whittle KR, Harvey EJ, Redfern SAT, Zaluzec NJ (2009) Ion irradiation of ternary pyrochlore oxides. Chem Mater 21:2746–2754
Lutze W, Ewing RC (eds) (1988) Radioactive waste forms for the future. North-Holland, Amsterdam
Morgan PED, Ryerson FJ (1982) A new "cubic" crystal compound. J Mater Sci Lett 1:351–352
Perevalov SA, Stefanovsky SV, Yudintsev SV, Mokhov AV, Ptashkin AG (2006) Leaching of neptunium from garnet – and murataite-based ceramics. Radiochim Acta 94:509–514
Ringwood AE, Kesson SE, Ware NG, Hibberson WO, Major A (1979) Immobilization of high level nuclear reactor wastes in SYNROC. Nature 278:219–223
Sickafus KE, Grimes RW, Valdez JA, Cleave A, Tang M, Ishimaru M, Corish SM, Stanek CR, Uberuaga BP (2007) Radiation-induced amorphization resistance and radiation tolerance in structurally related oxides. Nature Mater 6:217–223
Sobolev IA, Stefanovsky SV, Yudintsev SV, Omelianenko BI, Nikonov BS, Mokhov AV (1997) Study of melted synroc doped with simulated high level waste. Mater Res Soc Symp Proc 465:363–370
Stefanovsky SV, Yudintsev SV, Gieré R, Lumpkin GR (2004) Nuclear waste forms. In: Gieré R, Stille P (eds) Energy, waste and the environment: a geochemical perspective. Special publication, vol 236. Geological Society of London, London, pp 36–63
Stefanovsky SV, Yudintsev SV, Perevalov SA, Startseva IV, Varlakova GA (2007a) Leach resistance of murataite-based ceramics containing actinides. J Alloys Compds 444–445:618–620
Stefanovsky SV, Ptashkin AG, Knyazev OA, Dmitriev SA, Yudintsev SV, Nikonov BS (2007b) Inductive cold crucible melting of actinide-bearing murataite-based ceramics. J Alloys Compds 444–445:438–442
Stefanovsky SV, Varlakova GA, Startseva IV, Yuduntsev SV, Nikonov BS, Lapina MI (2009) Leach rates of uranium and thorium from murataite ceramics. Radiochim Acta 97:17–21
Strachan DM, Scheele RD, Buck EC, Kozelisky AE, Sell RL, Elovich RJ, Buchmiller WC (2008) Radiation damage effects in candidate titanates for Pu disposition: Zirconolite. J Nucl Mater 372:16–31
Subramanian MA, Aravamudan G, Subba Rao GV (1983) Oxide pyrochlores – a review. Progr Solid State Chem 15:55–143
Urusov VS, Rusakov VS, Yudintsev SV (2002) Valent state and structural position of Fe atoms in synthetic murataite. Dokl Earth Sci 384:461–465
Urusov VS, Organova NI, Karimova OV, Yudintsev SV, Stefanovsky SV (2005) Synthetic "murataites" as modular members of a pyrochlore-murataite polysomatic series. Dokl Earth Sci 401:319–325
Urusov VS, Organova NI, Karimova OV, Yudintsev SV, Ewing RC (2007) A modular model of the crystal structure of the pyrochlore-murataite polysomatic series. Crystallogr Rep 52:37–46
Weber WJ, Navrotsky A, Stefanovsky S, Vance ER, Vernaz E (2009) Materials science of high-level nuclear waste immobilization. MRS Bull 34:46–53
Yudintsev SV, Stefanovsky SV, Ewing RC (2007a) In: Krivovichev SV, Burns PC, Tananaev IG (eds) Structural chemistry of inorganic actinide compounds. Elsevier, Amsterdam, pp 457–490
Yudintsev SV, Stefanovsky SV, Nikonov BS, Maslakov KI, Ptashkin AG (2007b) Structural characterization of Pu-bearing murataite ceramic. J Alloys Compds 444–445:606–609
Zhang FX, Wang JW, Lian J, Lang MK, Becker U, Ewing RC (2008) Phase stability and pressure dependence of defect formation in $Gd_2Ti_2O_7$ and $Gd_2Zr_2O_7$ pyrochlores. Phys Rev Lett 100:045503
Zhang Y, Stewart MWA, Li H, Carter ML, Vance ER, Moricca S (2009) Zirconolite-rich titanate ceramics for immobilisation of actinides – waste form/HIP can interactions and chemical durability. J Nucl Mater 395:69–74

Lattice Distortion Upon Compression in Orthorhombic Perovskites: Review and Development of a Predictive Tool

Matteo Ardit, Michele Dondi, and Giuseppe Cruciani

1 Introduction

Perovskites are one of the most important class of materials belonging to the category "*minerals as advanced materials*". In fact, thousands of works every year are devoted to establish the countless properties of these compounds which were first discovered in 1839 as minerals in the Ural mountains and whose synthetic analogs find now application in many technological fields.

The perovskite archetype, with general formula $^{[XII]}A^{[VI]}BO_3$, is a compact structure with a cubic symmetry (s.g. *Pm*-3*m*) composed by an un-tilted framework of corner-sharing octahedra, which contain the *B* cations, and extra-framework dodecahedral sites in which are placed the *A* cations. Depending on the chemical nature of the *A* and *B* cations (i.e. size, valence, electronic configuration, etc.), and on the pressure and temperature conditions, the ideal cubic symmetry is often lowered, and the resulting structure is described by tilted octahedra and more distorted polyhedra, in which the coordination number of *A* cation is usually reduced (Mitchell 2002). In many cases the result of such tilting and distortion system is a perovskite with an orthorhombic symmetry (s.g. *Pbnm*). This kind of perovskites is characterized by peculiar properties (e.g. electric, magnetic, piezoelectric, optic) and therefore employed in several application fields, including solid oxide fuel cells (Ishihara 2009), multiferroics (Belik et al. 2009), antiferromagnetics (Komarek et al. 2008), magnetotransports (Yang et al. 2009), piezoelectric ceramics (Lim et al. 2010), photoluminescence studies (Zhydachevskii et al. 2006), etc.

Furthermore, in Earth sciences, perovskites became a striking theme since it was discovered that, at the high pressure and temperature conditions of the Earth's lower mantle, the pyroxene enstatite, $MgSiO_3$, transforms into a denser perovskite-structured

M. Ardit • G. Cruciani (✉)
Department of Earth Sciences, University of Ferrara, Via Saragat 1, 44100 Ferrara, Italy
e-mail: cru@unife.it

M. Dondi
CNR-ISTEC, Via Granarolo 64, 48018 Faenza, Italy

S.V. Krivovichev (ed.), *Minerals as Advanced Materials II*,
DOI 10.1007/978-3-642-20018-2_29,

polymorph (Navrotsky and Weidner 1989). From here, the importance to understand the relationships between external condition (such as *P* and *T*), structural properties, and chemistry of perovskite compounds becomes particularly essential to design a perovskite with specific properties for a practical application, and to suggest a reliable model on the deep Earth behavior.

In such an exciting context, many studies have been devoted to assess the high pressure behavior of the $GdFeO_3$-type perovskites (see references in Table 1), and some attempts to predict their evolution under pressure were made (Andrault and Poirier 1991; Thomas 1998; Zhao et al. 2004a).

From these works it is clear that the behavior under pressure of the orthorhombic perovskites is essentially associated to the relative compressibilities of the two polyhedra (AO_{12} and BO_6) controlling the tilting and distortion effects. The only two possible structural answers to the compression regime are: (1) the AO_{12} site is more compressible than the BO_6 octahedron, then volume reduction will lead to an increasing of the octahedral tilting; (2) the AO_{12} site is less compressible than the BO_6 octahedron, then the structure will evolve by decreasing the octahedral tilting raising its symmetry towards the cubic archetype.

The experimental methodology to obtain trustworthy bond distance values, and thus realistic polyhedral volumes, at non-ambient pressure, it is currently based on the high-pressure single-crystal X-ray diffraction. However, it is important to consider that single-crystal samples are not always available (or they require a very complex synthesis procedure). Hence, this technique suffers the major limitation of sample availability. Furthermore, in order to ensure adequate diffracted intensity (signal reduction is a common consequence of the tiny crystal size) and hydrostatic conditions, such experiments are generally performed at pressures of less than 10 GPa. On the other hand, a simpler way to achieve well crystalline compounds derives by the powder synthesis. However, high-pressure X-ray powder diffraction – although typically performed using area detectors and synchrotron radiation – may suffer of serious difficulties in reliable extraction of the diffracted intensities. This may lead to poor accuracy in Rietveld refined bond lengths and bond angles. Nevertheless, from these latter kind of experiments one can obtain high quality lattice parameters, since the positions of the diffraction maxima can be determined with best accuracy. Hence, it would be desirable to establish simple correlations between the variation of the unit-cell parameters under high-pressure and the compressibilities of AO_{12} and BO_6 polyhedra.

This work is aimed to give a useful tool that relates the lattice distortion of an orthorhombic perovskite under pressure to the relative compressibilities of the two polyhedra, by comparing several high pressure studies of $GdFeO_3$-type perovskites under high-pressure from the literature.

2 Cell Distortion Factor

The most popular parameter used to evaluate tilting and distortions of polyhedra in perovskite structures (including the orthorhombic $GdFeO_3$-type) is the well known *tolerance factor* (*t*) first introduced by Goldschmidt in 1926 on the basis of ideal ionic radii sums and most conveniently redefined by Sasaki et al. (1983) as the *observed tolerance factor* (t_{obs}) considering the observed bond distances:

$$t_{obs} = \langle ^{XII\ or\ VIII}A - O\rangle / \sqrt{2} \cdot \langle ^{VI}B - O\rangle$$

where $\langle ^{XII\ or\ VIII}A - O\rangle$ and $\langle ^{VI}B - O\rangle$ are the mean cation-anion distances based on 12- (or 8-) and 6-fold coordination of A and B cations, respectively.

A useful tool to evaluate the lattice departure of an orthorhombic perovskite from the ideal cubic model is the cell distortion factor (d). Introduced by Sasaki et al. in 1983 and by Vasylechko et al. in 1996, this parameter is calculated by comparing the unit cell axes through the relation $a \approx b \approx \sqrt{2}a_p$, and $c \approx 2a_p$ (a_p: the pseudocubic subcell parameter) as it follows:

$$d = \frac{\left[\left(\frac{a}{\sqrt{2}} - a_p\right)^2 + \left(\frac{b}{\sqrt{2}} - a_p\right)^2 + \left(\frac{c}{2} - a_p\right)^2\right]}{3{a_p}^2 \times 10^4}$$

where:

$$a_p = \frac{\left(\frac{a}{\sqrt{2}} + \frac{b}{\sqrt{2}} + \frac{c}{2}\right)}{3}$$

The cell distortion factor (we will refer to it hereafter as the *absolute* cell distortion factor) has been successfully used in many instances where the effects of changes in chemistry, symmetry and other features had to be evaluated in terms of lattice distortion variation (e.g. Sasaki et al. 1983; Vasylechko et al. in 1996; Moustafa et al. 2004).

A relatively well defined linear relationship (see Fig. 1) can be observed on a semi-logarithmic plot between the observed tolerance factor, based on the 8-fold A-site coordination ($^{VIII}t_{obs}$), and the cell distortion factor (d) for all of the orthorhombic perovskite compounds selected for this work (see next section) except those with very small lattice distortion (e.g. $CaGeO_3$ with a $d = 0.004$, very close to the value for a cubic structure). This correlation suggests that at least within the d-range from 0.2 to 8.2 the distortion of orthorhombic perovskites as measured by the departure of the unit cell from the cubic one and as evaluated by the increasing mismatch between the $\langle A - O\rangle$ and $\langle B - O\rangle$ distances are strictly connected.

Most recently an extended version of the d factor was developed by Ardit et al. (2010) in order to compare the lattice evolution under pressure of three isotypic perovskite compounds ($YM^{3+}O_3$, where M^{3+} = Cr, Al, Ti) by calculating the rate of the change of the *normalized* cell distortion factor ($d_{norm} = d\,/\,d_0$) with pressure, at constant temperature, via the equation:

$$d_{norm}(P) = \frac{1}{d_0} \cdot \left(\frac{\partial d}{\partial P}\right)_T$$

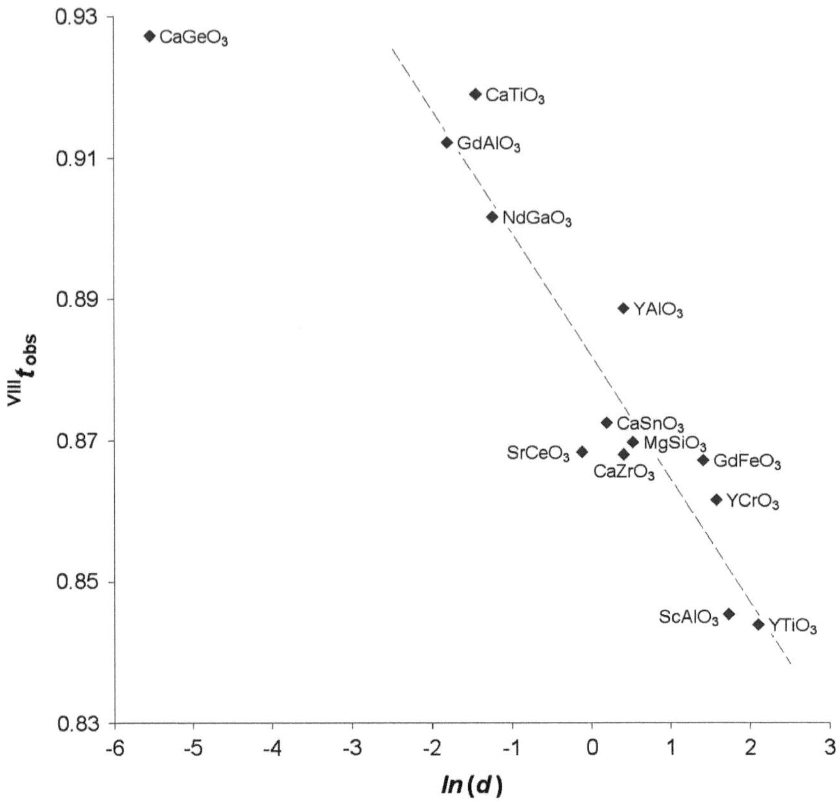

Fig. 1 Semilogarithmic plot of the cell distortion factor variation (*d*) as a function of the observed tolerance factor ($^{VIII}t_{obs}$: calculated by averaging the first eight *A*–O bond distances). Both *d* and $^{VIII}t_{obs}$ are dimensionless by definition

where d_0 is the initial absolute cell distortion factor calculated from the cell parameters at room pressure derived by equation of state fitting.

The $d_{norm}(P)$ values calculated for the Y-perovskite isotypes scale linearly with the $\langle B - O - B \rangle$ bond angles as well as with the Shannon ionic radii (Shannon 1976) of the *B* cations in the same structures. Further investigations are needed, in a wider set of data, to assess whether the same relationships apply as a more general behavior for orthorhombic perovskite under pressure.

3 Review of Orthorhombic Perovskites Under Pressure

In order to evaluate the applicability of the rate of change of the normalized lattice distortion parameter, $d_{norm}(P)$, to describe the behavior of orthorhombic perovskites upon compression, thirteen compounds were selected from the literature (see Table 1) by combining the following criteria: (i) the chosen compound must

Table 1 References of the selected compounds

Compound	High-pressure conditions	Ambient conditions
$ScAlO_3$	Ross 1998	Sinclair et al. 1979
$CaGeO_3$	Ross and Angel 1999	Sasaki et al. 1983
$CaTiO_3$	Ross and Angel 1999	Liu and Liebermann 1993
$CaZrO_3$	Ross and Chaplin 2003	Levin et al. 2003
$GdAlO_3$	Ross et al. 2004a	Ross et al. 2004a
$GdFeO_3$	Ross et al. 2004a	Ross et al. 2004a
$YAlO_3$	Ross et al. 2004b	Diehl and Brandt 1975
$CaSnO_3$	Zhao et al. 2004b	Zhao et al. 2004b
$SrCeO_3$	Knight et al. 2005	Knight and Bonanos 1995
$MgSiO_3$	Sugahara et al. 2006	Dobson and Jacobsen 2004
$YTiO_3$	Loa et al. 2007	Hester et al. 1997
$NdGaO_3$	Angel et al. 2007	Vasylechko et al. 2000
$YCrO_3$	Ardit et al. 2010	Cruciani et al. 2009

be studied both under compression and at ambient conditions; (ii) the high-pressure selected works must report, at least, the value of the cell parameters under compression, and the volumetric bulk modulus (in the case of multiple reports, priority was given to works in which the polyhedral bulk moduli values are also calculated); (iii) complete structural information (i.e. the refined atomic coordinates) must be available for the given crystal structure at ambient conditions. Table 2 reports a selection of crystallographic data for the chosen compounds. These data were considered in order to establish the correlations between structures at ambient conditions and their high-pressure evolution.

A comment is needed concerning the methods used for the polyhedral volume calculations for the *A* site in 12-fold coordination. Table 2 gives two columns, $V(AO_{12})_{convex}$ and $V(AO_{12})_{diff}$, where the $V(AO_{12})$ volumes are calculated in different ways.

The first method is based on software developed to calculate the volume, surface area and inter-edge angles for any convex polyhedron: e.g. *POLYVOL* (Swanson and Peterson 1980) or *IVTON* (Balic Zunic and Vickovic 1996). The second method to obtain the AO_{12} volume is by subtracting the octahedral volume, $V(BO_6)$, to the unit-cell volume, V_{cell}, through the relation:

$$V(AO_{12})_{diff} = (V_{cell}/Z) - V(BO_6)$$

where $Z = 4$ for $GdFeO_3$-type perovskites.

The *A*-site polyhedral volumes $V(AO_{12})$ were also calculated by multiplying the octahedral volumes to the $V(AO_{12})/V(BO_6)$ ratios determined directly from the atomic coordinates using the appropriate equations of Avdeev et al. (2007). It turned out that the values found with this last mode were exactly the same of those achieved by the second method and therefore not reported in Table 2. Comparing the volumetric values obtained with the first two methods, one can

Table 2 Unit cell parameters, lattice distortion parameters, mean bond lengths, mean bond angles, polyhedral volumes, tolerance factors, volumetric bulk modulus, ratio of the polyhedral compressibilities, and ratio of total estimated variation of bond valence of the selected perovskites

	a [Å]	b [Å]	c [Å]	V [Å^3]	d	$d_{norm}(P)$[a] [GPa^{-1}]	$^{XII}\langle A-O\rangle$ [Å]	$^{VIII}\langle A-O\rangle$ [Å]	$^{IV}\langle A-O\rangle$ [Å]	$\langle B-O\rangle$ [Å]	$\langle B-O-B\rangle$ [°]	$V(AO_{12})_{convex}$ [Å^3]	$V(AO_8)$ [Å^3]	$V(BO_6)$ [Å^3]	$V(AO_{12})_{diff}$ [Å^3]	t	$^{VIII}t_{obs}$	K_{T0}[a] [GPa]	$\beta(BO_6)/\beta(AO_{12})$[a]	M_A/M_B
$YTiO_3$	5.3381	5.6901	7.6130	231.24	8.173	0.040	2.806	2.439	3.542	2.044	141.9	48.82	24.40	11.36	46.45	0.829	0.844	163	-	0.98
$ScAlO_3$	4.9355	5.2313	7.2007	185.92	5.643	0.009	2.607	2.273	3.276	1.901	142.0	39.18	19.63	9.15	37.33	0.833	0.845	218	1.07	1.14
$YCrO_3$	5.2434	5.5242	7.5356	218.27	4.800	0.010	2.738	2.417	3.379	1.984	146.2	46.18	23.52	10.40	44.17	0.852	0.862	208	1.01	1.18
$GdFeO_3$	5.3511	5.6125	7.6711	230.38	4.064	−0.002	2.782	2.469	3.408	2.014	147.1	48.77	24.93	10.88	46.72	0.852	0.867	182	1.00	1.15
$YAlO_3$	5.1800	5.3300	7.3750	203.62	1.495	−0.042	2.655	2.401	3.163	1.911	151.8	43.16	22.55	9.30	41.60	0.889	0.889	192	1.18	1.42
$NdGaO_3$	5.4276	5.4979	7.7080	230.01	0.288	−0.109	2.756	2.526	3.216	1.981	154.0	48.67	25.70	10.36	47.14	0.883	0.902	177	1.20	1.18
$GdAlO_3$	5.2537	5.3030	7.4435	207.38	0.164	−0.121	2.658	2.457	3.058	1.905	156.4	43.86	23.44	9.21	42.63	0.901	0.912	191	1.14	1.44
$SrCeO_3$	6.0089	6.1483	8.5833	317.11	0.883	0.161	3.086	2.757	3.745	2.245	145.6	66.93	34.07	15.05	64.22	0.831	0.868	110	0.63	0.55
$MgSiO_3$	4.7780	4.9298	6.8990	162.50	1.681	0.014	2.473	2.205	3.009	1.793	147.0	34.37	17.60	7.67	32.95	0.905	0.870	254	0.87	0.70
$CaZrO_3$	5.5911	5.7613	8.0169	258.24	1.504	0.037	2.884	2.573	3.506	2.096	146.1	54.57	27.81	12.26	52.30	0.844	0.868	154	–	0.53
$CaSnO_3$	5.5142	5.6634	7.8816	246.14	1.208	0.041	2.836	2.537	3.435	2.056	147.5	52.07	26.68	11.58	49.96	0.856	0.872	163	0.81	0.56
$CaTiO_3$	5.3785	5.4419	7.6400	223.62	0.234	0.006	2.722	2.532	3.101	1.949	157.5	47.29	25.40	9.86	46.04	0.893	0.919	171	–	0.61
$CaGeO_3$	5.2607	5.2688	7.4452	206.36	0.004	−0.085	2.647	2.481	2.979	1.892	159.7	43.72	23.69	9.03	42.56	0.929	0.927	194	–	0.74

[a]Data quoted or calculated from the high-pressure studies of Table 1

M_A/M_B: ratio of total estimated variation of bond valence in a polyhedral site due to change of average bond distance (Zhao et al. 2004a). $M_i = \frac{R_i N_i}{B}\exp\left(\frac{R_0-R_i}{B}\right)$ where R_i = average bond length; N_i = coordination number of the cation site at ambient condition; R_0 = bond valence parameter (Brown and Altermatt 1985); and B = universal constant (0.37).

note that the volume calculated on a convex polyhedra, $V(AO_{12})_{convex}$, is systematically overestimated with respect to the volume obtained by difference, $V(AO_{12})_{diff}$.

It is noteworthy that such a mismatch in polyhedral volume calculations leads to a significant difference in the derived polyhedral bulk modulus, K_{AO12}. For instance, in the case of $YCrO_3$ and $YAlO_3$, the K_{YO12} polyhedral bulk moduli as derived from volumes calculated by convex polyhedra are equal to 254 and 190 GPa, respectively. On the other hand, when derived by calculating the cubic volume as difference of the cell volume and the octahedral volume, values of 200 GPa ($YCrO_3$) and 208 GPa ($YAlO_3$) are obtained for the K_{YO12}. This implies a substantial difference in evaluating the perovskite pressure evolution.

4 Pressure Evolution Trends

The relationships between the absolute lattice distortion parameter (*d*) at ambient conditions and the rate of change with pressure of its normalized form, $d_{norm}(P)$, are plotted in Fig. 2 for the selected perovskite compounds. This figure shows that $A^{3+}B^{3+}O_3$ (i.e. "3:3") and $A^{2+}B^{4+}O_3$ (i.e. "2:4") perovskites lie on two distinct logarithmic trends.

From these trends we infer several pieces of information: (i) 3:3 and 2:4 compounds are characterized by a different curve growth rate; (ii) the degree of lattice distortion at ambient condition for 2:4 perovskites are usually smaller than that of 3:3 compounds. The maximum *d* value for the 2:4 group is close to 1.50 (the one of $CaZrO_3$) while it reaches $d = 8.17$ for $YTiO_3$; (iii) if $SrCeO_3$ is excluded (see below), the maximum degree of lattice distortion upon compression levels to almost the same value for both 2:4 and 3:3 (maximum $d_{norm}(P) \approx 0.04$ for $CaSnO_3$ and $YTiO_3$).

$SrCeO_3$ and $MgSiO_3$ are two exceptional cases within the family of orthorhombic perovskites. According to Knight et al. (2005), $SrCeO_3$ behaves atypically with respect to other $A^{2+}M^{4+}O_3$ perovskites with M^{4+} = Sn , Ti, and Ge (Zr can be also included in this list) due to the strongest Ce–O bond length reduction under pressure. This is explained by the far largest ionic radius of Ce^{4+} (0.87 Å) compared to that of Sn^{4+} (0.69 Å), Ti^{4+} (0.605 Å), Ge^{4+} (0.53 Å), and Zr^{4+} (0.72 Å) with the same coordination number. Hence, as the Ce cation–anion bond length is the longest for the five elements considered, it will compress further before steric repulsion between the oxygen anions becomes too severe. A similar line of reasoning can be applied to justify the outlying behavior of $MgSiO_3$, which also appears anomalous although less than $SrCeO_3$. We speculate that these two structures are on opposite extremes as far as the main response to compression is achieved by compression or rotation of octahedra. While $SrCeO_3$ undergoes an exceptional compression of the Ce–O bond, on the other side, the short Si–O bond length in $MgSiO_3$ (due to the smallest ionic radius of 6-fold Si^{4+}, 0.40 Å)

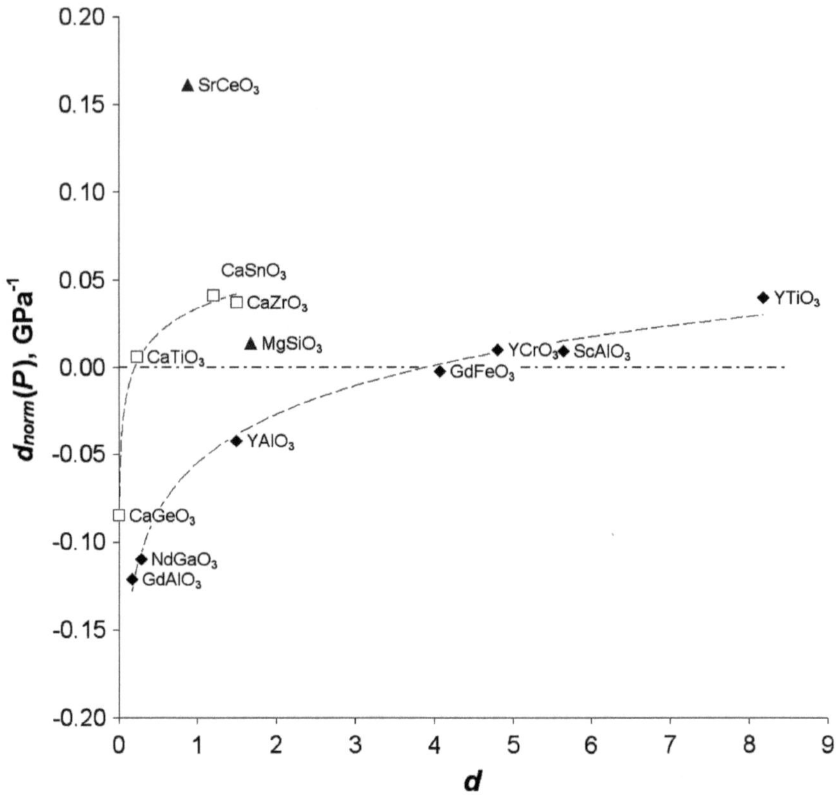

Fig. 2 Logarithmic relationships between the rate of the change of the normalized cell distortion factor, $d_{norm}(P)$, as a function of the cell distortion parameter, *d*. *Dash-dotted* line discriminates compounds with an increasing lattice distortion, $d_{norm}(P) > 0$, and compounds with decreasing lattice distortion, $d_{norm}(P) < 0$, upon compression

implies that the steric repulsion between the oxygen anions are active since the very beginning and the compression must be accounted exclusively by tilting of the SiO_6 octahedra.

When plotted against the observed tolerance factor ($^{VIII}t_{obs}$) the high-pressure lattice distortion rate identifies two well defined sub-parallel trends (see Fig. 3a). This suggests that the tolerance factor at room pressure has indeed a predictive significance for the degree of pressure-induced lattice distortion of perovskites. This contradicts what stated by Zhao et al. (2004a) who found no apparent correlation between their M_A/M_B parameter and the tolerance factor but confirms the speculations of Kunz and Brown (1995). On the plot of Fig. 3a $SrCeO_3$ behaves again like an outlier, deviating from the trends defined by the other compounds because of its comparably strongest lattice distortion under compression. It is noteworthy that, contrary to what stated by Knight et al. (2005), the small tolerance factor does not explain by itself the large degree of pressure distortion in $SrCeO_3$.

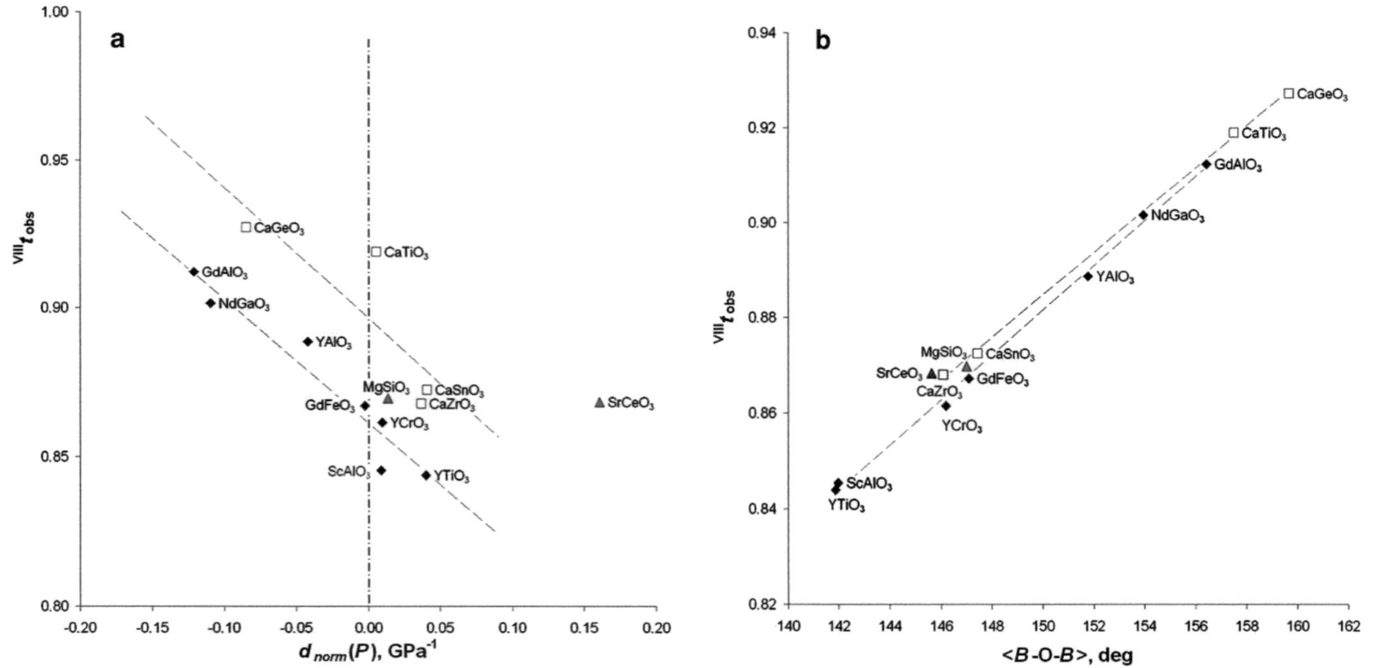

Fig. 3 (**a**) Observed tolerance factor, ${}^{VIII}t_{obs}$, vs. rate of change of the normalized cell distortion factor, $d_{norm}(P)$. *Dash-dotted* line discriminates compounds with an increasing lattice distortion, $d_{norm}(P) > 0$, and compounds with decreasing lattice distortion, $d_{norm}(P) < 0$, upon compression. (**b**) The observed tolerance factor, ${}^{VIII}t_{obs}$, as a function of the mean octahedral bond angles, $\langle B - \mathrm{O} - B \rangle$

In order to get further insights on the relations between the high-pressure lattice distortion rate and the tolerance factor at room pressure, the latter parameter is plotted in Fig. 3b as a function of the initial octahedral tilting, expressed by the mean bond angles at ambient pressure, $\langle B-\mathrm{O}-B\rangle$. The expected excellent correlation between $^{\mathrm{VIII}}t_{\mathrm{obs}}$ and $\langle B-\mathrm{O}-B\rangle$ shows that both the 3:3 and 2:4 perovskites considered here are aligned on a common trend and both range from low to high degree of initial octahedral tilting. This could be regarded as unexpected, given the systematically lower initial *d* factor of $A^{2+}B^{4+}\mathrm{O}_3$ perovskites compared to $A^{3+}B^{3+}\mathrm{O}_3$ (see Fig. 2); however it points to the same evidence of Fig. 2 about the presence of a common upper limit for the maximum rate of distortion with pressure, irrespective of the composition. In other words, it appears that while the tolerance factor reflects the degree of octahedral tilting, the absolute lattice distortion parameter in the pressure evolution of perovskites as expressed by $d_{norm}(P)$ accounts uniformly for the two distortion mechanisms (both the octahedral tilting and the initial compression), in different proportions for any given structure. It becomes more clear now that the tolerance factor of $\mathrm{SrCeO_3}$ is perfectly consistent with its initial octahedral tilting but the same parameter is not able, for this particular compound, to predict the high-pressure behavior as it successfully does for the other perovskites.

5 Conclusions

The high-pressure behavior of thirteen orthorhombic perovskites has been reviewed and rationalized with the help of a parameter newly introduced here, the rate of change with pressure of the normalized cell distortion factor, $d_{norm}(P)$. The correlations found between this parameter and others calculated from the crystallographic data refined for the given structures at ambient conditions, such as the unit cell (absolute *d* factor of Sasaki et al. 1983) or the bond distances (observed tolerance factor, $^{\mathrm{VIII}}t_{\mathrm{obs}}$) and angles ($\langle B-\mathrm{O}-B\rangle$), suggest that the pressure evolution of perovskites can be predicted on the basis of room pressure data, in particular the degree of cell distortion.

A more ambitious goal would be to develop a tool capable to predict the relative compressibility of the $B\mathrm{O}_6$ and $A\mathrm{O}_{12}$ polyhedra from the evolution of the cell parameters under pressure. In fact, as proposed and discussed by several authors (e.g. Andrault and Poirier 1991; Thomas 1998; Kunz and Brown 1995; Zhao et al. 2004a; etc.) the structural adjustments and the high-pressure behavior of perovskite structures are mainly dependent on the polyhedral volume (V_A/V_B) and polyhedral compressibility (β_B/β_A) ratios, respectively. On this base Zhao et al. (2004a) devised a general rule in which the compressibility ratio, β_B/β_A, is calculated from the ratio of the estimated variation of bond valence in the *A* and *B* polyhedral sites due to the change of average bond distance, M_A and M_B. This rule predicts that when $\beta_B / \beta_A = M_A / M_B < 1$, as in the case of 2:4 perovskites, the octahedral tilting and perovskite distortion will increase with pressure while the opposite will

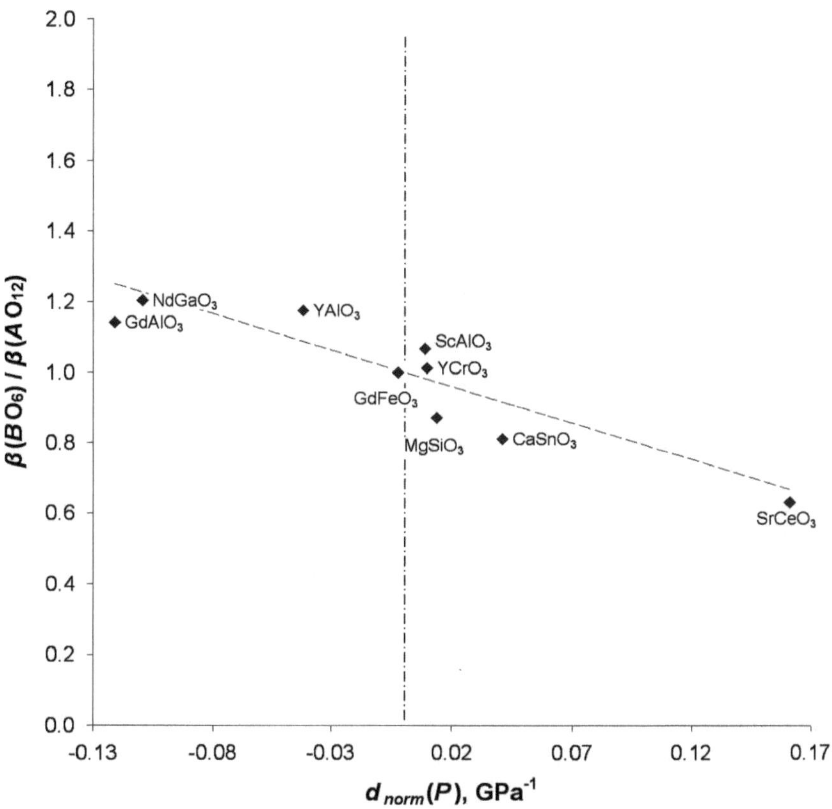

Fig. 4 Variation of the polyhedral compressibility ratio, $\beta(BO_6)$ / $\beta(AO_{12})$ with the changing rate of the normalized cell distortion factor, $d_{norm}(P)$. *Dash-dotted* line discriminates compounds with an increasing lattice distortion, $d_{norm}(P) > 0$, and compounds with decreasing lattice distortion, $d_{norm}(P) < 0$, upon compression

take place in 3:3 perovskites where β_B / $\beta_A = M_A$ / $M_B < 1$; no change would occur when β_B / $\beta_A = M_A$ / $M_B = 1$. The relationship between β_B / β_A and $d_{norm}(P)$ is reported in Figure 4 for the subgroup of $GdFeO_3$-type compounds reviewed in this work, whose elastic properties were available or could be derived from the literature. The clear-cut inverse correlation shows that the rate of change of lattice distortion with pressure scales with the relative compressibility of BO_6 and AO_{12} polyhedra such a way that β_B / β_A ratios > 1 are associated to a decreasing distortion upon compression, i.e. $d_{norm}(P) < 0$. Among the limited number of cases considered in the present review, only 3:3 peroskites occur in this group. For β_B / β_A ratios < 1 the polyhedral tilting and perovskite distortion increase, showing $d_{norm}(P) > 0$. It is noteworthy that both 2:4 and 3:3 perovskites belong to this latter group. Therefore, as a general trend, the rule formulated by Zhao et al. (2004a) is confirmed although the sharp equivalences '3:3 perovskites

= decreasing distortion at HP' and '2:4 perovskites = increasing distortion at HP' are clearly confuted. Possible effects from different electronic configurations (e.g. those of transition metals), justifying the presence of $YCrO_3$ and $YTiO_3$ in the $d_{norm}(P) > 0$ (see also Table 2), cannot be excluded, as anticipated by Zhao et al. (2004a). However, the relationship in Figure 4 appears valid for the whole group of structures considered, irrespective of the electronic configuration of the *B* cation. Furthermore, $ScAlO_3$ also belongs to the $d_{norm}(P) > 0$ group even though very close to the $d_{norm}(P) = 0$ (corresponding to $\beta_B / \beta_A = 1$ limit).

More studies of perovskites under pressure are obviously required to validate the above considerations and to turn the relationship in Figure 4 into a practical tool to predict the relative compressibility of the BO_6 and AO_{12} polyhedra in $GdFeO_3$-type perovskites simply from the variation of cell parameters under pressure.

In particular, more work is needed (and it is in progress) to find and test correlations which would allow to independently estimate the polyhedral bulk modulus of site *A* using room pressure data.

More work in progress is aimed to test the predictive capability of the relationships presented in this review. For instance, they predict that a perovskite with composition $YAl_{0.25}Cr_{0.75}O_3$ will neither increase nor decrease its tilting distortion under compression.

References

Andrault D, Poirier J (1991) Evolution of the distortion of perovskites under pressure: an EXAFS study of $BaZrO_3$, $SrZrO_3$ and $CaGeO_3$. Phys Chem Miner 18:91–105

Angel R, Zhao J, Ross N, Jakeways C, Redfern S, Berkowski M (2007) High-pressure structural evolution of a perovskite solid solution (La1-x, Ndx)GaO3. J Solid State Chem 180:3408–3424

Ardit M, Dondi M, Merlini M, Bouvier P, Cruciani G (2010) Elastic properties of $YCrO_3$ perovskite up to 60 GPa. Phys Rev B 82(064109):1–7

Avdeev M, Caspi E, Yakovlev S (2007) On the polyhedral volume ratios V_A/V_B in perovskites ABX_3. Acta Crystallogr B63:363–372

Balic Zunic T, Vickovic I (1996) *IVTON* – program for the calculation of geometrical aspects of crystal structures and some crystal chemical applications. J Appl Crystallogr 29:305–306

Belik A, Yusa H, Hirao N, Ohishi Y, Takayama-Muromachi E (2009) Structural properties of multiferroic $BiFeO_3$ under hydrostatic pressure. Chem Mater 21:3400–3405

Brown I, Altermatt D (1985) Bond-valence parameters obtained from a systematic analysis of the inorganic crystal structure database. Acta Crystallogr B41:244–247

Cruciani G, Ardit M, Dondi M, Matteucci F, Blosi M, Dalconi M, Albonetti S (2009) Structural relaxation around Cr^{3+} in $YAlO_3$–$YCrO_3$ perovskites from electron absorption spectra. J Phys Chem A113:13772–13778

Diehl R, Brandt G (1975) Crystal structure refinement of $YAlO_3$, a promising laser material. Mater Res Bull 10:85–90

Dobson D, Jacobsen S (2004) The flux growth of magnesium silicate perovskite single crystals. Am Miner 89:807–811

Goldschmidt V (1926) Geochemische Verteilungsgesetze der elemente. Kongelike norske videnskabers selskabs skrifter No8

Hester J, Tomimoto K, Noma H, Okamura F, Akimitsu J (1997) Electron density in $YTiO_3$. Acta Crystallogr B53:739–744

Ishihara T (2009) Perovskite oxide for solid oxide fuel cells. Fuel cells and hydrogen energy series. Springer, New York

Knight K, Bonanos N (1995) The crystal structures of some doped and undoped alkaline earth cerate perovskites. Mater Res Bull 30:347–356

Knight K, Marshall W, Bonanos N, Francis D (2005) Pressure dependence of the crystal structure of $SrCeO_3$ perovskite. J Alloy Compd 394:131–137

Komarek A, Streltsov S, Isobe M, Möller T, Hoelzel M, Senyshyn A, Trots D, Fernández-Díaz M, Hansen T, Gotou H, Yagi T, Ueda Y, Anisimov V, Grüninger M, Khomskii D, Braden M (2008) $CaCrO_3$: an anomalous antiferromagnetic metallic oxide. Phys Rev Lett 101 (167204):1–4

Kunz M, Brown I (1995) Out-of-center distortions around octahedrally coordinated d^0 transition metals. J Solid State Chem 115:395–406

Levin I, Amos T, Bell S, Farber L, Vanderah T, Roth R, Toby B (2003) Phase equilibria, crystal structures, and dielectric anomaly in the $BaZrO_3$–$CaZrO_3$ system. J Solid State Chem 175:170–181

Lim J, Zhang S, Jeon J-H, Shrout T (2010) (K, Na)NbO3-based ceramics for piezoelectric "hard" lead-free materials. J Am Cer Soc 93:1218–1220

Liu X, Liebermann R (1993) X-ray powder diffraction study of $CaTiO_3$ perovskite at high temperatures. Phys Chem Miner 20:171–175

Loa I, Wang X, Syassen K, Roth H, Lorenz T, Hanfland M, Mathis Y-L (2007) Crystal structure and the Mott–Hubbard gap in $YTiO_3$ at high pressure. J Phys Condens Mat 19 (406223):1–9

Mitchell R (2002) Perovskites: modern and ancient. Almaz Press, Ontario

Moustafa A, Ahmed Farag I, Salah L (2004) Structural characterization of substituted calcium titanate compounds $Ca_{1-x}La_xTi_{1-x}Fe_xO_3$. Egypt J Solids 27:213–222

Navrotsky A, Weidner D (1989) Perovskite: a structure of great interest to geophysics and materials science, vol 45, Geophysical Monograph. AGU, Washington DC

Ross N (1998) High pressure study of $ScAlO_3$ perovskite. Phys Chem Miner 25:597–602

Ross N, Angel R (1999) Compression of $CaTiO_3$ and $CaGeO_3$ perovskites. Am Miner 84:277–281

Ross N, Chaplin T (2003) Compressibility of $CaZrO_3$ perovskite: comparison with Ca-oxide perovskites. J Solid State Chem 172:123–126

Ross N, Zhao J, Burt J, Chaplin T (2004a) Equations of state of $GdFeO_3$ and $GdAlO_3$ perovskites. J Phys Condens Mat 16:5721–5730

Ross N, Zhao J, Angel R (2004b) High-pressure single-crystal X-ray diffraction study of $YAlO_3$ perovskite. J Solid State Chem 177:1276–1284

Sasaki S, Prewitt C, Liebermann R (1983) The crystal structure of $CaGeO_3$ perovskite and the crystal chemistry of the $GdFeO_3$-type perovskite. Am Miner 68:1189–1198

Shannon R (1976) Revised ionic radii and systematic studies of interatomic distances in halides and chalcogenides. Acta Crystallogr A32:751–767

Sinclair W, Eggleton R, Ringwood A (1979) Crystal synthesis and structure refinement of high pressure $ScAlO_3$ perovskite. Z Kristallogr 149:307–314

Sugahara M, Yoshiasa A, Komatsu Y, Yamanaka T, Bolfan-Casanova N, Nakatsuka A, Sasaki S, Tanaka M (2006) Reinvestigation of the $MgSiO_3$ perovskite structure at high pressure. Am Miner 91:533–536

Swanson D, Peterson R (1980) Polyhedral volume calculations. Can Miner 18:153–156

Thomas N (1998) A new global parameterization of perovskite structures. Acta Crystallorg B54:585–599

Vasylechko L, Akselrud L, Matkovskii A, Sugak D, Durygin A, Frukacz Z, Lukasiewicz T (1996) Crystal structure of the compound $Y_{0.5}Er_{0.5}AlO_3$. J Alloy Compd 242:18–21

Vasylechko L, Akselrud L, Morgenroth W, Bismayer U, Matkovskii A, Savytskii D (2000) The crystal structure of $NdGaO_3$ at 100 K and 293 K on synchrotron data. J Alloy Compd 297:46–52

Yang H, Cao Z, Shen X, Xian T, Feng W, Jiang J, Feng Y, Wei Z, Dai J (2009) Fabrication of 0–3 type manganite/insulator composites and manipulation of their magnetotransport properties. J Appl Phys 106(104317):1–7

Zhao J, Ross N, Angel R (2004a) New view of the high-pressure behaviour of $GdFeO_3$-type perovskites. Acta Crystallogr B60:263–271

Zhao J, Ross N, Angel R (2004b) Tilting and distortion of $CaSnO_3$ perovskite to 7 GPa determined from single-crystal X-ray diffraction. Phys Chem Miner 31:299–305

Zhydachevskii Ya, Galanciak D, Kobyakov S, Berkowski M, Kamińska A, Suchocki A, Ya Z, Durygin A (2006) Photoluminescence studies of Mn^{4+} ions in $YAlO_3$ crystals at ambient and high pressure. J Phys Condens Mat 18:11385–11396

Natural and Synthetic Layered Pb(II) Oxyhalides

Oleg I. Siidra, Sergey V. Krivovichev, Rick W. Turner, and Mike S. Rumsey

1 Introduction

Lead oxyhalides occur under variety of natural and technological conditions. They can be found as secondary minerals in oxidation zones of mineral deposits. For instance, Merehead quarry in England is the famous place for many findings of lead oxyhalides first described in 1923 by Spencer and Mountain (1923). Genesis of these deposits is still a topic of discussions. The most recent version points out that galena deposits were emplaced into limestones during the Triassic period (Turner 2006). Afterwards, they were locally exposed to the action of seawater. Oxidation of galena initiated deposition of manganate minerals from the seawater as well as adsorption of heavy metals from both seawater and local environment. A subsequent hydrothermal event heated the lead-manganate deposits causing conditions, which led to the formation of the suite of unusual secondary minerals – including a number of rare oxychlorides: kombatite $Pb_{14}(VO_4)_2O_9Cl_4$, sahlinite $Pb_{14}(AsO_4)_2O_9Cl_4$, asisite $Pb_7SiO_8Cl_2$, parkinsonite $Pb_7MoO_9Cl_2$, mendipite $Pb_3O_2Cl_2$, damaraite $Pb_3O_2Cl(OH)$, symesite $Pb_{10}(SO_4)O_7Cl_4(H_2O)$, blixite $Pb_8O_5(OH)_4Cl_2$, schwartzembergite $Pb_5IO_6H_2Cl_3$ and others (Siidra et al. 2008).

In general, formation and precipitation of lead oxysalts plays an important role in the transport of Pb from mines and mill tailings to the biosphere. Post and Buseck (1985) identified mendipite-related compounds in automobile exhaust gases. The adding of lead acetate to the fuel is no longer a problem in the most of European

O.I. Siidra (✉) • S.V. Krivovichev
Department of Crystallography, Geological Faculty, Saint-Petersburg State University, University emb. 7/9, St. Petersburg 199034, Russia
e-mail: siidra@mail.ru

R.W. Turner
The Drey, Allington Track, Allington, Salisbury SP4 0DD, Wiltshire, UK

M.S. Rumsey
Department of Mineralogy, Natural History Museum, Cromwell Road, London SW7 5BD, UK

S.V. Krivovichev (ed.), *Minerals as Advanced Materials II*,
DOI 10.1007/978-3-642-20018-2_30,

countries, where most of the gasoline is lead-free. However, in developing countries tetraethyl lead additives are widely used as a common antiknock octane booster. Several researchers identified lead oxyhalides in roadside soils (Edwards et al. 1992). Lead oxysalts – derivatives from tetragonal PbO are well-known corrosion products of lead artefacts and precipitates in natural gas production installations (Smith 1976). Phases belonging to lead oxosulfates form during production of lead acid batteries (Steele et al. 1997).

However, lead oxyhalides have not only environmental importance, but are also interesting from the viewpoint of material science.

Materials with anisotropic crystal structure exhibit a variety of physical properties that depend strongly on crystallographic direction, such as a refractive index, the permanent dipole moment in ferroelectrics, the magnetocrystalline anisotropy in ferromagnets, and carrier mobilities in ionic conductors. Our investigations (Siidra et al. 2007) revealed that the crystal structure of a new non-stoichiometric oxychloride is formed by nonconducting and conducting two dimensional blocks and can be considered to be a two-dimensional solid-state ionic conductor. The thickness of the nonconducting layer is about 1.5 nm, which allows to describe this oxychloride as a nanocondenser. Lead oxyhalides are also of considerable interest for fabrication of nanomaterials. Highly birefringent lead oxychlorides with the mendipite composition were synthesized (Sigman and Korgel 2005) and shown to exhibit greatly enhanced birefringence relative to the bulk material.

2 General Aspects on the Crystal Chemistry of Layered Lead Oxyhalides

The crystal structure of tetragonal modification of divalent lead oxide (litharge) (Boher et al. 1985) is based upon continuous layers formed by edge-sharing OPb_4 tetrahedra (Fig. 1). The interesting feature of this modification of PbO is the possible presence of weak Pb···Pb attractive interactions. Despite the layered character of its structure, the mineral has no cleavage parallel to (001). The nuclear magnetic resonance (NMR) investigations showed the presence of Pb···Pb weak interactions (Gabuda et al. 1999), which, however, are strong enough to inhibit the (001) cleavage. Pb···Pb distance between the layers is about 3.65 Å. The role of the close shell interactions seems to be very important in stabilizing some structure types (Pyykkö 1988).

To transform the [OPb] layer into any type of layers observed in lead oxyhalides, one has to excise certain blocks of OPb_4 tetrahedra from the former (Fig. 1) (Krivovichev et al. 2004). This procedure is known for the PbO derivative structures and has been also applied to oxychalcogenides, intermetallic and cluster compounds as well (Liao and Kanatzidis 1993).

Lead oxychlorides with PbO-derived blocks have first been described as synthetic compounds by Aurivillius (1982). Structural investigations revealed that the

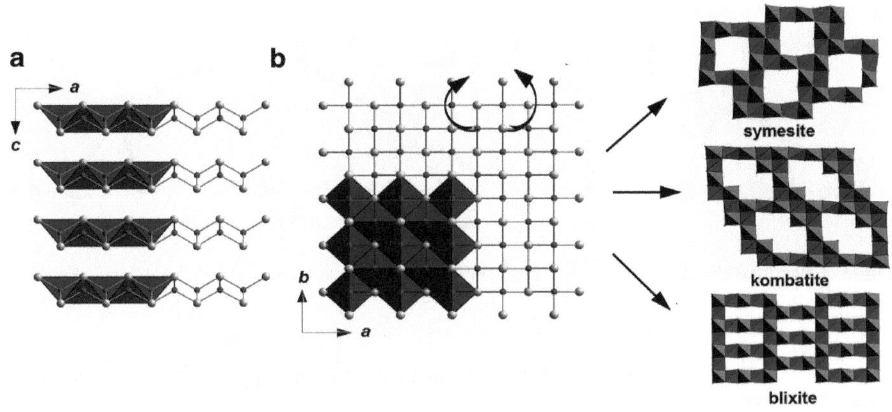

Fig. 1 General projection of the crystal structure of tetragonal PbO projected along the *b* (**a**) and *c* (**b**) axis. The PbO derivative structures can be obtained by the excision of certain blocks of OPb_4 tetrahedra. General projection of layers formed by OPb_4 tetrahedra in lead oxyhalides (*right*)

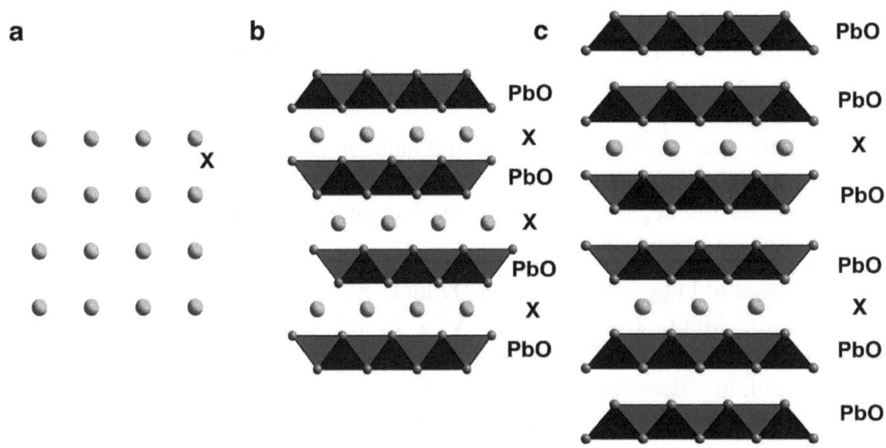

Fig. 2 The illustration of structural architecture of 1:1 (**b**) and 2:1 (**c**) in lead oxyhalides containing PbO blocks and tetragonal *X* sheets of halide ions (*X* = Cl, Br, I) (**a**)

structures of these phases consist of PbO-like blocks alternating with tetragonal sheets of halide ions, which we can designate as *X*. In the PbO-related phases reported in Aurivillius (1982), the stacking sequence of sheets is ...*X*|PbO|PbO|*X*|PbO|PbO|..., i.e. the (PbO):*X* ratio is 2:1 (Fig. 2a). The 2:1 structure has been observed in asisite and parkinsonite. The crystal structures of these minerals (Welch 2004; Lepore and Welch 2010) have disordered configuration. Another possible stacking sequence is 1:1 (...*X*|PbO|*X*|PbO|*X*...) with simple alternation of the PbO blocks and halogen sheets (Fig. 2b). This type of structure is specifically

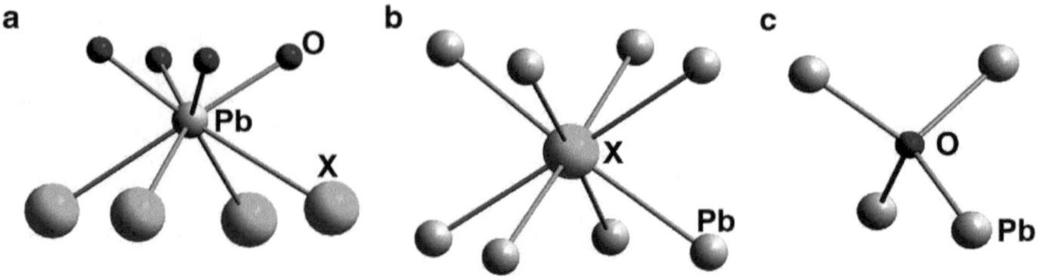

Fig. 3 Coordination of Pb (**a**), *X* (**b**), and O (**c**) atoms in the structure of ideal 1:1 lead oxyhalide with PbO blocks

characteristic for minerals and has been found in kombatite, sahlinite, symesite, schwartzembergite, and blixite.

In an ideal 1:1 structure, Pb atoms are coordinated by four O atoms in one coordination hemisphere and four halogen atoms in another (Fig. 3). The PbO_4X_4 polyhedron is a square antiprism. The halogen atoms have a cubic coordination, whereas O atoms have tetrahedral coordination, thus being central for the oxocentered OPb_4 tetrahedra. These tetrahedra are important structural units in a number of lead oxysalts and, in particular, in natural and synthetic lead oxyhalides (Siidra et al. 2008). The presence of oxocentered tetrahedra in crystal structure controls the anisotropy of such physical properties as thermal expansion, cleavage and optical properties.

Since PbO blocks are electroneutral and halogen sheets are negatively charged, the structure requires a charge-balance mechanism (Krivovichev et al. 2009) to compensate for the negative charge of the halogen sheets. This may be achieved through three different schemes:

1. Substitution of PbO groups in the PbO blocks by strong oxygen polyhedra – SO_4 (symesite), MoO_4 (parkinsonite), VO_4 (kombatite), AsO_4 (sahlinite), IO_4 (schwartzembergite), SiO_4 (asisite). The substitution may be either ordered or disordered. In ordered structures, vacancies in the PbO blocks have the form of a square (in symesite) (Fig. 1a), or a double square (butterfly-like vacancies in kombatite and sahlinite) (Fig. 1b). In the case of disordered arrangements, a number of superstructures have been observed using transmission electron microscopy.
2. $O^{2-} \leftrightarrow OH^-$ substitution mechanism in the PbO blocks. This mechanism is realized in the structure of blixite (Krivovichev and Burns 2006) where three oxygen anions are replaced by two hydroxyl anions. The vacancies in the PbO block have a form of an elongated rectangle (Fig. 1c).
3. Insertion of Pb atoms into the halogen sheet. The Pb atom can be inserted into the center of square formed by four halogen atoms. In this case, the Pb atom forms four Pb-halogen bonds in the halogen sheet, thus adopting a planar square coordination. This mechanism was first reported for synthetic compound $Pb_{31}O_{22}Br_{10}Cl_8$ (Krivovichev et al. 2006).

3 Structural Chemistry of Complex Lead Oxyhalides

Mereheadite is the example of a Pb oxychloride based upon PbO blocks. It was first described by Welch and coauthors (1998) as a new mineral from the Merehead quarry, Somerset, that was formed as a result of the interaction of brines with primary sulfides. The mineral was described as monoclinic, $C2/c$, $a = 5.680(2)$ Å, $b = 5.565(3)$ Å, $c = 13.143(9)$ Å, $\beta = 90.64(4)$º, $V = 415.4(8)$ Å^3. In their work, Welch et al. (1998) noted that the structure of the mineral had been determined and that it contains PbO blocks, which alternate along the c axis with layers of chlorine anions. Although, in this structural model, there are no positions for borate and carbonate groups, Welch et al. (1998) suggested that they replace chlorine in between the PbO blocks. The role of the OH groups has also been unclear. Thus, the structure model of mereheadite suggested by Welch et al. (1998) represents a substructure with disordered arrangements of borate, carbonate, and hydroxyl groups. It is well probable that the sample of mereheadite used by Welch et al. (1998) for single-crystal structure study represented a disordered variety of mereheadite.

The structural studies of layered Pb oxyhalides are challenging due to a number of factors, including X-ray absorption by heavy Pb^{2+} cations with diffuse electron shells, strong pseudosymmetries, poorly defined weak superstructure reflections, and twinning problems. Thus, the choice of crystal for X-ray diffraction study is dramatically important.

The structure of mereheadite contains 30 symmetrically independent Pb positions. 28 out of 36 sites belong to the PbO blocks, whereas two positions are located within the tetragonal sheets of the Cl^- anions according to the third mechanism (see above). These interlayer positions are slightly under-occupied and are complemented by the Pb sites of the PbO blocks. The coordination configurations of the Pb atoms of the PbO blocks are distorted versions of the square antiprism. In one half of the coordination hemisphere, they are coordinated by hard oxygen and OH^- anions, which number varies from three to four. Another coordination hemisphere invariably consists of four soft Cl^- anions located at the vertices of a distorted square. The Pb atoms in between the PbO blocks have almost planar square coordination of four Cl^- anions. The $PbCl_4$ squares are complemented by triangular borate and carbonate groups, so that a sevenfold coordination is achieved.

In Fig. 4 the layer in the crystal structure of mereheadite is viewed from the standpoint of oxocentered OPb_4 tetrahedra. The block can be described as a highly defect version of an ideal PbO block. It is noteworthy that the block contains square-shaped vacancies occupied by triangular groups. There are seven OH groups in the Pb-O/OH block. Mostly all of OH groups are coordinated by four Pb atoms each, whereas the OH6 group is coordinated by two Pb atoms only. Thus, the Pb-O/OH block in the crystal structure of mereheadite can be obtained from the ideal PbO block by the following list of procedures:

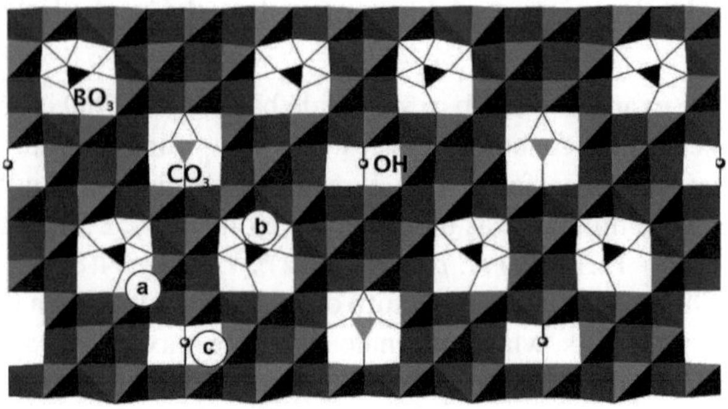

Fig. 4 General representation of the Pb-O/OH block in the crystal structure of mereheadite. The layer can be obtained by the following list of procedures: 1. removal of some PbO_4 groups (**a**); 2. insertion of triangular groups; 3. replacement of two O^{2-} anions by one OH^- anion

(a) Removal of some PbO_4 groups that results in formation of square-shaped vacancies;
(b) Insertion of triangular groups into these vacancies;
(c) Replacement of two O^{2-} anions by one OH^- anion with twofold coordination; this results in formation of the 1×2 elongated rectangular vacancy.

The structural formula that can be derived on the basis of the results of single-crystal structure determination is $Pb_{47}O_{24}(OH)_{13}Cl_{25}(BO_3)_2(CO_3)$. In a previous work (Welch et al. 1998), the simplified formula PbO(OH)Cl for mereheadite was proposed, which assumes that neither borate or carbonate are essential constituents of mereheadite and their presence in the mineral is due to their disordered replacements of Cl^- anions. However, it has been shown (Krivovichev et al. 2009) that this is not the case, as borate and carbonate groups have well-defined structural positions confined in the vacancies of the Pb-O/OH blocks. Thus mereheadite is an outstanding example of lead oxyhalides, where all three basic mechanisms are at work. Mereheadite is the first mineral, where additional Pb sites are located within the sheets of Cl^- anions. The only other example of such architecture is $Pb_{31}O_{22}Br_{10}Cl_8$ (Fig. 5) (Krivovichev et al. 2006). This compound is a high-temperature phase that has been obtained by rapid quenching of lead oxyhalide melt (Krivovichev et al. 2006). The mixture of PbO, $PbBr_2$, and $PbCl_2$ was loaded into a platinum crucible and kept at 780° for 1 h in air, followed by cooling to room temperature over 3 h. The product consisted of dark-green platy crystals of the new mixed lead oxyhalide. It is of interest that cooling of the same mixture down to room temperature at a lower cooling rate of a few degrees per hour results in the formation of transparent yellowish crystals of lead oxyhalide with the structure of mendipite. Thus, the new compound is a metastable high-temperature phase that can be obtained exclusively by rapid quenching of the lead oxyhalide melt. The structure of mixed lead oxyhalide (Fig. 5) is remarkable in many ways. It contains 31 symmetrically independent lead cations, 18 halide sites statistically occupied by Br- and Cl- ions, and 22 O positions. The lead cations of the Pb1-Pb30 sites have

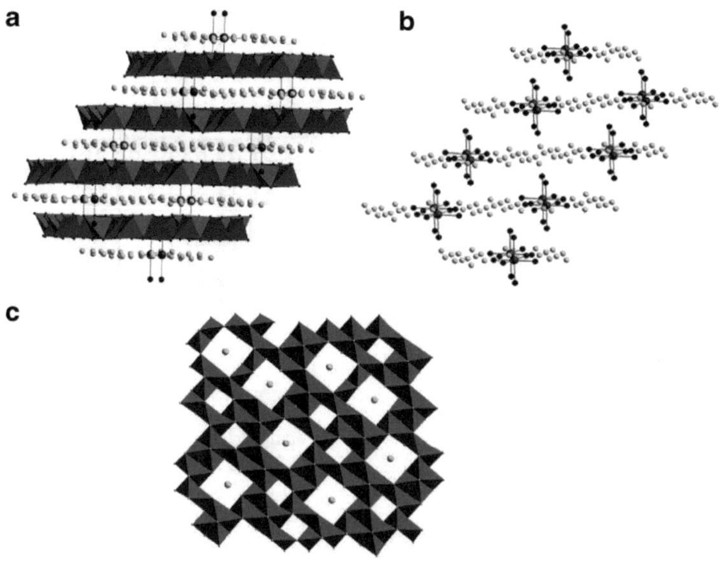

Fig. 5 General representation of the crystal structure of $Pb_{31}O_{22}Br_{10}Cl_8$ (**a**). The location and coordinadion of Pb31 atom by the halogen atoms in the halogen sheet (**b**). General structure of $[O_{22}Pb_{30}]$ block with the square vacancies filled by halogen atoms in the crystal structure of $Pb_{31}O_{22}Br_{10}Cl_8$ (**c**)

mixed oxyhalide coordination with oxygen and halide anions range from 2 to 4. Among the Pb atoms some are coordinated by two O atoms, some by three, and some by four. The Pb31 site (Fig. 5b) is coordinated solely by halogen anions, being at the center of PbX_6 octahedral units. All 22 oxygen anions are tetrahedrally coordinated by lead cations, thus forming oxocentered OPb_4 tetrahedra. The structure of mixed lead oxyhalide can be described as incorporation of $[PbX_6]$ halide units into a defect PbO matrix. The latter represents a two-dimensional $[O_{22}Pb_{30}]$ cationic layer of the OPb_4 tetrahedra that can be again derived from the [OPb] tetrahedral layer (Fig. 5c). In the structure of mixed lead oxyhalide, holes in the 2D PbO matrix correspond to either single tetrahedra or 2 × 2 blocks. The $[O_{22}Pb_{30}]$ cationic layers are parallel to the (102) plane and are separated by halogen anions, which form square pavements with halogen-halogen distances of 4 Å. The $[PbX_6]^{4-}$ octahedra are located between the layers in such a way that membered 2 × 2 holes in the PbO matrix are exactly above and below these units. The lead oxide $[O_{22}Pb_{30}]$ block in the structure of mixed lead oxyhalide is remarkable in its exceptional topological complexity. It consists of 22 symmetrically independent OPb_4 tetrahedra. From the chemical viewpoint, the appearance of such complexity should be ascribed to the incorporation of octahedral halide clusters into the metal oxide matrix that induces modification of the latter in a complex way.

To discuss the topology of the tetrahedral layers in lead oxyhalides, and the outstanding complexity of the layers in $Pb_{31}O_{22}Br_{10}Cl_8$ in more details, we shall use the approach which we have called "method of square lattices" (Siidra et al.

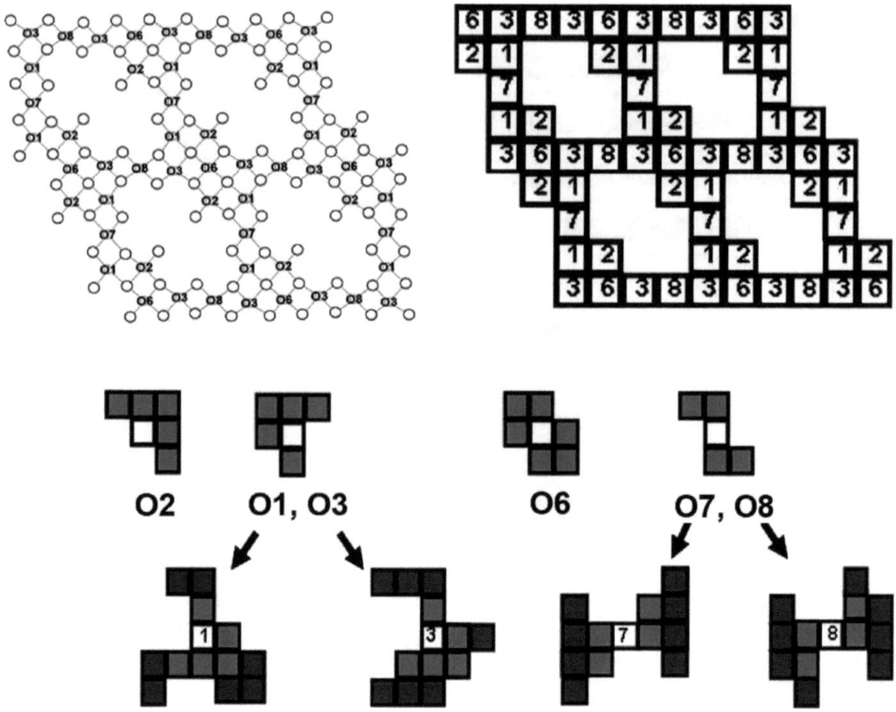

Fig. 6 Topological structure of the $[O_9Pb_{14}]$ 2D layer in the structure of kombatite (*above*) and first and second coronas (local coordinations) of central OPb_4 tetrahedra (*below*). See the text for details

2006). Within this approach, a single OPb_4 tetrahedron is symbolized by a square. Thus, the [OPb] layer in tetragonal PbO corresponds to a two-dimensional layer of black squares that fill the plane without gaps and overlaps. In turn, tetrahedral layers in the PbO derivative structures correspond to two dimensional arrangements of black and white squares, where the latter symbolize vacancies.

In Fig. 6 the two-dimensional array of squares that represents the arrangement of OPb_4 tetrahedra within the $[O_9Pb_{14}]$ layer in the structure of kombatite and sahlinite is illustrated. Each black square is labeled by a number that corresponds to the designation of the oxygen site at the center of the OPb_4 tetrahedron. The topological function of a tetrahedron within the layer can be visualized by investigation of the local coordination of a given square by the adjacent squares, or by the other words by all squares with which it has common points.

The arrangement of light-grey squares around the central square is designated as the first corona. In turn, the second corona, is defined as a set of black squares that surround the first corona, and so on. The schemes of first coronas for all of the six tetrahedra present in the $[O_9Pb_{14}]$ block are depicted in Fig. 6b. It is obvious that $O1Pb_4$ and $O3Pb_4$, and also $O7Pb_4$ and $O8Pb_4$ have the same first coronas consisting

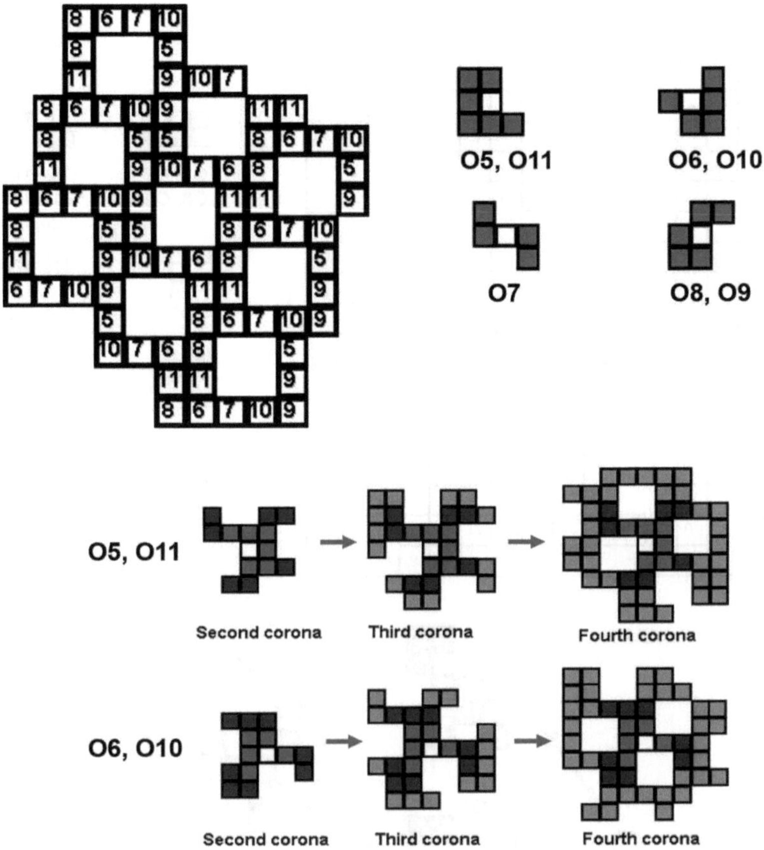

Fig. 7 Topological structure of the $[O_7Pb_{10}]$ 2D layer in the structure of symesite (*above*) and first (*right*), second, third and fourth coronas (*below*) of central OPb_4 tetrahedra (*below*). See the text for details

of five tetrahedra in the first case and four tetrahedra in the second. To further investigate whether topological functions of the tetrahedra are different, one has to examine their second coronas designated by black. Obvious, that despite the fact that the first coronas of these two pairs of tetrahedra are identical, their second coronas are different, and therefore the topological functions of the tetrahedra are different.

The two-dimensional array of squares that represents the arrangement of OPb_4 tetrahedra within the $[O_7Pb_{10}]$ in the structure of symesite is shown in Fig. 7. During the analysis of its structure it was revealed that even the second coronas are the same for some pairs of OPb_4 tetrahedra. So it was necessary to investigate their third coronas. Interestingly that $O5Pb_4$ and $O11Pb_4$, and also $O6Pb_4$ and $O10Pb_4$ have the same third and fourth coronas consisting of 13 and 20 tetrahedra accordingly. Thus $O5Pb_4$ and $O11Pb_4$, and also $O6Pb_4$ and $O10Pb_4$ are topologically the same. Note that crystallographicaly they are different.

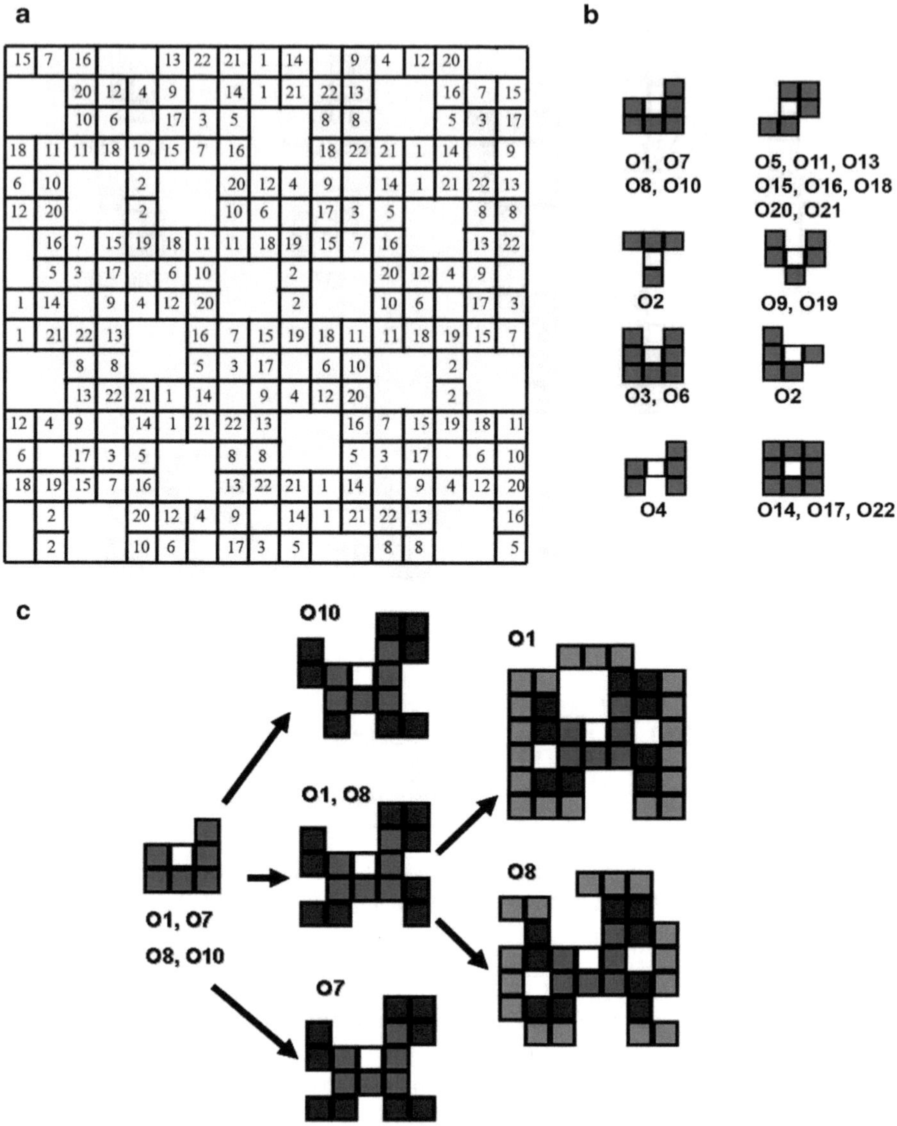

Fig. 8 Topological structure of the $[O_{22}Pb_{30}]$ 2D layer in the structure of $Pb_{31}O_{22}Br_{10}Cl_8$ (*above*) and first, second and third coronas for some pairs of OPb_4 tetrahedra (*below*). See the text for details

Many of the first coronas are common for several oxocentered tetrahedra in the $[O_{22}Pb_{30}]$ block (Fig. 8) in the structure of mixed lead oxyhalide $Pb_{31}O_{22}Br_{10}Cl_8$. For instance, the $O1Pb_4$, $O7Pb_4$, $O8Pb_4$, and $O10Pb_4$ tetrahedra have the same coronas consisting of six tetrahedra arranged around the central one in the same way. To further investigate whether topological functions of the tetrahedra we need to examine their second coronas. The tetrahedra centered by the O5, O11, O13,

O15, O16, O18, O20, and O21 atoms are distinguished by their second coronas, whereas their first coronas are the same. Obviously, that for the most of tetrahedra the second coronas are different. And therefore the topological functions of the tetrahedra are different. The situation is more complicated for the $O1Pb_4$ and $O8Pb_4$ tetrahedra because they have identical first and second coronas. However, their third coronas are different, and therefore their topological functions within the sheet are inequivalent. To our knowledge, this is the first example of a structure where third coronas are necessary to reveal topological differences between single tetrahedra. Thus all 22 symmetrically independent tetrahedra in the $[O_{22}Pb_{30}]$ block have unique functions in the topology of this unit. This topological complexity is exceptional.

The model of squares proposed to describe different levels of complexity in PbO derivative structures is rather simple. In particular, it is especially suitable for modeling self-organization of complex topologies using cellular automata that have attracted much attention within the past few years (Krivovichev 2004).

The Kombat mine in Namibia is another excellent place containing many of structurally beautiful Pb minerals, in particular, kombatite that was named in honor of this deposit. During the recent single-crystal study of Pb minerals associations from Kombat two new layered lead oxyhalides were identified.

The first is a Mn-containing unnamed mineral with the formula $\{[Pb_4Mn_2O](BO_3)_8\}[Pb_{32}O_{18}]Cl_{14}$ and remarkable crystal structure (*Pmmn*, $a = 12.7591(19)$ Å, $b = 27.1690(40)$ Å, $c = 11.5159(11)$ Å, $V = 3992.01(9)$ Å^3). It also belongs to the 1:1 type and consists of alternating Pb-O blocks and Pb-Cl sheets. The formation of this structure involves new and unusual mechanism of charge-balance compensation. The manganese atoms have quite usual coordination (Fig. 9) consisting of five O and one Cl atoms forming octahedron. The Pb^{2+} cation coordination is rather typical showing the location of stereochemically active lone-electron pair.

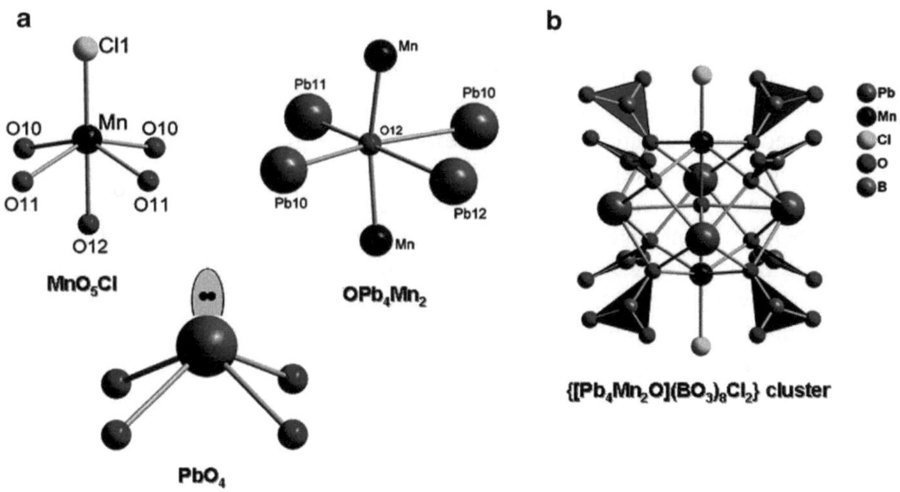

Fig. 9 Local coordinations of atoms in the crystal structure of unnamed $\{[Pb_4Mn_2O](BO_3)_8\}$ $[Pb_{32}O_{18}]Cl_{14}$ mineral

Fig. 10 General projection of the $[O_{18}Pb_{32}]$ layer in the crystal structure of unnamed $\{[Pb_4Mn_2O](BO_3)_8\}$ $[Pb_{32}O_{18}]Cl_{14}$ mineral

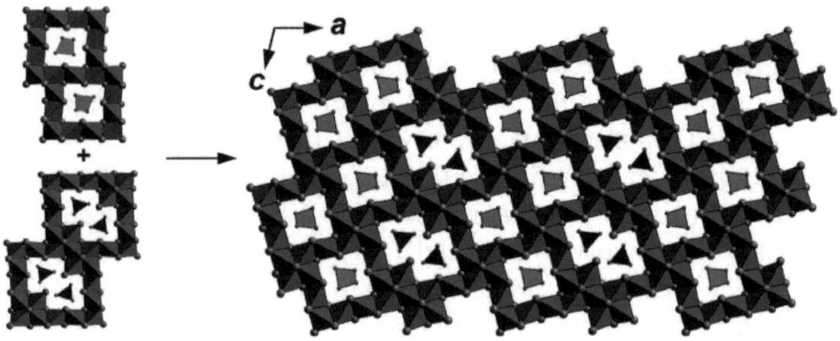

Fig. 11 General projection of the $Pb_{32}O_{21}$ layer formed the kombatite and symesite modules in the crystal structure of unnamed $Pb_{32}O_{21}(AsO_4)_2(VO_4)_2Cl_{10}$ mineral

Boron atoms form BO_3 triangles. The remarkable feature of the structure is the formation of heterometallic oxocentered OPb_4Mn_2 octahedron by the O12 atom. Oxocentered octahedra are found in numerous number of different oxysalts prominent in structural diversity (Krivovichev 2008; Siidra et al. 2009). The formation of heterometallic oxocentered clusters causes the presence of cross-like vacancies in the PbO-defficient layer with the composition $[O_9Pb_{16}]$ (Fig. 10). The layer is formed from complex chains built up from edge-sharing OPb_4 tetrahedra. Due to the low charge of the central O^{2-} anion, oxocentered tetrahedra may link through edges as well as through corners (Krivovichev and Filatov 1999), which is observed in this topology. As a result, structural complexity of the oxocentered tetrahedral units is higher than that of other tetrahedral units known in chemistry, for example, of silicate anions.

Another new unnamed Pb vanadate-arsenate oxyhalide with the composition $Pb_{32}O_{21}(AsO_4)_2(VO_4)_2Cl_{10}$ from Kombat mine has a rather big monoclinic cell ($C2/c$, $a = 23.1394(37)$ Å, $b = 22.6835(36)$ Å, $c = 12.3895(20)$ Å, $\beta = 102.090(3)°$ $V = 6358.79(45)$ Å^3). Its crystal structure again belongs to the 1:1 type. It consists of alternating PbO double layers and chlorine sheets. The vanadate and arsenate groups locate within the OPb block. The structure of the layer is formed by combination of symesite and kombatite modules as shown in Fig. 11. It is very likely that other combinations of modules are possible as well that may provide a unique sequence of phases.

Topological, structural and crystallographic complexity make the layered lead oxyhalides one of the most structurally complex family of inorganic compounds known to date. The reasons for this complexity may be seen as follows. The first reason is the presence of stereochemically active lone electron pairs on lead cations which play a role of additional ligands. The lone electron pair is active when the structure contains strong Lewis bases, e.g. additional O^{2-} anions. Therefore, in Pb (II) compounds with additional O atoms, lone electron pairs on the lead cations are stereochemically active and induce distortion and complication of the structure. Another possible reason could be a quite rare mechanism of incorporation of lead atoms into halogen layer and its modification a complex way.

Acknowledgements This work was financially supported by Russian President grant (MK-1645.2009.5), Deutsche Forschungsgemeinschaft (DE 412/45-1) and Russian Federal Programme "Scientific Cadres for Innovative Russia" (state contract # 02.740.11.0326).

References

Aurivillius B (1982) On the crystal structure of a number of non-stoichiometric mixed lead oxide halides composed of PbO like blocks and single halogen layers. Chem Scr 19:97–107

Boher P, Garnier P, Gavarri JR (1985) Monoxyde quadratique PbO alpha (I): description de la transition structurale ferroelastique. J Solid State Chem 57:343–350. doi:10.1016/0022-4596 (85)90197-5

Edwards R, Gillard RD, Williams PA, Pollard AM (1992) Studies of secondary mineral formation in the PbO-H_2O-HCl system. Miner Mag 56:53–65

Gabuda SP, Kozlova SG, Terskikh VV, Dybowski C, Neue G, Perry DL (1999) ^{207}Pb NMR study of novel Pb-Pb chemical bonding in lead monoxides, α-PbO and β-PbO. Chem Phys Lett 305:353–358. doi:10.1016/S0009-2614(99)00407-8

Krivovichev SV (2004) Crystal structures and cellular automata. Acta Crystallogr A60:257–262. doi:10.1107/S0108767304007585

Krivovichev SV (2008) Minerals with antiperovskite structure: a review. Z Kristallogr 223:109–113. doi:10.1524/zkri.2008.0008

Krivovichev SV, Burns PC (2006) The crystal structure of $Pb_8O_5(OH)_2Cl_4$, a synthetic analogue of blixite? Can Miner 44:515–522. doi:10.2113/gscanmin.44.2.515

Krivovichev SV, Filatov SK (1999) Metal arrays in structural units based on anion-centered metal tetrahedra. Amer Miner 84:1099–1106

Krivovichev SV, Armbruster T, Depmeier W (2004) Crystal structures of $Pb_8O_5(AsO_4)_2$ and $Pb_5O_4(CrO_4)$, and review of PbO-related structural units in inorganic compounds. J Solid State Chem 77:1321–1332. doi:10.1016/j.jssc.2003.11.005

Krivovichev SV, Siidra OI, Nazarchuk EV, Burns PC, Depmeier W (2006) Exceptional topological complexity of lead oxide blocks in $Pb_{31}O_{22}X_{18}$ (X = Br, Cl). Inorg Chem 45:3846–3848. doi:10.1021/ic060166m

Krivovichev SV, Turner R, Rumsey M, Siidra OI, Kirk CA (2009) The crystal structure and chemistry of mereheadite. Miner Mag 73:75–89. doi:10.1180/minmag.2009.073.1.103

Lepore GO, Welch MD (2010) The crystal structure of parkinsonite, nominally $Pb_7MoO_9Cl_2$: a naturally occurring Aurivillius phase. Miner Mag 74:269–275. doi:10.1180/minmag.2010.074.2.269

Liao J-H, Kanatzidis MG (1993) Quaternary rubidium copper tin sulfides ($Rb_2Cu_2SnS_4$, $A_2Cu_2Sn_2S_6$ (A = Na, K, Rb, Cs), $A_2Cu_2Sn_2Se_6$ (A = K, Rb), potassium gold tin sulfides, $K_2Au_2SnS_4$, and $K_2Au_2Sn_2S_6$. Syntheses, structures, and properties of new solid-state chalcogenides based on tetrahedral $[SnS_4]^{4-}$ units. Chem Mater 5:1561–1569. doi:10.1021/cm00034a029

Post J, Buseck PR (1985) Quantitative energy dispersive analysis of lead halide particles from the Phoenix urban aerosol. Environ Sci Technol 19:682–685. doi:10.1021/es00138a004

Pyykkö P (1988) Relativistic effects in structural chemistry. Chem Rev 88:563–594. doi:10.1021/cr00085a006

Sigman MB Jr, Korgel BA (2005) Strongly birefringent $Pb_3O_2Cl_2$ nanobelts. J Am Chem Soc 127:10089–10095. doi:10.1021/ja051956i

Siidra OI, Krivovichev SV, Depmeier W (2006) Method of square lattices for the description of structural topologies of minerals and inorganic compounds derivatives of tetragonal PbO (litharge). Vestn S-Peterb U Geol 3:18–26

Siidra OI, Krivovichev SV, Depmeier W (2007) Structure and mechanism of the ionic conductivity of the nonstoichiometric compound $Pb_{2+x}OCl_{2+2x}$. Dokl Phys Chem 414:128–131. doi:10.1134/S0012501607060024

Siidra OI, Krivovichev SV, Filatov SK (2008) Minerals and synthetic Pb(II) compounds with oxocentered tetrahedra: review and classification. Z Kristallogr 223:114–126. doi:10.1524/zkri.2008.0009

Siidra OI, Krivovichev SV, Depmeier W (2009) Crystal structure of $Pb_6O[(Si_6Al_2)O_{20}]$. Glass Phys Chem 35:406–410. doi:10.1134/S1087659609040099

Smith WH (1976) Lead contamination of the roadside ecosystem. J Air Pollut Control Assoc 26:753–766

Spencer LJ, Mountain ED (1923) New lead-copper minerals from the Mendip Hills, Somerset, England. Miner Mag 20:67–92

Steele IM, Pluth JJ, Jr Richardson JW (1997) Crystal structure of tribasic lead sulfate ($3PbO{\cdot}PbSO_4{\cdot}H_2O$) by X-rays and neutrons: an intermediate phase in the production of lead acid batteries. J Solid State Chem 132:173–181. doi:10.1006/jssc.1997.7440

Turner R (2006) A mechanism for the formation of the mineralized Mn deposits at Merehead Quarry, Cranmore, Somerset, England. Miner Mag 70:629–653. doi:10.1180/0026461067060359

Welch MD (2004) Pb-Si ordering in sheet-oxychloride minerals: the super-structure of asisite, nominally $Pb_7SiO_8Cl_2$. Miner Mag 68:247–254. doi:10.1180/0026461046820185

Welch MD, Criddle AJ, Symes RF (1998) Mereheadite, $Pb_2O(OH)Cl$: a new litharge-related oxychloride from Merehead Quarry, Cranmore, Somerset. Miner Mag 62:387–393

Tetradymite-Type Tellurides and Related Compounds: Real-Structure Effects and Thermoelectric Properties

Oliver Oeckler

1 Introduction

Tellurium is one of the rarest elements in the lithosphere. Consequently, tellurium minerals are extremely rare. Most of them are tellurides of noble metals, bismuth or lead. Their crystal chemistry is complicated. Tetradymite, Bi_2Te_2S, is a textbook example for a typical telluride (Harker 1934). The mineral exhibits a layered crystal structure with hexagonal layers of the elements in a sequence Te-Bi-S-Bi-Te that resembles a cubic close-packed arrangement of Te and S with Bi centering the octahedral voids. Five atom layers form a slab that corresponds to a cutout of a ternary variant of the rocksalt type. The slabs are stacked to form a rhombohedral structure with van der Waals gaps between the slabs; however, the distances Te-Te between the slabs are significantly shorter than the sum of the van der Waals radii, which indicates a certain degree of covalence. Numerous solid solutions with tetradymite-type structures are known. The chalcogenide positions can contain various amounts of S, Se and Te in a more or less disordered fashion, whereas additional cations such as Pb, Sb, Au, Cu and others may – at least to a certain degree – replace Bi.

In contrast to their uncommonness in nature, tellurides are very important materials for diverse technical applications even though their manufacturing is rather costly. The addition of small amounts of Te improves the properties of some alloys, and the element is sometimes used for doping semiconductors. CdTe is used in photodiodes and some thin-film solar cells, whereas TeO_2 glasses are well known for their high refractive index. However, the most important application are thermoelectric materials that can reversibly convert thermal and electrical energy. Although numerous research projects have focused on the development of new thermoelectrics, materials based on PbTe and Bi_2Te_3 still dominate the market. They are used in Peltier elements for cooling and, less frequently, heating devices as

O. Oeckler (✉)
Department Chemie, Ludwig-Maximilians-Universität München, Munich, Germany
e-mail: oliver.oeckler@gmx.de

S.V. Krivovichev (ed.), *Minerals as Advanced Materials II*,
DOI 10.1007/978-3-642-20018-2_31, © Springer-Verlag Berlin Heidelberg 2012

well as for electric power generation in thermoelectric generators. Whereas these have been used in radionuclide batteries for decades, the discussion about renewable energies and sustainable approaches to reduce CO_2 emission has led to ambitious plans to use thermoelectric generators for the generation of electric power from waste heat. Recently, the first prototype thermoelectric generators for the application in automobiles have been presented. They reduce the gasoline consumption significantly; however, the efficiency of thermoelectric generators is still very low. Furthermore, the scarcity and toxicity, and consequently the high costs, of tellurium and its compounds will probably limit future applications to more specialized devices where smaller amounts of energy need to be converted.

Tellurides also play a key role concerning data storage in rewritable media such as CD-RW, DVD-RW, DVD-RAM or BD-RE and non-volatile RAM devices. The recording layers consist of tellurides, the most prominent ones being $Ge_2Sb_2Te_5$ and other germanium antimony tellurides (GST materials) or quaternary silver-indium-antimony-tellurium alloys (AIST materials) (Wuttig and Yamada 2007; Bensch and Wuttig 2010). During the recording process in optical storage media, laser irradiation produces amorphous marks in a crystalline layer. Erasing means recrystallization and is achieved by irradiation with reduced laser power. In RAM devices, 'resistance' heating is induced by electric currents. The crystalline phase is metastable and energetically very close to the amorphous one so that little enthalpy exchange is involved in the write-erase process. Furthermore, the crystalline phases are highly disordered and exhibit simple basic structures that are accessible from the amorphous modifications via very short diffusion pathways.

2 Structural Chemistry of Layered Pnicogen Tellurides

2.1 Stable Phases

The crystal structures of binary pnicogen tellurides with normal oxidation states are closely related to the rocksalt type. Whereas As_2Te_3 exhibits a rather complicated orthorhombic layered structure, Sb_2Te_3 and Bi_2Te_3 are isotypic with tetradymite, S formally being replaced by additional Te (cf. Fig. 1). Sb-rich and Bi-rich binary tellurides form homologous series of compounds $(M_2Te_3)_m(M_2)_k$ (M = Sb, Bi) (Kifune et al. 2005; Bos et al. 2007), which in addition to tetradymite blocks contain A7-type layers of M atoms similar to those in the structure of the elements. These tetradymite blocks can be 'enlarged' by formally adding YTe (Y = Ge, Sn, Pb) (Konstantinov et al. 2001). The resulting homologous series $(YTe)_n(M_2Te_3)_m$ contain rocksalt-type slabs whose thickness depends on n and m, with Te occupying anion positions including those adjoining the van der Waals gaps between the blocks and Y and M occupying the cation positions, usually in a disordered fashion.

The number of stable compounds in the homologous series $(YTe)_n(M_2Te_3)_m$ depends on the element combination. The series with Y = Ge are the most

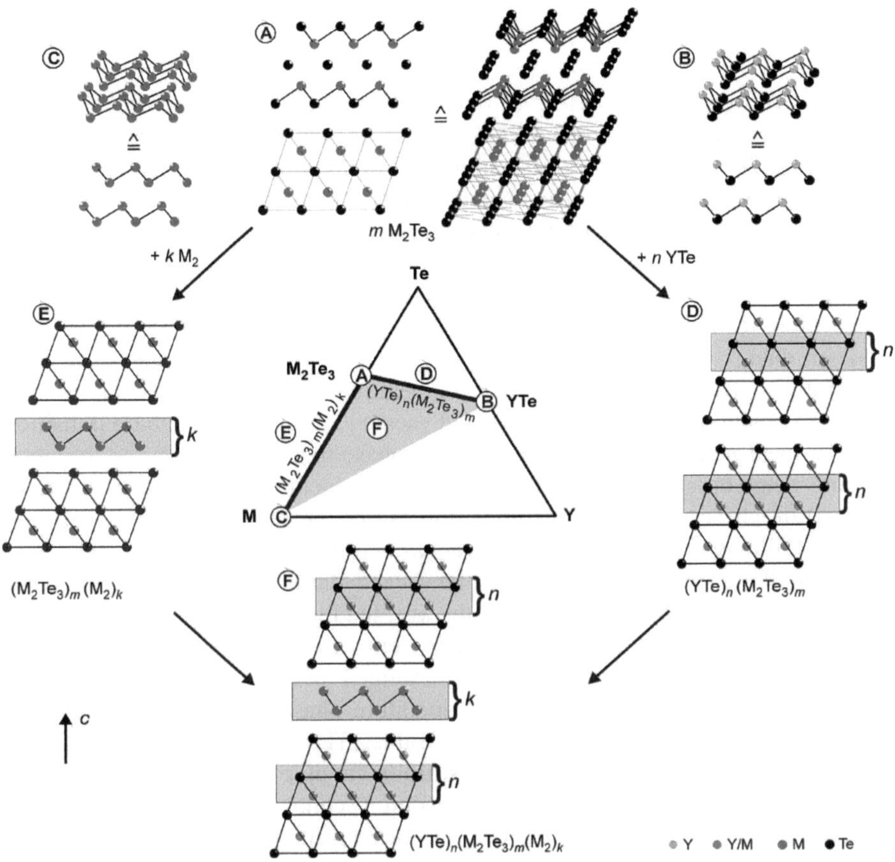

Fig. 1 Layered compounds in the ternary systems Y – M – Te (Y = Ge, Sn, Pb, M = Sb, Bi). The structures shown are drawn for Y = Ge and M = Sb; however, the other systems are similar (exception: PbTe and SnTe exhibit an undistorted rocksalt-type structure not shown in the figure); cf. text for further explanations

comprehensive ones, whereas the compounds YM_2Te_4 and $Y_2M_2Te_5$ dominate the other phase diagrams. In general, there are more discrete compounds for M = Bi than for M = Sb. Although all stable compounds are characterized by cation disorder, no significant amounts of cation vacancies or anti-site defects have been reported.

For $n >> m$, large parts of the structures become similar to the compounds YTe and finally YTe precipitates. The compounds YTe themselves exhibit simple structures (at ambient pressure). The rhombohedral structure of GeTe is a binary variant of the A7 (gray arsenic) structure type in which cation-anion alternation leads to the loss of centrosymmetry. At 432°C a second-order phase transition takes place, the high-temperature phase exhibits the rocksalt type (Chattopadhyay et al. 1987). SnTe crystallizes with the rocksalt type at room temperature but transforms to the rhombohedral structure at −175°C (Iizumi et al. 1975), whereas no phase transitions have been reported for rocksalt-type PbTe. Numerous phase transitions at high pressures have also been reported.

2.2 Metastable Phases

In addition to the stable phases, a number of metastable ones is known. As the metastable phases are essential in GST phase-change materials, a great deal of recent research has focused on germanium antimony tellurides which, in addition, exhibit interesting thermoelectric characteristics. The metastable phases of phase-change materials are produced by sputtering techniques and are not accessible by conventional solid-state synthesis. Whereas for AIST phases there is no corresponding thermodynamically stable quaternary phase, some metastable crystalline modifications of GST materials $(GeTe)_n(Sb_2Te_3)$ (i. e. those with $n \geq 3$) used in data storage devices correspond to phases that are stable at high temperatures. These rocksalt-type high-temperature phases contain Te anions, and the cation position is shared by Ge, Te and vacancies. Whereas the phase transition in GeTe itself is displacive, a reconstructive phase transition is required to obtain the long-range ordered layered phases stable at ambient conditions from the high-temperature phase. Formally, this transition can be viewed as a vacancy diffusion process. Cation vacancies arrange in layers as two-dimensional defects, and due to relaxation of the surrounding Te atoms the vacancy layer becomes a van der Waals gap. As this involves long diffusion pathways, the high-temperature phase can be quenched. As illustrated in Fig. 2, vacancy layers are also present in quenched bulk material, however, they are finite and arranged perpendicular to all <111> directions of the original cubic high-temperature phase. The average metrics of these quenched phases remains close to cubic. The diffusion process leads to trigonal structures; however, as this distortion involves microstrain in twinned crystals, the stresses impede rapid diffusion (Schneider et al. 2010a). Upon annealing, two-dimensionally extended vacancy layers, i. e. van der Waals gaps, are formed. These are parallel, although not equidistant unless the samples are annealed for very long times in order to reach the thermodynamical minimum.

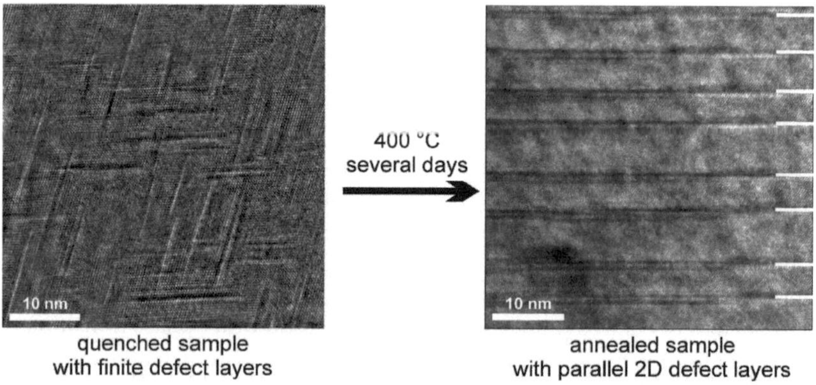

Fig. 2 HRTEM images of $(GeTe)_{12}(Sb_2Te_3)$: the domain structure of quenched samples develops into a layered structure with parallel defect layers, i.e. van der Waals gaps, upon annealing. After short-time annealing, the parallel gaps are not equidistantly spaced

Various micro- and nanostructures can be obtained by different thermal treatment. In addition, the domain structure depends on the stoichiometry. With increasing GeTe content, the percentage of cation vacancies in the high-temperature phases decreases. As a consequence, quenching leads to a more pronounced the domain structure as the diffusion pathways to form two-dimensional vacancy layers increase.

The quenched pseudocubic phases are obviously metastable, however, this does not mean that long-periodically ordered phases are always stable. Whereas the stable phases in the system Ge-Sb-Te correspond to the homologous series $(GeTe)_n(Sb_2Te_3)_m$ and $(Sb_2Te_3)_m(Sb_2)_k$, metastable phases $(GeTe)_n(Sb_2Te_3)_m$ $(Sb_2)_k$ with $n, m, k \neq 0$ can be obtained by quenching melts and subsequently annealing the ingots at moderate temperatures (Schneider et al. 2009; Schneider and Oeckler 2009). Their crystal structures combine the blocks known from both homologous series, i. e. A7-type Sb layers are formally inserted between 'enlarged' tetradymite-type blocks. Probably a process related to spinodal decomposition takes place during quenching, and lamellar concentration gradients evolve into these long-periodic structures by subsequent short-range diffusion at moderate temperatures. The thermodynamically stable state corresponds to Ge precipitation, which requires both nucleation and long-range diffusion and thus does not take place without long-time annealing.

3 Challenges for Structure Determination

The establishment of structure–property relationships in tellurides requires a detailed knowledge of the crystal structure. Both in stable and metastable compounds, disorder on various scales plays an important role. Local disorder phenomena can best be elucidated by analysis of diffuse X-ray scattering, especially if the relaxation around defects is very pronounced and two-body short-range correlations dominate. However, disorder models can be very complex and the optimization of simulations of diffuse scattering is far from straightforward. The refinement of statistical disorder parameters based on experimentally measured diffuse intensities still poses demanding conceptual and computational challenges. High-resolution transmission electron microscopy yields valuable information on domain structures and also on intermediate-range ordering. Concerning concentration gradients of certain elements, the advent of scanning transmission electron microscopy (STEM) employing high-angle annular dark-field detectors (HAADF) has been a breakthrough. Additionally, defect ordering can directly be imaged using this technique as such phenomena of course mean pronounced Z contrast. Concerning phase transitions, space- and time-resolved diffraction studies using synchrotron radiation and very fast and sensitive detectors are necessary to clarify the mechanism of diffusion processes.

Even 'standard' single-crystal structure analysis from Bragg data, either those of completely ordered phases or those representing the average structure of disordered compounds poses numerous problems. Elements like Sb and Te (and in some compounds, additionally In or Sn) cannot be distinguished by conventional X-ray diffraction. As their neutron scattering lengths are also very similar, resonant X-ray diffraction using synchrotron radiation with wavelengths close to the absorption edges of the elements is the method of choice. It is a future goal to exploit the effect of resonant scattering on diffuse intensities. Furthermore, it has been shown that very similar structure factors for non-congruent and chemically very different structures of layered tellurides can result from almost homometric structure models and not just from similar form factors (Schneider et al. 2010b). Ambiguities arising from this effect can only be resolved by very precise chemical analysis. Stacking disorder in layered compounds further complicates the situation, especially if powder data are analyzed.

4 Thermoelectric Properties

The efficiency of thermoelectric generators is directly related to the figure of merit $ZT = S^2\sigma T/\kappa$, which means that the Seebeck coefficient S and the electrical conductivity σ should be high at high temperatures T, whereas the thermal conductivity κ should be low. The optimization of ZT is a challenge as all relevant properties are related to the charge carrier concentration and cannot be changed independently. High Seebeck coefficients are characteristic for semiconductors with a rather low electrical conductivity whereas typical metals exhibit low values of S. Therefore, semimetals or small-bandgap semiconductors like many tellurides are the best compromise to obtain a high power factor $S^2\sigma$. According to the Wiedemann-Franz law, the electronic part κ_e of the thermal conductivity is proportional to σ so that κ can only be effectively reduced by influencing its phononic part κ_p, which means that phonon scattering must be enhanced. This can be achieved by introducing so called ‘rattling’ atoms or, more efficiently, by nanostructuring. Nanostructures can be generated by small precipitates, domain boundaries and a broad range of real-structure effects. The most prominent examples for such an approach in bulk materials are again tellurium-based materials which have been extensively reviewed in the literature (Sootsman et al. 2009; Shevelkov 2008).

An intriguing approach to prepare nanostructured tellurides involves the use of phase transitions. The highly disordered high-temperature phases of GST materials are a prime example. Quenching leads to partial vacancy ordering, and the associated twinning yields microdomain structures whose domain sizes and relative orientations can be tuned by varying the composition or thermal treatment. Whereas the stable layered modifications of compounds $(GeTe)_n(Sb_2Te_3)$ exhibit ZT values around 0.2, surprisingly high values have been measured for nanostructured materials obtained by quenching the corresponding melts. The ZT value increases from $n = 3$ ($ZT = 0.5$ at 450°C) to $n = 12$ ($ZT = 1.3$ at 450°C) as the domain structure becomes more pronounced (Schneider et al. 2010c). κ_p is quite low in these materials, however, κ_e is less favorable due to the rather high value of σ. The combination of high values for σ and S is unusual and leads to the high values of ZT. In contrast to other promising thermoelectrics, nanostructured GST materials are single-phase, which means that in this case it is more straightforward to understand structure–property relationships on the atomic and the nanoscale than it is for materials that contain precipitates.

It is interesting to note that GST materials corroborate the hypothesis that the metastable crystalline modifications of phase-change materials for data storage are promising precursors for efficient thermoelectrics. The optimum compositions for GST materials with high ZT values correspond to those used in BluRay discs or non-volatile RAM devices. For both applications, the combination of low thermal conductivity and intermediate electrical conductivity is required. During the writing process in storage media, very small spots need to be heated which is only possible if κ is low, whereas for RAM devices the material must be conductive. An intermediate range of σ is required for 'resistance heating' during write-erase cycles. In thermoelectrics, intermediate values of σ are a compromise to achieve reasonably high Seebeck coefficients.

5 Geo-Inspired Thermoelectrics?

Multinary tellurides have been shown to exhibit very promising thermoelectric properties; however, optimization usually requires substitution or doping experiments. It can be very tedious work to find out which dopants are tolerated without phase separation. The chemical compositions known from minerals may constitute interesting hints concerning possible dopants and existing solid solution series. Moreover, compounds found as minerals are sufficiently long-time stable and may provide intriguing new crystal structures which can then be systematically investigated. As many tellurides can be synthesized from the elements in silica glass ampoules, it is no problem that the actual measurements of physical properties usually cannot be carried out using natural material as many telluride minerals are so rare. The multinary systems found in nature might inspire research in unusual but promising classes of compounds and last but not least lead to a plethora of interesting 'new' thermoelectrics. Tetradymite-type or related building blocks in highly disordered and nanostructured systems may be just a starting point.

Acknowledgements The author thanks his coworkers Matthias Schneider, Tobias Rosenthal, Thorsten Schröder, Felix Fahrnbauer and Philipp Urban for their contributions to this area of research. We thank all students who worked on research projects in our group and acknowledge Thomas Miller, Christian Minke, Dr. Markus Döblinger and Dr. Christian Stiewe for performing numerous measurements. Prof. Dr. Wolfgang Schnick has supported our work for many years. We further thank the Deutsche Forschungsgemeinschaft for supporting this study (grant OE/530/1-1).

References

Bensch W, Wuttig M (2010) Optische speichermedien (optical storage media). Chem unserer Zeit 44:92–107. doi:10.1002/ciuz.201000493

Bos JWG, Zandbergen HW, Lee MH, Ong NP, Cava RJ (2007) Structures and thermoelectric properties of the infinitely adaptive series $(Bi_2)_m(Bi_2Te_3)_n$ Phys. Rev B 75:195203. doi:10.1103/PhysRevB.75.195203

Chattopadhyay T, Boucherle JX, von Schnering HG (1987) Neutron diffraction study on the structural phase transition in GeTe. J Phys C Solid State Phys 20:1431–1440. doi:10.1088/0022-3719/20/10/012

Harker D (1934) The crystal structure of the mineral tetradymite, Bi_2Te_2S. Z Kristallogr 89:175–181

Iizumi M, Hamaguchi Y, Komatsubara KF, Kato Y (1975) Phase transition in SnTe with low carrier concentration. J Phys Soc Jpn 38:443–449. doi:10.1143/JPSJ.38.443

Kifune K, Kubota Y, Matsunaga T, Yamada N (2005) Extremely long period-stacking structure in the Sb–Te binary system. Acta Cryst B 61:492–497. doi:10.1107/S0108768105017714

Konstantinov PP, Shelimova LE, Avilov ES, Kretova MA, Zemskov VS (2001) Thermoelectric properties of nGeTe.mSb_2Te_3 layered compounds. Inorg Mater 37:662–668. doi:10.1023/A:1017613804472

Schneider MN, Oeckler O (2009) $GeSb_4Te_4$ – a new 9P-type phase in the system Ge-Sb-Te. Z Anorg Allg Chem 636:137–143. doi:10.1002/zaac.200900453

Schneider MN, Seibald M, Oeckler O (2009) A new series of long-range ordered metastable phases in the system M-Sb-Te (M = Ge, Ag). Dalton Trans 2004–2011. doi:10.1039/b814000f

Schneider MN, Urban P, Leineweber A, Döblinger M, Oeckler O (2010a) The influence of stress and strain on the kinetic stability and phase transitions of cubic and pseudocubic Ge-Sb-Te materials. Phys Rev B 81:184102. doi:10.1103/PhysRevB.81.184102

Schneider MN, Seibald M, Lagally P, Oeckler O (2010b) Ambiguities due to almost homometric structure models and stacking disorder concerning the structure determination of antimony tellurides. J Appl Cryst 43:1012–1020. doi:10.1107/S0021889810032644

Schneider MN, Rosenthal T, Stiewe C, Oeckler O (2010) From phase-change materials to thermoelectrics? Z Kristallogr 225:463–470. doi:10.1524/zkri.2010.1320

Shevelkov AV (2008) Chemical aspects of the design of thermoelectric materials. Russ Chem Rev 77:1–19. doi:10.1070/RC2008v077n01ABEH003746

Sootsman JR, Chung DY, Kanatzidis MG (2009) New and old concepts in thermoelectric materials. Angew Chem Int Ed 48:8616–8639. doi:10.1002/anie.200900598

Wuttig M, Yamada N (2007) Phase-change materials for rewriteable data storage. Nat Mater 6:824–832. doi:10.1038/nmat2009

Rare-Earth Metal(III) Fluoride Oxosilicates Derivatized with Alkali or Alkaline-Earth Elements

Marion C. Schäfer and Thomas Schleid

1 Introduction

While searching for new and effective materials as host lattices for luminescence (Blasse and Grabmaier 1994; Dieke 1968; Höppe 2009; Jüstel et al. 1998; Ronda 2008), synthetic oxosilicates containing rare-earth metal(III) cations and fluoride anions got into the focus of our work (Schleid et al. 2011). In some cases, the charge compensation is achieved using alkali metals such as sodium to cesium or alkaline-earth metals such as calcium to barium. By studying the quinary systems AF-MF$_3$-M_2O$_3$-SiO$_2$ (A^+ = alkali metal, M^{3+} = rare-earth metal), several new representatives could be obtained in the last years such as A_5M_4F[SiO$_4$]$_4$ (A^+ =Na, K, M^{3+} = Y, La) (Merinov et al. 1980; Schäfer and Schleid 2010), $K_5Eu_2F[Si_4O_{13}]$ (Chiang et al. 2007), $K_9M_3F_2Si_{12}O_{32}$ (M = Sm – Gd) (Tang et al. 2008), A_2MFSi$_4$O$_{10}$ (A^+ = K, Rb, M^{3+} = Sc, Y) (Kolitsch and Tillmanns 1994; Schäfer and Schleid 2007) and $Rb_3Sc_2F_5Si_4O_{10}$ (Schäfer 2010; Schäfer and Schleid 2011). The ratios of rare-earth metal to fluorine change from 4:1 in the fluoride-poor compounds A_5M_4F[SiO$_4$]$_4$ via 2:1 in $K_5Eu_2F[Si_4O_{13}]$, 3:2 in $K_9M_3F_2Si_{12}O_{32}$ and 1 : 1 in A_2MFSi$_4$O$_{10}$ to 2:5 in the fluoride-rich example $Rb_3Sc_2F_5Si_4O_{10}$. Thereby, the oxosilicates $K_2ScFSi_4O_{10}$ and $Cs_2YFSi_4O_{10}$ are *not* isostructural. Interestingly, the crystal structures of the four compositions $K_5Eu_2F[Si_4O_{13}]$, $K_2ScFSi_4O_{10}$, $Cs_2YFSi_4O_{10}$ and $Rb_3Sc_2F_5Si_4O_{10}$ have substantial similarities in their constitutions. The basic building units are always the typical $[SiO_4]^{4-}$ tetrahedra of course, along with $[M(O,F)_6]^{n-}$ octahedra with a variable ratio of fluoride and oxide ligands. Both building units join together in each case to form three-dimensional octahedral-tetrahedral frameworks. The structural relationships are shown in the first part of this short report.

M.C. Schäfer (✉) • T. Schleid
Institut für Anorganische Chemie, Universität Stuttgart, D-70569 Stuttgart, Germany
e-mail: schaefer@iac.uni-stuttgart.de

S.V. Krivovichev (ed.), *Minerals as Advanced Materials II*,
DOI 10.1007/978-3-642-20018-2_32, © Springer-Verlag Berlin Heidelberg 2012

In contrast, only one structure type is known for rare-earth metal(III) fluoride oxosilicates derivatized with alkaline-earth metal cations. Starting from the mixed-valent europium fluoride oxosilicate $Eu_2^{II}Eu_3^{III}F[SiO_4]_3$ (Wickleder et al. 2002) with *apatite*-type structure, the divalent europium can be substituted by the alkaline-earth metals calcium, strontium and barium. The first representative $Ba_2La_3F[SiO_4]_3$: Yb^{3+} was reported by Gong et al. (2005). Meanwhile, the series $Ca_2M_3F[SiO_4]_3$ (M^{3+} = La – Nd, Sm, Eu), $Sr_2M_3F[SiO_4]_3$ (M^{3+} = La – Nd, Sm) and $Ba_2M_3F[SiO_4]_3$ (M^{3+} = La – Nd, Sm, Eu) (Schäfer and Schleid 2008; Schäfer 2010) could be synthesized by a special preparation technique within a high-frequency furnace. Moreover, the isotypic hydroxide-containing compounds $AE_2M_3(OH)[SiO_4]_3$ (AE^{2+} = Ca – Ba, M^{3+} = Y, La, Ce, Nd, Sm, Gd, Dy, Er, Lu) are already known since 1968 (Ito 1968).

2 Alkali-Metal Rare-Earth Metal(III) Fluoride Oxosilicates

The fluoride poor *soro*-oxosilicate $K_5Eu_2FSi_4O_{13}$ (Chiang et al. 2007) crystallizes monoclinically in the space group $P2_1/m$ (no. 11) with a = 718.50(2), b = 579.81(2), c = 1816.75(6) pm, β = 92.248(2)° and two formula units per unit cell. In this structure, all four crystallographically independent $[SiO_4]^{4-}$ tetrahedra are connected to form a four-membered *catena*-$[Si_4O_{13}]^{10-}$ fragment with Si–O distances from 158 to 167 pm (Fig. 1, *left*). Both crystallographically different Eu^{3+} cations show one F^- and five O^{2-} anions (d(Eu–F) = 249–255 pm, d(Eu–O) =222–228 pm) as first-sphere neighbours. These $[EuFO_5]^{8-}$ octahedra are linked via their F^- anions to form $[Eu_2FO_{10}]^{18-}$ dimers with Eu–F–Eu angles of 178° (Fig. 1, *right*). The F^- anions themselves reside in octahedral environments of two Eu^{3+} and four K^+ cations.

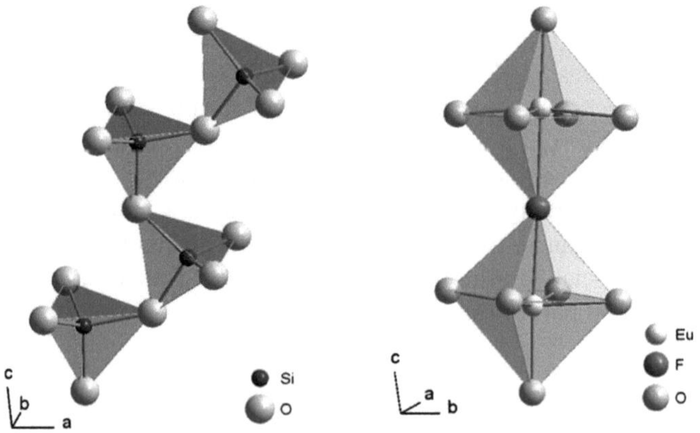

Fig. 1 *Catena*-$[Si_4O_{13}]^{10-}$ fragments (*left*) and dimeric $[Eu_2FO_{10}]^{18-}$ units (*right*) in $K_5Eu_2F[Si_4O_{13}]$

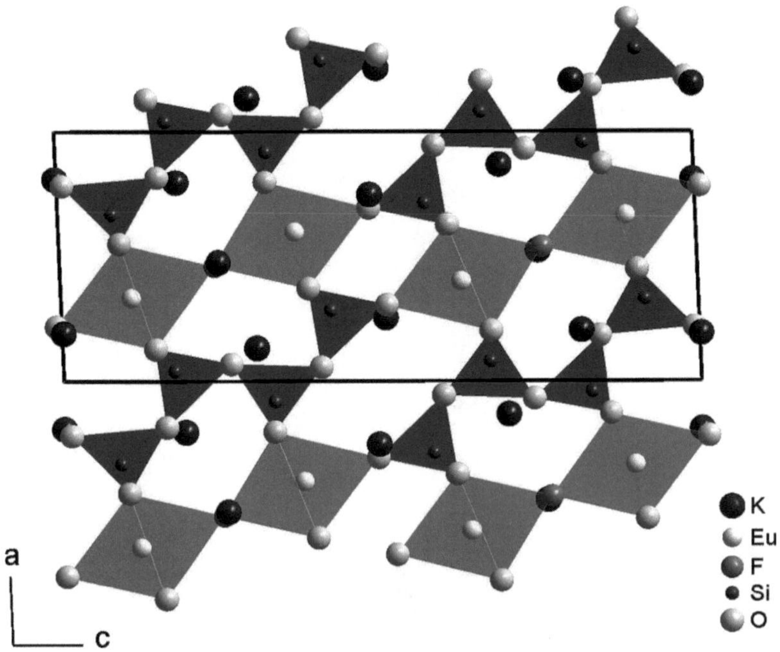

Fig. 2 View at the monoclinic crystal structure of $K_5Eu_2F[Si_4O_{13}]$ along [010]

The dimeric $[Eu_2FO_{10}]^{18-}$ units are linked via all O^{2-} anions to a total of six *catena*-$[Si_4O_{13}]^{10-}$ fragments to form a three-dimensional framework (Fig. 2). Five crystallographically different K^+ cations occupy the resulting cages of the crystal structure, which results in the formation of $[(K1)F_2O_{10}]^{21-}$, $[(K2)FO_{10}]^{20-}$, $[(K3)FO_{10}]^{20-}$, $[(K4)O_{11}]^{21-}$ and $[(K5)O_{12}]^{23-}$ polyhedra with d(K – F) = 265–290 pm and d(K – O) = 275–379 pm.

Both structures of the composition $A_2M\text{FSi}_4\text{O}_{10}$ represent *ino*-oxosilicates. The first one, $K_2ScFSi_4O_{10}$ (Kolitsch and Tillmanns 1994), crystallizes tetragonally in the space group *I*4/*m* (no. 87) with a = 1120.7(2), c = 816.6(2) pm and four formula units per cell. The topological arrangement in $K_2ScFSi_4O_{10}$ is isotypic with that in the mineral *narsarsukite* $Na_2(Ti,Fe^{III})(O,F)Si_4O_{10}$ (Kolitsch and Tillmanns 1994; Peacor and Buerger 1962). The crystallographically unique $[SiO_4]^{4-}$ tetrahedra are linked to infinite tubular chains $^1_\infty\{[Si_4O_{10}]^{4-}\}$ along [001] (Fig. 3, *left*), built up by four-membered rings. The Si–O distances range between 157 and 164 pm in these unbranched vierer double chains according to the nomenclature of Liebau (1985). The unique Sc^{3+} cations are surrounded by two F^- and four O^{2-} anions with d(Sc–F) = 200–208 pm and d(Sc–O) = 207 pm (4×). The *trans*-$[ScF_2O_4]^{7-}$ octahedra get fused via both apical F^- anions to exactly linear infinite chains $^1_\infty\{[ScF^v_{2/2}O^t_{4/1}]^{6-}\}$, which run along [001] (Fig. 3, *right*). One of the two crystallographically different F^- anions is coordinated only in a linear fashion by two Sc^{3+} cations, while the second one has an octahedral environment consisting of two apical Sc^{3+} and four equatorial K^+ cations. Each of the infinite chains

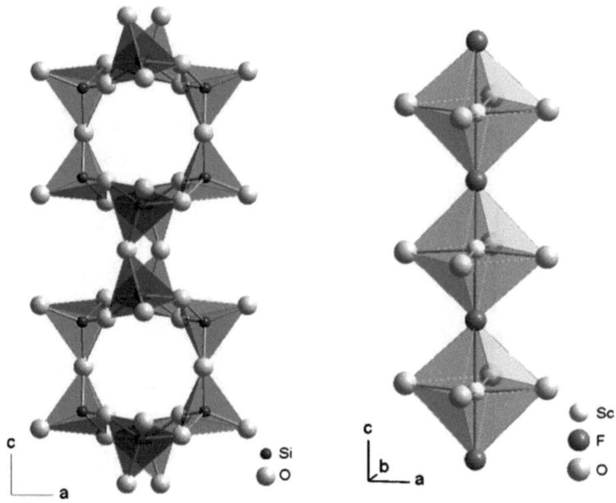

Fig. 3 Infinite tubular chains $^1_\infty\{[Si_4O_{10}]^{4-}\}$ (*left*) and linear infinite chains $^1_\infty\{[ScF^{v}_{2/2}O^{t}_{4/1}]^{6-}\}$ (*right*), both along [001] in $K_2ScFSi_4O_{10}$

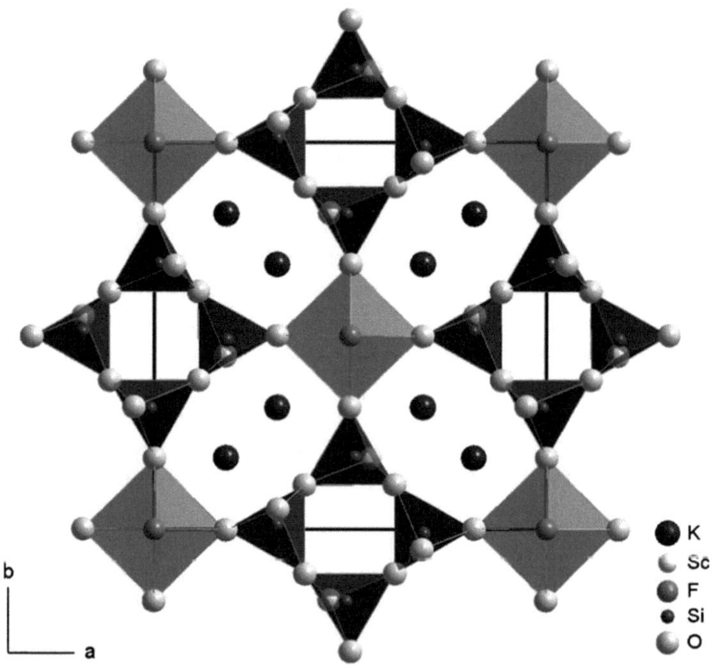

Fig. 4 View at the tetragonal crystal structure of $K_2ScFSi_4O_{10}$ along [001]

$^1_\infty\{[ScF^{v}_{2/2}O^{t}_{4/1}]^{6-}\}$ connects via all four equatorial oxygen atoms to four oxosilicate chains $^1_\infty\{[Si_4O_{10}]^{4-}\}$. In the resulting octahedral-tetrahedral framework (Fig. 4), the K^+ cations reside in the hexagonal channels between both types of chains as $[KFO_{10}]^{20-}$ polyhedra with d(K–F) = 293 and d(K–O) = 281–338 pm.

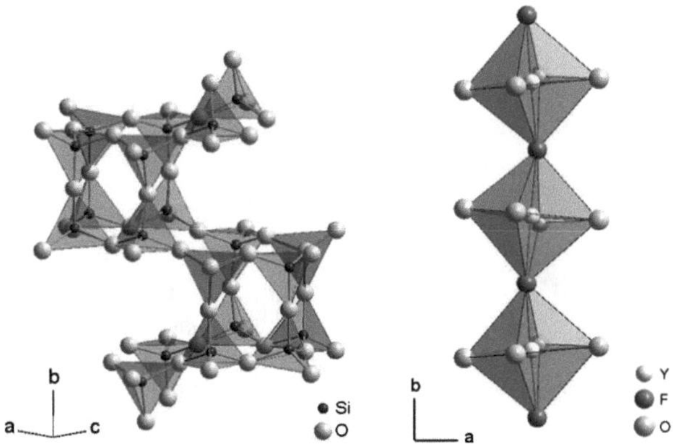

Fig. 5 Infinite tubular chains $^1_\infty\{[Si_4O_{10}]^{4-}\}$ (*left*) and infinite chains $^1_\infty\{[YF^v_{2/2}O^t_{4/1}]^{6-}\}$ (*right*), both along [010] in $Cs_2YFSi_4O_{10}$

The second structure representing the formula $A_2MFSi_4O_{10}$, $Cs_2YFSi_4O_{10}$ (Schäfer and Schleid 2007), crystallizes orthorhombically in the space group *Pnma* (no. 62) with $a = 2239.75(9)$, $b = 884.52(4)$, $c = 1198.61(5)$ pm and eight formula units per unit cell. The characteristic building unit of $Cs_2YFSi_4O_{10}$ also comprises infinite tubular chains $^1_\infty\{[Si_4O_{10}]^{4-}\}$ of vertex-sharing $[SiO_4]^{4-}$ tetrahedra along [010], now consisting of eight-membered half-open cube-shaped oxosilicate cages (Fig. 5, *left*). According to the classification of Liebau (1985), this silicate chain is a loop-branched vierer double chain. The four crystallographically different Si^{4+} cations occupy general sites 8*d* with Si − O-distances ranging from 157 to 165 pm. Because of the rigid structure of this kind of oxosilicate chain the bridging Si–O–Si angles vary extremely between 128 and 167°. The crystallographically unique Y^{3+} cations are again surrounded by two F^- and four O^{2-} anions (d(Y − F) = 122 pm, d(Y − O) = 221–225 pm). These slightly distorted *trans*-$[YF_2O_4]^{7-}$ octahedra share both apical F^- anions as common vertices to form nearly linear infinite chains $^1_\infty\{[YF^v_{2/2}O^t_{4/1}]^{6-}\}$ along [010] with Y–F–Y angles of 169 and 177° (Fig. 5, *right*). The four- and fivefold environments of the two different F^- anions are recruited of two Y^{3+} and two or three Cs^+ cations, respectively. Each of the $^1_\infty\{[YF^v_{2/2}O^t_{4/1}]^{6-}\}$ chains connects via all four equatorial O^{2-} anions to three neighbouring oxosilicate chains to erect a three-dimensional framework (Fig. 6). The resulting hexagonal and octagonal channels along [010] are occupied by four crystallographically different Cs^+ cations being 10-, 12-, 13- and 14-fold coordinated by O^{2-} and F^- anions with coordination polyhedra like $[(Cs1)O_{10}]^{19-}$, $[(Cs2)F_2O_{10}]^{21-}$, $[(Cs3)FO_{12}]^{24-}$ and $[(Cs4)F_2O_{12}]^{25-}$ and seperations of d(Cs − F) = 360–371 pm and d(Cs − O) = 309–390 pm. Meanwhile, compounds with the same structure type and formula $Cs_2MFSi_4O_{10}$ could also be obtained for M^{3+} = Pr, Er and Yb (Schäfer 2010).

The fourth compound $Rb_3Sc_2F_5Si_4O_{10}$ (Schäfer 2010; Schäfer and Schleid 2011) crystallizes orthorhombically in the space group *Pnma* (no. 62) with $a = 962.13(5)$, $b = 825.28(4)$, $c = 1838.76(9)$ pm and four formula units per unit cell. Compared with both modifications of the *ino*-oxosilicates $A_2MFSi_4O_{10}$, the crystal structure has one formula unit of RbF and one of ScF_3 in excess.

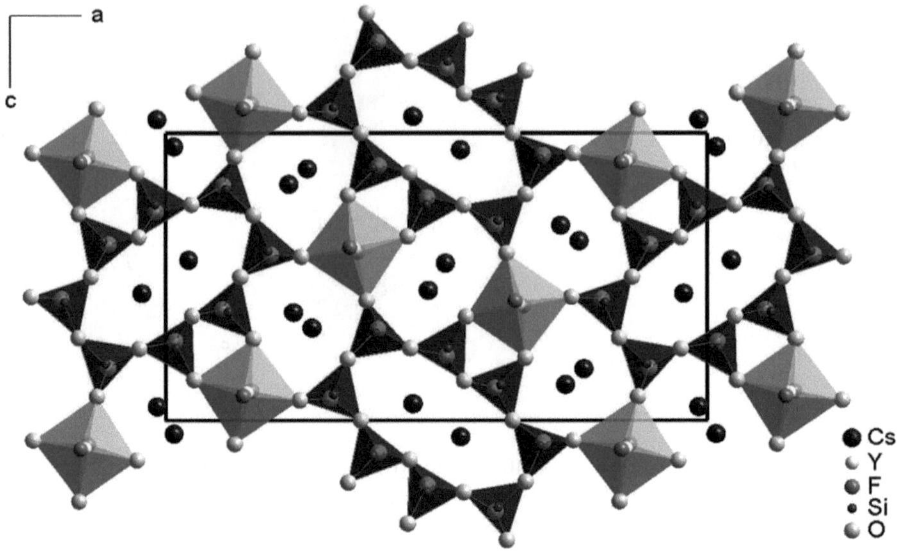

Fig. 6 View at the orthorhombic crystal structure of $Cs_2YFSi_4O_{10}$ along [010]

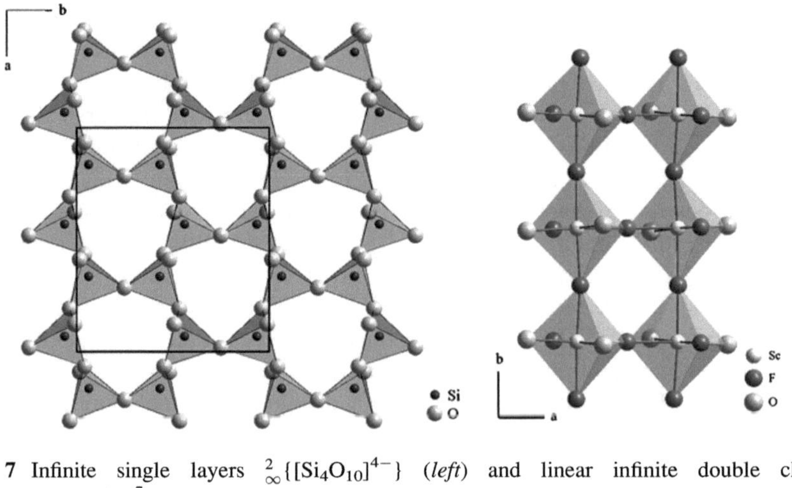

Fig. 7 Infinite single layers ${}^{2}_{\infty}\{[Si_4O_{10}]^{4-}\}$ (*left*) and linear infinite double chains ${}^{1}_{\infty}\{[Sc_2F^{t}_{2/1}F^{v}_{6/2}O^{t}_{4/1}]^{7-}\}$ (*right*) in $Rb_3Sc_2F_5Si_4O_{10}$

In consequence of the minimized SiO_2 content, a higher condensation of the $[SiO_4]^{4-}$ tetrahedra takes place to constitute layers in the *phyllo*-oxosilicate $Rb_3Sc_2F_5Si_4O_{10}$. The resulting layer topology is based on distorted six-membered rings, which are linked to corrugated, unbranched vierer single layers parallel to the (001) plane (Fig. 7, *left*). At the current state of knowledge, this up and down conversion in a layer topology seems to be unknown in the oxosilicate chemistry,

Fig. 8 View at the orthorhombic crystal structure of $Rb_3Sc_2F_5Si_4O_{10}$ along [010].

but an identical tetrahedral layer is found, for example, in the borophosphate $Mg(H_2O)_2[B_2P_2O_8(OH)_2] \cdot H_2O$ (Ewald et al. 2005). In $Rb_3Sc_2F_5Si_4O_{10}$, the Si–O distances range from 158 to 165 pm, the O–Si–O angles from 103 to 114° and the Si–O–Si angles from 125 to 145°. The Sc^{3+} cations are coordinated octahedrally by four F^- and two O^{2-} anions (d(Sc–F) = 200–208 pm, d(Sc–O) = 202–205 pm). These *cis*-$[ScF_4O_2]^{5-}$ octahedra become connected via both apical F^- anions to form slightly undulated chains $^1_\infty\{[ScF^t_{2/1}F^v_{2/2}O^t_{2/1}]^{4-}\}$ along [010] with Sc – F–Sc angles of 170° and 172°. Furthermore, the single chains $^1_\infty\{[ScF^t_{2/1}F^v_{2/2}O^t_{2/1}]^{4-}\}$ are condensed via one of the equatorial F^- anions with exactly linear Sc–F–Sc bridges to generate $^1_\infty\{[Sc_2F^t_{2/1}F^v_{6/2}O^t_{4/1}]^{7-}\}$ double chains (Fig. 7, *right*). Here, the four crystallographically different fluoride anions have five- and sixfold cationic coordinations. The linkage to the oxosilicate layer occurs again via both equatorial O^{2-} anions to form the tetrahedral-octahedral framework (Fig. 8).

The resulting unique cavities host the three crystallographically independent Rb^+ cations with coordination polyhedra $[(Rb1)F_6O_5]^{15-}$, $[(Rb2)F_6O_6]^{17-}$ and $[(Rb3)F_6O_7]^{19-}$ and distances such as d(Rb – F) = 282–352 pm and d(Rb – O) = 296–373 pm.

In conclusion, the basic silicate units of the four reported alkali-metal rare-earth metal(III) fluoride oxosilicates always contain motifs of four corner-shared $[SiO_4]^{4-}$ tetrahedra. Their further condensation features increase from oligomers in the fluoride-poor oxosilicate $K_5Eu_2F[Si_4O_{13}]$ via different infinite chains in both 1:1 structure types of the composition $A_2MFSi_4O_{10}$ to even infinite layers in the fluoride-rich oxosilicate $Rb_3Sc_2F_5Si_4O_{10}$. The cross-linkage of the different $[MF_nO_{6-n}]^{m-}$ octahedra to vertex-sharing fragments or single and double chains takes place only via the F^- anions and the degree of condensation rises with increasing F^- content within the compounds. The connectivity to the final three-dimensional octahedral-tetrahedral frameworks occurs only via the oxide anions of the octahedral building units.

3 Alkaline-Earth Metal Rare-Earth Metal(III) Fluoride Oxosilicates

The *neso*-oxosilicates of the formula type $AE_2M_3F[SiO_4]_3$ (AE^{2+} = Ca – Ba, M^{3+} = La – Nd, Sm, Eu) (Schäfer and Schleid 2008; Schäfer 2010) all crystallize isotypically with the mineral *apatite* $Ca_5(OH,F)[PO_4]_3$ (Hughes et al. 1989) in the hexagonal space group $P6_3/m$ (no. 176). The evolution of their lattice parameters (Table 1) obeys the lanthanide contraction in all three series. The *apatite*-type structure contains isolated $[SiO_4]^{4-}$ tetrahedra (d(Si–O) = 158–164 pm), threefold coordinated F^- anions and two possible *Wyckoff* sites, 6 *h* and 4*f*, for the AE^{2+} and M^{3+} cations. In detail, the cations are surrounded by one F^- and six O^{2-} anions as pentagonal bipyramids on site 6 *h* and by nine O^{2-} anions as tricapped trigonal prisms on site 4 *f*.

Table 1 Lattice parameters of the $AE_2M_3F[SiO_4]_3$ representatives with *apatite*-type structure (hexagonal, $P6_3/m$; Z = 2)

Formula	*a*/pm	*c*/pm	*c*/*a*	V_m/cm^3 mol^{-1}
$Ca_2La_3F[SiO_4]_3$	964.92(3)	708.96(2)	0.735	172.13
$Ca_2Ce_3F[SiO_4]_3$	960.47(3)	705.50(3)	0.735	169.71
$Ca_2Pr_3F[SiO_4]_3$	956.59(2)	702.81(2)	0.735	167.70
$Ca_2Nd_3F[SiO_4]_3$	953.57(4)	699.78(3)	0.734	165.93
$Ca_2Sm_3F[SiO_4]_3$	947.00(3)	695.25(2)	0.734	162.59
$Ca_2Eu_3F[SiO_4]_3$	943.97(3)	692.45(3)	0.734	160.90
$Sr_2La_3F[SiO_4]_3$	973.80(3)	723.09(2)	0.743	178.81
$Sr_2Ce_3F[SiO_4]_3$	967.77(2)	718.97(2)	0.743	175.59
$Sr_2Pr_3F[SiO_4]_3$	967.46(5)	715.76(3)	0.740	174.70
$Sr_2Nd_3F[SiO_4]_3$	960.94(3)	711.60(2)	0.740	171.34
$Sr_2Sm_3F[SiO_4]_3$	957.59(3)	704.81(3)	0.736	168.53
$Ba_2La_3F[SiO_4]_3$	987.62(5)	734.05(3)	0.743	186.70
$Ba_2Ce_3F[SiO_4]_3$	985.21(8)	729.02(6)	0.740	184.52
$Ba_2Pr_3F[SiO_4]_3$	982.78(6)	723.16(4)	0.736	182.14
$Ba_2Nd_3F[SiO_4]_3$	980.27(3)	718.21(3)	0.733	179.97
$Ba_2Sm_3F[SiO_4]_3$	977.02(9)	711.05(7)	0.728	176.99
$Ba_2Eu_3F[SiO_4]_3$	973.89(4)	705.30(3)	0.724	174.44

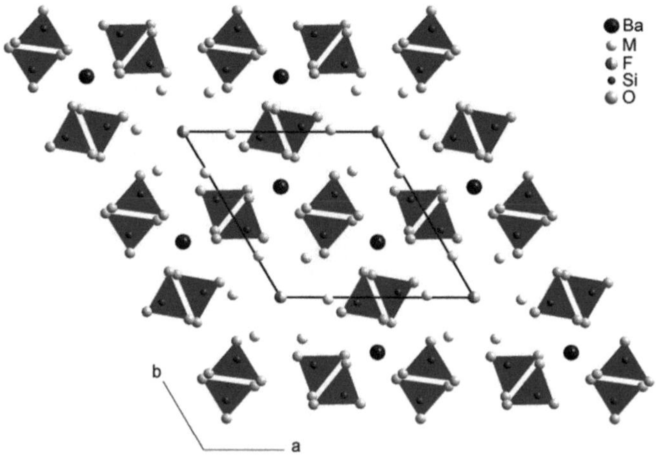

Fig. 9 View at the hexagonal crystal structure of the apatites $Ba_2M_3F[SiO_4]_3$ along [001]

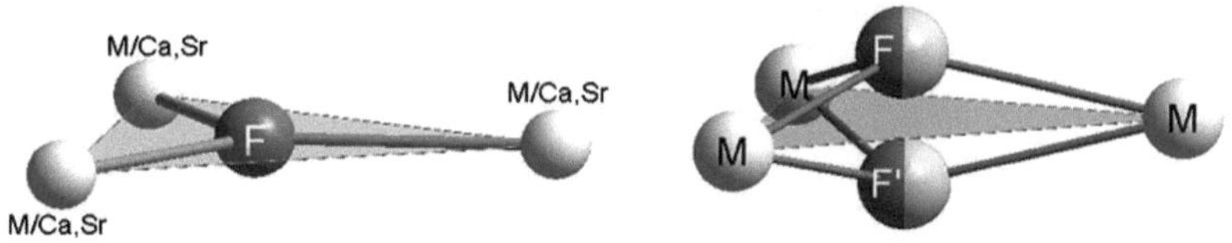

Fig. 10 Triangular coordination of the F^- anions in the apatites $AE_2M_3F[SiO_4]_4$ (AE = Ca and Sr) (*left*) and $Ba_2M_3F[SiO_4]_4$ (*right*)

In the calcium and strontium compounds, both possible cationic sites are mixed occupied by AE^{2+} and M^{3+} cations (d(Ca/M–O) = 234–286 pm, d(Ca/M–F) = 224–236 pm; d(Sr/M–O) = 233–289 pm, d(Sr/Mr–F) = 233–240 pm). In contrast, only the M^{3+} cations reside at the 6 h site (d(M–O) = 239–307 pm, d(M–F) = 243–250 pm) and the Ba^{2+} cations occupy the 4f site (d(Ba–O) = 239–284 pm) exclusively in all barium compounds (Fig. 9). While the F^- anions (site 2a) are coordinated trigonal planar by three cations from the 6 h site in $Ca_2M_3F[SiO_4]_3$ (M^{3+} = La–Nd, Sm, Eu) and $Sr_2M_3F[SiO_4]_3$ (M^{3+} = La–Nd, Sm) (Fig. 10, *left*, and Fig 11), they move out of their triangular plane to the 4e site with an occupation probability of 50% in the $Ba_2M_3F[SiO_4]_3$ series (M^{3+} = La–Nd, Sm, Eu) (Fig. 10, *right*, and Fig. 12).

Meanwhile, phase-pure samples of all $AE_2La_3F[SiO_4]_3$:Eu^{3+} representatives (AE^{2+} = Ca – Ba) could be synthesized, which always show the typical red luminescence for Eu^{3+} under UV excitation. However, the complete spectroscopic investigations (excitation, emission, life time and quantum efficiency) are still work in progress.

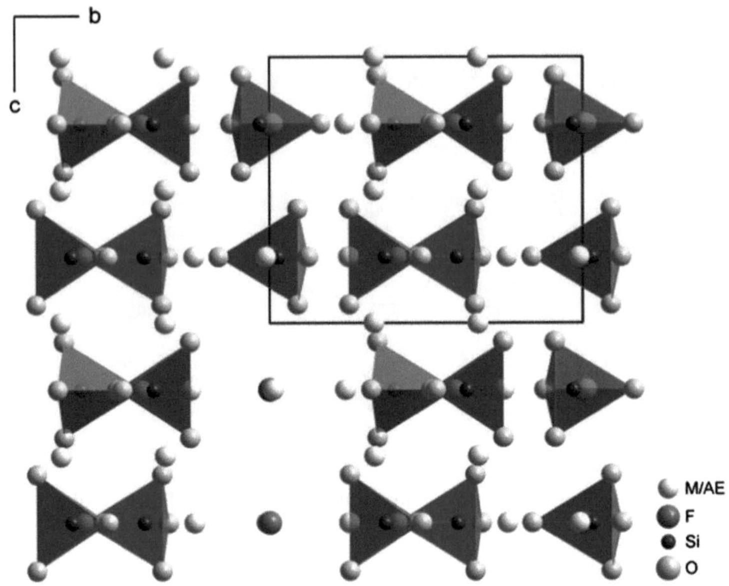

Fig. 11 View at the crystal structure of the apatites $AE_2M_3F[SiO_4]_3$ along [001], for a better look at the F^- anions some $[SiO_4]^{4-}$ tetrahedra are omitted

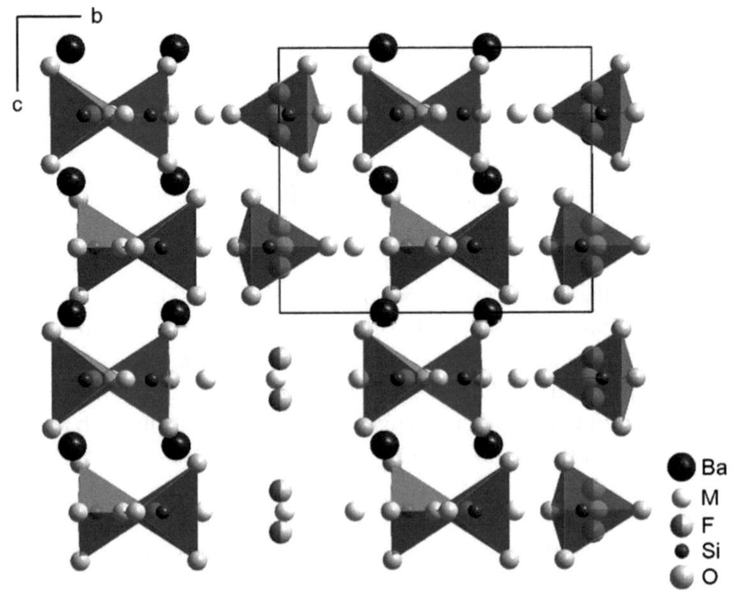

Fig. 12 View at the crystal structure of the apatites $Ba_2M_3F[SiO_4]_3$ along [001], for a better look at the F^- anions some $[SiO_4]^{4-}$ tetrahedra are omitted

Acknowledgements We gratefully acknowledge the financial support of the *State of BadenWürttemberg* (Stuttgart, Germany) and the *Deutsche Forschungsgemeinschaft* (Bonn, Germany) for the promotion within the DFG priority program 1166 (Lanthanoid-Specific Functionalities in Molecule and Material).

References

Blasse G, Grabmaier BC (1994) Luminescent materials. Springer, Heidelberg

Chiang PY, Lin TW, Dai JH, Chang BC, Lii KH (2007) Flux Synthesis, crystal structure, and luminescence properties of a new Europium fluoride-silicate: $K_5Eu_2FSi_4O_{13}$. Inorg Chem 46:3619–3622

Dieke GH (1968) Spectra and energy levels of rare-earth ions in crystals. Intersience Publ, New York

Ewald B, Öztan Y, Prots Yu, Kniep R (2005) Structural patterns and dimensionality in magnesium borophosphates: the crystal structures of $Mg_2(H_2O)[BP_3O_9(OH)_4]$ and $Mg(H_2O)_2[B_2P_2O_8(OH)_2] \cdot H_2O$. Z Anorg Allg Chem 631:1615–1621

Gong X, Lin Y, Chen Y, Huang Z, Huang Y, Luo Z (2005) Syntheses, structure, and characterization of crystal $La_6Ba_4(SiO_4)_6F_2$, a promising laser host. Chem Mater 17:1135–1138

Höppe HA (2009) Recent developments in the field of inorganic phosphors. Angew Chem Int Ed 48:3572–3582

Hughes JM, Cameron M, Croxley KD (1989) Structural variations in natural F, OH, and Cl apatites. Amer Mineral 74:870–876

Ito J (1968) Silicate apatites and oxyapatites. Amer Mineral 53:890–907

Jüstel T, Nicol H, Ronda C (1998) New developments in the field of luminescent materials for lighting and displays. Angew Chem Int Ed 37:3085–3103

Kolitsch U, Tillmanns E (1994) The structural relation between the new synthetic silicate K2ScFSi4O10 and narsarsukite, Na2(Ti, Fe(3+))(O, F)Si4O10. Eur J Mineral 16:143–149

Liebau F (1985) Structural chemistry of silicates. Springer, Heidelberg

Merinov BV, Maksimov BA, Ilyukhin VV, Belov NV (1980) The crystal structure of $(NaYSiO_4)_4NaF$. Dokl Akad Nauk SSSR 255:348–351 (in russian)

Peacor DR, Buerger MJ (1962) The determination and refinement of the structure of narsarsukite, $Na_2TiOSi_4O_{10}$. Amer Mineral 47:539–556

Ronda C (2008) Luminescence. Weinheim, Viley-VCH

Schäfer MC (2010) Synthese, Charakterisierung und Eigenschaften multinärer Oxosilicate der Selten-Erd-Metalle. Dissertation, Universität Stuttgart (2010)

Schäfer MC, Schleid Th (2007) Synthese und Kristallstruktur des Fluorid-*ino*-Oxosilicats $Cs_2YFSi_4O_{10}$. Z Anorg Allg Chem 633:1018–1023

Schäfer MC, Schleid Th (2008) $Ba_2Ce_3F[SiO_4]_3$: Ein ausgeordnetes Fluorid-Oxosilicat im Apatit-Typ. Z Anorg Allg Chem 634:2074–2074

Schäfer MC, Schleid Th (2010) $K_5La_4F[SiO_4]_4$: a fluoride-poor quinary lanthanum neso-oxosilicate. Z Anorg Allg Chem 636:2069–2069

Schäfer MC, Schleid Th (2011) Synthesis and crystal structure of the fluoride-rich rubidium scandium fluoride oxosilicate $Rb_3Sc_2F_5Si_4O_{10}$. Z Anorg Allg Chem 637, in press

Schleid Th, Müller-Bunz H, Janka O (2011) Geo-inspired phosphors based on rare-earth metal(III) fluorides with complex oxoanions: I. Fluoride oxocarbonates and oxosilicates. In: Krivovichev SV (ed) Minerals as advanced minerals II. Springer, Heidelberg

Tang MF, Chiang PY, Su YH, Jung YC, Hou GG, Chang BC, Lii KH (2008) Flux Synthesis, crystal structures, and luminescence properties of salt-inclusion lanthanide silicates: $[K_9F_2][Ln_3Si_{12}O_{32}]$ (Ln = Sm, Eu, Gd). Inorg Chem 48:8985–8989

Wickleder C, Hartenbach I, Lauxmann P, Schleid Th (2002) $Eu_5F[SiO_4]_3$ und $Yb_5S[SiO_4]_3$: Gemischtvalente Lanthanoid-Silicate mit Apatit-Struktur. Z Anorg Allg Chem 628:1602–1606

Geo-Inspired Phosphors Based on Rare-Earth Metal(III) Fluorides with Complex Oxoanions: I. Fluoride Oxocarbonates and Oxosilicates

Thomas Schleid, Helge Müller-Bunz, and Oliver Janka

1 Introduction

The quest for luminescent materials with high energy efficiency in illumination devices has become a serious issue for saving our planet earth from global warming. On the search for new and effective solids as host lattices for luminescence (Dieke 1968; Blasse and Grabmaier 1994; Jüstel et al. 1998; Ronda 2008; Höppe 2009), synthetic oxosilicates containing rare-earth metal(III) cations and fluoride anions got into the focus of our work, because hard anions (Pearson 1963) such as oxide and fluoride seem to be most suitable to preserve the applied energy and prevent it from being transformed into heat by lattice-vibration processes. During the everlasting fight photons versus phonons, the transfer of energy originating from external UV radiation to luminescent cations needs to prevail over any kind of non-radiative fading. The most simple combination of both anionic species, oxide O^{2-} and fluoride F^{-}, is realized in the rare-earth metal oxide fluorides with the composition *MOF* (M^{3+} = Y, La; Ce – Nd, Sm – Lu), which are not easy to prepare. A straightforward method to obtain them phase-pure and on the gram-scale results from the thermal decomposition of *bastnaesite*-type fluoride oxocarbonates MF[CO_3] at reasonable low temperatures. Moreover, they can serve as active educts for the synthesis of fluoride oxosilicates, where all oxide anions are bonded to silicon central atoms in isolated or vertex-shared $[SiO_4]^{4-}$ tetrahedra and thus get in close proximity to each another. Therefore high coordination numbers most suitable for M^{3+} cations can be achieved in a few prototypic mineral-related compounds such as $La_3F_3[Si_3O_9]$, $Y_3F[Si_3O_{10}]$, $Er_4F_2[Si_2O_7][SiO_4]$ and $Eu_5F[SiO_4]_3$ (≡ $(Eu^{II})_2(Eu^{III})_3F[SiO_4]_3$).

T. Schleid (✉) • H. Müller-Bunz • O. Janka
Institut für Anorganische Chemie, Universität Stuttgart, D-70569 Stuttgart, Germany
e-mail: schleid@iac.uni-stuttgart.de

S.V. Krivovichev (ed.), *Minerals as Advanced Materials II*,
DOI 10.1007/978-3-642-20018-2_33, © Springer-Verlag Berlin Heidelberg 2012

2 From *Bastnaesite*-Type Fluoride Oxocarbonates to Rare-Earth Metal(III) Oxide Fluorides

Bulk and M^{3+}-doped *bastnaesite*-type LaF[CO_3] (Donnay and Donnay 1953) can be prepared from aqueous solutions, containing La^{3+}, F^- and $[CO_3]^{2-}$ ions along with up to 3% M^{3+} dopant (M = Eu or Tb) if desired (Janka and Schleid 2009). Its thermal decomposition at 450°C yields volatile CO_2 and single-phase LaOF (trigonal, $R\bar{3}m$) as solid residue, which shows brilliant red (with $M = Eu^{3+}$) or green luminescence (with $M = Tb^{3+}$). The same routine leads to amorphous YF[CO_3], which shows similar behaviour in terms of pyrolysis. Just like in the case of crystalline LaF[CO_3] it can be thermally decomposed to YOF under CO_2 evolution and the crystal structure of the obtained solid phase becomes subject to the actual temperature control. At 450°C the trigonal γ-phase ($R\bar{3}m$), up to 650°C the tetragonal β-phase ($P4/nmm$) and above 850°C the cubic α-phase ($Fm\bar{3}m$) are formed as exclusive modifications.

From the topological point of view the *bastnaesite*-type crystal structure of LaF[CO_3] (a = 717.85(6) pm, c = 982.68(9) pm, c/a = 1.369, Z = 6, hexagonal, $P\bar{6}2c$, no. 190; Janka 2010) shows some resemblance to the *tysonite*-type arrangement of the binary LaF_3 (a = 718.80(6) pm, c = 735.94(6) pm, c/a = 1.024, Z = 6, trigonal, $P\bar{3}c1$, no. 165). In both cases, the crystallographically unique La^{3+} cations are coordinated by 9 + 2 anions. Consisting of F^- anions exclusively in LaF_3 (d(La–F) = 242–264 *plus* 301 pm; Müller-Bunz and Schleid 1999) they split up into 3 F^- and 6 + 2 O^{2-} anions with La–F distances ranging from 241 to 243 pm and values of 254–258 *plus* 309 pm for the La–O separations (Janka and Schleid 2009; Grice et al. 2007) in LaF[CO_3] (Fig. 1). The La^{3+} cations form hexagonal layers with both crystallographically different F^- anions in LaF[CO_3] and the two sorts of threefold coordinated F^- anions in LaF_3 within the corresponding (001) planes (Fig. 2). All fluoride anions exhibit thereby coordination numbers of three, but whereas $(F1)^-$ resides in the same plane as the lanthanum triangles,

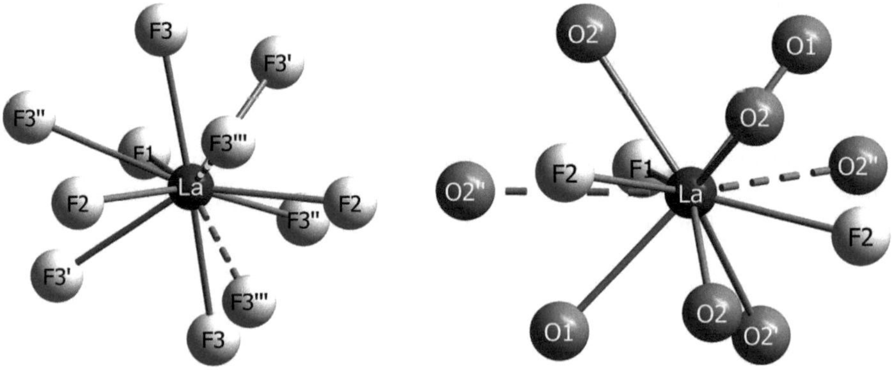

Fig. 1 Polyhedra of coordination about the La^{3+} cations in *tysonite*-type LaF_3 (*left*) and *bastnaesite*-type LaF[CO_3] (*right*)

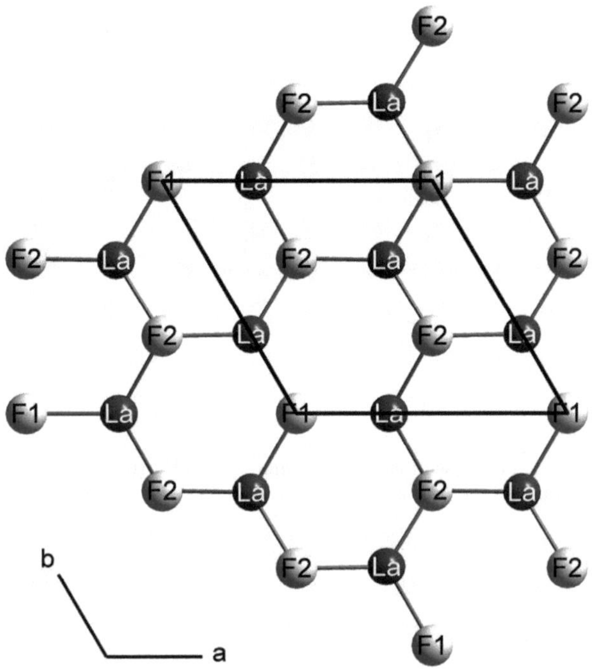

Fig. 2 Hexagonal $^2_\infty\{[La(F1)_{1/3}(F2)_{2/3}]^{2+}\}$ layer parallel to the (001) plane of *tysonite*-type LaF_3, *bastnaesite*-type $LaF[CO_3]$ and *bastnaesite*-related $La_3F_3[Si_3O_9]$

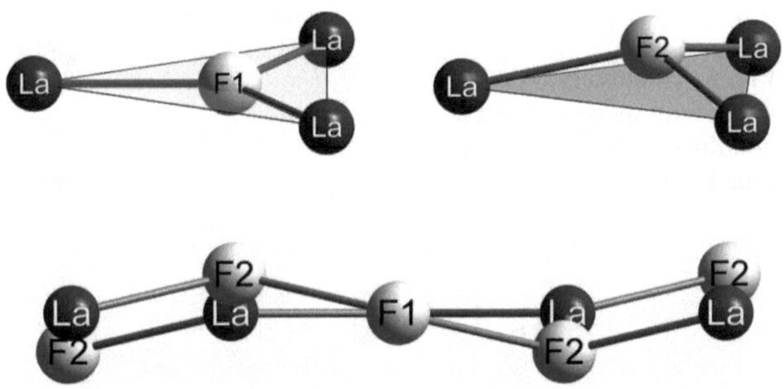

Fig. 3 La^{3+} coordination of the two different layer-forming F^- anions (*top*) within the undulated $^2_\infty\{[La(F1)_{1/3}(F2)_{2/3}]^{2+}\}$ layer of *tysonite*-type LaF_3, *bastnaesite*-type $LaF[CO_3]$ and *bastnaesite*-related $La_3F_3[Si_3O_9]$

$(F2)^-$ is coordinated in a slightly non-planar fashion, shifted away by 46 pm in LaF_3 and 32 pm in $LaF[CO_3]$ from the equilateral $(La^{3+})_3$ planes (Fig. 3). A view of the complete crystal structures (Fig. 4) indicates that two topologically different layers of oxocarbonate anions $[CO_3]^{2-}$ (d(C–O) = 132–134 pm, Fig. 5) appear, separating the hexagonal $^2_\infty\{[FLa_{3/3}]^{2+}\}$ monolayers in hexagonal $LaF[CO_3]$, which can thus be written as $^2_\infty\{[La(F1)_{1/3}(F2)_{2/3}]^{2+}\}$. The same formalism holds for trigonal LaF_3 (Zalkin and Templeton 1985), where the complex $[CO_3]^{2-}$ anions of $LaF[CO_3]$ have formally just to be replaced layerwise by a doubled number of non-complex F^- anions (Fig. 4).

Single-phase γ-LaOF (a = 405.1(1) pm, c = 2003.8(7) pm, c/a = 4.946, Z = 6, trigonal, $R\bar{3}m$, no. 166; Zachariasen 1951) results as solid product after the thermal decomposition of $LaF[CO_3]$ at 450°C, while β-LaOF (a = 409.1(1) pm,

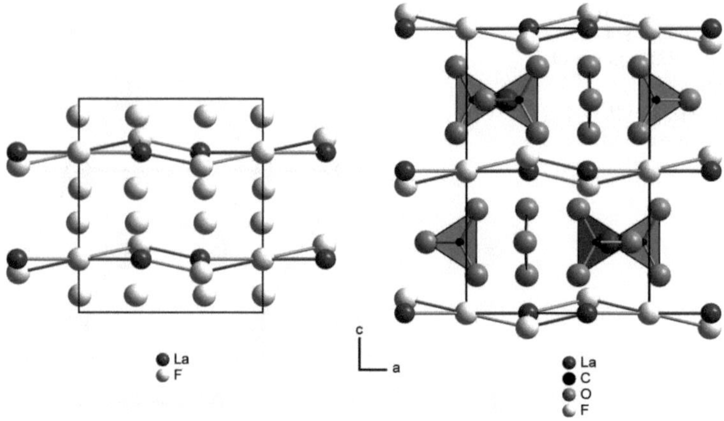

Fig. 4 (110) sections of *tysonite*-type LaF_3 (trigonal, *left*) and *bastnaesite*-type $LaF[CO_3]$ (hexagonal, *right*)

Fig. 5 Isolated $[CO_3]^{2-}$ anion with its complete La^{3+} coordination in *bastnaesite*-type $LaF[CO_3]$

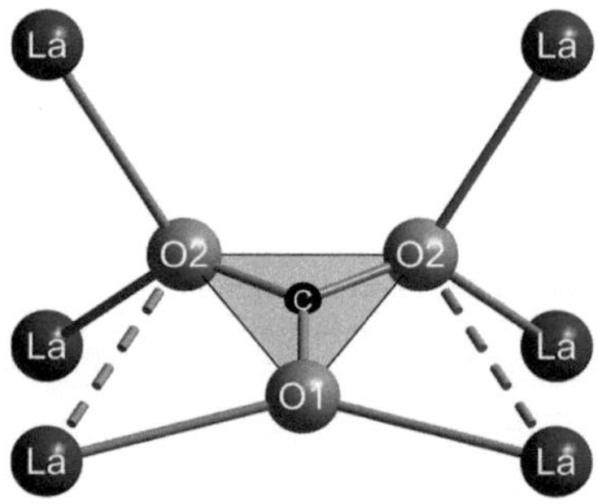

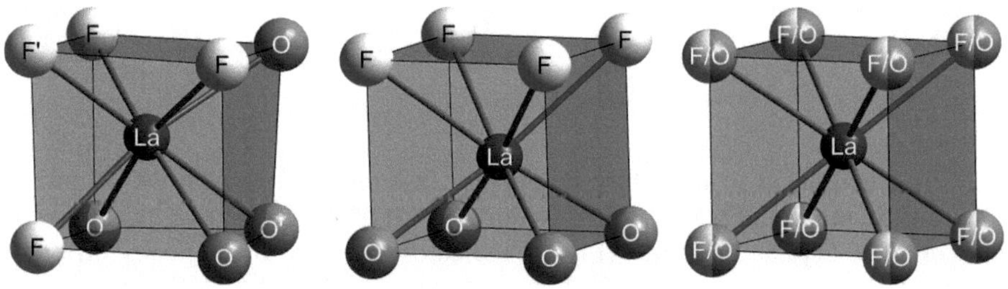

Fig. 6 Polyhedra of coordination about the La^{3+} cations in trigonal γ-LaOF (*left*), tetragonal β-LaOF (*mid*) and cubic α-LaOF (*right*)

$c = 583.6(2)$ pm, $c/a = 1.427$, $Z = 2$, tetragonal, $P4/nmm$, no. 129; Zachariasen 1951) and α-LaOF ($a = 575.6(3)$ pm, $Z = 4$, cubic, $Fm\bar{3}m$, no. 225; Klemm and Klein 1941) emerge at higher temperatures (650°C and 850°C, respectively). In all three LaOF modifications the unique La^{3+} cations are surrounded cube-like by eight anions (four O^{2-} and four F^-, Fig. 6) in an ordered way for γ-LaOF

(d(La–O) = 240–242 pm, d(La–F) = 256–258 pm) and β-LaOF (d(La–O) = 241.5 pm, d(La–F) = 261.8 pm, 4× each), but as disordered real cube for α-LaOF (d(La–O/F) = 249.2 pm, 8× each).

3 From Rare-Earth Metal(III) Oxide Fluorides to Fluoride Oxosilicates

Fluoride-derivatized rare-earth metal(III) oxosilicates are available from appropriate M_2O_3:MF_3:SiO_2, M_2O_3:MOF:SiO_2 or MOF:SiO_2 mixtures at elevated temperatures. For *bastnaesite*-related LaF(SiO_3) (≡ $La_3F_3[Si_3O_9]$, hexagonal, $P\bar{6}2c$; Müller-Bunz and Schleid 1999) UV-luminescence has already been proven with Ce^{3+}-doped samples (Oskam et al. 2002). *Thalenite*-type $Y_3F[Si_3O_{10}]$ (monoclinic, $P2_1/n$; Schleid and Müller-Bunz 1998) can be doped with medium-size lanthanoids (M = Sm – Er) and exhibits not only the expected visible, but also IR-luminescence (Schäfer et al. 2011). In contrast, $Er_4F_2(Si_3O_{11})$ (≡ $Er_4F_2[Si_2O_7][SiO_4]$, triclinic, $P\bar{1}$; Müller-Bunz and Schleid 2001) has meanwhile even an analogous yttrium counterpart as $Y_4F_2[Si_2O_7][SiO_4]$ host for lanthanoid dopants, but so far no successful doping in single-phase samples was possible. A very promising *apatite*-type candidate contains di- and trivalent europium simultaneously: $Eu_5F(Si_3O_{12})$ (≡ $(Eu^{II})_2(Eu^{III})_3F[SiO_4]_3$, hexagonal, $P6_3/m$; Wickleder et al. 2002). Therefore it should be possible to dope isotypic samples of *i.e.* $Ba_2La_3F[SiO_4]_3$ with both suitable di- and trivalent lanthanoid cations. From the structural point of view more or less planar $[FM_3]^{8+}$ triangles and regular $[SiO_4]^{4-}$ tetrahedra are present in all four examples, which occur either isolated or condensed. The cations $[FM_3]^{8+}$ share vertices to form layers ($^2_\infty\{[FLa_{3/3}]^{2+}\}$ just like in *tysonite*-type LaF_3 and *bastnaesite*-type LaF[CO_3]) or edges to build up dimers ($^0_\infty\{([FEr_{2/1}Er_{2/2}]^{5+})_2\}$) if necessary, while the tetrahedral $[SiO_4]^{4-}$ anions use common corners, whenever they need to condense at all.

4 *Bastnaesite*-Related $La_3F_3[Si_3O_9]$

Not only due to its reduced formula, LaF(SiO_3) namely, hexagonal $La_3F_3[Si_3O_9]$ (Müller-Bunz and Schleid 1999) should exhibit a *bastnaesite*-related crystal structure. Even the space group ($P\bar{6}2c$, no. 190) for LaF[CO_3] (a = 709.38(7) pm, c = 485.91(5) pm, c/a = 0.685, Z = 6) and $La_3F_3[Si_3O_9]$ (a = 708.32(3) pm, c = 1089.48(6) pm, c/a = 1.538, Z = 2) is identical and indeed, the crystallographically unique La^{3+} cations are coordinated by 9 + 2 anions in both cases (Fig. 7). They even show the same diversification into 3F^- and 6 + 2 O^{2-} anions with quite similar intervals for the La–F (241–243 pm in LaF[CO_3] versus 243–251 pm in $La_3F_3[Si_3O_9]$) and La–O distances (254–258 *plus* 309 pm

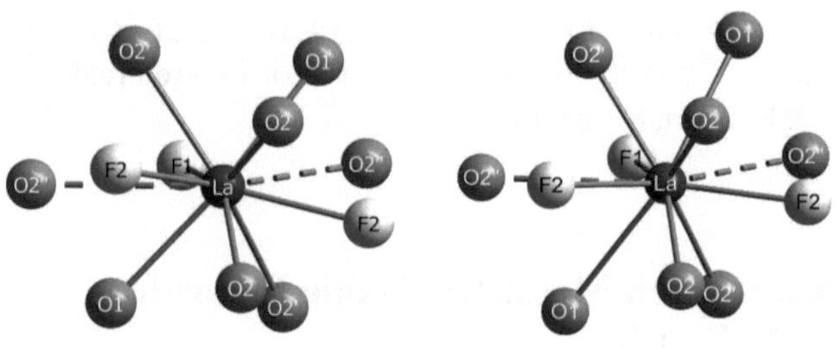

Fig. 7 Polyhedra of coordination about the La^{3+} cations in *bastnaesite*-type $LaF[CO_3]$ (*left*) and *bastnaesite*-related $La_3F_3[Si_3O_9]$ (*right*)

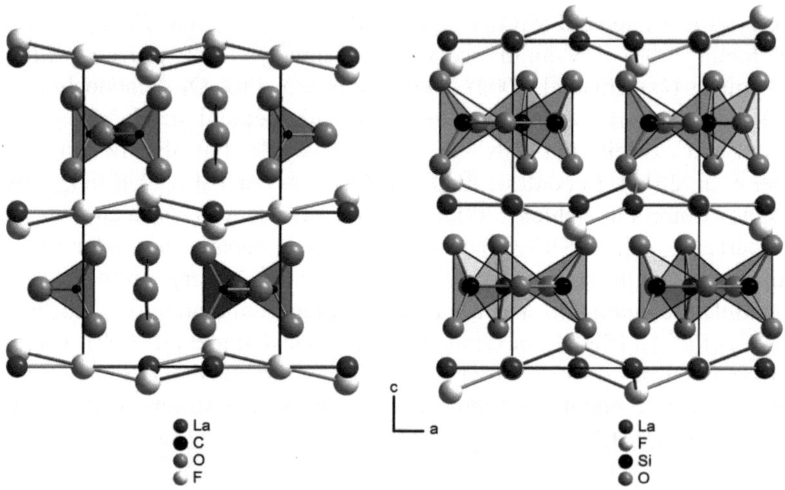

Fig. 8 (110) sections of the hexagonal crystal structures of *bastnaesite*-type $LaF[CO_3]$ (*left*) and *bastnaesite*-related $La_3F_3[Si_3O_9]$ (*right*)

in $LaF[CO_3]$ versus 249–291 *plus* 322 pm in $La_3F_3[Si_3O_9]$). Moreover, hexagonal cationic monolayers $^2_\infty\{[FLa_{3/3}]^{2+}\}$, best written according to $^2_\infty\{[La(F1)_{1/3}(F2)_{2/3}]^{2+}\}$ as already described for *tysonite*-type LaF_3 and *bastnaesite*-type $LaF[CO_3]$ (Fig. 3), persist topologically equivalent in the crystal structure of $La_3F_3[Si_3O_9]$. The measure of the hexagonal *ab*-plane shrinks slightly from LaF_3 ($a = 718.80(6)$ pm) via $LaF[CO_3]$ ($a = 717.85(6)$ pm) towards $La_3F_3[Si_3O_9]$ ($a = 708.32(3)$ pm). Of all threefold coordinated fluoride anions, $(F1)^-$ always resides in the same plane as the $(La^{3+})_3$ triangles, while $(F2)^-$ shows slightly non-planar coordination figures with displacements of 32 pm in $LaF[CO_3]$, 46 pm in LaF_3 and 80 pm in $La_3F_3[Si_3O_9]$ from the corresponding lanthanum planes. The same trend is reflected by the individual *c*/*a*-ratios of $c/a = 1.024$ with $c = 735.94(6)$ pm for LaF_3 as lowest, followed by $c/a = 1.369$ with $c = 982.68(9)$ pm for $LaF[CO_3]$ as intermediate and $c/a = 1.538$ with $c = 1089.48(6)$ pm for $La_3F_3[Si_3O_9]$ as largest value (Fig. 8). This does not come as a real surprise, since double layers of non-complex

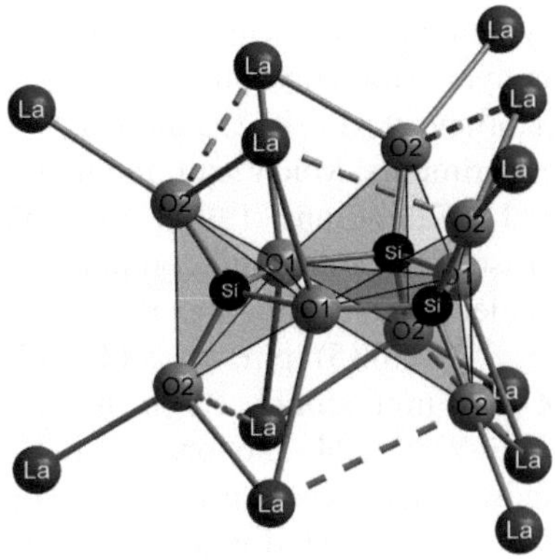

Fig. 9 Discrete *cyclo*-oxotrisilicate anion $[Si_3O_9]^{6-}$ with its complete La^{3+} coordination in *bastnaesite*-related $La_3F_3[Si_3O_9]$

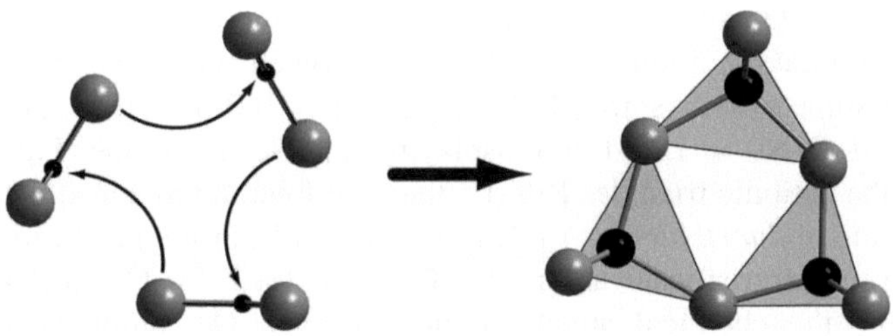

Fig. 10 Formal cyclotrimerization of three $[CO_3]^{2-}$ anions to a single $[Si_3O_9]^{6-}$ anion for the hypothetical transformation of *bastnaesite*-type $LaF[CO_3]$ into *bastnaesite*-related $La_3F_3[Si_3O_9]$

F^- anions in LaF_3 are replaced with monolayers of erected complex $[CO_3]^{2-}$ anions in $LaF[CO_3]$, which again find their own substitution in complex *cyclo*-oxotrisilicate anions $[Si_3O_9]^{6-}$ (Fig. 9) in $La_3F_3[Si_3O_9]$. Every substitutional step is accompanied by a certain lattice expansion in the *c*-direction of almost 34% for the LaF_3-$LaF[CO_3]$ and about 48% for the LaF_3-$La_3F_3[Si_3O_9]$ transformation.

A final view at the $LaF[CO_3]$-$La_3F_3[Si_3O_9]$ couple reveals that only a formal cyclotrimerization (Fig. 10) of three $[CO_3]^{2-}$ anions (d(C–O) = 132–134 pm, $\angle$(O–C–O) = 120°) to a single $[Si_3O_9]^{6-}$ anion (d(Si–O) = 159–169 pm, $\angle$(O–Si–O) = 105–109° *plus* 121°, $\angle$(Si–O–Si) = 134°) needs to take place between the hexagonal cationic monolayers ${}^{2}_{\infty}\{[FLa_{3/3}]^{2+}\}$ in order to transform the *bastnaesite*-type structure of $LaF[CO_3]$ into the topological arrangement of $La_3F_3[Si_3O_9]$ combined with an almost 11% widening of the *c*-axis (or 5.5% per monolayer of complex anions).

5 *Thalenite*-Type $Y_3F[Si_3O_{10}]$

For quite a long time, the mineral *thalenite* was misinterpreted as an yttrium oxodisilicate $Y_2[Si_2O_7]$ (Nauman et al. 1957; Semenov 1963), sometimes with a certain content of crystal water ($Y_2[Si_2O_7] \cdot {}^1/_3\ H_2O$, Ito and Johnson 1968), but the pyroanionic character of its oxosilicate unit has never been doubted. The true nature as hydroxide-containing *catena*-oxotrisilicate of the composition $Y_3(OH)[Si_3O_{10}]$ (Kornev et al. 1972) could only be proven with single-crystal X-ray diffraction experiments. Much later, the first synthetic fluoride-analogue $Y_3F[Si_3O_{10}]$ (Schleid and Müller-Bunz 1998) was able to be prepared, followed by a short series of isotypic lanthanoid compounds $M_3F[Si_3O_{10}]$ (M = Dy – Tm; Müller-Bunz and Schleid 2000; Schäfer et al. 2011; see Table 1). The crystal structure of $Y_3F[Si_3O_{10}]$ (a = 730.38(5) pm, b = 1112.47(8) pm, c = 1037.14(7) pm, β = 97.235(6)°, Z = 4; monoclinic, $P2_1/n$, no. 14) comprises discrete triangular $[FY_3]^{8+}$ cations (d(F–Y) = 224–236 pm, ∠(Y–F–Y) = 110–134°), where the central F^- anions show a displacement of only 20 pm from their (Y1,Y2,Y3)-plane (Fig. 11, *left*). The three crystallographically independent Y^{3+} cations are all coordinated by a single F^- anion (d(Y–F) = 224–236 pm) and six or six *plus* one O^{2-} anions (d(Y–O) = 221–273 pm *plus* 293 pm) resulting in mono – or bicapped trigonal prismatic coordination figures (Fig. 12). All oxygen atoms belong to isolated *catena*-oxotrisilicate anions $[Si_3O_{10}]^{8-}$ (d(Si–O) = 160–165 pm, ∠(O–Si–O) = 98–116°, ∠(Si–O–Si) = 133–139°) displaying a horseshoe shape (Fig. 11, *right*). Along with the cationic triangles $[FY_3]^{8+}$ they are lined up to constitute the whole crystal structure of *fluorthalenite* $Y_3F[Si_3O_{10}]$ (Fig. 13). From the chemical point of view, the *cyclo*-oxotrisilicate anions $[Si_3O_9]^{6-}$ of $La_3F_3[Si_3O_9]$ just need to be opened up by the chemical attack of an additional O^{2-} anion generating the

Table 1 Lattice parameters of the *fluorthalenite*-type representatives $M_3F[Si_3O_{10}]$ (M = Dy – Tm; monoclinic, $P2_1/n$; Z = 2)

Formula	a/pm	b/pm	c/pm	β/deg	V_m/cm³/mol
$Dy_3F[Si_3O_{10}]$	734.06(6)	1116.55(9)	1040.62(8)	97.281(7)	127.371(9)
$Ho_3F[Si_3O_{10}]$	730.91(6)	1111.68(9)	1037.83(8)	97.238(7)	125.945(9)
$Er_3F[Si_3O_{10}]$	727.89(6)	1107.02(9)	1035.21(8)	97.209(7)	124.591(9)
$Tm_3F[Si_3O_{10}]$	725.04(6)	1102.43(9)	1032.57(8)	97.185(7)	123.280(9)

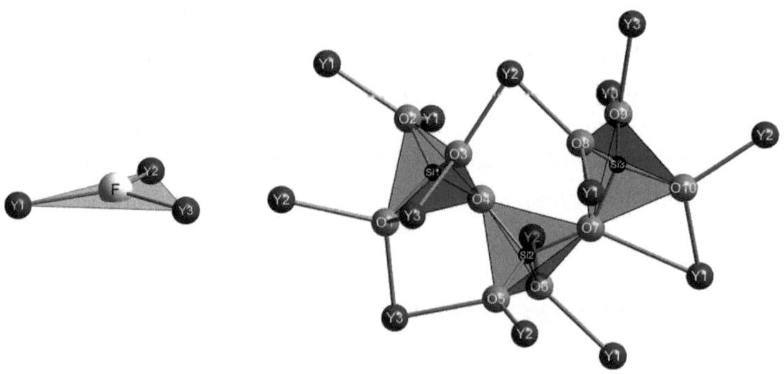

Fig. 11 Isolated triangular $[FY_3]^{8+}$ cation (*left*) and discrete horseshoe-shaped *catena*-oxotrisilicate anion $[Si_3O_{10}]^{8-}$ with its complete Y^{3+} coordination (*right*) in *thalenite*-type $Y_3F[Si_3O_{10}]$

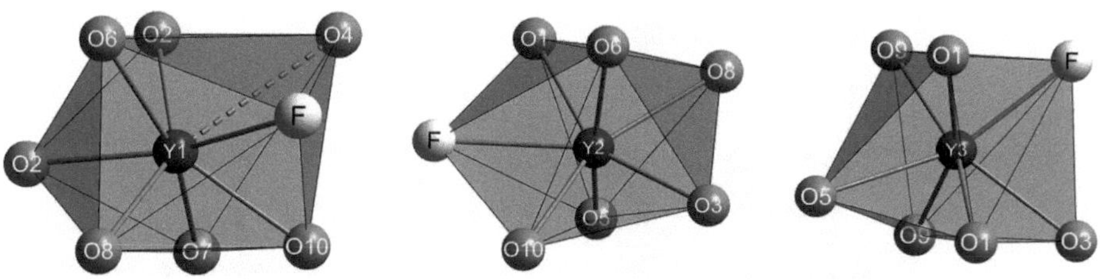

Fig. 12 Polyhedra of coordination about the Y^{3+} cations in *thalenite*-type $Y_3F[Si_3O_{10}]$

Fig. 13 Monoclinic crystal structure of *thalenite*-type $Y_3F[Si_3O_{10}]$ as viewed along [100]

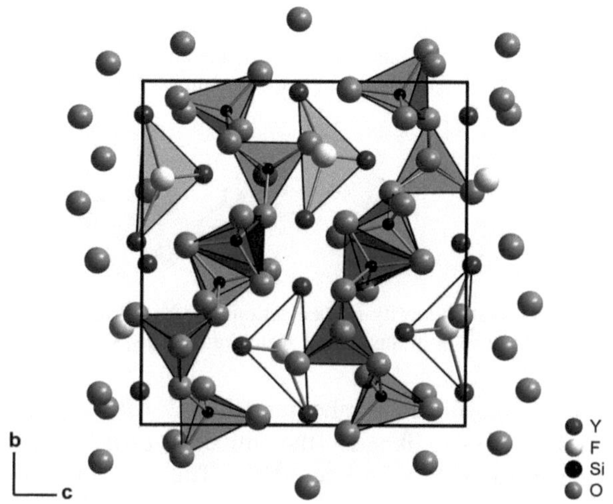

catena-oxotrisilicate anion $[Si_3O_{10}]^{8-}$ in $Y_3F[Si_3O_{10}]$ undergoing an acid–base reaction according to Lewis (Lewis 1926).

6 Two Different Oxosilicate Anions in $Y_4F_2[Si_2O_7][SiO_4]$

Another Lewis acid–base reaction of an additional O^{2-} anion tears the *catena*-oxotrisilicate anion $[Si_3O_{10}]^{8-}$ in $Y_3F[Si_3O_{10}]$ apart into two fragments. This has been demonstrated for the first time with the example of $Er_4F_2(Si_3O_{11})$ ($\equiv Er_4F_2[Si_2O_7][SiO_4]$; Müller-Bunz and Schleid 2001), but meanwhile the isotypic yttrium analogue $Y_4F_2[Si_2O_7][SiO_4]$ could be prepared as well and is waiting for experiments with suitable lanthanoid dopants. Just like in $Er_4F_2[Si_2O_7][SiO_4]$ ($a = 648.51(5)$ pm, $b = 660.34(5)$ pm, $c = 1324.43(9)$ pm, $\alpha = 87.449(4)°$, $\beta = 85.793(4)°$, $\gamma = 60.816(3)°$, $Z = 2$; triclinic, $P\bar{1}$, no. 2; Müller-Bunz and Schleid 2001), there are two different oxosilicate anions in the crystal structure of $Y_4F_2[Si_2O_7][SiO_4]$ ($a = 649.87(5)$ pm, $b = 661.96(5)$ pm, $c = 1329.78(9)$ pm, $\alpha = 87.418(4)°$, $\beta = 85.702(4)°$, $\gamma = 60.854(3)°$; Schäfer 2010), namely a pyroanionic bitetrahedral oxodisilicate unit $[Si_2O_7]^{6-}$ (d(Si–O) = 161–167 pm, $\angle$(O–Si–O) = 98–117°) and

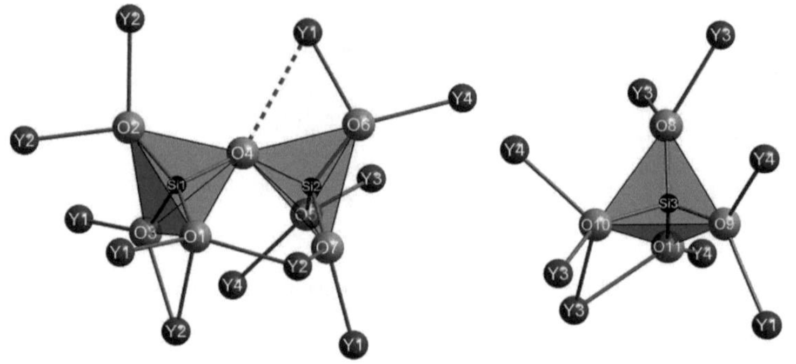

Fig. 14 Discrete bitetrahedral oxodisilicate anion $[Si_2O_7]^{6-}$ (*left*) and isolated *ortho*-oxosilicate anion $[SiO_4]^{4-}$ (*right*) both with their complete Y^{3+} coordination in the crystal structure of $Y_4F_2[Si_2O_7][SiO_4]$

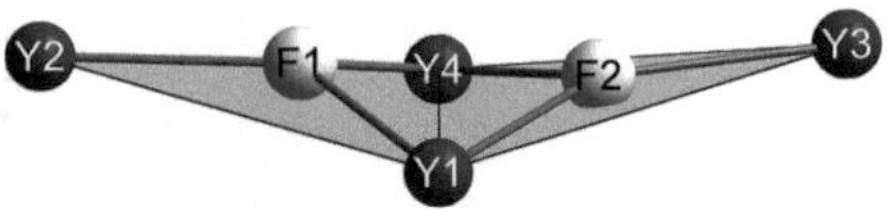

Fig. 15 Discrete butterfly-shaped cationic $[F_2Y_4]^{10+}$ unit as dimer of two $[FY_3]^{8+}$ triangles fused via a common edge in the crystal structure of $Y_4F_2[Si_2O_7][SiO_4]$

an *ortho*-oxosilicate tetrahedron $[SiO_4]^{4-}$ (d(Si–O) = 162–168 pm, ∠(O–Si–O) = 100–116°) just like in the mineral *orthite*, $Al_2CaFeCeO(OH)[Si_2O_7][SiO_4]$ (Rumanova and Nikoleva 1959). The first one consists of two vertex-shared $[SiO_4]^{4-}$ tetrahedra in eclipsed conformation exhibiting a bridging (Si–O–Si)-angle of 131°. Together with these two different anionic building blocks ($[Si_2O_7]^{6-}$ and $[SiO_4]^{4-}$, Fig. 14), discrete cationic $[F_2Y_4]^{10+}$ entities (Fig. 15) complete the crystal structure of $Y_4F_2[Si_2O_7][SiO_4]$ (Fig. 16). For the formation of the latter, two almost plain $[FY_3]^{8+}$ triangles are fused via a common edge resulting in a butterfly-shaped $[F_2Y_4]^{10+}$ dimer (d(F–Y) = 222–240 pm, ∠(Y–F–Y) = 102–149°), where the central F^- anions are only shifted by 22 or 28 pm away from their $(Y^{3+})_3$ planes and both triangular planes include a dihedral angle of 162°. Two out of four crystallographically distinct Y^{3+} cations have only one, but the remaining have two F^- anions in their irregular coordination spheres, which are always filled up with oxygen atoms (d(Y–O) = 222–254 pm *plus* 286 pm) to form mono – and bicapped trigonal prisms again (Fig. 17).

7 Mixed-Valent *Apatite*-Type $(Eu^{II})_2(Eu^{III})_3F[SiO_4]_3$

The final fragmentation of vertex-sharing $[SiO_4]^{4-}$ tetrahedra is accomplished in $Eu_5F(Si_3O_{12})$ (≡ $(Eu^{II})_2(Eu^{III})_3F[SiO_4]_3$), where only isolated *ortho*-oxosilicate units $[SiO_4]^{4-}$ (Fig. 18, *right*) remain eventually in accord with the isotypic

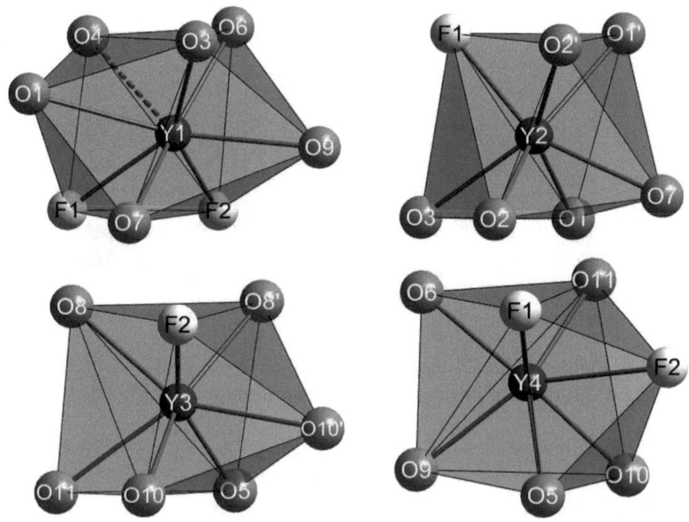

Fig. 16 Polyhedra of coordination about the Y^{3+} cations in the crystal structure of $Y_4F_2[Si_2O_7][SiO_4]$

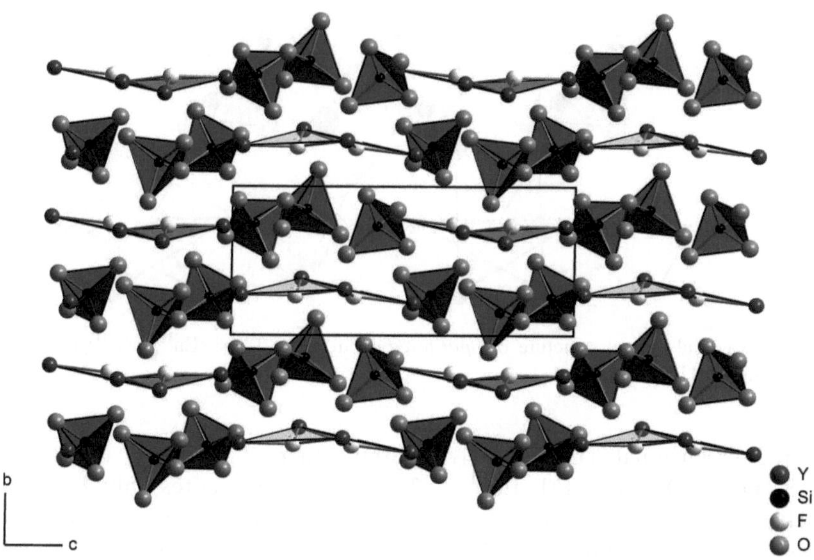

Fig. 17 Triclinic crystal structure of $Y_4F_2[Si_2O_7][SiO_4]$ as viewed along [100]

fluorapatite-type architecture of $Ca_5F[PO_4]_3$ (Hughes et al. 1989). In the crystal structure $Eu_5F[SiO_4]_3$ (a = 954.79(8) pm, c = 704.16(6) pm, c/a = 0.737, Z = 2; hexagonal, $P6_3/m$, no. 176; Wickleder et al. 2002) planar $[FEu_3]^{8+}$ triangles (Fig. 18, *left*) bearing trivalent europium recruit the second building block. Since

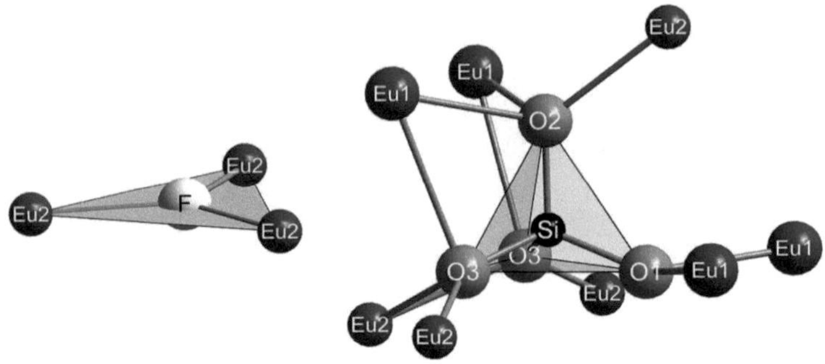

Fig. 18 Isolated triangular $[FEu_3]^{8+}$ cation (*left*) and discrete *ortho*-oxosilicate anion $[SiO_4]^{4-}$ with its complete Eu^{2+}/Eu^{3+} coordination (*right*) in *apatite*-type $Eu_5F[SiO_4]_3$ (= $(Eu^{II})_2(Eu^{III})_3F[SiO_4]_3$)

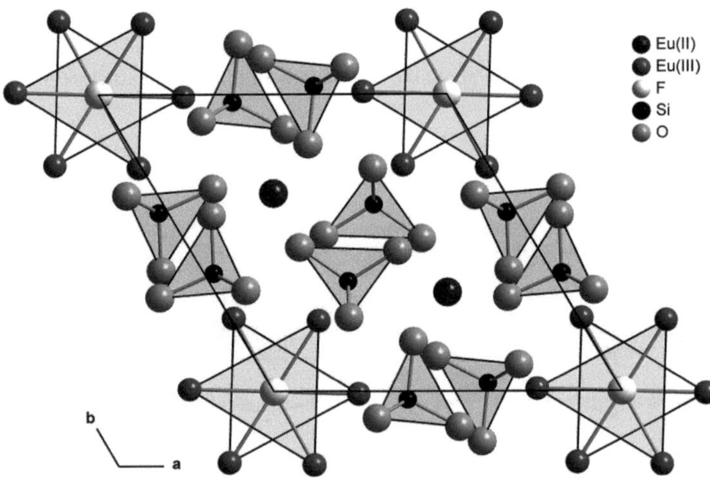

Fig. 19 Hexagonal crystal structure of *apatite*-type $Eu_5F[SiO_4]_3$ (= $(Eu^{II})_2(Eu^{III})_3F[SiO_4]_3$) as viewed along [001]

the three-dimensional $\{Eu_3F[SiO_4]_3\}^{4-}$ arrangement (d(Si–O) = 162–163 pm, ∠(O–Si–O) = 105–112°; $d(F–Eu^{III})$ = 231.5 pm, 3×) requires only 2 × 2 positive charges for electroneutrality, the remaining two europium cations in the *fluorapatite*-structure (Fig. 19) and – formula $Eu_5F[SiO_4]_3$ need to be only divalent according to $(Eu^{II})_2(Eu^{III})_3F[SiO_4]_3$. In contrast to their trivalent counterparts, which are sevenfold coordinated by one F^- and six O^{2-} anions as pentagonal bipyramids ($d(Eu^{III}–F)$ = 231.5 pm, $d(Eu^{III}–O)$ = 238–267 pm), they have contact to nine oxygen atoms exclusively ($d(Eu^{II}–O)$ = 242–285 pm) defining a tricapped trigonal prism (Fig. 20). The dark red colour of this fluoride oxosilicate indicates dynamic mixed-valance behaviour, however, driven by polaron activities. Thus intervalence charge transfer processes (Robin and Day 1967) severely hamper an

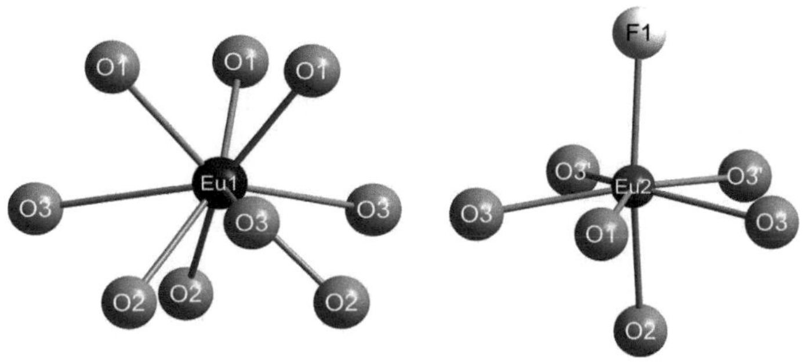

Fig. 20 Polyhedra of coordination about the Eu^{2+} (*left*) and Eu^{3+} cations (*right*) in *apatite*-type $Eu_5F[SiO_4]_3$ (= $(Eu^{II})_2(Eu^{III})_3F[SiO_4]_3$)

efficient luminescence of both Eu^{2+} and Eu^{3+}. Promising *apatite*-type candidates as hosts for luminescent materials containing innocent di – and trivalent cations simultaneously may result from combinations of alkaline-earth and rare-earth metals (Gong et al. 2005) in their regular oxidation states for the formula type $AE_2M_3F[SiO_4]_3$ (AE^{2+} = Ca – Ba, M^{3+} = Y, Gd, Lu and La; Schäfer and Schleid 2008), where it should be possible to dope these samples with both suitable di – or trivalent lanthanoid cations (Schäfer 2010). A short state-of-the-art survey on this topic is given in the following contribution (Schäfer and Schleid 2011) of this book on "Minerals as Advanced Materials II".

Acknowledgements We gratefully acknowledge the financial support of the *State of BadenWürttemberg* (Stuttgart, Germany) and the *Deutsche Forschungsgemeinschaft* (Bonn, Germany) for the promotion within the DFG priority program 1166 (Lanthanoid-Specific Functionalities in Molecule and Material).

References

Blasse G, Grabmaier BC (1994) Luminescent materials. Springer, Heidelberg

Dieke GH (1968) Spectra and energy levels of rare-earth ions in crystals. Intersience Publ, New York

Donnay G, Donnay JDH (1953) The crystallography of *bastnaesite*, *parisite*, *roentgenite*, and *synchisite*. Amer Miner 38:932–963

Grice JD, Maisonneuve V, Leblanc M (2007) Natural and synthetic fluoride carbonates. Chem Rev 107:114–132

Gong X, Lin Y, Chen Y, Huang Z, Huang Y, Luo Z (2005) Syntheses, structure, and characterization of crystal $La_6Ba_4(SiO_4)_6F_2$, a promising laser host. Chem Mater 17:1135–1138

Höppe HA (2009) Recent developments in the field of inorganic phosphors. Angew Chem Int Ed 48:3572–3582

Hughes JM, Cameron M, Croxley KD (1989) Structural variations in natural F, OH, and Cl apatites. Amer Miner 74:870–876

Ito J, Johnson H (1968) Synthesis and study of *yttrialite*. Amer Miner 53:1940–1952

Janka O (2010) Synthese und Charakterisierung von multinären Selten-Erd-Metall(III)-Fluoriden und Untersuchungen zu deren optischem Verhalten. Dissertation, Universität Stuttgart

Janka O, Schleid Th (2009) Facile synthesis of *bastnaesite*-type $LaF[CO_3]$ and its thermal decomposition to LaOF for bulk and Eu^{3+}-doped samples. Eur J Inorg Chem 2009:357–362

Jüstel T, Nicol H, Ronda C (1998) New developments in the field of luminescent materials for lighting and displays. Angew Chem Int Ed 37:3085–3103

Klemm W, Klein HA (1941) Lanthanoxyfluorid. Z Anorg Allg Chem 248:167–171

Kornev AN, Batalieva NG, Maksimov BA, Ilyukhin VV, Belov NV (1972) Crystalline structure of *thalenite* $Y_3(Si_3O_{10})(OH)$. Dokl Akad Nauk SSSR 202:1324–1327

Lewis GN (1926) Valence and the nature of the chemical bond. Chem Catal Comp, New York

Müller-Bunz H, Schleid Th (1999) $La_3F_3[Si_3O_9]$: Das erste Fluoridsilicat aus dem ternären System $LaF_3/La_2O_3/SiO_2$. Z Anorg Allg Chem 625:1377–1383

Müller-Bunz H, Schleid Th (2000) Darstellung und Aufbau der Lanthanoidfluorid-*catena*-Trisilicate $M_3F[Si_3O_{10}]$ (M = Dy, Ho, Er) im *Fluorthalenit*-Typ ($Y_3F[Si_3O_{10}]$). Z Anorg Allg Chem 626:845–852

Müller-Bunz H, Schleid Th (2001) $Er_4F_2[Si_2O_7][SiO_4]$: Das erste Selten-Erd-Fluoridsilicat mit zwei verschiedenen Silicat-Anionen. Z Anorg Allg Chem 627:218–223

Nauman H, Sverdrup T, Saebo P (1957) X-ray powder patterns for mineral identification, III. Silicates. Norske Vidensk Acad Oslo 6:16–54

Oskam KD, Kaspers KA, Meijerink A, Müller-Bunz H, Schleid Th (2002) Luminescence of $La_3F_3[Si_3O_9]$:Ce^{3+}. J Lumin 99:101–105

Pearson RG (1963) Hard and soft acids and bases. J Amer Chem Soc 85:3533–3539

Robin MB, Day P (1967) Mixed valence chemistry – a survey and classification. Adv Inorg Chem Radiochem 10:247–422

Ronda C (2008) Luminescence. Viley-VCH, Weinheim

Rumanova IM, Nikoleva TV (1959) Crystal structure of *orthite*. Sov Phys Crystallogr 4:789–795

Schäfer MC (2010) Synthese, Charakterisierung und Eigenschaften multinärer Oxosilicate der Selten-Erd-Metalle. Dissertation, Universität Stuttgart

Schäfer MC, Schleid Th (2008) $Ba_2Ce_3F[SiO_4]_3$: Ein ausgeordnetes Fluorid-Oxosilicat im *Apatit*-Typ. Z Anorg Allg Chem 634:2074–2074

Schäfer MC, Schleid TH (2011) Rare-earth metal(III) fluoride oxosilicates derivatized with alkali or alkaline-earth elements. In: Krivovichev SV (ed) Minerals as advanced minerals II. Springer, Heidelberg (this book)

Schäfer MC, Petter M, Zhang S, Wickleder C, Schleid Th (2011) Luminescence properties of $Y_3F[Si_3O_{10}]$:M^{3+} (M = Eu, Tb, Er) with *thalenite*-type host lattice and crystal structure of $Tm_3F[Si_3O_{10}]$. Solid State Sci 13: (in press)

Schleid Th, Müller-Bunz H (1998) Einkristalle von $Y_3F[Si_3O_{10}]$ im *Thalenit*-Typ. Z Anorg Allg Chem 624:1082–1084

Semenov EI (1963) Mineralogy of rare earths. Izv. Akad. Nauk SSSR, Moscow

Wickleder C, Hartenbach I, Lauxmann P, Schleid Th (2002) $Eu_5F[SiO_4]_3$ und $Yb_5S[SiO_4]_3$: Gemischtvalente Lanthanoid-Silicate mit *Apatit*-Struktur. Z Anorg Allg Chem 628:1602–1606

Zachariasen WH (1951) Crystal chemical studies of the 5*f*-series of elements, XIV. Oxyfluorides, *X*OF. Acta Crystallogr 4:231–236

Zalkin A, Templeton DH (1985) Refinement of the trigonal crystal structure of lanthanum trifluoride with neutron diffraction data. Acta Crystallogr B41:91–93

$REECa_4O(BO_3)_3$ (REECOB): New Material for High-Temperature piezoelectric applications

R. Möckel, M. Hengst, J. Götze, and G. Heide

1 Introduction

Especially in terms of the modern environmental consciousness with demand for more efficient, cleaner and more ecological machinery, processes has to be improved. This is mainly essential for combustion engines, coal fired electrical plants as well as gas heating installations etc. Processes have to be measured and controlled in situ in order to increase efficiency by manipulating parameters. Doing so, sensors which can resist and work under high temperature/high pressure conditions are strongly required. Unfortunately, most piezoelectric sensor materials known today (like quartz, $Bi_4Ti_3O_{12}$-and PZT-ceramics) can not proceed at elevated temperatures above 600°C. Others that can, are either expensive and energy-consuming in production ($GaPO_4$) or only available in sufficient size as naturally occurring minerals (tourmaline) with fluctuating properties. In the last few years, the new material REECOB ($REECa_4O(BO_3)_3$ with REE = rare earth elements: Gd, Y, La, Sm, Nd) emerged to be a promising candidate for high temperature applications, displaying constant piezoelectric properties up to 1,200°C. This material has made a steep career in optical applications (laser host material, nonlinear optics) since the mid-1990s but good properties for sensing applications are only known for a few years (Shimizu et al. 2004; Markiewicz et al. 2006). Investigations on ultra high temperature properties are relatively new (Zhang et al. 2008a, b, c). Unfortunately, thermomechanical data of the material are very rare and partially contradictional.

In addition to temperature stability, the requirements to piezoelectrical sensor materials are: preferably low temperature dependence of dielectric and piezoelectric properties, an electrical resistivity above 10^6 Ωcm and mechanical as well as

R. Möckel (✉) • M. Hengst • J. Götze • G. Heide
Institute of Mineralogy, TU Bergakademie Freiberg, Brennhausgasse 14, D-09596 Freiberg, Germany
e-mail: robert.moeckel@mineral.tu-freiberg.de

S.V. Krivovichev (ed.), *Minerals as Advanced Materials II*,
DOI 10.1007/978-3-642-20018-2_34,

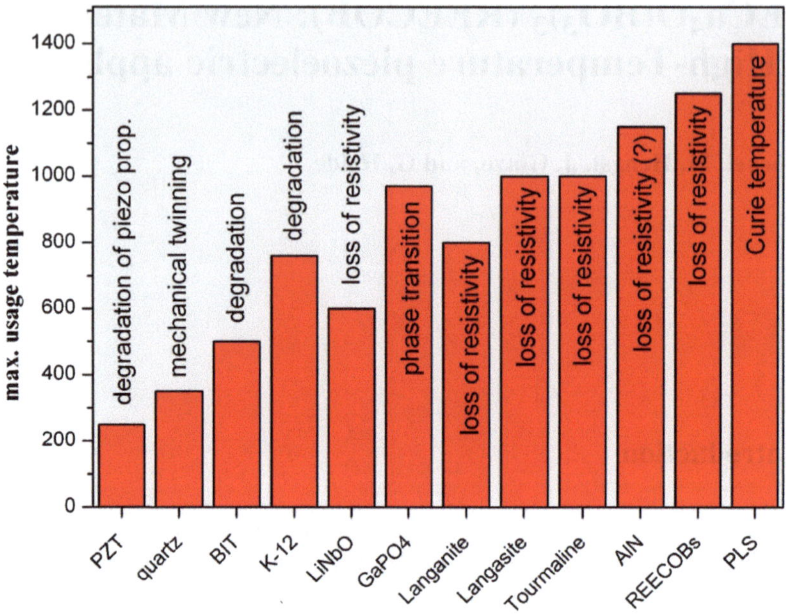

Fig. 1 Comparison of different piezoelectric materials, their maximum usage temperature and cause of restriction. Abbreviations: *PZT* lead-zirconate-titanate-ceramics, *BIT* Bi4Ti3O12-based ceramics, *K-12* modified *BIT*-ceramic, Langanite/Langasite - $La_3Ga_5SiO_{14}$/$La_3Ga_{5.5}Nb_{0.5}O_{14}$, *PLS* perovskite-like layer structures (Data from Zhang et al. 2008b; Damkanovic 1998; Turner et al. 1994)

chemical stability. Limiting factors are also: low Curietemperature, degradation of electrical resistivity or piezoelectric properties at elevated temperatures, and of course phase transitions. Figure 1 gives some piezoelectric sensor materials, their maximum usage temperature and cause of restriction.

Although preliminary tests on the growth of $LaCa_4O(BO_3)_3$ and also $YCa_4O(BO_3)_3$ (LaCOB and YCOB, respectively) led to satisfying results, the present study will focus on $GdCa_4O(BO_3)_3$.

2 Materials and Methods

2.1 *Crystal Growth*

Polycrystalline REECOB was synthesized by simple solid-state reaction. The starting materials were stoichiometrically mixed following the formular

$$REE_2O_3 + 8CaCO_3 + 3B_2O_3 \rightarrow 2REECa_4O(BO_3)_3 \quad (1)$$

A 2 wt.% excess of boron oxide was added to compensate the loss of adsorbed water. Sintering was performed in a two-step method: 1,000°C/10 h, afterwards grinding and heating again for 10 h at 1,200°C. The reaction was proven to be completed by powder X-ray diffraction (XRD). Crystal growth using the Czochralski method was performed using a Cyberstar Oxypuller 05–03 pulling device with growth (pulling) rates of 1–3 mm/h and rotation rates of 6 rpm under nitrogen atmosphere. Growth directions were [010] and [001], whereby the [010]-direction seemed to lead to better results in terms of less cracks. Crystals dimensions reached lengths of 50 mm and a diameter of 15 mm. More details of crystal growth are described elsewhere (Möckel et al. 2009).

2.2 *Thermal Expansion*

Samples for thermal expansion measurements were cut from GdCOB single crystals. Measurements in [010]- and [001]-direction were made from the upper crystal parts (neck) with a diameter of 5 mm and lengths between 10 and 27 mm of the related growth directions. Additionally, samples of the [100]- and [101]-direction were cut from [010]-single crystal, with lengths of 12–14 mm. Lengths of the sample material were in a 20% range of the calibration sticks (Al_2O_3-ceramics) as recommended by the manufacturer. Measurements were performed using a Netzsch DIL 409 PC equipment. Heating rate was kept constant at 5K/min from 25°C to 1,200°C. Contact pressure of the pushrod was 25 cN. Measurements were repeated at least three times to minimize errors.

Additionally, high-temperature X-ray diffraction on grounded powder from GdCOB-single crystal was performed in a temperature range of 30–1,300°C, 2-theta range of 10–80°, with scanspeed of 0.3 s per step and a stepwidth of 0.0143°(2-theta). The device used was a Bruker AXS D8 Discover, working with CoKα radiation (1.7889 Å) and Göbel mirror.

3 Discussion

Crystals were grown successfully by the Czochralski method mostly without cracks and other macroscopical defects. However, there have been failed experiments, when the crystal cracked. The cooling time was therefore increased to 20 h immediately (melting point of 1,490°C leads to a cooling rate of 1.2K/min). Crystals in [001]-direction have a circular cross section, whereas the [010]-direction led to a parallelogram formed by the facets ($20\bar{1}$) and (101).

The thermal expansion coefficient is a second rank tensor, which can be calculated for any direction by the formula:

$$\alpha(\vec{\mathbf{n}}) = \vec{\mathbf{n}} \cdot {}^2\vec{\alpha} \cdot \vec{\mathbf{n}} \tag{2}$$

Where

$$^2\vec{\alpha} = \begin{pmatrix} \alpha_{11} & 0 & \alpha_{13} \\ 0 & \alpha_{22} & 0 \\ \alpha_{31} & 0 & \alpha_{33} \end{pmatrix} \tag{3}$$

For symmetrical reasons α_{31} equals α_{13} (see Paufler 1986 for example), which leads to

$$^2\vec{\alpha} = \begin{pmatrix} \alpha_{11} & 0 & \alpha_{13} \\ 0 & \alpha_{22} & 0 \\ \alpha_{31} & 0 & \alpha_{33} \end{pmatrix} \tag{4}$$

From Eq. 2, it follows that the tensor elements α_{11}, α_{22}, and α_{33} equate the thermal expansion coefficients of the crystallographic a, b, and c directions (α_a, α_b, α_c), respectively and α_{13} correlates with the diagonal between a- and c-axis (ac- or [101]-direction) and can be achieved from:

$$\vec{\alpha}_{13} = \alpha_{ac} - (\alpha_a + \alpha_c)/2 \tag{5}$$

Therefore, four independent directions have to be measured in order to describe the tensor completely. In Fig. 2 the results of these measurements are shown.

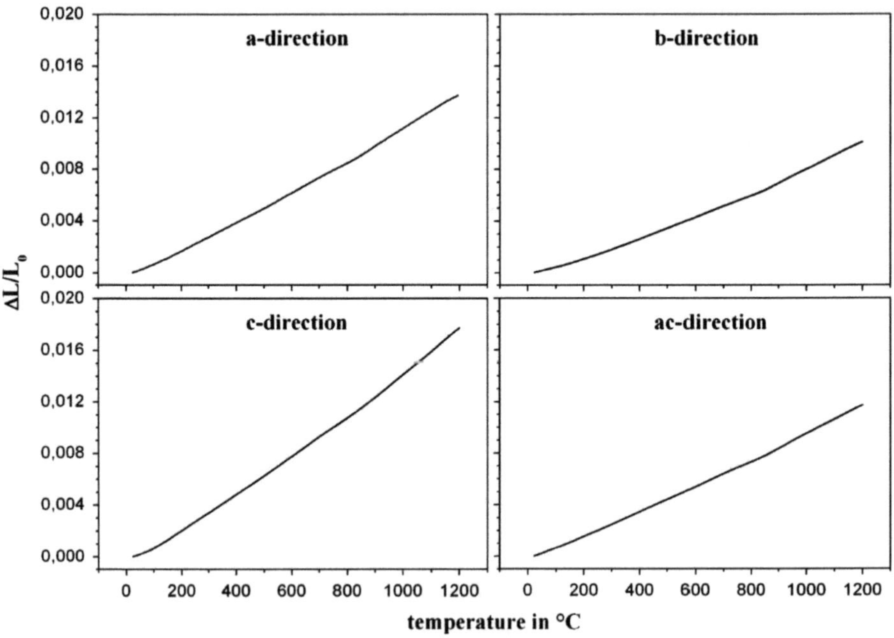

Fig. 2 Relationship between thermal expansion and temperature for the measured crystaographic directions (for values see Table 1)

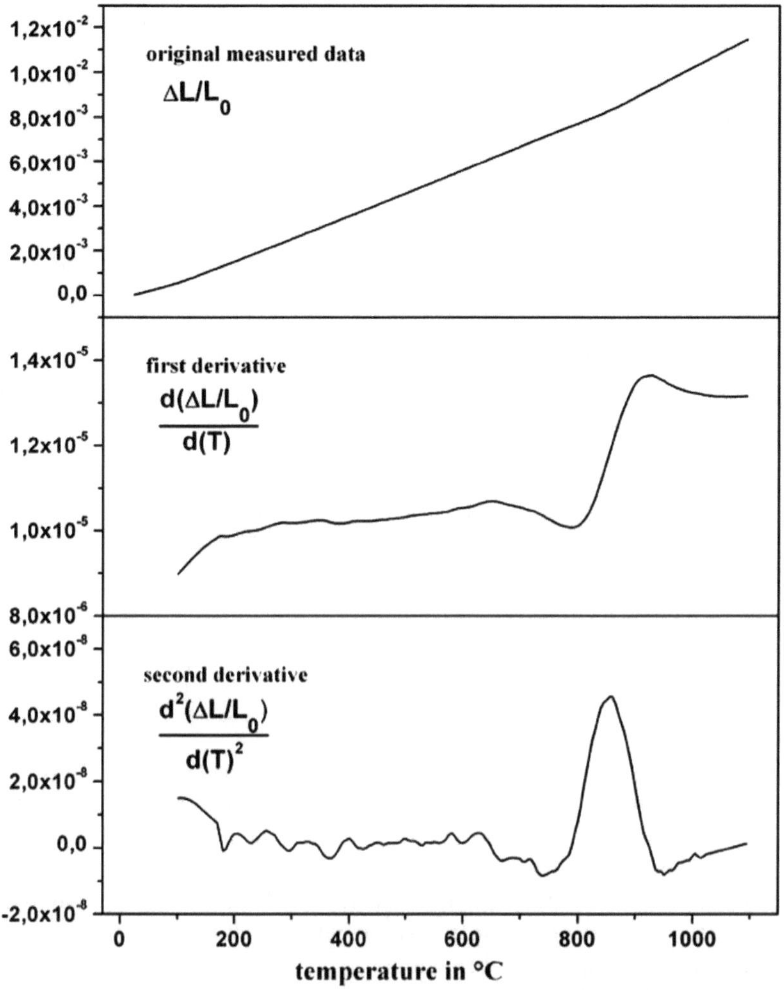

Fig. 3 Results of thermal expansion measurement with originally measured data, first (slightly smoothed) and second derivative, displaying the maximum change in the slope at 855°C (α_{ac}-measurement)

The thermal expansion increases almost linearly in all directions until around 830–860°C, where a slight change in the slope appears. Afterwards, in the measured range, thermal expansion is almost linear, again. Thermal coefficients were calculated from the slope of the curve, which is represented by the first derivative of the measured curve, where two different values can clearly be seen. The second derivative, which in turn displays the change of slope, was used to determine the point of change, represented by a maximum peak (Fig. 3). In the case of the crystallographic a-, b-, and ac-directions the peak was relatively sharp, with values around 850°C (852 ± 13°C, n = 21). Exclusively, the thermal expansion of the

Table 1 Thermal expansion coefficients in different crystallographic directions

	<850°C, in $10^{-6}K^{-1}$	>850°C, in $10^{-6}K^{-1}$
α_a	11.04	14.10
α_b	8.59	11.70
α_c	14.54	17.87
α_{ac}	10.36	13.30

c-direction showed very little variation in the slope, with a very broad peak in the second derivative. Nonetheless, the thermal expansion coefficient of the c-direction increases about 23% in the range of 850–1,200°C (in contrast to an increase of around 28% of a and ac-direction, and 36% in b-direction, see also Table 1).

Following equation 5, $\alpha_{13} = -2.43$ (-2.69 for temperatures above 850°C), resulting in the second-rank tensors:

$$^{2}\vec{\alpha}_{<850^\circ C} = \begin{pmatrix} 11.04 & 0 & -2.43 \\ 0 & 8.59 & 0 \\ -2.43 & 0 & 14.54 \end{pmatrix} \tag{6}$$

and

$$^{2}\vec{\alpha}_{<850^\circ C} = \begin{pmatrix} 14.10 & 0 & -2.69 \\ 0 & 11.70 & 0 \\ -2.69 & 0 & 17.87 \end{pmatrix} \tag{7}$$

High-temperature X-ray diffraction (HT-XRD) provided similar results in terms of the point of change, which lies between 800°C and 900°C. Unfortunately, Rietveld refinement of the data led to unsatisfying achievements for thermal expansion coefficients from an atomic scale, due to unsufficient peak profiles. Therefore, calculated lattice parameters revealed large errors. Nonetheless, measurements detected no phase transition which would have been a possible explanation for the temperature behaviour.

Surprisingly, this effect has been rarely discussed yet. Solely Mougel et al. (1998) reported the anomaly using HT-XRD and single crystal determination with values similar to those presented here (at least for the temperature range below 850°C). Most publicated measurements, also on other members of the REECOB family, do either not exceed the point of turn (e.g. Shujun et al. 2000; Ge et al. 2007; Wei et al. 2008) or ignore it (Zhou et al. 2003).

4 Conclusion

Crystal growth has been performed successfully for $GdCa_4O(BO_3)_3$ single crystals, being a promising material for high temperature sensing. Thermal expansion measurements revealed a change in thermal expansion coefficients at around 850°C.

Below and above this point thermal expansion is almost linear. Anisotropy and the change in thermal expansion might be a reason why cooling down of the as-grown crystal is somewhat difficult. We suppose a possible restriction for high temperature application due to this fact as well, but further work will be necessary, especially on other REECOB-materials.

Acknowledgements This work would not have been possible without the kind help of a bunch of people, who will be thanked for: Dr. A. Erb for crystal orientation, Dr. M. Kurnikova for HT-XRD measurements as well as Dr. V. Geist for mathematical advice.

References

Damkanovic D (1998) Materials for high temperature pieroelectric transducers. Curr Opin Solid State Mater Sci 3(5):469–473

Ge WW, Zhang HJ, Wang JY, Jiang MH, Sun SQ, Ran DG, Xia HR, Boughton RI (2007) Thermal properties of monoclinic crystal Er^{3+}:Yb^{3+}:$Ca_4YO(BO_3)_3$. J Appl Crystallogr 40:125–132

Markiewicz E, Pawlaczyk C, Klos A, Hofman W, Pajaczkowska A (2006) Temperature behavior of piezoelectric gadolinium calcium oxoborate $GdCa_4O(BO_3)_3$ crystal. Phys Stat Sol (A) Appl Mater 203:372–378

Möckel R, Hengst M, Reuther C, Götze J (2009) Synthesis of $Ca_4GdO(BO_3)_3$ single crystals using Czochralski method. J Siber Fed Univ Eng Technol 2:400–408

Mougel F, Kahn-Harari A, Aka G, Pelenc D (1998) Structural and thermal stability of Czochralski grown GdCOB oxoborate single crystals. J Mater Chem 8:1619–1623

Paufler P (1986) Physikalische kristallographie. Akademie-Verlag, Berlin

Shimizu H, Kodama K, Takeda H, Nishida T, Shikida T, Okamura S, Shiosaki T (2004) Evaluation of material constants and temperature properties in lanthanum calcium oxoborate $LaCa_4O(BO_3)_3$ single crystals. Jap J Appl Phys 43(9B):6716–6720

Shujun Z, Hua Y, Zhenxiang C, Xuesong L, Huanchu C (2000) Crystal growth, thermal and optical properties of $SmCa_4O(BO_3)_3$ crystal. J Cryst Growth 208:482–486

Turner R, Furier P, Newnham R, Shrout T (1994) Materials for high temperature acoustic and vibration sensors: a review. Appl Acoust 41:299–324

Wei B, Lin Z, Zhang L, Wang G (2008) Growth, thermal and polarized spectral properties of Nd^{3+}-doped $Gd_{1-x}La_xCa_4O(BO_3)_3$ ($x = 0.16$ and 0.33) crystals. Cryst Growth Des 8:186–191

Zhang S, Fei Y, Chai BHT, Frantz E, Snyder DW, Jiang X, Shrout TR (2008a) Characterization of piezoelectric single crystal $YCa_4O(BO_3)_3$ for high temperature applications. Appl Phys Lett 92(20):202905 (3 pages)

Zhang S, Frantz E, Xia R, Everson W, Randi J, Snyder DW, Shrout TR (2008b) Gadolinium calcium oxyborate piezoelectric single crystals for ultrahigh temperature (>1000°C) applications. J Appl Phys 104(8):084103 (7 pages)

Zhang S, Shrout T, Fei Y, Chai B, Frantz E, Snyder D (2008c) High temperature piezoelectric single crystal $ReGd_4O(BO_3)_3$ for sensors. 2008 IEEE international frequency control symposium, FCS, pp 316–319

Zhou J, Zhong Z, Xu J, Luo J, Hua W, Fan S (2003) Bridgman growth and characterization of nonlinear optical single crystals $Ca_4GdO(BO_3)_3$. Mater Sci Eng B97:283–287

"Shock Wave" Synthesis of Oxygen-Bearing Spinel-Type Silicon Nitride γ-$Si_3(O,N)_4$ in the Pressure Range from 30 to 72 GPa with High Purity

T. Schlothauer, M.R. Schwarz, M. Ovidiu, E. Brendler, R. Moeckel E. Kroke, and G. Heide

1 Introduction

Silicon nitride, Si_3N_4 is a hard, refractory ceramic substance first synthesized more than 100 years ago. In the 1990s, its so-called α-modification was discovered to occur also as an exotic extraterrestrial mineral in certain chondrites and termed Nierite (Lee et al. 1995). The ambient-pressure modifications α-Si_3N_4 and β-Si_3N_4 have hexagonal (trigonal) lattice symmetry. β-Si_3N_4 crystallizes in the space group $P6_3/m$ (Goodmann and O'Keefe 1980), α-Si_3N_4 in the space group P31c (Zhou et al. 1995). Both types consist of SiN_4 tetrahedra and N_3Si groups in a close-to-planar arrangement. In 1999 a high pressure modification of Si_3N_4 with spinel-type structure (space group Fd-3 m) and sixfold coodination of Si to nitrogen was discovered (ZERR et al. 1999). The γ-modification can be quenched to ambient conditions and shows a remarkable metastability and a high bulk modulus (Kuwabara et al. 2008; Jiang et al. 2001; Sekine et al. 2001; Sekine 2002). The excellent thermal and mechanical properties qualify γ-Si_3N_4 as an interesting candidate for high performance ceramics.

Typical static HP/HT conditions for the phase transformation from β-Si_3N_4 to γ-Si_3N_4 are ~15 GPa and 1,800°C (Schwarz 2003). It is possible to synthesize γ-Si_3N_4 under dynamic loading conditions (He et al. 2000; Schwarz 2003; Deribas et al. 2004). For this method, the peak shock pressures are usually higher than 20 GPa with temperatures >2,000 K. Almost all of these studies used the crystalline low pressure phases α- and/or β-Si_3N_4 as starting materials.

A thorough literature survey of the works published on the dynamic synthesis of γ-Si_3N_4 shows that it was with this method apparently impossible to obtain spinel-type material that was completely free of other phases (α – and β-Si_3N_4 or

T. Schlothauer (✉) • M.R. Schwarz • M. Ovidiu • E. Brendler • R. Moeckel
• E. Kroke • G. Heide
Institute of Mineralogy, TU Bergakademie Freiberg, Brennhausgasse 14,
D-09596 Freiberg, Germany
e-mail: Thomas.schlothauer@mineral.tu-freiberg.de

S.V. Krivovichev (ed.), *Minerals as Advanced Materials II*,
DOI 10.1007/978-3-642-20018-2_35,

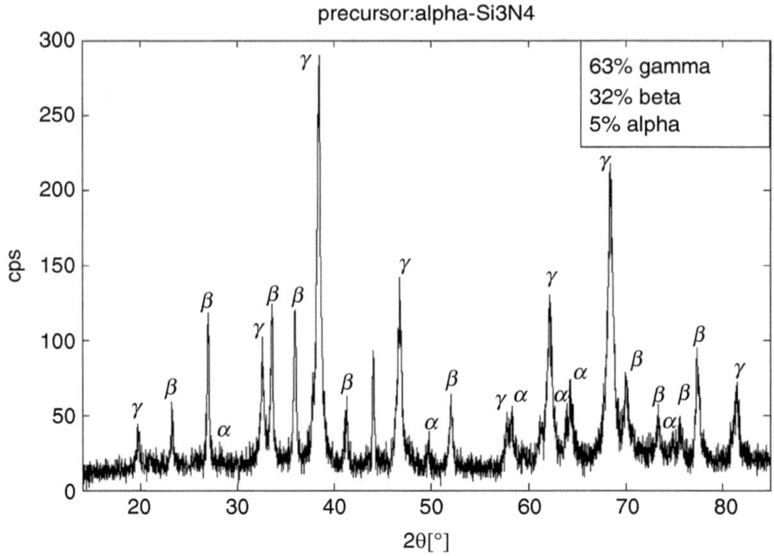

Fig. 1 Mixture of three Si_3N_4-phases after shock, precursor: α-Si_3N_4, p = 72 GPa, T = 5,700 K, T/p = 79.47 K/GPa

amorphous parts). For all types of further analysis low pressure phases with similar chemical compositions like the spinel-type nitrides are absolutely disruptive (Fig. 1). Although the pressure was relatively high (>70 GPa) the resulted high-pressure phase is polluted with α- and β-Si_3N_4.

Further problems may be a predicted ternary Si_2N_2O-phase with defect spinel structure could exist in the pressure range between 11 and 16 GPa, and a decomposition into a phase assemblage of stishovite and γ-Si_3N_4 at static pressures up to 40 GPa, and "a possible Si_2N_2O phase with corundum-type structure beyond 80 GPa". All this phases are able to falsify the structural and chemical measurements. This is especially for this case very important if it is necessary to analyze for example the oxygen values in our shocked samples.

Actually it is impossible to obtain pure spinel phases of those nitrides. In every case the appearance of more than one phase is common (Sekine et al. 2002). But how to analyze a complex mixture of two, three or four nanocrystalline phases (including amorphous parts)? With the XRD the identification of the different phases is possible, the IR (bondings etc.) is very difficult and the NMR (coordination of the Si) is completely impossible. In the last case occurs an "overlay" of the $Si^{[4]}$ from the spinel structure with the $Si^{[4]}$ from the low pressure phases.

For this reason, systematic synthesis of spinel-structured high-pressure silicon-oxonitride-phases with high purity is essential. It is necessary to evaluate and develop the best synthesis conditions, the best precursors and also the best system of sample recovery.

Different amorphous Si-N-(O)-(H) precursors and inert-gas techniques were employed for this study. The shocked products from different amorphous Si-N-(O)-(H) precursors were investigated by X-ray powder diffraction (XRD), high-resolution scanning electron microscopy (HR-SEM) and energy-dispersive X-ray spectroscopy (EDX), as well as infrared (IR) and nuclear magnetic resonance (NMR) spectroscopy.

2 Shock Wave Method and Loading Conditions

For our experiments the flyer plate method with an explosive plane-wave-generator was used (Fig. 2). The flyer plate velocity was calculated from the mass ratio of flyer and charge and specific properties of the explosive after DeCarli and Meyers (1981). Practical measurements with the contact pin assemblage showed small differences between predicted and measured velocities of the flyer plate. The differences are ca. 0.1 km/s. The flyer plate material was steel St37, the sample container was made from ARMCO-iron. The similarity in the behaviour under shock wave loading between these two materials permits the approximation for the determination of the particle velocity:

$$u_p = V_{FP}/2 \tag{1}$$

The calculation of shock velocity, pressure, internal energy and all other shock parameters within the sample container follows Batsanov (1994).

The precursor powder was mixed with spherical copper powder (size 0.02–0.063 mm, ECKA Granulate Velden GmbH, Germany). The precursor/Cu-weight

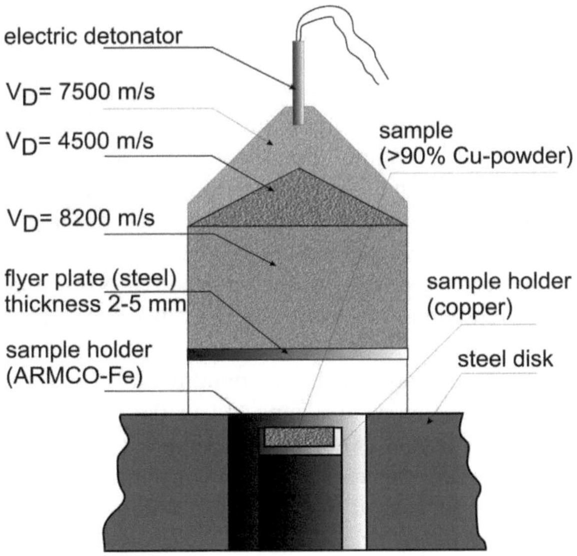

Fig. 2 Synthesis apperature with active plane wave lens, used commonly at the TU Bergakademie Freiberg

ratio was in every case smaller than 0.1. It is hence possible to approximate the pressures in the sample as those for pure (porous) Cu-Powder, as established by Milyavskii (2000). The porosity factor

$$k = \left(V_{\text{sample-chamber}}/V_{\text{copper}}\right) \tag{2}$$

in the experiments was fixed at 2, corresponding to 50 vol.% porosity. The lower boundary of this ratio is given by the mathematical expression

$$k \leq 1 + \frac{2}{\Gamma} \tag{3}$$

after Milyavskii (2000) with Γ being the Grüneisen-coefficient of the metal powder (Cu = 1.96). In the case of Cu-powder the porosity limit is thus $k = 2.02$. The calculation for the shock temperatures follows Sharp and DeCarli (2006) using the pressure dependence of the Grüneisen-parameter after Davison (2008).

Because in every case under shock loading occurs a rapid increase of the temperature, a mathematical help will ne necessary to describe the relative temperature in relation to the pressure, the so called "relative temperature". It was calculated by

$$Trel = \frac{Tschock}{p} \tag{4}$$

and its value was given in K/GPa.

3 Starting Materials

In order to adjust different levels of oxygen content, amorphous Si-N-H precursors stored under different conditions were used. Three experiments were conducted with silicon nitride imide that had been stored under inert conditions (hereafter abbreviated with Si_2N_2NH). Two further experiments were conducted with partially hydrolysed silicon nitride imide that had been stored in ambient air (~12 Mass-% O, hereafter abbreviated with Si_2N_2NH:O).

For one experiment silicon diimide $Si(NH)_2$, synthesized from $SiCl_4$ and anhydrous ammonia was used. The oxygen values were further varied with the type of Cu-powder (reduced Cu- or Cu-granules with an oxide scale) and the sample preparation being either performed in air or in a glove box under N_2-atmosphere.

The samples were compacted in the sample holder with a static press to the necessary density. After the experiment, the sample ampoules were recovered, the copper matrix was removed with concentrated nitric acid, and the residual ceramic powder was washed with destilled water and separated by centrifugation several times and finally dried.

4 Sample Analysis

The precursors and the shocked samples were characterized with XRD, FT-IR- and NMR-spectroscopy.

For the quantification of the oxygen values, the hot gas extraction method, with LECO TC-436 at 2,500°C in helium atmosphere with carbon crucibles was applied. XRD was performed on a RD7 (Seifert) with CuKα1-radiation in Bragg-Brentano-geometry. The powders were pressed into a sample holder with a flat mould (diameter 10 mm, thickness: 0.8 mm). Measurements were conducted with automatic divergency aperture (irradiated area 6 mm) and a vertical aperture of 6 mm. The diffraction patterns were collected in the range from 13° to 85° with 2θ-steps of 0.02° and a measurement time of 2 s per step.

^{29}Si MAS and CP/MAS NMR, ^{1}H MAS NMR were measured on a Bruker Avance 400 WB using a 4 mm CP/MAS probehead (shocked samples) or 7 mm CP/MAS-Probehead (precursor) at 79.51 MHz (^{29}Si) and 400.23 MHz(1 H). The rotation frequency was 4–6 kHz (precursors) and 10 kHz (shocked samples) for the Si- measurements and 15 kHz for 1 H-. The ^{1}H MAS NMR spectra were scaled for equal intensity of the broad peak at 4.7 ppm. Chemical shifts are reported relative to TMS (= 0 ppm).

Quantitative and qualitative IR-spectroscopy was made with the Thermo Electron Corporation NICOLET 380 FT-IR. For the IR- phase analysis, 300 mg dried KBr were mixed with 0.5 mg shocked sample. For the IR-pellets itself 80 mg of this mixture were used. The calculations were made with the Non-Negative-Linear-Square-Method (NNLS) and a MATLAB-software package after Xu et al. (2004).

5 Precursor Characterization

The IR-spectrum of Si_2N_2NH:O shows a sharp peak at 492 cm^{-1} and a pronounced peak at 939 cm^{-1}(Fig. 8). These strong absorptions are assigned to the Si–N stretching and Si–N deformation mode, respectively (description here after ZHU et al. 2007). The two weak peaks at 1,401 cm^{-1} and 1,630 cm^{-1} are possibly the rocking and bending bands of N–H. The broad peak around 3,368 cm^{-1} is an overlap of N–H stretching vibrations and O–H vibrations.

It is noteworthy that some of the shock experiments yielded products that were also in an amorphous state. These show similar infrared adsorption bands like Si_2N_2NH:O. The intensity of the hydrogen related peaks at 1,400 and 1,600 cm^{-1} has become smaller and additionally a strong peak at 1,078 cm^{-1}, possibly the stretching mode of Si-O.

The NMR-spectrum of the material is also given in Fig. 6. It shows the typical broad peak of the [Si-(N,O)]$_4$ in tetrahedral coordination in an amorphous structure.

Sample SN46 was prepared from a under oxygen poor conditions synthesized precursor and was completely prepared in the glove box. The copper for this experiment was also stored under N_2-atmosphere and it was purchased in a reduced state from the supplier.

The oxygen values for the following experiments were changed by using the two kinds of silicon nitride imide (Si_2N_2NH and Si_2N_2NH:O) and non-reduced copper powder. The sample compaction was either performed under air or in the glove box (c.f. Table 1).

It should be noted that the shock loading-unloading history of SN48 is different from the other samples. Here the reflected shock wave from the bottom of the Cu-sample holder penetrated the head end cap of the sample container. The result was a small hole with a diameter of 1 mm.

The massive decompression of the hot Cu-powder combined with this process lead to an evaporation of the metal and a more or less complete loss of the sample. Here the criterion of limited porosity is not fullfilled and a temperature calculation is not possible. The value of 12,000°C is a theoretical number, resulting from different experimental studies of this case (Belyakov et al. 1977).

Another limiting factor is the ratio sample/Cu-powder. Because we calculate the pressures with the Hugoniot-parameters of Cu (the equivalent parameters for the samples are completely unknown), this ratio should never been greater than 10 M-%. In addition to porosity, also the other loading parameters (flyer plate mass, charge mass) were kept approximately constant at $p \approx 35$ GPa (peak shock pressure at the passage of the incident shock wave) and $T \approx 6{,}300$°C (temperature/pressure-coefficient $T/p = 175$ K/GPa).

6 Influence of p-T-Conditions

The FT-IR-analysis turned out to be a quick and comfortable way for the identification of γ-silicon nitride-based materials, and a valuable method to identify the best conditions for the synthesis.

Figure 3 shows the spectra of products obtained by shocking the same precursor (Si_2N_2NH:O) prepared under the same conditions (under air), but with different porosities of the copper matrix. For the shock wave experiments different charge masses were used. Only the sample loaded to a high T/p-coefficient shows the characteristical octahedral lattice vibrations between 600 and 850 cm^{-1} which are also typical for spinel-type oxides.

7 Sample Characterisation After Shock Loading

7.1 FE-SEM

The nano-powders show grain sizes between 10 and 50 nm (Fig. 4). The single grains of the γ-Si-O-N appear to be connected, possibly an effect of the thin carbon films that had been deposited on the powders to avoid electrostatic charging effects.

Table 1 Loading conditions and sample preparation of six shock wave experiments

Sample	Precursor	Prec./ Cu [−%]	k	p [GPa]	T [K]	T/p [K/GPa]	Cu-powder	Atmosphere	Sample mass [g]	Sample loadings
SN46	$Si(NH)_2$	3.76	2.00	36	6,300	176	Reduced	N_2	0.138	Synth. prec.+ reduced Cu + N_2-Atm.
SN47	Si_2N_2NH	8.55	2.00	36	6,300	176	Reduced	Air	0.693	Commercial prec. (stored in glovebox) + red. Cu + air
SN48	Si_2N_2NH	8.28	2.00	36	> 12,000	Esc.	Reduced	N_2	0.027	Commercial prec. (stored in glovebox) + red. Cu + N_2 Atm.
SN49	Si_2N_2NH:O	8.77	2.00	36	6,300	176	Reduced	Air	0.844	Commercial prec. (stored in Air with 11.4% O) + reduced Cu + air
SN50	Si_2N_2NH:O	9.30	2.00	36	6,200	175	Not reduced	Air	0.733	Commercial prec. (stored in air with 11.4% O) + Cu + air
SN51	Si_2N_2NH	8.45	2.00	36	6,200	175	Not reduced	N_2	0.862	Commercial prec. (stored in Glovebox with 11.4% O) + Cu + N_2

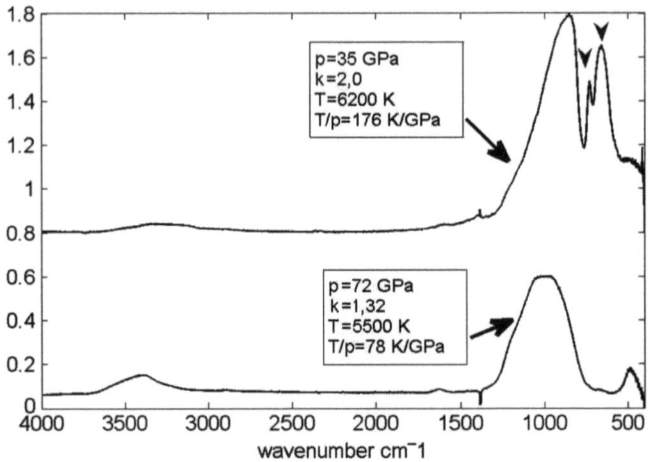

Fig. 3 Comparison between two different shock conditions with the same precursor: medium pressure, high temperature (*top*) and high pressure, low temperature (*bottom*) with the octahedral lattice vibrations (*marker*)

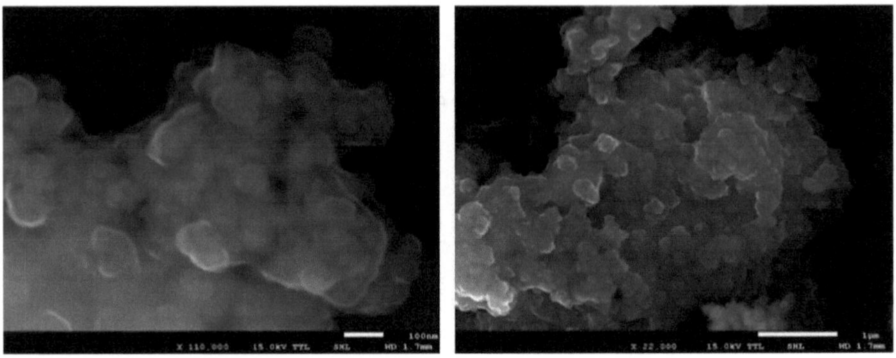

Fig. 4 FE-SEM-microphotographs of the sample SM-51

During evacuation of the FE-SEM sample chamber, a strong degassing of the samples was evident, possibly an effect the high specific surface area of the fine powdered materials.

7.2 *X-ray-Diffraction (XRD)*

The X-ray diffraction patterns (Fig. 5) demonstrate the successful synthesis of a single phase with spinel structure and a low amorphous background. Other phases such as α-Si_3N_4 and β-Si_3N_4 are below the detection limit. In some patterns reflections from the aluminium sample holder are visible (SN46 and SN48). With exception of a continuously small peak shift, all spectra are very similar.

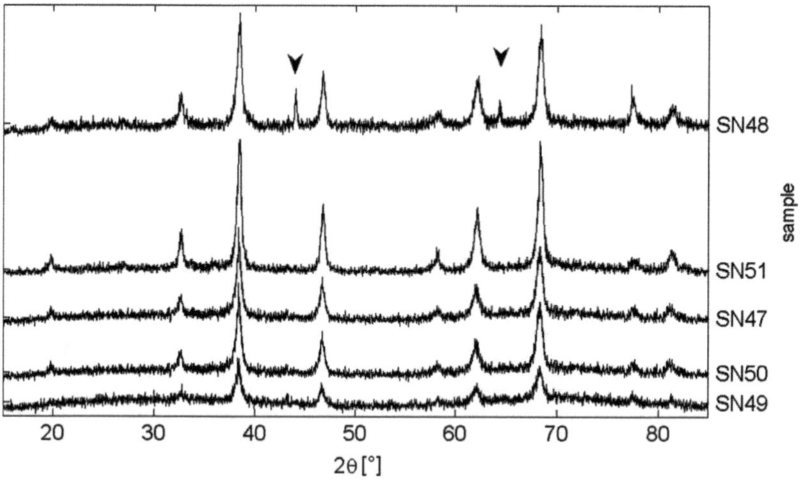

Fig. 5 X-ray-diffraction pattern of the five samples in the order of increasing crystallinity (*bottom* to *top*), *arrows*: Al from the sample mould

7.3 FT-IR-Spectroscopy

Figure 6 shows the IR-spectra of six samples in the order of increasing cristallinity. IR-spectra of Si-bearing oxide spinel modifications were described by Reynard (1993) and Jeanloz (1980). The spectrum of regular γ-Si_3N_4 shows 4 characteristic bands. The tetrahedral modes are at 855–951 cm^{-1} (Si-N-Si stretching) and at 1,066–1,083 cm^{-1} (Si-O bending only in the samples SN46 and SN48), the first octahedral SiN_6-lattice mode occurs at 661–674 cm^{-1}, the second is visible between 727 and 734 cm^{-1}. The overlap of N–H stretching vibrations and O–H vibrations at ca. 3,368 cm^{-1} is weakly visible.

This FT-IR-spectra permit us to write the samples in direction of increasing crystalline order in the following way: SN49 → SN50 → SN47 → SN51 → SN48.

7.4 NMR-Spectroscopy

The NMR-spectra of all products show a sharp signal at −49 ppm, that is nearly identical to the shift in the low pressure phases of Si_3N_4. (Fig. 7, sample SN48 could not be measured because of its very low mass) It corresponds to Si in tetrahedral coordination to nitrogen (SiN_4). A second sharp signal can be found at −221 ppm, corresponding to the octahedral SiN_6 unit (i.e. $Si^{[6]}$). The value for this latter signal by another group is −225 ppm (Sekine et al. 2001).

All $Si^{[4]}$-peaks are characterised by a pronounced upfield tail of almost triangular shape. The intensity of the $Si^{[4]}$ principal signal at −49 ppm increases continuously

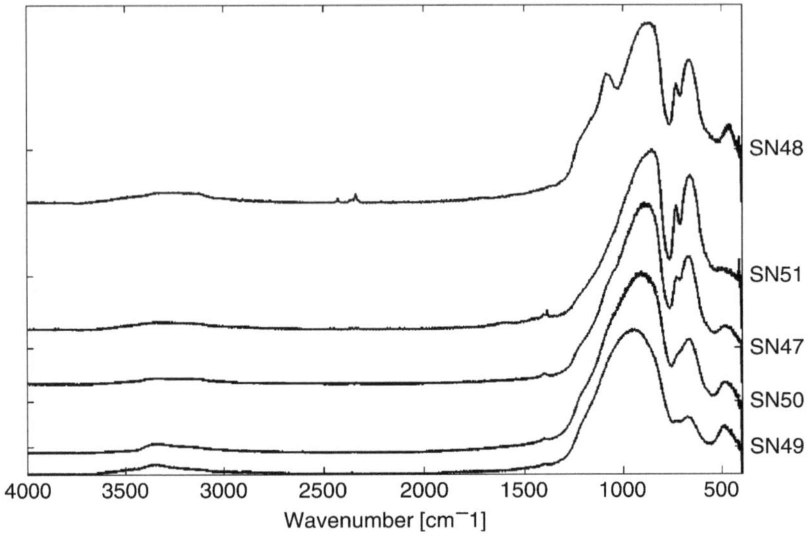

Fig. 6 FT-IR-spectra of the five samples with the for spinel phases typical double peak at 660 and 750 cm^{-1} and with decreasing values of amorpheous substance

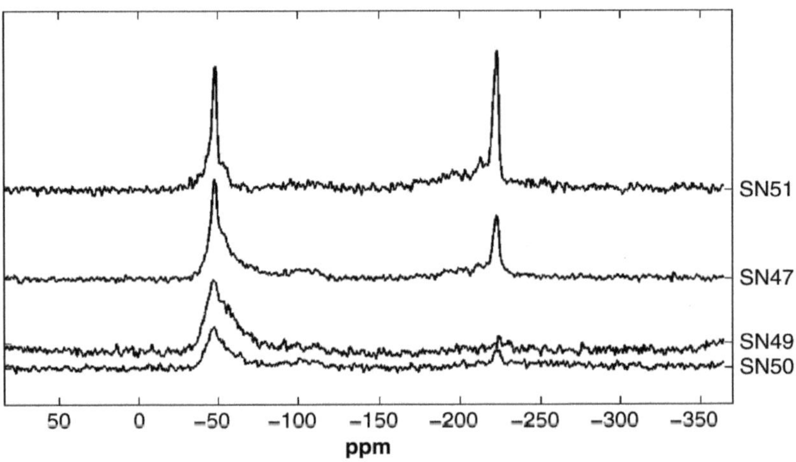

Fig. 7 NMR-spectra of four samples (except sample SN48) in order of increasing crystallinity

and is narrowing from sample to sample. The same is notable for the peak of the $Si^{[6]}$ - environment. In sample SN49 it is not clearly visible. In the sample SN51 the ratio between $Si^{[4]}$ and $Si^{[6]}$ is contrary to all other samples. Here the octahedral $Si^{[6]}$ signal shows a higher intensity than the tetrahedral one, as expected for spinels (see following Section 'Discussion') and a small but well-defined peak on its left side.

In addition the samples SN47, 49 and 50 show a broad signal around−100 ppm, which is typically associated with SiO_4-surroundings bearing different numbers of

OH-groups (−110 ppm Q^4, at −99 ppm Q^3 with 1 OH-group,−91 ppm Q^2 with 2 OH-groups).

Surface hydrolysis during acid treatment of the nanopowders to remove the copper matrix could be responsible for their presence. Ranking of the samples according to their increasing degree of local crystalline order gives the following: SN50 > SN49 > SN47 > SN51.

8 Discussion

It is clear from XRD characterization, that the only crystalline phase in the obtained product is of spinel-type. However, the investigations of FTIR and NMR spectroscopy revealed also disordered or amorphous sample fractions. For comparison, IR- an NMR-spectra of a sample that was entirely amorphous (SN29) after the shock treatment is shown in Fig. 8, together with the corresponding signals of the starting material Si_2N_2NH:O.

The intensity of the hydrogen related peaks at 1,400 and 1,600 cm^{-1} has become smaller and additionally a strong peak at 1,078 cm^{-1}, possibly the stretching mode of Si-O. The FT-IR-spectrum is similar the spectrum of amorpheous Si-O-N, except the sharp peak at 1,078 cm^{-1}. This peak was a stretching mode of Si-O in composition with hydrolysis. This effect was erased by shocking. This spectrum is similar to the spectra of the sample SN49 with its very low cristallinity and low spinel content.

For this phases the quantitative FT-IR- analysis with the NNLS- method was tested in principle. This method will be successful to determine the contents on amorphous substance, but for this method a permanent working under N2-atmosphere will be unavoidable (partially hydrolysis of the powdered KBr!).

The spectrum of the amorphous shock product shows two broad signals. The chemical shift of their maxima coincide with those of the Si[4] signal at −49 ppm and the Q-groups within the spinel-type materials. However, its profile appears to be more Gaussian and it remains unclear whether the spectra of the other samples

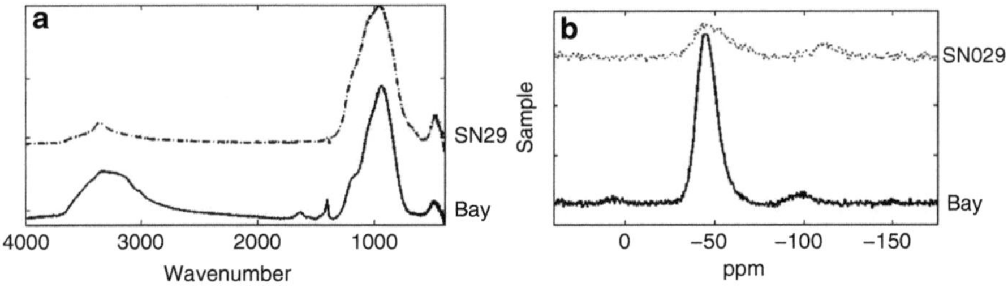

Fig. 8 Comparision between FT-IR-spectrum (**a**) and NMR-spectrum (**b**) of amorphous Si_2N_2NH:O (lower spectrum) and shocked Si_2N_2NH:O at 62 GPa and ca. 4,000 K (67.4 K/GPa, preparation in air)

can be simply thought as a superposition of a similar amorphous phase plus sharp peaks of a well-defined spinel structure.

NMR and also the FT-IR have often been used to determine cation disorder which is identical with the degree of inversion in spinels (Hafner 1961; Hafner and Laves 1961; Brun, Hafner 1962a, b). The applied models are however directly applicable in the present case, because in the γ-Si-N:O phase only one type of cation occurs, the Si^{4+}, while the introduction of oxygen into the anion sublattice causes silicon vacancies.

In normal cases, one-third of the ions occupy the tetrahedral positions in the lattice, the two-thirds on the octahedral ones. Thus the intensity of the sixfold coordinated Si should be higher than it appears within the present NMR spectra – even if there was a considerable fraction of vacancies. Only sample SN51 shows a NMR-spectrum according to these theoretical expectations. Here, the $Si^{[6]}$- peak is in deed higher than the $Si^{[4]}$-peak.

The XRD – diffractograms shows little indications for a high content of amorphous substances, such as an elevated background. Assuming a spinel phase with almost zero occupation of octahedral sites would be unphysical and does not coincide with the results of Rietveld refinement not reported in this paper. However, relaxation times for silicon nitride materials are known to be extremely long. Especially for highly crystalline environments with no or almost no hydrogen species in vicinity. The delay time for the acquisition of the discussed spectra has been kept constant at 60 s. Thus, only those fractions of the material with a smaller relaxation time than 60 s, i.e. only a certain, more disordered fraction has been probed, while material with a higher degree of crystalline order – apparently including most of the $Si^{[6]}$ species remained 'invisible'. Experiments with a series of relaxation times (not shown) have been conducted and indeed indicate an intensity gain of these sharp peaks with increasing relaxation time.

A possible explanation for these apparently findings is that the synthesized material contains different regions that have an overall long range order but may contain domains with different degrees of local disorder that is severe enough to cause rapid decrease of relaxation time, as well as tailing and broadening in the NMR-signals and IR absorption bands.

For all the synthesized crystalline samples a constant pressure and temperature was used (with exception of SN48), but under different conditions of sample preparation. The fraction of maintained γ-Si-O-N should depend on other factors, for example oxygen content, use of oxidized or reduced copper, atmosphere and more.

The samples with a high degree of ordering according to IR-spectroscopy are sample SN51 and the ultra-high temperature sample SN48. Table 1 shows that these two samples were prepared in the glovebox under N_2-atmosphere, but one with reduced copper and one with ordinary copper powder. Thus the atmosphere for the sample preparation is presumably a more critical factor with respect to enhanced cristallinity. In other words: The presence of higher oxygen contents seems to decrease crystalline order, which is in agreement with the low thermal stability of stishovite the high pressure SiO_2 and the tendency of SiO_2 to preferentially form amorphous materials upon intense shock loading.

9 Conclusions

The application of amorphous precursors for the synthesis of pure high pressure phase shows promising results. Under the used conditions no α- and β-phases in the products were found. By avoiding crystalline phases as starting materials we "jump" directly from the amorphous state to the spinel type material with a reconstructive phase transition with low activation barrier.

The evaluation of the best conditions to synthesize γ-$Si(O,N)_4$-phases shows the best crystallinity not in the samples with the highest pressure alone. Correctly, the method of dynamic loading should be named "dynamic high pressure-high temperature-synthesis", because the temperature plays a very important role. The degree of waste heat, introduced in the system consisting of Cu-powder and precursor with the shock wave, controls the yield of high pressure phases too. In particular, the T/p-ratio which is a function of the external loading and the porosity of the Cu-powder appears to be a valuable parameter.

For a better understanding of the difficult structure in the synthesized nanomaterials further investigations with more measurements methods are indispensable. This includes neutron diffraction, HR-TEM- analysis with EELS, DTA-MS and with the FE-SEM and more. But this measurements are completely impossible without pure high pressure phases.

Acknowledgments The development of a shock wave synthesis laboratory would not be possible without many helpful hands. At first we thank the Dr. Erich-Krüger-Foundation for her generous financially support. Further many thanks to Prof. Dr. Paul S. DeCarli for his fruitful theoretical and practical help and the working group "Spreng- und Treibmittel" on the BAM Berlin for her important calculations. Our work would be completely impossible without the practical permanent benefit by the team of Dr. Klaus Grund from the "Lehr- ud Forschungsbergwerk Reiche Zeche Freiberg".

References

Batsanov SS (1994) Effects on explosions on materials. Springer, New York

Belyakov GV, Rodionov VN et al (1977) Heating of porous material under impact compression. Comb Expl Shock Waves 13(4):524–528

Brun E, Hafner S (1962a) Die elektrische Quadrupolaufspaltung von Al^{27} in Spinell $MaAl_2O_4$ und Korund Al_2O_3 I Paramagnetische Kernresonanz von Al^{27} und Kationenverteilung in Spinell. Z Kristallogr 117:37–62

Brun E, Hafner S (1962b) Die elektrische Quadrupolaufspaltung von Al^{27} in Spinell $MaAl_2O_4$ und Korund Al_2O_3 II Das Ionenmodell zur Deutung der gemessenen Quadrupolaufspaltung. Z Kristallogr 117:63–78

Davison L (2008) Shock wave and high pressure phenomena: fundamentals of shock wave propagation in solids. Springer, Heidelberg/Berlin

DeCarli PS, Meyers MA (1981) Design of uniaxial strain shock recovery experients. Shock waves and high strain-rate phenomena. In: Meyers MA, Murr LE (eds) Metals: concepts and applications. Plenum Press, New York, pp 341–373

Deribas A, Silvestrov V, Yunoschev A (2004) Shock-wave synthesis of cubic phase of silicon nitride Si_3N_4. Mater Sci Forum 465–466:113–116
Goodmann P, O'Keefe M (1980) The space group of β-Si_3N_4. Acta Crystallogr B36:2891–2893
Hafner S (1961) Ordnung/Unordnung und Ultrabsorption IV Die Adsorption einiger Minerale mit Spinellstruktur. Z Kristallogr 115:331–358
Hafner S, Laves F (1961) Ordnung/ Unordnung und Ultrabsorption III Die Systeme $MgAl_2O_4$ – Al_2O_3 und $MgAl_2O_4$ – $LiAl_5O_8$. Z Kristallogr 115:330–331
He H, Sekine T, Kobayashi T, Hirosaki H, Suzuki I (2000) Shock-induced phase transition of β-Si_3N_4 to c-Si_3N_4. Phys Rev B62:11412–11417
Jeanloz R (1980) Infrared spectra of olivine polymorphs: α, β phase and spinel. Phys Chem Miner 5:327–341
Jiang JZ, Kragh F et al (2001) Hardness and thermal stability of cubic silicon nitride. J Phys Condens Mater 13:L515–L520
Kuwabara A, Matsunaga K et al (2008) Lattice dynamics and thermodynamical properties of silicon nitride polymorphs. Phys Rev B78:064104
Lee MR, Russel SS, Arden JW, Pillinger CT (1995) Nierite (Si_3N_4), a new mineral from ordinary and enstatite chondrites. Meteor Planet Sci 30:387–398
Milyavskii VV (2000) Empirical estimates of shock compression parameters for porous materials. High Temp 38:215–223
Reynard B (1993) Infrared reflectivity of γ-(spinel)Ni_2SiO_4. Eur J Miner 5:31–35
Schwarz MR (2003) High pressure synthesis of novel hard materials: spinel-Si_3N_4 and derivates, Ph. D. Thesis, Darmstadt
Sekine T (2002) Shock synthesis of cubic silicon nitride. J Amer Ceram Soc 85:113–116
Sekine T, Matsunaga K et al (2001) High-temperature metastability of cubic spinel Si_3N_4. Appl Phys Lett 79:2719–2721
Sekine T, He H et al (2002) Spinel type sialon, spinel type silicon oxinitride and methods for producing their powders. U.S. P. A. Publ. US, Natl Inst Res Tsukuba-shi (JPN)
Sharp TG, DeCarli PS (2006) Shock effects in meteorites. In: Lauretta DS, Hym Jr (eds) Meteorites and the early solar system II. University of Arizona Press, Tucson, pp 653–677
Xu ZCB, Cornilsen PDC, Pennington WC, Wood JR, Hwang JY (2004) Quantitative Mineral Analysis by FTIR Spectroscopy. The Internet Jounal of Vibrational Spectroscopy 5(1), http://www.ijvs.com/volume5/edition1/section2.html
Zerr A, Miehe G et al (1999) Synthesis of cubic silicon nitride. Nature 400:340–342
Zhou D, Mitchell T (1995) Rotation domains in α-silicon nitride Phil Mag A72:1131–1140
Zhu L, Chen L et al (2007) Nitridation of silica to an α-Silicon-nitride nanorod using $NaNH_2$ in the autoclave at 700°C. J Amer Ceram Soc 90:1243–1245

Decomposition of Aluminosilicates and Accumulation of Aluminum by Microorganisms on Fumarole Fields of Tolbachik Volcano (Kamchatka Peninsula, Russia)

S.K. Filatov, L.P. Vergasova, and R.S. Kutuzova

1 Introduction

Although aluminum is the significant constituent of weathered minerals, its precipitating by microorganisms is not well studied. Moreover, the participation of microorganisms in biogenic aluminum hydroxides formation is still has no wide experimental justification (Sneath et al. 1997). The present paper gives such facts of the microbiological origin of aluminum sediments on a recent volcano.

The Kluchevskaya group of volcanoes is the biggest volcanic center in the Kurile-Kamchatka region (Klychevskoi, Tolbachik, and Bezimyannyi volcanoes). Fissure eruptions of volcanic products with volume exceeding 1 km^3 have been occurring here for several centuries.

Last eruption of the Tolbachik volcano took place in 1975–1976 and was named the Great Fissure Tolbachik Eruption (GFTE, (Great Fissure ... 1984)). It is represented by the Northern and Southern breaches (NB and SB, correspondingly). In 1 year and a half, new cinder cones were formed in this area. Their heights are up to 300 m. Lave flows covered an area of about 45 km^2, and a new caldera originated at the top of Ploskiy Tolbachik volcano, as a result of the foundering of 1 km^3 of rock. The total amount of erupted substance was estimated as 2.2 km^3. The GFTE was the largest basalt eruption in the Kurile-Kamchatka Volcanic belt, and one of the six largest Fissure eruptions worldwide.

S.K. Filatov (✉)
Saint Petersburg State University, Department of Crystallography, St. Petersburg, Russia
e-mail: filatov.stanislav@gmail.com

L.P. Vergasova
Institute of Volcanology and Seismology RAS, Laboratory of Posterupted Processes, Petropavlovsk-Kamchatsky,
Kamchatka Peninsula, Russia

R.S. Kutuzova
State Scientific University of Agriculture Microbiology RAAS,
Saint Petersburg, Pushkin, Russia

S.V. Krivovichev (ed.), *Minerals as Advanced Materials II*,
DOI 10.1007/978-3-642-20018-2_36,

Investigation of phenomena accompanying the volcanic activity is one of the important tasks of modern volcanology. Fumarole processes are of special interest, for they are closely related to metasomatism and genesis of minerals and ores. Since the onset of the GFTE (from September, 15, 1975 at the NB and December, 10, 1975 at the SB), the fumarole activity has been developing at lava streams and limited crater zones of slag cones. It can be characterized by temperature heterogeneity, significant transformation of the erupted rock under influence of volcanic gases, and numerous formations of various minerals and ores.

Fumarole activity following the major eruption provides valuable material for detailed mineralogical and crystal chemical researches. Thirty new mineral species were discovered by our group of volcanologists from the Institute of Volcanology and Seismology Russian Academy of Sciences and the Saint Petersburg State University (Vergasova and Filatov 1993, etc.). In particular Cu- and Al-ores forming minerals were revealed and investigated. Al-enriched sediments have aluminum concentrations in the range 25–40 wt. % of Al_2O_3, while original basalt volcanic rocks contain twice less of Al_2O_3 (13–17 wt. %).

Thus, for the first time the volcanism is considered as a source of high aluminous substance. Predominantly, sediments enriched by aluminum consist of new mineral – lesukite, $Al_2(OH)_5Cl{\cdot}2H_2O$ (Vergasova et al. 1997), less gibbsite $Al(OH)_3$, and amorphous Al-containing phase (Kutuzova et al. 2004; Vergasova et al. 2004; Filatov et al. 2004), whose chemical composition is similar to that of bauxite.

Study of the natural process of bauxite formation connected to volcanic activity is essential in view of three reasons: (1) it provides evidences of a leading role of desilification followed by fixation of aluminum *in situ*; (2) it proves the fact of development of primary minerals under the direct influence of volcanic gases, probably formation of these minerals is a result of crystallization of primary alumogel, and (3) in previous our researches the attention was also drawn to a possible influence of the microbiological factor along with the influence processes on the destruction of the erupted rock and subsequent genesis of minerals (Vergasova et al. 1999).

Here we summarize the results of investigations in the aria of recent bauxite formation with participation of microorganisms that can accumulate aluminum – a chemical element without variable valence. The activity of the Al-precipitating microorganisms was discovered as a component of a modern post eruptive volcanism. New evidences of the microbiological factor of bauxite formation on the recent volcano are represented. The mineral covers were removed from the microorganism structures and more clear pictures of them were obtained. Due to this, the identification of the aluminum-precipitating microorganisms was made more reliable.

2 Description of Bauxite (Protobauxite) Sediments

Bauxite (lesukite)-bearing sediments are formed in zone of low-temperature transformations of pyroclastic erupted rock (not above 50°C). The fine-dispersive sediments have intensive yellow-orange, yellow-brown, or yellow-cream color distinguish on an otherwise dark volcanic surface. The lesukite is cubic mineral with *a* parameter about 20Å.

The lesukite sediments were noticed on GFTE immediately after termination of the First and Third cones' eruptions at Northern Break in 1976 (Vergasova et al. 1997). The sediments also sporadically occurred on the SB.

The main Al-enriched sediments were observed in the crater zone of the First GFTE cone. In the probes of 1995, lesukite and halite were presented there. Under dynamic thermal conditions, a stable formation of protobauxite sediments has tendency to expand with time.

In 2000, about 200–250 m^2 of lesukite (bauxite) field were observed on the eastern edge of the crater of the First cone and extend further 5–10 m (in some places, up to 30 m) along the eastern external slope of the cone. Thickness of the coating varies from about 1–35 cm. Their color varies from cream to bright orange and brown; the texture varies from macroporous and highly wet rock to finely dispersed and absolutely dry powder. X-ray phase identification showed the presence in the samples of lesukite, gibbsite (first occurrence on the GFTE) and amorphous phase.

A vertical section (30 cm) of the orange "lesukite" sediments was made at the bauxite field of the First cone. The highest layer (7–8 cm, yellow-orange, N 2000/15) was presented by gibbsite, neighboring layer (10–12 cm, orange, N 2000/16) by lesukite, next one (5 cm, grayish-yellow, N 2000/17) – lesukite, and last layer (5–7 cm, reddish-brown, N 2000/18) – lesukite. Just under the last layer of the sediments there was a dense and viscous dark red plate consisting of transformed, highly oxidized slag-ashes pyroclastic material.

3 Experimental

Lesukite samples were taken at the First cone of the NB of GFTE and studied by X-ray powder diffraction (XRD), high-temperature XRD (HTXRD), differential thermal analyses (DTA), thermal gravimetry (TG), and microbiological methods.

X-ray powder diffraction data were collected at room temperature using the DRON 2.0 diffractometer (CuKα and CoKα, 35 kV, 20 mA), 5–75° (2θ). The powder samples were prepared from water suspension with germanium or potassium diphthalate as internal standards. The cubic cell parameter was refined by least-squares method.

HTXRD data were collected using DRON 3.0 diffractometer (CuKα radiation) with a high temperature KRV-1100 chamber under abundant atmosphere.

The sample of lesukite was investigated also in vacuum (down to 10^{-7} bar) and forevacuum (down to 10^{-5} bar) at room temperature, diffractometer Siemens D5000, vacuum chamber, CuKα radiation.

DTA-TG data were recorded on a "MOM" derivatograph with 0.6 g/min heating rate, 5°C temperature intervals, and 2 mg step of weights scale. Platinum crucibles were used with corundum as a standard. Since at high temperatures, the volatile constituents of lesukite escape intensively, some experiments were carried out with a platinum multi-plate holder. This allowed one to have a thin friable layer of a sample (1–1.5 mm) and simplified evaporation of gases.

For microbiological researches, to confirm the living nature of the ore-impregnated structure, we got the accumulating cultures of aluminum-precipitating microorganism on synthetic mineral media with presence of aluminum lactate (Al^{3+} = 8.7 and 86.5 mg/l) or aluminum sulfate (Al^{3+} = 8.1 and 40.5 mg/l) at pH 3.5–4. The presence of iron-containing compounds on the surface of the microorganism cells was revealed with the reaction "Prussian blue," (blue staining), and the presence of aluminum was determined by the reaction with aluminon/aurintricarboxylic (red staining). When working with the lesukite samples with an admixture of pyroclastic particles containing up to 18% iron, a particular emphasis was put upon the iron-precipitating organisms. Iron hinders the identification of aluminum on cells, because Fe^{3+} is responsible for the color appearing in the reaction with aluminon; the latter is used as test for aluminum. To be sure of the capability of the microorganisms to precipitate aluminum, samples of ralstonite, $(Mg, Al)_2(F, OH)_6{\cdot}nH_2O$, containing 20–24% Al_2O_3 and an insignificant admixture of iron (less than 0.7%) were used.

The methodic of detection of aluminum-precipitating microorganism on fixing preparation (glass plates) with further microscopic investigation and photographing was developed. Microscopes Axiolab-Zeiss and Axiostar plus-Zeiss were used with main working zoom 100 × 10 and 40 × 10.

4 Results and Discussion

4.1 *High-Temperature Investigations*

According to the HTXRD previous data (Vergasova et al. 1997), with heating from room temperature, the cubic cell of lesukite is compressed. The rate of the compression increases above 85 ± 5°C and proceeds until destabilization (145 ± 5°C). DTA curve has three endothermic peaks at 160, 195, and 340°C (Fig. 1). Energy absorption at 160°C indicates the destabilization of lesukite. At this stage of heating, 13 wt. % is lost. At 195°C, lost of the weight is 17%, whereas at 340°C, it is 37 wt. %. Stepwise loss of weight is caused by the presence of four volatile components: hygroscopic water (in wet samples), structural water, $(OH)^-$ groups, and Cl^- ions. Two sets of experiments have been carried out by HTXRD, and TG, both at 20–90 and 20–140°C temperature intervals, to obtain more detailed data about temperature transformations of lesukite and to investigate a reversibility of the process (Filatov et al. 2004). Series of XRD experiments in vacuum were made also to study the thermal behavior of the mineral in different atmospheres (Filatov et al. 2005).

HTXRD data. 20–90°C. The results of HTXRD investigation are shown in Fig. 2. It is clearly seen (Fig. 2a) that, within the 20–90 ± 5°C (filled circles), the

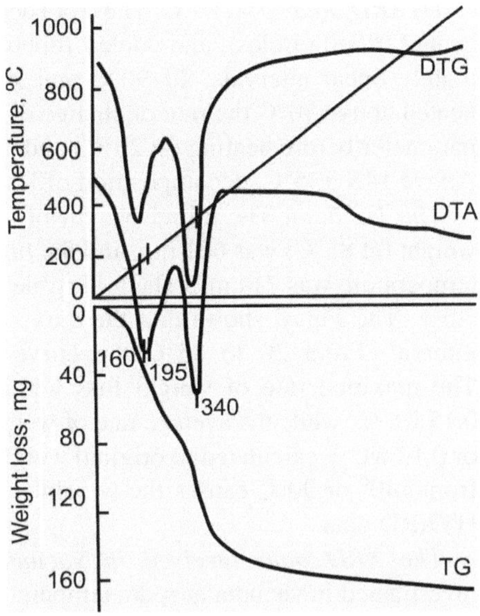

Fig. 1 DTA curves of lesukite after (Vergasova et al. 1997)

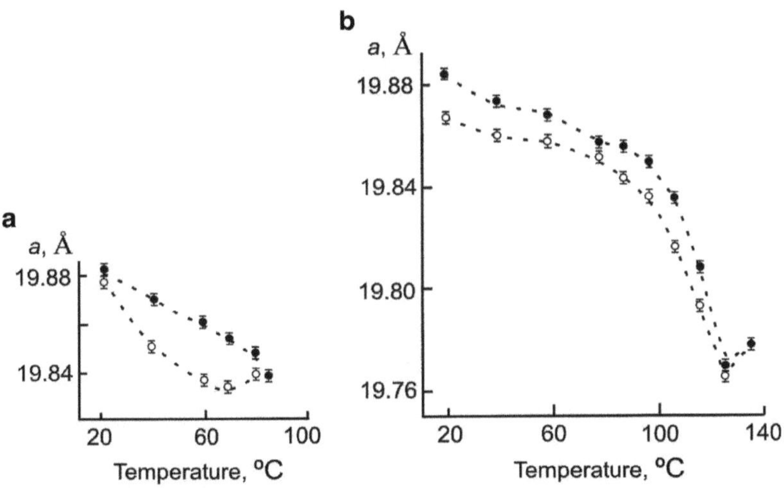

Fig. 2 Temperature dependence of the *a* cell lesukite parameter after (Filatov et al. 2004). (**a**) 20–90°C, (**b**) 20–140°C, ● – heating data, ○ – cooling data

dependence of the *a* cell parameter on temperature is almost linear. When cooled, starting from 90°C (open circles), the dependence curve is parabolic. However, the values of the *a* parameter at 20°C, before heating and after cooling, are practically identical. Hence, we conclude that the thermal transformation of the lesukite is reversible in the 20–90°C interval. This is also confirmed by TG data (see below).

HTXRD data. 20–140°C. The *a* (*T*) dependence in this temperature interval when heated (filled circles), and cooled (open circles) can be divided into two approximately linear intervals: 20–90°C and 90–130°C (Fig. 2b). When the sample was heated above 90°C the rate of change of the *a* parameter increases. Values of the *a* parameter before heating (at 20°C) and after cooling to at 20°C differed by 0.02 Å. Above 145 ± 5°C, decomposition of lesukite takes place.

The TG data. The initial weight of the sample used was 714 mg. Its minimal weight (at 85°C) was 661 mg, and the final weight after cooling from 90°C to room temperature was 714 mg. The reversible loss of weight in the interval 20–90°C was 7.5%. The Fig. 3 shows that the curve of weight change has a bend in 65–85°C interval. From 20 to 85°C the curve of weight change is slightly S-shaped. The maximal rate of weight loss within 40–60°C interval is 0.85 mg/grad., or 0.15 wt. %, while the average rate of weight loss within this interval is 0.62 mg/grad., or 0.12 wt. % calculated to original weight. Dropping the sample temperature either from 140° or 90°C causes the weight to increase immediately, which agreed with HTXRD data.

The XRD data received in vacuum. The lesukite, $Al_2(OH)_5Cl{\cdot}2H_2O$, was investigated in vacuum at room temperature (Filatov et al. 2005). It was shown, that in case of atmosphere conditions lesukite dehydrates and becomes decomposed at 145°C, but in vacuum conditions (10^{-6}–10^{-7} bar) its dehydration and amorphisation occur even at room temperature; dehydration is accompanied by decrease of the cubic cell parameter (a = 19.82 Å) by 1 Å (5%). So, vacuum treatment may be used for "pumping out" the weak-bonded water from the matter, like heating. Dehydration in vacuum may be preceded at lower temperatures than in atmosphere conditions. This fact may be of interest in technology and in geology of the Moon or other planets that have no atmosphere.

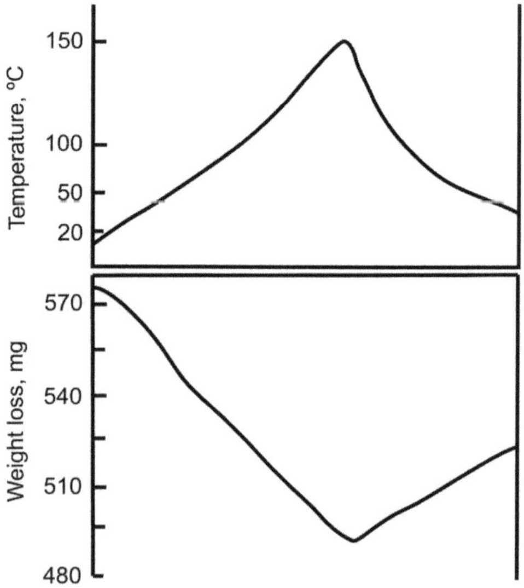

Fig. 3 The curves of lesukite weight losses during heating up to 140°C and cooling to room temperature after (Filatov et al. 2004)

The nature of the thermal behavior of lesukite. The values of the *a* parameter at 20°C (before heating) and after cooling from 90°C, are nearly identical (Fig. 2a). Hence, we conclude that the thermal transformation of lesukite is reversible in the 20–90°C interval. This is also confirmed by the equality of the initial weight of the sample and its weight after cooling from 90°C (714 mg). In the interval 20–140°C the change of lesukite unit cell parameter is not completely reversible. When the temperature increases from room temperature to 140°C the loss of weight is 15%, and the reversible weight loss is 7.5%.

We assume that such thermal behavior of lesukite can be explained by step-by-step loss of H_2O. The first (reversible) step (20–85°C) proceeds in accordance with the following scheme:

$$Al_2(OH)_5Cl \cdot 2H_2O \leftrightarrow Al_2(OH)_5Cl \cdot H_2O + H_2O \uparrow .$$

The reversible loss of weight is 7.5 wt. %, which is in good agreement with values of one H_2O molecule in lesukite, which amounts to approximately 8 wt. %.

Further heating causes the second irreversible step of weight loss:

$$Al_2(OH)_5Cl \cdot H_2O \rightarrow Al_2(OH)_5Cl + H_2O \uparrow .$$

Heating lesukite above 145 ± 5°C, probably, causes loss of Cl^- atoms and $(OH)^-$ groups that results in decomposition of lesukite. The process of reversible dehydration-hydration of lesukite takes place at low temperature. This means that molecular water is located in voids of the crystal structure, for example between layers analogous to that in gibbsite.

4.2 The Dynamic of Recent Bauxite Formation

In 2000, about 200–250 m^2 of lesukite (bauxite) field were observed on the eastern edge of the First cone crater and extend further 5–10 m (in some places, up to 30 m) along the eastern external slope of the cone. Thickness of the coating varies from about 1 to 35 cm. Their color varies from cream to bright orange and brown; the texture varies from macroporous and highly wet rock to finely dispersed and absolutely dry powder. X-ray phase identification showed the presence in the samples of lesukite, gibbsite (first occurrence on the GFTE) and amorphous phase.

The vertical section (30 cm) of the orange "lesukite" sediments was made at the bauxite field of the First cone. The highest layer (7–8 cm, yellow-orange, sample N T2000/15) was presented by gibbsite; neighboring layer (10–12 cm, orange, N T2000/16) by lesukite; next one (5 cm, grayish-yellow, N T2000/17) – lesukite; and last layer (5–7 cm, reddish-brown, N T2000/18) – by lesukite. Just under the last layer of the sediments there was a dense and viscous dark red plate consisting of transformed, highly oxidized slag-ashes pyroclastic material.

Thus X-ray phase investigation of yellow finely dispersed sample N T2000/15 selected from the surface showed that it is gibbsite, instead of lesukite. Diffraction peaks of lesukite in the samples NN T2000/15– T2000/18 selected from the cross-section have considerably lower intensity than those of samples taken from another surface spots of the field. Probably, lesukite from samples T2000/16 – 18 is already being transformed into gibbsite or amorphous phase, and eventually, all lesukite samples will contain mainly gibbsite.

The discovery of the gibbsite indicates that under certain physical-chemical conditions, the process of "bauxitisation" of the fumarolic precipitate (represented by lesukite) may occur. Among the lesukite encrustations, an amorphous phase enriched in aluminum (alumogel), was also found. Hence, lesukite that is formed under direct participation of volcanic gases can be thought of as a bauxite proto-mineral. The major role in the processes of the protobauxite substance formation belongs to microorganisms.

4.3 Microbiological Investigations

The process of formation of the protobauxite substance on modern volcano can be divided on two stages: (1) desilification of the aluminosilicate erupted rocks, and (2) accumulation of free alumina. We suppose that microorganisms (inserted by chance on cooling volcano) take part at the both stages under suitable temperature and chemical regimes. The first stage (desilification) may be a result of the combine action of the temperature, fumarole gases and microorganisms, which isolate acids and decompose aluminosilicate material this way. At the second stage (accumulation of free alumina) Al-precipitating microorganism plays leading role.

According to its morphological properties the microorganism was conventionally attributed to group *Metallogenium-Siderococcus* (Kutuzova et al. 2006). In the end of 70 years of the last century in sod-podzol the peculiar and still insufficiently investigated microorganisms of this group were discovered to have ability to accumulate iron and manganese as well as aluminum on its cell structures (Aristovskaya 1980). Originally the ferromanganese bacteria of *Metallogenium-Siderococcus* group were discovered in silt deposits of ore-bearing lakes as active mineralization agents (Perfiliev and Gabe 1964).

Below we describe such microorganism revealed on recent volcano. The probes of the culture of aluminum-precipitating microorganism were taken from fumarole deposits of the First GFTE cone (see Paragraph 2) and got on synthetic mineral media mentioned in Paragraph 3. Acid medium, presence of Al^{3+} ion, that is toxic for many microorganisms, addition of ampicillin (the antibiotic for a wide range of bacteria), – all this enables creation of elective conditions for this microorganism that yields to the banal micro flora in speed of developmental growth (Kutuzova et al. 2004, 2006; Vergasova et al. 2004).

At the initial phase of accumulating culture development (1–1.5 month) the aurintricarboxylic acid reveals the mineralized oval formations of red color lying freely or encapsulated in streaks of aluminum, from which they disengage themselves as far as the latter is taken away. Dominating form of microorganisms disengaged from sediments are small oval bodies reaching 1–3 μm (or smaller) in diameter with more compact central part and transparent circlet that looks like mucous capsule (Fig. 4a). It turned out that the supposed capsule is not the mucous formation but is able to transform into radial spread of extremely thin filaments that gradually transform into the chains of small oval bodies that further form solid mass at the center of growth (Fig. 4b). Actually, these are micro colonies, trichospherical and zoogleal according to the name of kin structures at *Metallogenium* (Aristovskaya 1980). Micro colonies give start to the next development cycle by casting the radial chains of filaments or by breaking up to oval bodies. According to Perfiliev and Gabe (1964), a cell growing out by one or several radial sprouts from

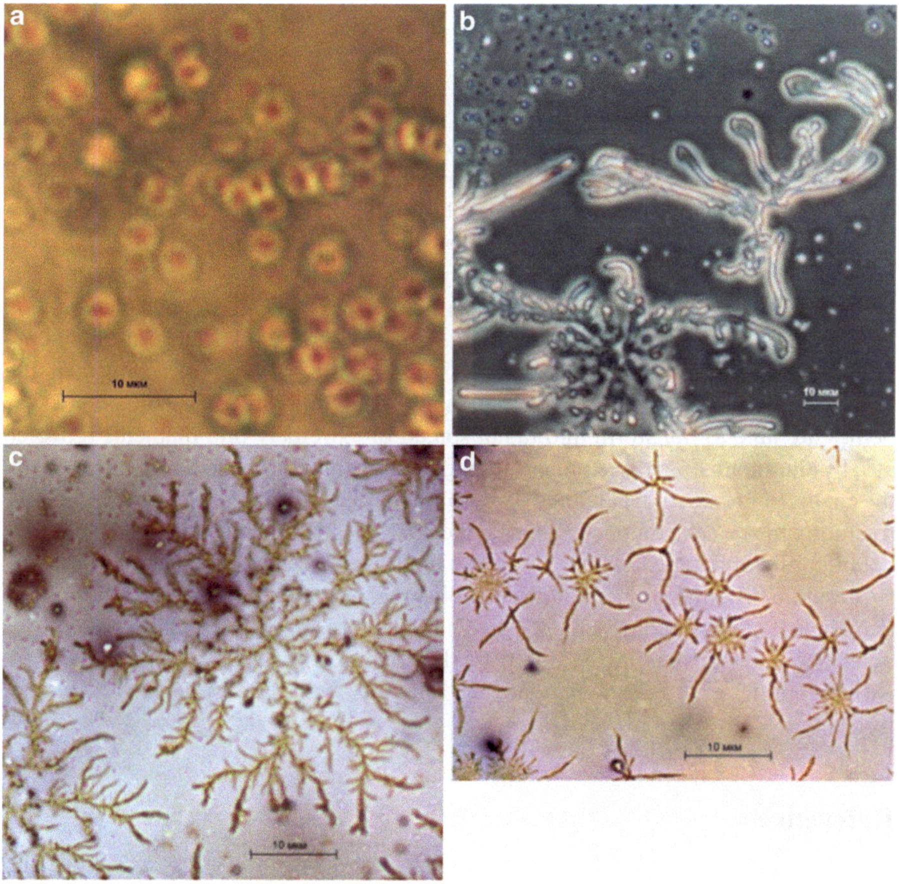

Fig. 4 The Al-precipitating microorganisms of *Metallogenium-Siderococcus* group from fumarole field of Tolbachik volcano. **a**, **b** – original appearance, **c**, **d** – the same after removal Al-containing mineral covers

the chain of small bodies, forms colonies – "spiders" typical for *Metallogenium.* Further increase of colony leads to lengthening of radial sprouts, their branching and thickening for the account of Fe-Mn precipitation, and in our case, of aluminum.

Trichospherical micro colonies (10–70 μm) is the most effective and typical stage of the microorganism development. As a rule, in the central part of the colony aluminum precipitation is mostly significant while it decreases to its boundaries.

Comparison of the microorganism, discovered in the volcanic material, with *Metallogenium* of lake silts (Perfiliev and Gabe 1964) and the results of the other authors investigations (Aristovskaya 1980; Dubinina 1970), as well as the description of *Metallogenium,* presented in Bergey's "Manual of Determinative Bacteriology", 1997, indicates their similarity in functional aspect – precipitation of iron and manganese as well as aluminum, and in morphological aspect – the same small oval bodies and radial spread of extremely thin filaments.

5 Conclusions

For the first time the recent volcanism is considered as a source of high aluminous substance. Rigid basalt volcanic alumosilicate glass of Tolbachik volcano (1975–1976) transforms due to a microbiological factor into a soft bauxite-like material which is ore on aluminum: a new mineral – lesukite, $Al_2(OH)_5Cl.2H_2O$; gibbsite, $Al(OH)_3$; and Al-enriched amorphous phase.

In the bauxite-like sediments of the Tolbachik volcano, aluminum-precipitating microorganism was found, first in geological processes, and attributed to *Metallogenium-Siderococcus* group. Dominant forms of the microorganism are small oval bodies covered by camouflaging Al-containing minerals (Fig. 4a). These bodies are able to transform into radial spread of extremely thin filaments (covered also by aluminum mineral sediments) that gradually transform into new generations of the chains of the small oval bodies (Fig. 4b). After the Al-mineral covers removal, the microorganisms have more clear appearance (Fig. 4c, d). And due to this, the possible doubt in microbiological nature of the bauxite formations is going done.

Lesukite, as a main mineral of recent volcanogenic bauxite-like sediments, has law temperature of the decomposition (145°C) in comparison with a melting point of the volcanic alumosilicate glass (≥1,000°C). For this reason, lesukite permits to extract easily aluminum from the compound, and this way, from basalt volcanic rocks.

References

Aristovskaya T (1980) Microbiology of soil-forming processes. Nauka, Leningrad (In Russian)

Dubinina GA (1970) Untersuchungen uber die Morphologie von Metallogenium und die Beziehungen zu Mycoplasma. Z Allg Mikrobiol 965:309–320

Fedotov SA (ed) (1984) Great fissure tolbachik eruption. Nauka, Moscow (in Russian)

Filatov SK, Vergasova LP, Stepanova EL, Kutuzova RS, Poliyakova IG (2004) Microbiological transformation of basalts of the Great Fissure Tolbachik Eruption of 1975–1976 (Kamchatka) into lesukite, $Al_2(OH)_5Cl{\cdot}2H_2O$, and other phases of bauxites. Proc Russ Miner Soc 133:1–11 (in Russian)

Filatov SK, Paufler P, Vergasova LP, Levin AA, Yanson OM, Meyer DC (2005) Modeling of the hypergenesis process on volcano-surface with use of vacuum [by example of lecukite $Al_2(OH)_5Cl{\cdot}2H_2O$]. Proc Russ Miner Soc 134:104–109 (in Russian)

Kutuzova RS, Vergasova LP, Filatov SK (2004) Transformation of igneous rocks by microbial biocenosis on the First cinder cone of the Great Tolbachik fissure eruption. Volcanol Seismol, 133:46–54 (in Russian)

Kutuzova RS, Vergasova LP, Filatov SK (2006) Contribution of Metallogenium-Siderococcus bacteria to the postvolcanic transformation of igneous material (Kamchatka). Eurasian Soil Sci 39:298–306

Perfiliev BV, Gabe DR (1964) Investigation of bacteria accumulating Mn-Fe in bed silts by method of microbal landscaping. The part of microorganisms in formation of ferromanganese lake ores. Nauka, Moskow-Leningrad (in Russian)

Sneath PHA et al (1997) Bergey's manual of determinative bacteriology. In: Holt G, Krieg NR (eds) Aerobic chemolithotrophic bacteria and associated organisms. Williams and Wilkins, Mir, Baltimore/Moscow

Vergasova LP, Filatov SK (1993) Minerals of volcanic exhalations – special genetic group. Proc Russ Miner Soc 122:68–76 (in Russian)

Vergasova LP, Stepanova EL, Serafimova EK, Filatov SK (1997) Lesukite – $Al_2(OH)_5Cl{\cdot}2H_2O$ – a new mineral from volcanic exhalation. Proc Russ Miner Soc 126:104–110 (in Russian)

Vergasova LP, Stepanova EL, Zelenskii ME, Filatov SK (1999) Bauxite formation in recent volcanic zones. In: Proceedings of the IX congress of Russian mineralogical society, St. Petersburg, Russia, pp 215–216 (in Russian)

Vergasova LP, Filatov SK, Kutuzova RS, Zelenskii ME (2004) The origin of bauxites: the Great Tolbachik fissure eruption, Kamchatka. Volcanol Seismol 133:48–53 (in Russian)

Biogenic Crystal Genesis on a Carbonate Rock Monument Surface: The Main Factors and Mechanisms, the Development of Nanotechnological Ways of Inhibition

Olga V. Frank-Kamenetskaya, Dmitriy Yu. Vlasov, and Olga A. Shilova

1 Introduction

The problem of the biofilm mineralization on the carbonate rock surfaces involves the increased attention of the scientists (Gadd 2007). The urgency of this problem is explained by the importance of understanding of rock transformation by the lithobiotic microbial communities (microfungi, lichens, algae and bacteria) in various environment conditions as well as the necessity of new solutions for rock monuments protection.

The formation of metabiogenic minerals involves following processes: the microbial biofilm growth; the mechanical action of organisms on stone surface; biochemical action of organisms on rock including the producing acids and other metabolites and chemical reactions of these aggressive products with stone: pitting and etching of mineral surfaces, mineral substitution and dissolution reactions, secondary (metabiogenic) mineral formation; patina (organic – inorganic composite system) formation. The mechanisms of patina generation and the role of micro-organisms in these processes are unsufficiently investigated.

The present work is focused on the calcium oxalate and gypsum rich patina which are widespread on cultural heritage monument surfaces. Destructive gypsum crystallization on marble and limestone monuments often results in irreplaceable losses (Fig. 1).

O.V. Frank-Kamenetskaya (✉)
Department of Crystallography, St. Petersburg State University, 199034 St. Petersburg, Russia
e-mail: ofrank-kam@mail.ru

D.Y. Vlasov
Labotary of Lower Plant, Institute of Silicate Chemistry, Russian Academy of Sciences, 199034 St. Petersburg, Russia

O.A. Shilova
Labotary of Inorganic Sinthesis, Labotary of Lower Plant, Institute of Silicate Chemistry, Russian Academy of Sciences, 199034 St. Petersburg, Russia

S.V. Krivovichev (ed.), *Minerals as Advanced Materials II*,
DOI 10.1007/978-3-642-20018-2_37, © Springer-Verlag Berlin Heidelberg 2012

Fig. 1 Gypsum-rich patina on Manufacturer Demidov house in Saint Petersburg, arch. O. Monferan

Fig. 2 Micromycetes colonies on Carrara marble sculpture (Saint Petersburg) in a year after treatment by organotin and quaternary ammonium biocides

The most widespread countermeasures to biodestruction of the stone monuments are the chemical biocide treatment and laser cleaning of monument surface. The main defects of these approaches are the short time durability (Fig. 2), probable destruction of monument surface and toxicity (by biocide treatment). So, the creation of new scientific concepts of ecological long-lived protection of monuments is required.

The aims of this work are the following.

1. To study the factors and the mechanisms of oxalate and gypsum-rich patina formation and in particular to investigate: the morphogenesis peculiarities of forming crystals, the comparative role of microorganisms and environment in crystallization processes, the dependence of biogenic crystallization on peculiarities of the bedrock.

2. To develop long selfcleaning inorganic-organic coatings for carbonate rock monument protection: to investigate the bioactivity of different photocatalysts and nanodiamond as perspective long-term "soft biocides" and to carry out the synthesis and testing of the selfcleaning inorganic-organic films with anti-microbial properties.

The experiments and field works were carried out by the scientists of Saint Petersburg State University in cooperation with the specialists of Silicate Chemistry Institute, Russian Academy of Science (Vlasov and Frank-Kametskaya 2005, 2006; Timasheva et al. 2007; Frank-Kamenetskaya et al. 2009; Rousakov et al. 2010; Barinova et al. 2010; Shilova et al. 2009).

2 Materials and Methods

The rock samples with the oxalate and sulphate patina were found on the surface of marble and limestone monuments which are exposed under the intensive influence of anthropogenic factors in different climatic conditions (Saint Petersburg, Russia; Chersonesus, Ukraine; Oslo, Norway; Oerebru, Sweden). The samples for biological analyses were collected from the same stone objects as for mineralogical ones. The non-destructive techniques were applied for isolation of microorganisms from the stone surface. The simulation experiments on crystallization of calcium oxalates under the microfungal influence have been carried out (Table 1).

The acid producing by microscopic fungi (stone destructors) by chromatography-mass spectrometry method has been studied. The phase composition of patina and synthesized materials was analyzed by X-Ray powder diffraction method. The study of morphological features of metabiogenic and synthesized crystals (calcium oxalates and gypsum) and biological components of the patina using scanning electron microscopy (SEM) was carried out. The variations of chemical composition of the gypsum

Table 1 Condition of experiments on crystallization of calcium oxalates under the microfungi influence

Experiment	Micromycetes species (isolated from damaged marble)	Method of inoculation (exposition, days)	pH	
			Days	Values
1	Aspergillus niger	Spores (5, 15, 25, 60)	0–25	6.20–5.49
			25–60	5.49–7.49
	Penicillium oxalicum		0–15	6.20–5.82
			15–60	5.82–8.97
	Penicillium vitale		0–15	6.20–5.55
			15–60	5.55–9.14
2	Aspergillus niger Penicillium cyclopium	Mycelia and spores (2, 4, 6, 9, 11, 13, 17)	0–17	5–6

crusts were investigated by inductively coupled plasma-mass spectrometry (ICP-MS) and electron microprobe (EMP) analysis (across the whole thickness). The species composition of lithobiotic microbial community was identified using special mycological diagnostic methods.

The opportunities of application of inorganic and organic photocatalysts which active in various ranges of waves lengths for protection of rock monuments against biocrystallization were investigated for the first time. The inorganic photocatalysts on the base of TiO_2, organic photocatalysts (phthalocyanines and diphthalocyanines of different metals) and detonation nanodiamonds as promising selfcleaning bioactive materials were tested. The pure cultures of biodestructors (micromycetes) on the Chapek agar (pH 5.2) (as basic medium) with $CaCO_3$ (2 wt %, pH = 6.5) and test substances (in proportions from 1 to 10wt %) have been used in these experiments. The type of colony, pigmentation (coloration), dissolution of calcium carbonate, growth speed (colony diameter) and micromorphology peculiarities were applied as the test criterions on the bioactive properties of substances. The measurements were carried out every 48 h. The special attention has been given to the opportunity of photocatalytic materials fixing on the rock monument surface. For this aim the meso- and nanoporous composite coatings: epoxy-silicate with different biactive dopants (photocatalysts, nanodiamonds) and epoxy-titanate were prepared by sol–gel-derived method (Fig. 3) and tested in field and laboratory conditions. Use as one of precursors the epoxy rezin has allowed to spend drying in the open air (annealing is not needed). Produced coatings satisfy the following important demands: low-temperature synthesis, high vapor penetration, transparency (invisible), colorlessness and absence of shine.

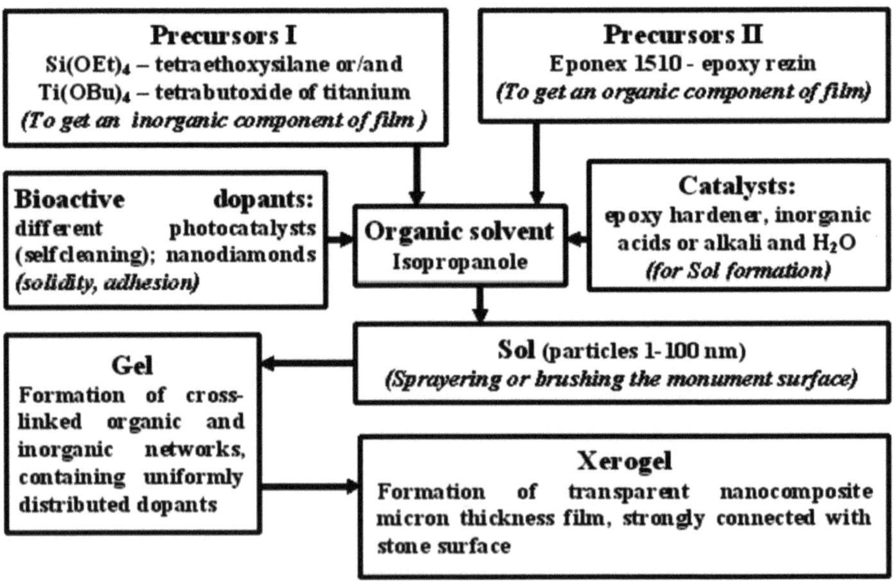

Fig. 3 Flow diagram of sol–gel synthesis of organic – inorganic coatings

3 Results and Discussion

3.1 Biogenic Crystal Genesis on the Surface of Carbonate Rock

3.1.1 Calcium Oxalate Crystallization

Calcium oxalates: whewellite $Ca(C_2O_4){\cdot}H_2O$ *and weddellite* $Ca(C_2O_4){\cdot}(2 + x)H_2O)$, in detail described on the surface of marble monuments in Mediterranean (Del Monte and Sabbioini 1987; Burford et al. 2003 and other), now are revealed in the biofolms from a limestone monuments surface in Chersonesus (Crimea, Ukraine). The X-ray diffraction analysis showed that whewellite and weddelite occur in biofilms in various ratios. On the scanning electron microscopy pictures of biofilms we can see the lichen thalluses and the mycelia of fungi as well as various oxalate crystals (Fig. 4). The majority of crystals which were found on the rock surface have dipyramidal or dipyramidal-prismatic habit typical for tetragonal weddellite (Fig. 4a, b). Among them there are crystals with rounded dissolution forms (Fig. 4c).

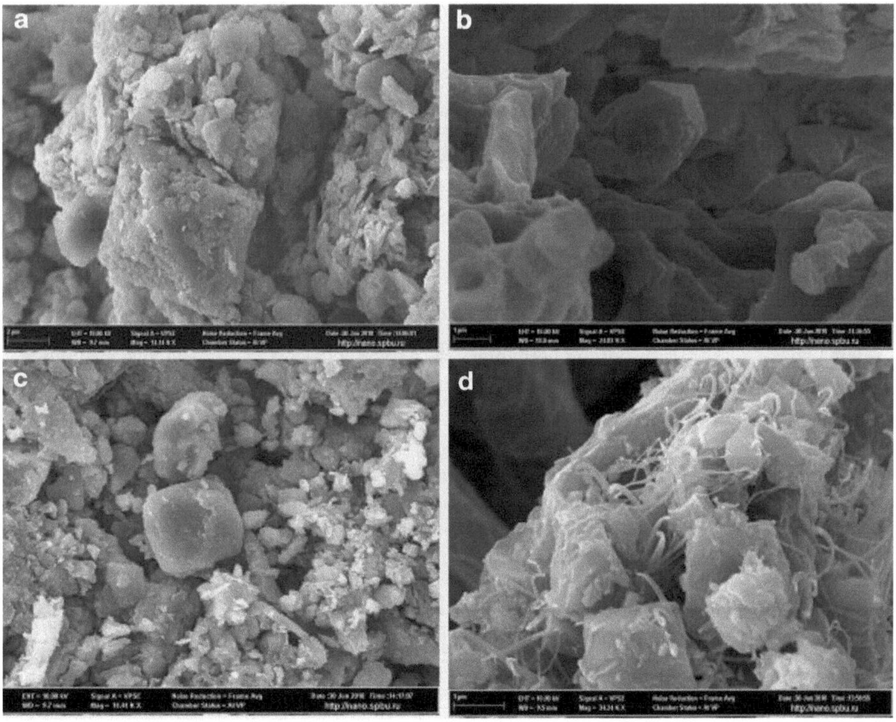

Fig. 4 SEM images of film weddellite crystals: "ideal" single crystals (**a**, **b**), with rounded dissolution forms (**c**), surrounded by mycobacteria (**d**). Aggregates with whewellite crystals are seen. Chersones, Ukraina

The length of the crystallites lies between 5 and 20 μm. Besides we see the crystalline aggregates with monoclinic whewellite. Biological research on the Chersonesus monuments has allowed to detect the same genera of lichens (*Lecanora* and *Caloplaca)* connected with oxalate patina producing. Such lichens were found before on oxalate patina on the surface of marble monuments in Mediterranean (Table 2). Besides some new species of lichens connected with Chersonesus monuments deterioration have been established. The study shows that, besides lichens, there are numerous of micromycetes (no less than 14 species) in Chersonesus oxalate patina. The dominated species belong to the genus *Cladosporium*, which is well known as a destructor of carbonate rocks in various conditions. Consequently we can propose that micromycetes can take part in the process of oxalate patina formation.

The results of simulation experiments supported this assumption. The carboxylic acid producing activity was fixed for the 18 micromycetes strains isolated from carbonate rock substrates. It was shown that the acid composition produced by fungi changes with the age of a culture. Besides it was revealed that calcium oxalate crystallization can be initiated by microscopic fungi, producing oxalic acid. The excretion of citric acid resulted in formation of small calcium citrates crystals which then quickly dissolve. So, the phase composition of fungal crystallization products is defined by variable content of organic acids in fungal culture.

Table 2 Species of microorganisms connected with oxalate patina

Chersonesos ruins (results of field work)		Piza (Burford et al. 2003)	
Lichens	Micromycetes	Lichens (Whewellite)	Lichens (Weddellite)
Botryolepraria lesdanii	Alternaria alternate	Acarospora rugulosa	Acarospora rugulosa
Diplocia canescens	Aspergillus niger	Acarospora smargdula	Aphyllophorales spp.
Caloplaca lacteal	Cladosporium cladosporioides	Aspicila alpine	Aspicilia calcarea
Collema sp.	Cladosporium herbarum	Aspicilia calcarea	Caloplaca aurantia
Lecanora pruinosa	Cladosporium sphaerospermum	Aspicilia radiosa	Caloplaca flavescens
Verrucaria sp.	Epicoccum nigrum	Caloplaca flavescens	Hypogymnia physodes
	Fusarium oxysporum	Hypogymnia physodes	Geastrum spp.
	Fusarium solani	Lecanora atra	Hysterangium Crassum
	Mucor racemosus	Lecanora rupicola	Lecanora atra
	Mucor ramosissimus	Lecidea inops	Lecanora rupicola
	Penicillium chrysogenum	Lecidea lactea	Lecidea inops
	Penicillium purpurogenum	Ochrolechia parella	Lecidea lactea
	Rhizopus stolonifer	Parmelia conspersa	Ochrolechia parella
	Ulocladium charatrum	Parmelia subrudecta	

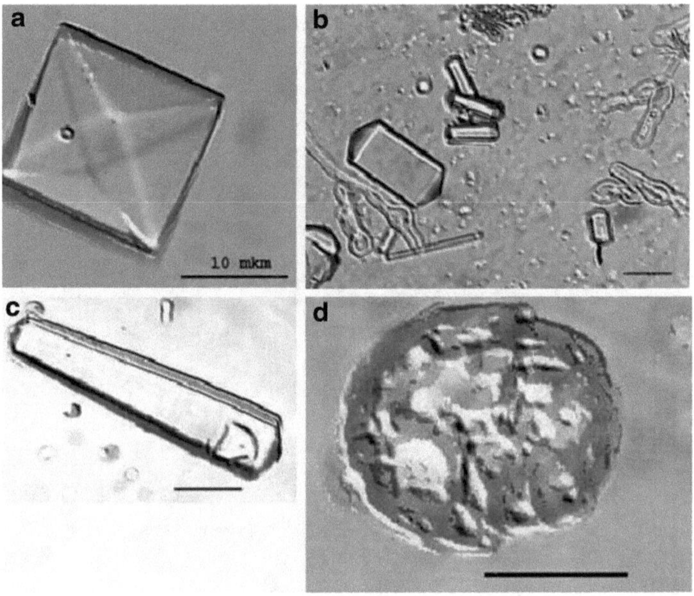

Fig. 5 The morphology of calcium oxalate crystals formed in fungal culture: "ideal" tetragonal crystals of weddellite (**a**, **b**), splitting weddelite crystals (**c**), spherical aggregate with whewellite (**d**)

The main stages of calcium oxalate crystallization in fungal culture in laboratory condition are following (Fig. 5):

1. Dipyramidal and dipyramidal-prismatic single crystals of weddellite. The size of the crystals is from 5 to 50 μm.
2. Splitting of Weddellite crystals.
3. Dumbell-like and than spherical aggregates of oxalate crystals.

The absence of monoclinic whewellite single crystals indicates that the whewellite have been raised from weddellite on stage of aggregates formation. Summarizing the results of this part of the work we can conclude that the crystallization of biofilm calcium oxalates (weddellite and whewellite) typically initiated by the different microorganisms (lichens of genera Lecanora and Caloplaca as well as numerous microscopic fungi producing oxalic acid).

3.1.2 Gypsum Crystallization

The crystals of *gypsum* $CaSO_4 \cdot 2H_2O$ on the carbonate rock surface described in many papers (Kramar and Mirtič 2008; Siegesmund et al. 2007 and other). The obtained results demonstrate that a gypsum-rich patina (black crusts of various thicknesses) develops on different marble and limestone monuments in all regions

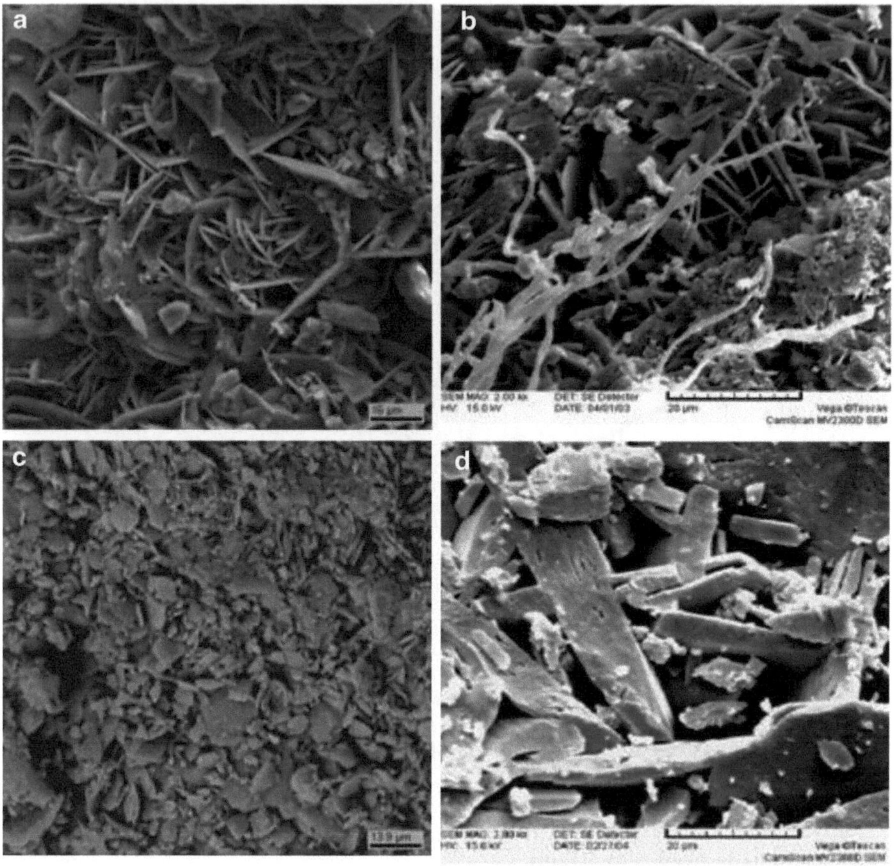

Fig. 6 SEM images of sulphated patina. Gypsum crystals on the surface of different carbonate rock: Carrara compact marble (**a**), Ruskeala heterogeneous marble (**b**), stratified Putilovskii limestone (**c**), porous Pudostskii travertine (**d**). Saint-Peterburg, Russia

of our field research. The X-Ray diffraction shows that the quantity of gypsum in sulphate crusts considerably varies. According to the SEM data the most gypsum crystals have lamellar habit (Fig. 6). The size of the individual crystals (on average) varies from 15 to 20 μm. The lamellar crystals may aggregate into rosettes, which indicate a high number of nucleation centers (Fig. 6a). In hollows and cracks lamellar crystals may up to 30–40 μm in size (Fig. 6b). Among them there are crystals with rounded (dissolution) forms (av. size of 10 μm) (Fig. 6c). Besides we see more large acicular and prismatic crystals with the strong development of the second pinacoid up to 170 μm in length (Fig. 6d).

On the whole we can conclude that the morphological features of the gypsum crystals developed on the surface of the carbonate rocks in urban environment indicate that the processes of growth and dissolution replace each other, which is typical for metasomatic crystallization. The degree of carbonatic rocks sulphation and the size of gypsum crystals depend on the local conditions, fissuring and porosity of the rock, structure/texture peculiarities and mineral composition of carbonate bedrock. A complicated relief of the monument promotes the moisture accumulation on the rock surface. For this reason, the high artistic monuments with a complex relief (made of homogeneous Carrara marble) as well as the monuments made of highly porous travertine suffer most from the sulphation. Dissolution, nucleation and growth of crystals take place in the presence of a numerous microorganisms such as bacteria, microfungi, algae and lichens (Fig. 7).

Three main stages of the gypsum-rich patina formation on the surface of carbonate rock monuments were noticed:

1. The initial stage, when a thin biofilm is formed on the rock surface. Beneath the biofilm, tiny gypsum crystals occur.

Fig. 7 SEM image of microfungi in gypsum crust

2. The intermediate stage is characterized by numerous large gypsum crystals. Between them, thin films of microfungi, algae cells and other microorganisms occur.
3. The intensive stage is reached as soon as a layer of gypsum crystals is formed. The microbial community develops beneath this gypsum layer.

Summarizing, metasomatic crystallization of gypsum occurring on the carbonate rocks surface in the urban environment connected mainly with the presence of essential amount sulfuric gas in an atmosphere. Additionally sulfates may be formed on the monument surface due to the sulfur bacteria (Sand 1997).The catalytic mechanism of the SO_2 oxidation is not clear yet. The formation of gypsum rich patina is always contemporaneous with the activity of many microorganisms.

These microorganisms produce aggressive metabolites and could catalyze SO_2 oxidation. Besides, iron which is according to our data presents very often on stone surfaces could catalyze SO_2 oxidation also. Results of our simulation experiments have shown that at the presence of air oxygen gypsum crystallization may start in a no saturated solution if SO_4^{2-} ions are absent (in presence only SO_3^{2-} ions). The experiments on catalytic mechanisms of the microorganism influence on gypsum crystallization are in progress.

3.2 Inorganic-Organic Coatings for Carbonate Rock Monument Protection

According to the experiment results the majority of the tested inorganic and organic photocatalysts (active in various ranges of wave lengths) suppressed the development of micromycetes isolated from the damaged rock monuments. It was fixed the wide spectrum of micromicetes reactions on the stressful influences of Photocatalysts. Photocatalyst effect is expressed in a stop or inhibiting of colonies growth and change of micromorphological attributes. The highest dependent on light antimicrobial effect was detected in the variants with the "photosens" (the mixture of Na-salts of dee-, three- and tetrasulfophthalocyanine of hydroxyaluminium) (Fig. 8). "Photosens" penetrates in fungal mycelium, produces blue coloration and induces the anomaly growth of the fungal cells.

Besides, the possibility of the application of inorganic-organic sol–gel matrix by the creation of nanocomposite protection coatings with antimicrobial properties was demonstrated. It was shown that the adding of photocatalysts and nanodiamond in the epoxy-silicate sol–gel matrixes increased the film protective activity (Fig. 9).

The photocatalytic activity of epoxy-titanate sol–gel film (alkali catalysis) is similar to Degussa P-25. Long-live nanocomposite coatings show inhibiting effect on microbial colonies growth (especially algae and micromycetes) in the field experiments (more than 5 years period of observation). According to film testing results in extreme condition (Antarctica, Russian Polar Station "Bellingshausen") incorporation of bioactive dopants increase the solidity of film.

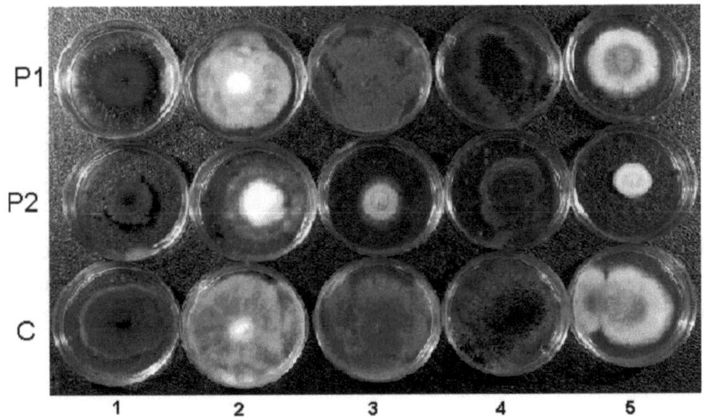

Fig. 8 The example of inhibition of micromycetes activity by organic photocatalyst "photosens": C – control (without "photosens"), P1 – "photosens" 1 h light, P2 – "photosens" 2 h light (repeatedly). The species of microfungi: *1 Ulocladium chartarum*; *2 Chaetomium globosum*; *3 Penicillium spinulosum*; *4 Aspergillus niger*; *5 Aspergillus versicolor*

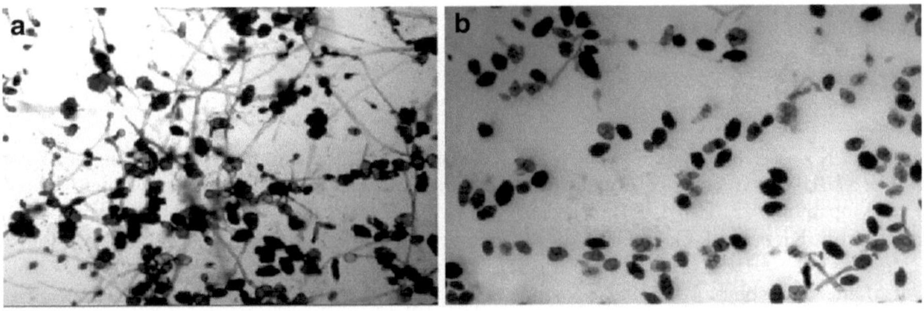

Fig. 9 Antimicrobial activity of nanocomposite protection coatings with nanodiamond against the active biodestructor – Ulocladium chartarum: (**a**) epoxy-silicate sol–gel matrixes without dopants: germination of fungal spores; (**b**) epoxy-silicate sol–gel matrixes with adding of nanodiamond (concentration in sol – 0.2%): full inhibiting of fungal spores germination

4 Conclusion

The study results demonstrate that the biogenic crystal genesis on a carbonate rock monuments surface is multifactor metasomatic process connecting with the activity of microorganisms, environment and bedrock peculiarities. This phenomenon is one of the main causes responsible for the carbonate rocks deterioration in urban environment. The crystallization of calcium oxalates and gypsum is contemporaneous with the activity of many microorganisms but the role of microorganisms in consideration processes is very different.

Crystallization of biofilm calcium oxalates (weddellite and whewellite) is typically initiated by a numerous microscopic fungi producing oxalic acid. Gypsum crystallization depends mainly on the local conditions, promoting moisture accumulation: fissuring and porosity of the bedrock and complicated relief of the monument. The catalytic mechanism of microorganism influence on this destructive process is not clear yet and experiments improving its understanding are in progress.

Long-term selfcleaning ecological organic–inorganic (silicate and titanate) coatings with antimicrobial properties for the carbonate rock monument protection were synthesized and tested. It was shown that the majority of the tested inorganic and organic photocatalysts can suppress the development of micromycetes isolated from carbonate bedrock. Adding of photocatalysts and nanodiamond into the epoxy-silicate sol–gel matrixes increase the film protective activity and the solidity of film in extreme conditions. The photocatalytic activity of epoxy-titanate sol–gel film (alkali catalysis) is similar to Degussa P-25.

Acknowledgments This study was supported by Russian Foundation for Basic Research (Project 09-05-01062-a).The authors are grateful to the Marina S. Zelenskaya, Aleksandr M. Marugin, Juliya V. Plotkina, Dmitri Gimelbrand, Irina V. Knauf, Mariya A. Timasheva, Alexander Mosyagin and Alexey Rousakov (Saint Petersburg State University) and Tamara Chamova (Institute of Silicate chemistry RAS) for their active collaboration in this work.

References

Barinova KV, Vlasov DYu, Schiparev SM, Zelenskaya MS, Rusakov AV, Frank-Kamenetskaya OV (2010) Production of organic acids by micromycetes from the rock substrates. Mycol Phytopathol 44(2):137–142 (in Russian)

Burford EP, Kierans M, Gadd GM (2003) Geomycology: fungi in mineral substrata. Mycologist 17:98–107

Del Monte M, Sabbioini C (1987) A study of the patina called scialbatura on imperial Roman marble. Stud Conserv 32:114–121

Frank-Kamenetskaya OV, Vlasov DYu, Zelenskaya MS, Knauf IV, Timasheva MA (2009) Decaying of the marble and limestone monuments in the urban environment: case studies from Saint Petersburg, Russia. Studia Universitatis Babes-Bolyai Geologia 54(2):17–22

Gadd GM (2007) Geomycology: biogeochemical transformations of rocks, minerals, metals and radionuclides by fungi, bioweathering and bioremediation. Mycol Res 111:3–49

Kramar S, Mirtič B (2008) Characterization of black crustsof Robba´s fountain statues, Ljubljana (Slovenia). RMZ – Mater Geoenviron 55(4):490–504

Rousakov AV, Frank-Kamenetskaya OV, Zelenskaya MS, Vlasov DYu, Gimelbrand DE, Knauf IV, Plotkina JV (2010) The first find of calcium oxalates in biofilms on the surface of archeological monuments of Tauric Chersonesos limestone (Crimea, Ukraine). Russian Mineralogical Society Notes (in Russian)

Sand W (1997) Microbial mechanisms of deterioration of inorganic substrates. A general mechanistic overview. Int Biodeterior Biodegrad 40(4):183–190

Shilova OA, Khamova TV, Vlasov DYu, Ryabusheva YuV, Mikhal'chuk VM, Frank-Kamenetskaya OV, Marygin AM, Dolmatov VYu (2009) Sol-gel synthesis and investigation nanocomposite protective biostable coatings. In: The second nanotechnology international forum rusnanotech'09 pp 342–344

Siegesmund S, Torok A, Hupers A, Muller C, Klemm W (2007) Mineralogical, geochemical and microfabric evidences of gypsum crusts: a case study from Budapest. Environ Geol 52:385–397

Timasheva MA, Frank-Kamenetskaya OV, Vlasov DYu (2007) The morphology and generation features of gypsum crystals on carbonate rocks surface in urban environment. Russ Mineral Soc Notes 5:98–104 (in Russian)

Vlasov DYu, Frank-Kametskaya OV (2005) Biodeterioration of rock monuments in urban environment: physical-chemical methods of control and preservation. In: Proceedings of the international conference solar renewable energy news research and applications, Firenze, pp 151–154

Vlasov DYu, Frank-Kametskaya OV (2006) Natural rock decaying in the urban environment. Trans St Petersburg Nat Soc 96:156–170 (in Russian)

Formation and Stability of Calcium Oxalates, the Main Crystalline Phases of Kidney Stones

Alina R. Izatulina, Yurii O. Punin, Alexandr G. Shtukenberg, Olga V. Frank-Kamenetskaya, and Vladislav V. Gurzhiy

1 Introduction

Wide spread of oxalate mineralization in living organisms is a strong reason for its intense worldwide study during recent decades. Calcium oxalates can be found in human body (stones of urinary system, calcifications in lungs, crystals in a bone marrow etc.) (Socol et al. 2003; Zuzuk 2005; Korago 1992), in body of animals (stones of cats urinary system), and in plants (Korago 1992). Most often calcium oxalates occur as a part of pathogenic formations of the human urinary system (Korago 1992). The part of oxalate kidney stones ranges from 50% to 75% depending of the geographical region. Our collection of kidney stones that were removed in Saint-Petersburg hospitals includes 263 samples, 81% of which consists of calcium oxalate minerals fully or as a part.

Oxalate monomineral stones are usually formed by whewellite – $CaC_2O_4.H_2O$, calcium oxalate monohydrate, or (less often) by weddellite – $CaC_2O_4.2H_2O$, calcium oxalate dihydrate (Zuzuk 2005). Bimineral stones formed by both calcium oxalates minerals also occur often enough (Fig. 1). At the normal body temperature (37°C), whewellite is a stable phase, whereas weddellite is metastable (Zuzuk 2005).

This paper is aimed to find out conditions, which control formation of calcium oxalate mineralization in human urinary system. Three main issues are addressed in the study: (1) Conditions that favor the calcium oxalate mineralization; (2) Stability of calcium oxalate dihydrate (weddellite); (3) Relationships between cell parameters and amount of zeolitic water in the weddellite crystal structure. The modeling experiments were used as a main approach to the problem.

A.R. Izatulina (✉) • Y.O. Punin • A.G. Shtukenberg • O.V. Frank-Kamenetskaya • V.V. Gurzhiy
Department of Crystallography, Saint Petersburg State University,
199034 St. Petersburg, Russia
e-mail: alina.izatulina@mail.ru

S.V. Krivovichev (ed.), *Minerals as Advanced Materials II*,
DOI 10.1007/978-3-642-20018-2_38,

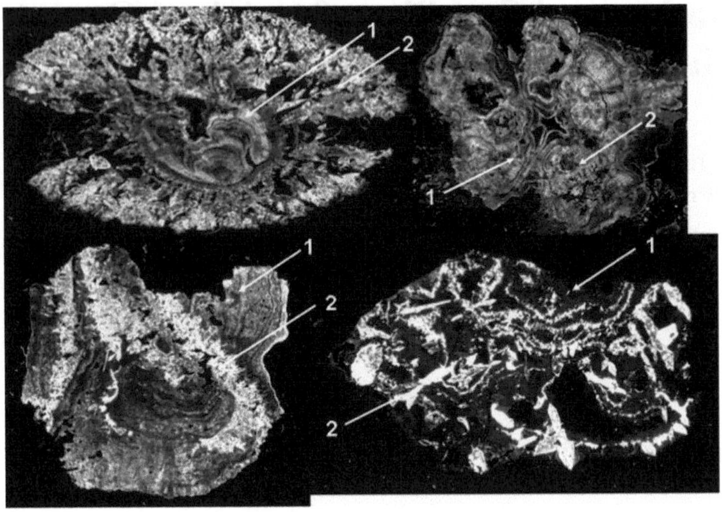

Fig. 1 Bimineral oxalate stones (*1* whewellite, *2* weddellite)

Table 1 Composition of initial solutions in modeling experiments (mmol/L) (I = 0.3 NaCl, pH = 4.0–8.5 (NaOH))

Components	Concentration
$CaCl_2.6H_2O$	1.70–5.00
$MgSO_4.7H_2O$	5.30–11.00
NH_4Cl	0.00–6.60
$KHSO_4$	21.70–69.00
$(NH_4)_2C_2O_4.H_2O$	0.20–0.42
$(NH_4)_2HPO_4$	6.50–24.60
$NaH_2PO_4.2H_2O$	6.50–8.40
$KHCO_3$	0.00–33.00

2 Experimental

There were two types of experiments. In the first, composition of physiological solution (urine) on all inorganic components (Table 1) was simulated. We used solutions with the minimum, average and maximum concentrations of components from a physiological range. The constant ionic strength of solution $I = 0.3$ (mmol/L) has been maintained by adding of NaCl (Fig. 2).

In the second case, solution contained only oxalate and calcium ions (supersaturation $\gamma = 7–50$). Except the factors denoted above, effect of crystallization rate, solution exposure, and the $Ca^{2+}/C_2O_4^{2-}$ ratio, were also studied.

Lattice constants and amount of zeolitic water in the weddellite crystal structure were determined by single crystal X-ray analysis. Six weddellite crystals extracted from kidney stones were analyzed on the STOE IPDS II and Bruker Smart APEX II diffractometers equipped with flat X-ray detectors. Crystal structures were solved

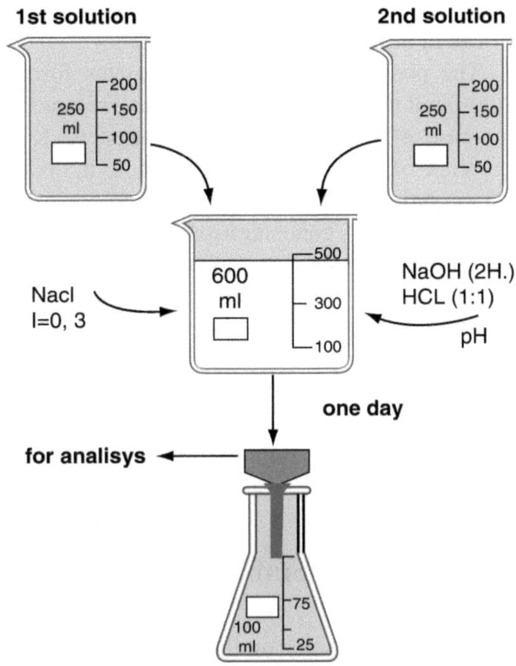

Fig. 2 The general scheme of modeling experiments

by the direct methods and refined in the *I*4/*m* space group using SHELXL-97 program (Sheldrick 2008). The absorption correction was introduced analytically taking into account the crystals shape.

3 Conditions of Calcium Oxalates Genesis

As shown by thermodynamic calculations (Elnikov et al. 2007; Levkovskiy and Levkovskiy 2006) hydroxylapatite is the most stable phase corresponding to the pH and composition ranges observed for physiological solution. The probability to find calcium oxalate in equilibrium with hydroxylapatite in physiological conditions is very low (Elnikov et al. 2007). Results of our modeling experiments (Table 1) do not contradict the data of thermodynamic calculations. Calcium oxalate was not found in a wide range of concentration and pH. Instead we obtained a series of other minerals often found in kidney stones: brushite $CaHPO_4{\cdot}2H_2O$, struvite $NH_4MgPO_4.6H_2O$, hydroxylapatite $Ca_5(PO_4)_3OH$ and whitlockite $Ca_3(PO_4)_2$, and also X-ray amorphous calcium phosphate. The hydrated forms of calcium oxalate (analogs of whewellite $CaC_2O_4.H_2O$ and weddellite $CaC_2O_4.2H_2O$) were absent in the experimental sediments. This agrees with the data on combined phosphates and calcium oxalates crystallization (Levkovskiy and Levkovskiy 2006; Rakin and Katkova 2003), in which calcium oxalate also has not been obtained. Thus, there is obvious contradiction between results of thermodynamic calculation and a wide distribution of calcium oxalates in renal stones.

The possibility of calcium oxalates formation in human urinary system was suggested to be due to the following reasons: acidation of urine up to pH 4.5–6.0; increase of oxalate-ion concentration to 1.0–1.5 mmol/L (oxalatourea disease); presence of certain organic substances in urea. Indeed, calcium oxalates are formed (in the ratio 1:5 to phosphates) in analogs of a physiological solution with increase of oxalate ions concentration to 1.5 mmol/L (that five times above the norm). It confirms relationship between oxalate urolithiasis and oxalatourea desease. The amount of precipitated oxalates increases (up to the ratio 1:4 to phosphates) with addition of bacteria-viral associates (rotavirus, a hepatitis virus, coli) to the same solution. However, phosphate phases prevail in sediments in all our experiments without exceptions. We also did not reveal promotion of oxalate formation by low acidity – in the range pH = 4.0–7.5 oxalates were not found. Thus, the problem of determination of oxalate urolithiasis conditions demands the further research.

4 Weddellite Stabilizing Factors

According to results of thermodynamic calculations the stable calcium oxalate phase in physiological conditions is whewellite (Elnikov et al. 2007). Nevertheless weddellite frequently occurs in oxalate stones. About 68% of stones in our collection contain weddellite, and 17% are monomineral weddellite stones. Rhythmic alternation of whewellite and weddellite zones, that is often observed, points to the periodic sharp changes of stone formation conditions (Fig. 1).

In accordance with the literature data (Marcovic et al. 1988; Bretherton and Rodgers 1998; Rakin and Katkova 2003; Gardner 1978; Werness et al. 1981; Garside et al. 1982; Guo et al. 2002) of variables can affect crystallization of various calcium oxalates: temperature, pH, $Ca^{2+}/C_2O_4^{\ 2-}$ ratio, various organic and inorganic components, bacteria-viral associates, and crystallization time. However, experimental data are contradictory, so that the same variable can work in opposite. At the same time the most of authors consider the calcium oxalate dihydrate to be the first precipitating phase. Kept in solution or dried at temperatures above 36°C for hours, days or weeks it completely transforms into calcium oxalate monohydrate (Marcovic et al. 1988; Zuzuk 2005).

Determination of the factors causing formation of metastable weddellite, was carried out for the system $CaCl_2 - (NH_4)_2C_2O_4 - H_2O$. We varied time of precipitation, crystallization rate, temperature, pH, $Ca^{2+}/C_2O_4^{\ 2-}$ ratio, and also added various inorganic and organic additives (Table 2).

Results of these experiments in many respects do not correspond to expectations: based on both: theoretical considerations and literature data. First of all, even for the shortest time dictated by our experimental setup (60 min from sedimentation to X-ray experiment) of keeping sediment with a solution, weddellite in a sediment aren't found out.

Table 2 Weddellite stabilizing factors

Factors		Phases
Temperature	From 20 to 58°C	Whewellite
Time of crystallization	40 min–1 day	Whewellite
pH	4.0–8.5	Whewellite
Ca/C_2O_4 ratio	From 1 to 10	Whewellite
Crystallization speed	10^{-6} mol/l·s – 10^{-4} mol/l·s	Whewellite
Additives	HPO_4^{2-}	Whewellite
	Mg^{2+}	Whewellite weddellite
	CO_3^{2-}	Whewellite weddellite
	Aminoacids (glycine, alanine, proline, glutamic acid)	Whewellite weddellite
	Protein (gelatine and ovalbumine)	Whewellite weddellite
	Bacteria-viral associates	Whewellite weddellite

On contrary, according to the literature data, transformation of metastable weddellite into stable whewellite in a solution requires from 1 h (Rakin and Katkova 2003) to about 1 week (Bretherton and Rodgers 1998). Further, reported in some papers (Bretherton and Rodgers 1998; Gardner 1978; Werness et al. 1981) increase of water content in oxalate phases with decreasing of crystallization temperature was not found as well, high crystallization rate, which often favors the growth of metastable phases and, in particularly, of weddelite (Rakin and Katkova 2003; Werness et al. 1981), has no effect in our experiments as well. The same relates to effect of the increase of the Ca/C_2O_4 ratio and wide pH variations, which in contradiction to the literature data (Bretherton and Rodgers 1998; Garside et al. 1982), also have not led to weddellite formation. Thus, all of the factors listed above cannot be considered as water content controlling factors in crystallizing calcium oxalates.

A number of authors put forward the assumption that formation of calcium oxalates with different water contents is controlled by the impurities. In particular, an impurity that breaks crystallization of whewellite promotes sedimentation of weddellite (Rakin and Katkova 2003; Guo et al. 2002). On the other hand, inhibitors can equally poison growth of both mono – and dihydrated calcium oxalates (Guo et al. 2002). As our experiments have shown, different well known inhibitors operate in different ways. Hydrophosphate-ion does not lead to weddellite formation, whereas Mg^{2+} and CO_3^{2-} ions promote its crystallization. Such powerful growth inhibitors as amino acids and proteins stabilize calcium oxalate dihydrate. Apparently, the similar action of bacteria-viral associates is reduced to a producing of inhibiting organic compounds.

It is also worth noting that in the course of sediments drying calcium oxalate dihydrate remains only in the sediments produced in the presence of organic components. Thus, the physiological solution composition has to be considered as a primary factor defining crystallization of calcium oxalates with different amounts of water.

Fluctuations of components concentrations that stabilize weddellite in a physiological solution will lead to alternation of oxalate phases in uroliths (phase zonal structure) that is often observed. However, that we were never able to get weddellite without admixture of whewellite – both phase always crystallize together, and whewellite predominates. On the other hand, as it was mentioned above, monomineral weddellite kidney stones are quite common.

5 Structural Water in Weddellite

The certain quantity of additional water in the structure of metastable weddellite could be the possible reason for its stability. It is well known that the correct chemical formula of weddellite should be written as $CaC_2O_4.(2+x)H_2O$.

The first determination of weddellite crystal structure was carried out by Sterling (Sterling 1965). Then Tazzoli and Domenegetti (Tazzoli and Domeneghetti 1980) refined its crystal structure and splited position of "zeolitic" water into two crystallographically independent positions. The results presented in given paper, supplement the data obtained by previous researchers. The unit cell parameters, experiment details, and parameters of the structure refinement for six weddellite crystals are presented in Table 3. Selected bond lengths for all compounds are listed in Table 4.

Mineral weddellite crystallizes in the tetragonal space group *I*4/*m*. In the structure of weddellite (Fig. 3) calcium atoms form six bonds with oxygen atoms belonging to oxalic groups and two to oxygen atoms from water molecules. Each calcium polyhedron shares an edge with two adjacent Ca polyhedra forming chains of square antiprisms that extend along the *c* axis. These chains are linked by the oxalic groups into the 3-D framework forming in the crystal structure two types of channels oriented along the [001] direction . The channels differ with the inner diameter. Larger channels with the inner diameter of about 4.5 Å are occupied by the "zeolitic" water molecules, which occupy two alternative crystallographically independent positions W3 and W31 with the distance W3 – W31 = 0.6 Å. The analysis of the single crystal X-ray data for the studied six crystals along with literature data has shown that the amount of "zeolitic" water in the structure of weddellite affects unit cell dimentions, especially *a* which varies in the range 12.336– 12.371 Å (Fig. 4). Unit cell parameters increase as the channel diameter increases that is likely caused by the presence of significant water amount in the weddelite crystal structure (Fig. 4).

For example, the distance between the neighbor molecules of water W1 located within one layer parallel to a (001) plane, ranges within 3.211– 3.287 Å. Determination of the unit cell parameter *a* of weddellite phases, stabilized by different additives, has allowed us to find amount of "zeolitic" water in structures of these phases (Table 5).

These data suggest that the stable weddellite crystals from kidney stones are characterized by $x = 0.13–0.37$ a.p.f.u., and most of points fall in the middle of the interval.

Table 3 The unit cell parameters experiment details, and parameters of the structure refinement for six weddellite crystals studied

Sample	Smp1	Smp2	Smp3	Smp4	Smp5	Smp6
a (Å)	12.3363 (13)	12.3443 (14)	12.3462(11)	12.3543 (13)	12.3567(6)	12.363(2)
c (Å)	7.3448(8)	7.3599(8)	7.3535(7)	7.3547(9)	7.3573(3)	7.3460(17)
V (Å^3)	1117.8(2)	1121.5(2)	1120.88(18)	1122.5(2)	1123.37(9)	1122.7(4)
Space group	I4/m	I4/m	I4/m	I4/m	I4/m	I4/m
μ (mm^{-1})	1.084	1.082	1.083	1.081	1.082	1.082
Z	8	8	8	8	8	8
D_{calc} (g/cm^3)	1.976	1.983	1.982	1.980	1.991	1.992
Diffractometer	Stoe IPDS II	Stoe IPDS II	Stoe IPDS II	Stoe IPDS II	Bruker Smart Apex II	Stoe IPDS II
Radiation	MoKα	MoKα	MoKα	MoKα	MoKα	MoKα
Total Ref.	6,579	5,254	6,609	6,502	7,437	5,150
Unique Ref.	1,026	815	1,032	1,030	1,382	815
2θ range, °	4.66–63.90	4.66–58.72	4.66–63.82	6.44–63.80	4.66–71.84	4.66–58.62
Unique $\lvert F_o \rvert \geq 4\sigma_F$	834	646	909	901	1,196	683
R_{int}	0.056	0.050	0.043	0.050	0.028	0.075
R_σ	0.033	0.034	0.022	0.025	0.020	0.042
R_1 ($-F_o\rvert \geq 4\sigma_F$)	0.047	0.034	0.043	0.052	0.027	0.041
wR_2 ($-F_o\rvert \geq 4\sigma_F$)	0.084	0.064	0.078	0.092	0.073	0.077
R_1 (all data)	0.066	0.054	0.054	0.063	0.033	0.057
wR_2 (all data)	0.089	0.069	0.082	0.097	0.076	0.081
S	1.095	0.999	1.078	1.092	1.066	1.098
ρ_{min}, ρ_{max}, e/Å^3	–0.535, 0.505	–0.422, 0.460	–0.397, 0.491	–0.596, 0.495	–0.423, 0.512	–0.424, 0.366

Note: $R_1 = \Sigma\lvert\lvert F_o\rvert - \lvert F_c\rvert\rvert/\Sigma\lvert F_o\rvert$; $wR_2 = \{\Sigma[w(F_o^2 - F_c^2)^2]/\Sigma[w(F_o^2)^2]\}^{1/2}$; $w = 1/[\sigma^2(F_o^2) + (aP)^2 + bP]$, where $P = (F_o^2 + 2F_c^2)/3$; $s = \{\Sigma[w(F_o^2 - F_c^2)]/(n - p)\}^{1/2}$ where n is the number of reflections and p is the number of refined parameters

Table 4 Selected bond lengths for the crystal structures of six weddellite crystals

Compound	Smp1	Smp2	Smp3	Smp4	Smp5	Smp6
Ca1-W1	2.392(3)	2.390(2)	2.391(3)	2.392(3)	2.3916(13)	2.394(3)
Ca1-O2	2.4449(18)	2.4461(16)	2.4453(16)	2.4458(19)	2.4470(8)	2.4450(18)
Ca1-W2	2.448(3)	2.446(3)	2.449(3)	2.457(3)	2.4506(14)	2.452(3)
Ca1-O1	2.4528(16)	2.4571(14)	2.4568(14)	2.4574(17)	2.4594(7)	2.4568(16)
Ca1-O1	2.4945(15)	2.5010(13)	2.4977(13)	2.4997(16)	2.5000(7)	2.4970(15)
C1-C1	1.549(4)	1.557(4)	1.549(3)	1.548(4)	1.5476(19)	1.552(4)
C1-O2	1.244(3)	1.242(2)	1.245(2)	1.246(3)	1.2490(11)	1.244(3)
C1-O1	1.251(2)	1.249(2)	1.251(2)	1.252(3)	1.2521(11)	1.252(3)
W3–W1	3.08(9)	3.10(5)	3.06(7)	3.09(5)	3.04(2)	3.08(3)
W3–W31	0.60(19)	0.57(14)	0.54(11)	0.62(9)	0.61(3)	0.66(8)
W1–W1	3.223(4)	3.239(4)	3.238(4)	3.245(5)	3.253(2)	3.253(4)
W2–W2	2.936(6)	2.933(6)	2.941(6)	2.943(7)	2.947(3)	2.927(7)

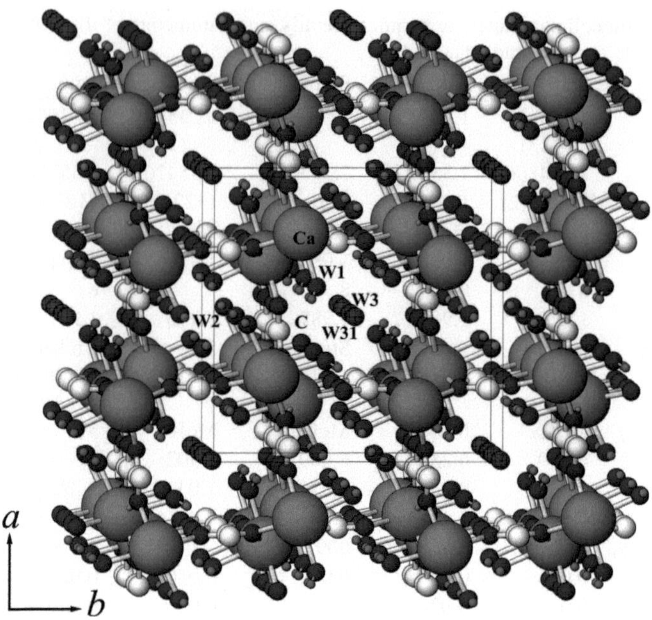

Fig. 3 Crystal structure of weddellite. Ca, C and O atoms are shown by *grey*, *light grey* and *dark grey* balls, respectively

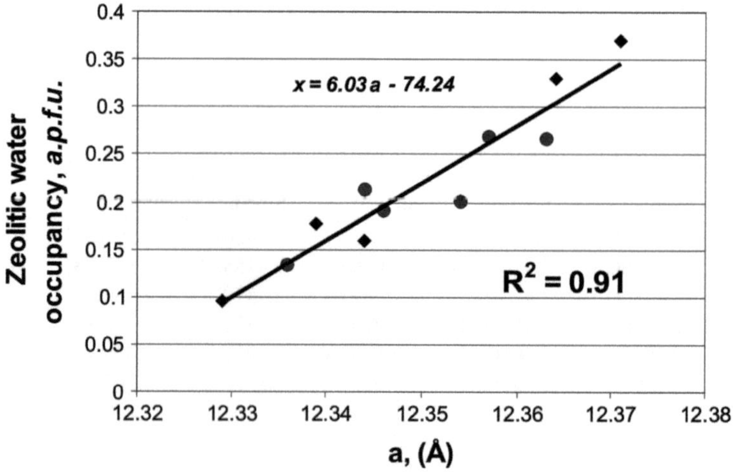

Fig. 4 Crystal structure of weddellite. Ca, C and O atoms are shown by *grey*, *light grey* and *dark grey* balls, respectively

Table 5 Amount of "zeolitic" water (x) in the weddellite crystals grown in the presence of different additives

Additives	Ratio whewellite: weddellite	a, Å	x
Mg^{2+}	3:1	12.365 (2)	0.31
CO_3^{2-}	5:1	12.357 (1)	0.26
Ovalbumin	5:2	12.349 (1)	0.21
Bacteria coli	5:2	12.344 (1)	0.18
Gepatite	5:2	12.351 (1)	0.23
Rotavirus	5:3	12.344 (1)	0.18

6 Conclusions

1. The increase of oxalates-ions concentration and the occurrence of organic substance in a physiological solution promote formation of calcium oxalates in human urinary system.
2. Presence of magnesium ions, carbonate-ions, glutamic acids, a glycine, ovalbumin and viruses in urina stabilizes the phase of calcium oxalate dihydrate.
3. Variations of unit cell parameter a in urolithic as well as in synthesized samples show that the weddelite stability can be defined by amount of additional "zeolitic" water molecules.

Acknowledgments This work was supported by Russian Foundation for Basic Research (grant # 10-05-00881-a)

References

Bretherton T, Rodgers A (1998) Crystallization of calcium oxalate in minimally diluted urine. J Cryst Growth 192:448–455

Elnikov VYu, Rosseeva EV, Golovanova OA, Frank-Kamenetskaya OV (2007) Thermodynamic and experimental modeling of the formation of major mineral phases of urolith. Russ J Inorg Chem 52(2):150–157

Gardner GL (1978) Effect of pyrophosphate and phosphonate anions on the crystal growth kinetics of calcium oxalates hydrates. J Phys Chem 82(8):864–870

Garside J, Lj B, Mullin JW (1982) The effect of temperature on the precipitation of calcium oxalate. J Cryst Growth 57(2):233–240

Guo S, Ward MD, Wesson JA (2002) Direct visualization of calcium oxalate monohydrate (COM) crystallization and dissolution with atomic force microscopy (AFM) and the role of polymeric additives. Langmuir 18:4284–4291

Korago AA (1992) Introduction to biomineralogy. Nedra, St.Petersburg (In Russian)

Levkovskiy NS, Levkovskiy SN (2006) Urolithiasis (ethiology, patogenesis, prediction and prophylaxis). Military medical academy of S.M. Kirov, St.Petersburg (In Russian)

Marcovic M, Lj K, Furedi-Milhofer H, Skrtic D (1988) Precipiation of calcium oxalates from high ionic strenght solutions VII. J Cryst Growth 88:118

Rakin VI, Katkova VI (2003) Diffusion-kinetic crystallization systems. Syktyvtar (In Russian)

Sheldrick GM (2008) A short history of SHELX. Acta Crystallogr 64:112–222

Socol EV, Nigmatulina EN, Maksimova NV (2003) Calcium oxalate spherulites in kidney stones: morphology and formation conditions. Chem Sustain Dev 11:547–558 (In Russian)
Sterling C (1965) Crystal-structure analysis of weddellite. Acta Crystallogr 18:917–921
Tazzoli V, Domeneghetti C (1980) The crystal structures of whewellite and weddellite: re-examination and comparison. Amer Miner 65:327–334
Werness G, Bergert JH, Smith LH (1981) Crystalluria. J Cryst Growth 53(1):166–181
Zuzuk FV (2005) Mineralogy of uroliyths. Lviv national university Lviv (In Ukranian)

Index

S.V. Krivovichev (ed.), *Minerals as Advanced Materials II*,
DOI 10.1007/978-3-642-20018-2, © Springer-Verlag Berlin Heidelberg 2012

MIX
Papier aus verantwortungsvollen Quellen
Paper from responsible sources
FSC® C105338

If you have any concerns about our products,
you can contact us on
ProductSafety@springernature.com

In case Publisher is established outside the EU,
the EU authorized representative is:
Springer Nature Customer Service Center GmbH
Europaplatz 3, 69115 Heidelberg, Germany

Printed by Libri Plureos GmbH
in Hamburg, Germany